T0253530

Lecture Notes in Artificial Intelligence   11108

Subseries of Lecture Notes in Computer Science

LNAI Series Editors

Randy Goebel
University of Alberta, Edmonton, Canada
Yuzuru Tanaka
Hokkaido University, Sapporo, Japan
Wolfgang Wahlster
DFKI and Saarland University, Saarbrücken, Germany

LNAI Founding Series Editor

Joerg Siekmann
DFKI and Saarland University, Saarbrücken, Germany

More information about this series at http://www.springer.com/series/1244

Min Zhang · Vincent Ng
Dongyan Zhao · Sujian Li
Hongying Zan (Eds.)

Natural Language Processing and Chinese Computing

7th CCF International Conference, NLPCC 2018
Hohhot, China, August 26–30, 2018
Proceedings, Part I

 Springer

Editors
Min Zhang
Soochow University
Suzhou
China

Sujian Li
Peking University
Beijing
China

Vincent Ng
The University of Texas at Dallas
Richardson, TX
USA

Hongying Zan
Zhengzhou University
Zhengzhou
China

Dongyan Zhao
Peking University
Beijing
China

ISSN 0302-9743   ISSN 1611-3349   (electronic)
Lecture Notes in Artificial Intelligence
ISBN 978-3-319-99494-9   ISBN 978-3-319-99495-6   (eBook)
https://doi.org/10.1007/978-3-319-99495-6

Library of Congress Control Number: 2018951640

LNCS Sublibrary: SL7 – Artificial Intelligence

© Springer Nature Switzerland AG 2018
This work is subject to copyright. All rights are reserved by the Publisher, whether the whole or part of the material is concerned, specifically the rights of translation, reprinting, reuse of illustrations, recitation, broadcasting, reproduction on microfilms or in any other physical way, and transmission or information storage and retrieval, electronic adaptation, computer software, or by similar or dissimilar methodology now known or hereafter developed.
The use of general descriptive names, registered names, trademarks, service marks, etc. in this publication does not imply, even in the absence of a specific statement, that such names are exempt from the relevant protective laws and regulations and therefore free for general use.
The publisher, the authors and the editors are safe to assume that the advice and information in this book are believed to be true and accurate at the date of publication. Neither the publisher nor the authors or the editors give a warranty, express or implied, with respect to the material contained herein or for any errors or omissions that may have been made. The publisher remains neutral with regard to jurisdictional claims in published maps and institutional affiliations.

This Springer imprint is published by the registered company Springer Nature Switzerland AG
The registered company address is: Gewerbestrasse 11, 6330 Cham, Switzerland

Preface

Welcome to the proceedings of NLPCC 2018, the 7th CCF International Conference on Natural Language Processing and Chinese Computing. Following the highly successful conferences in Beijing (2012), Chongqing (2013), Shenzhen (2014), Nanchang (2015), Kunming (2016), and Dalian (2017), this year's NLPCC was held in Hohhot, the capital and the economic and cultural center of Inner Mongolia. As a leading international conference on natural language processing and Chinese computing organized by CCF-TCCI (Technical Committee of Chinese Information, China Computer Federation), NLPCC 2018 served as a main forum for researchers and practitioners from academia, industry, and government to share their ideas, research results, and experiences, and to promote their research and technical innovations in the fields.

There is nothing more exciting than seeing the continual growth of NLPCC over the years. This year, we received a total of 308 submissions, which represents a 22% increase in the number of submissions compared with NLPCC 2017. Among the 308 submissions, 241 were written in English and 67 were written in Chinese. Following NLPCC's tradition, we welcomed submissions in eight key areas, including NLP Fundamentals (Syntax, Semantics, Discourse), NLP Applications, Text Mining, Machine Translation, Machine Learning for NLP, Information Extraction/Knowledge Graph, Conversational Bot/Question Answering/Information Retrieval, and NLP for Social Network. Unlike previous years, this year we intended to broaden the scope of the program by inviting authors to submit their work to one of five categories: applications/tools, empirical/data-driven approaches, resources and evaluation, theoretical, and survey papers. Different review forms were designed to help reviewers determine the contributions made by papers in different categories. While it is perhaps not surprising to see that more than 88% of the submissions concern empirical/data-driven approaches, it is encouraging to see three resources and evaluation papers and one theoretical paper accepted to the conference. Acceptance decisions were made in an online PC meeting attended by the general chairs, the Program Committee (PC) chairs, and the area chairs. In the end, 70 submissions were accepted as full papers (with 55 papers in English and 15 papers in Chinese) and 31 as posters. Six papers were nominated by the area chairs for the best paper award. An independent best paper award committee was formed to select the best papers from the shortlist. The proceedings include only the English papers accepted; the Chinese papers appear in *ACTA Scientiarum Naturalium Universitatis Pekinensis*.

We were honored to have four internationally renowned keynote speakers — Charles Ling, Joyce Chai, Cristian Danescu-Niculescu-Mizil, and Luo Si — share their views on exciting developments in various areas of NLP, including language communication with robots, conversational dynamics, NLP research at Alibaba, and megatrends in AI.

We could not have organized NLPCC 2018 without the help of many people:

- We are grateful for the guidance and advice provided by TCCI Chair Ming Zhou, General Chairs Dan Roth and Chengqing Zong, and Organizing Committee Co-chairs Dongyan Zhao, Ruifeng Xu, and Guanglai Gao.
- We would like to thank Chinese Track and Student Workshop Co-chairs Minlie Huang and Jinsong Su, as well as Evaluation Co-chairs Nan Duan and Xiaojun Wan, who undertook the difficult task of selecting the slate of accepted papers from the large pool of high-quality papers.
- We are indebted to the 16 area chairs and the 232 reviewers. This year, we operated under severe time constraints, with only a month between the submission deadline and the notification date. We could not have met the various deadlines during the review process without the hard work of the area chairs and the reviewers.
- We thank ADL/Tutorial Co-chairs Wenliang Chen and Rui Yan for assembling a tutorial program consisting of six tutorials covering a wide range of cutting-edge topics in NLP.
- We thank Sponsorship Co-chairs Kam-Fai Wong and Ming Zhou for securing sponsorship for the conference.
- Publication Co-chairs Sujian Li and Hongying Zan spent a tremendous amount of time ensuring every little detail in the publication process was taken care of and truly deserve a big applause.
- We thank Dan Roth, Xuanjing Huang, Jing Jiang, Yang Liu, and Yue Zhang for agreeing to serve in the best paper award committee.
- Above all, we thank everybody who chose to submit their work to NLPCC 2018. Without their support, we could not have put together a strong conference program.

Enjoy the conference as well as Hohhot' vast green pastures and natural sceneries!

July 2018   Vincent Ng
                                                                       Min Zhang

Organization

NLPCC 2018 is organized by Technical Committee of Chinese Information of CCF, Inner Mongolia University and the State Key Lab of Digital Publishing Technology.

Organizing Committee

General Chairs

Dan Roth — University of Pennsylvania, USA
Chengqing Zong — Institute of Automation, Chinese Academy of Sciences, China

Program Co-chairs

Min Zhang — Soochow University, China
Vincent Ng — University of Texas at Dallas, USA

Area Chairs

NLP Fundamentals
Nianwen Xue — Brandeis University, USA
Meishan Zhang — Heilongjiang University, China

NLP Applications
Bishan Yang — Carnegie Mellon University, USA
Ruifeng Xu — Harbin Institute of Technology, China

Text Mining
William Yang Wang — University of California, Santa Barbara, USA
Ping Luo — Institute of Computing Technology, Chinese Academy of Sciences, China

Machine Translation
Fei Huang — Facebook, USA
Derek Wong — University of Macau, Macau, SAR China

Machine Learning for NLP
Kai-Wei Chang — University of California, Los Angeles, USA
Xu Sun — Peking University, China

Knowledge Graph/IE
Yun-Nung (Vivian) Chen — National Taiwan University, Taiwan
Wenliang Chen — Soochow University, China

Conversational Bot/QA
Jianfeng Gao — Microsoft AI and Research, USA
Haofen Wang — Gowild.cn, China

NLP for Social Network

Wei Gao	Victoria University of Wellington, New Zealand
Bo Wang	Tianjin University, China

Organization Co-chairs

Guanglai Gao	Inner Mongolia University, China
Ruifeng Xu	Harbin University of Technology, China
Dongyan Zhao	Peking University, China

ADL/Tutorial Chairs

Wenliang Chen	Soochow University, China
Rui Yan	Peking University, China

Student Workshop Chairs

Minlie Huang	Tsinghua University, China
Jinsong Su	Xiamen University, China

Sponsorship Co-chairs

Kam-Fai Wong	The Chinese University of Hong Kong, SAR China
Ming Zhou	Microsoft Research Asia, China

Publication Chairs

Sujian Li	Peking University, China
Hongying Zan	Zhengzhou University, China

Publicity Chairs

Wanxiang Che	Harbin Institute of Technology, China
Qi Zhang	Fudan University, China
Yangsen Zhang	Beijing University of Information Science and Technology, China

Evaluation Chairs

Nan Duan	Microsoft Research Asia
Xiaojun Wan	Peking University, China

Program Committee

Wasi Ahmad	University of California, Los Angeles, USA
Xiang Ao	Institute of Computing Technology, Chinese Academy of Sciences, China
Deng Cai	Shanghai Jiao Tong University, China

Hailong Cao	Harbin Institute of Technology, China
Kai Cao	New York University, USA
Rongyu Cao	Institute of Computing Technology, CAS, China
Shaosheng Cao	Ant Financial Services Group, China
Yixuan Cao	Institute of Computing Technology, CAS, China
Ziqiang Cao	Hong Kong Polytechnic University, SAR China
Baobao Chang	Peking University, China
Kai-Wei Chang	UCLA, USA
Yung-Chun Chang	Graduate Institute of Data Science, Taipei Medical University, Taiwan, China
Berlin Chen	National Taiwan Normal University, Taiwan, China
Boxing Chen	Alibaba, China
Chen Chen	Arizona State University, USA
Chengyao Chen	Hong Kong Polytechnic University, SAR China
Dian Chen	blog.csdn.net/okcd00, China
Gong Cheng	Nanjing University, China
Li Cheng	Xinjiang Technical Institute of Physics and Chemistry, Chinese Academy of Sciences, China
Hongshen Chen	JD.com, China
Hsin-Hsi Chen	National Taiwan University, Taiwan, China
Muhao Chen	University of California Los Angeles, USA
Qingcai Chen	Harbin Institute of Technology Shenzhen Graduate School, China
Ruey-Cheng Chen	SEEK Ltd., Australia
Wenliang Chen	Soochow University, China
Xu Chen	Tsinghua University, China
Yidong Chen	Xiamen University, China
Yubo Chen	Institute of Automation, Chinese Academy of Sciences, China
Yun-Nung Chen	National Taiwan University, Taiwan, China
Zhumin Chen	Shandong University, China
Wanxiang Che	Harbin Institute of Technology, China
Thilini Cooray	Singapore University of Technology and Design, Singapore
Xinyu Dai	Nanjing University, China
Bhuwan Dhingra	Carnegie Mellon University, USA
Xiao Ding	Harbin Institute of Technology, China
Fei Dong	Singapore University of Technology and Design, Singapore
Li Dong	University of Edinburgh, UK
Zhicheng Dou	Renmin University of China, China
Junwen Duan	Harbin Institute of Technology, China
Xiangyu Duan	Soochow University, China
Jiachen Du	Harbin Institute of Technology Shenzhen Graduate School, China

Jinhua Du	Dublin City University, Ireland
Matthias Eck	Facebook, USA
Derek F. Wong	University of Macau, Macau, SAR China
Chuang Fan	Harbin Institute of Technology Shenzhen Graduate School, China
Chunli Fan	Guilin University of Electronic Technology, China
Yang Feng	Institute of Computing Technology, Chinese Academy of Sciences, China
Yansong Feng	Peking University, China
Guohong Fu	Heilongjiang University, China
Michel Galley	Microsoft Research, USA
Jianfeng Gao	Microsoft Research, Redmond, USA
Qin Gao	Google LLC, USA
Wei Gao	Victoria University of Wellington, New Zealand
Niyu Ge	IBM Research, USA
Lin Gui	Aston University, UK
Jiafeng Guo	Institute of Computing Technology, CAS, China
Jiang Guo	Massachusetts Institute of Technology, USA
Weiwei Guo	LinkedIn, USA
Yupeng Gu	Northeastern University, USA
Xianpei Han	Institute of Software, Chinese Academy of Sciences, China
Tianyong Hao	Guangdong University of Foreign Studies, China
Ji He	University of Washington, USA
Yanqing He	Institute of Scientific and Technical Information of China, China
Yifan He	Alibaba Inc., USA
Yulan He	Aston University, UK
Zhongjun He	Baidu Inc., China
Yu Hong	Soochow University, China
Dongyan Huang	Institute for Infocomm Research, Singapore
Fei Huang	Alibaba DAMO Research Lab, USA
Guoping Huang	Tencent AI Lab, China
Jiangping Huang	Chongqing University of Posts and Telecommunications, China
Ruihong Huang	Texas AM University, USA
Shujian Huang	Nanjing University, China
Ting-Hao Huang	Carnegie Mellon University, USA
Xiaojiang Huang	Microsoft, China
Xuanjing Huang	Fudan University, China
Zhiting Hu	Carnegie Mellon University, USA
Junyi Jessy Li	University of Texas at Austin, USA
Jingtian Jiang	Microsoft AI Research, USA
Shengyi Jiang	Guangdong University of Foreign Studies, China
Wenbin Jiang	Baidu Inc., China
Yuxiang Jia	Zhengzhou University, China

Peng Jin	Leshan Normal University, China
Chunyu Kit	City University of Hong Kong, SAR China
Fang Kong	Soochow University, China
Lingpeng Kong	Carnegie Mellon University, USA
Lun-Wei Ku	Academia Sinica, Taiwan, China
Man Lan	East China Normal University, China
Yanyan Lan	Institute of Computing Technology, CAS, China
Ni Lao	SayMosaic, USA
Wang-Chien Lee	The Penn State University, USA
Shuailong Liang	Singapore University of Technology and Design, Singapore
Xiangwen Liao	Fuzhou University, China
Bin Li	Nanjing Normal University, China
Binyang Li	University of International Relations, China
Changliang Li	Institute of automation, Chinese Academy of Sciences, China
Chen Li	Microsoft, USA
Chenliang Li	Wuhan University, China
Fei Li	Wuhan University, China
Hao Li	Rensselaer Polytechnic Institute, USA
Hongwei Li	ICT, China
Junhui Li	Soochow University, China
Liangyue Li	Arizona State University, USA
Maoxi Li	Jiangxi Normal University, China
Peifeng Li	Soochow University, China
Peng Li	Institute of Information Engineering, CAS, China
Piji Li	The Chinese University of Hong Kong, SAR China
Ru Li	Shanxi University, China
Sheng Li	Adobe Research, USA
Shoushan Li	Soochow University, China
Bingquan Liu	Harbin Institute of Technology, China
Jiangming Liu	University of Edinburgh, UK
Jing Liu	Baidu Inc., China
Lemao Liu	Tencent AI Lab, China
Qun Liu	Dublin City University, Ireland
Shenghua Liu	Institute of Computing Technology, CAS, China
Shujie Liu	Microsoft Research Asia, Beijing, China
Tao Liu	Renmin University of China, China
Yang Liu	Tsinghua University, China
Yijia Liu	Harbin Institute of Technology, China
Zhengzhong Liu	Carnegie Mellon University, USA
Wenjie Li	Hong Kong Polytechnic University, SAR China
Xiang Li	New York University, USA
Xiaoqing Li	Institute of Automation, Chinese Academy of Sciences, China
Yaliang Li	Tencent Medial AI Lab, USA

Yuan-Fang Li	Monash University, Australia
Zhenghua Li	Soochow University, China
Ping Luo	Institute of Computing Technology, CAS, China
Weihua Luo	Alibaba Group, China
Wencan Luo	Google, USA
Zhunchen Luo	PLA Academy of Military Science, China
Qi Lu	Soochow University, China
Wei Lu	Singapore University of Technology and Design, Singapore
Chen Lyu	Guangdong University of Foreign Studies, China
Yajuan Lyu	Baidu Company, China
Cunli Mao	Kunming University of Science and Technology, China
Xian-Ling Mao	Beijing Institute of Technology, China
Shuming Ma	Peking University, China
Yanjun Ma	Baidu, China
Yue Ma	Université Paris Sud, France
Fandong Meng	Tencent, China
Haitao Mi	Alipay US, USA
Lili Mou	AdeptMind Research, Canada
Baolin Peng	The Chinese University of Hong Kong, SAR China
Haoruo Peng	UIUC, USA
Nanyun Peng	University of Southern California, USA
Longhua Qian	Soochow University, China
Tao Qian	Wuhan University, China
Guilin Qi	Southeast University, China
Yanxia Qin	Harbin Institute of Technology, China
Likun Qiu	Ludong University, China
Xipeng Qiu	Fudan University, China
Weiguang Qu	Nanjing Normal University, China
Feiliang Ren	Northerstern University, China
Yafeng Ren	Guangdong University of Foreign Studies, China
Huawei Shen	Institute of Computing Technology, Chinese Academy of Sciences, China
Wei Shen	Nankai University, China
Xiaodong Shi	Xiamen University, China
Wei Song	Capital Normal University, China
Aixin Sun	Nanyang Technological University, Singapore
Chengjie Sun	Harbin Institute of Technology, China
Weiwei Sun	Peking University, China
Yu Su	University of California Santa Barbara, USA
Duyu Tang	Microsoft Research Asia, China
Zhi Tang	Peking University, China
Zhiyang Teng	Singapore University of Technology and Design, Singapore
Jin Ting	Hainan University, China
Ming-Feng Tsai	National Chengchi University, Taiwan, China

Yuen-Hsien Tseng	National Taiwan Normal University, Taiwan, China
Zhaopeng Tu	Tencent AI Lab, China
Bin Wang	Institute of Information Engineering, Chinese Academy of Sciences, China
Chuan Wang	Google Inc., USA
Di Wang	Carnegie Mellon University, USA
Haofen Wang	Shenzhen Gowild Robotics Co. Ltd., China
Kun Wang	Alibaba, China
Longyue Wang	Dublin City University, Ireland
Quan Wang	Institute of Information Engineering, Chinese Academy of Sciences, China
Xiaojie Wang	Beijing University of Posts and Telecommunications, China
Zhiguo Wang	IBM Watson Research Center, USA
Zhongqing Wang	Soochow University, China
Zhongyu Wei	Fudan University, China
Hua Wu	Baidu, China
Yunfang Wu	Peking University, China
Tong Xiao	Northestern University, China
Yanghua Xiao	Fudan University, China
Rui Xia	Nanjing University of Science and Technology, China
Wayne Xin Zhao	RUC, China
Chenyan Xiong	Carnegie Mellon University, USA
Deyi Xiong	Soochow University, China
Shufeng Xiong	Wuhan University, China
Jun Xu	Institute of Computing Technology, CAS, China
Kun Xu	IBM T.J. Watson Research Center, USA
Endong Xun	Bejing Language and Cultural University, China
Jie Yang	Singapore University of Technology and Design, Singapore
Liang Yang	Dalian University of Technology, China
Liner Yang	Tsinghua University, China
Liu Yang	University of Massachusetts Amherst, USA
Yating Yang	The Xinjing Technical Institute of Physics and Chemistry, CAS, China
Zi Yang	Google, USA
Oi Yee Kwong	The Chinese University of Hong Kong, SAR China
Peifeng Yin	IBM Almaden Research Center, USA
Wenpeng Yin	University of Pennsylvania, USA
Bei Yu	Syracuse University, USA
Dong Yu	Beijing Language and Culture University, China
Junjie Yu	Soochow University, China
Mo Yu	IBM Research, USA
Zhengtao Yu	Kunming University of Science and Technology, China
Xiangrong Zeng	Institute of Automation, Chinese Academy of Sciences, China

Ying Zeng	Peking University, China
Feifei Zhai	Sogou Inc., China
Chengzhi Zhang	Nanjing University of Science and Technology, China
Dongdong Zhang	Microsoft Research Asia, China
Fan Zhang	University of Pittsburgh, USA
Fuzheng Zhang	MSRA, China
Min Zhang	Tsinghua University, China
Peng Zhang	Tianjin University, China
Qi Zhang	Fudan University, China
Weinan Zhang	Shanghai Jiao Tong University, China
Xiaodong Zhang	Peking University, China
Xiaowang Zhang	Tianjin University, China
Yongfeng Zhang	Rutgers University, USA
Yue Zhang	Singapore University of Technology and Design, Singapore
Hai Zhao	Shanghai Jiao Tong University, China
Jieyu Zhao	University of California, Los Angeles, USA
Sendong Zhao	Harbin Institute of Technology, China
Tiejun Zhao	Harbin Institute of Technology, China
Guoqing Zheng	Carnegie Mellon University, USA
Deyu Zhou	Southeast University, China
Guangyou Zhou	Central China Normal University, China
Hao Zhou	Bytedance AI Lab, China
Junsheng Zhou	Nanjing Normal University, China
Ming Zhou	Microsoft Research Asia, China
Muhua Zhu	Alibaba Inc., China

Organizers

Organized by

China Computer Federation, China

Supported by

Asian Federation of Natural Language Processing

Hosted by

Inner Mongolia University

State Key Lab of Digital Publishing Technology

In Cooperation with:

Lecture Notes in Computer Science

Springer

 Springer

ACTA Scientiarum Naturalium Universitatis Pekinensis

Sponsoring Institutions

Primary Sponsors

AI Strong

JINGDONG

Diamond Sponsors

Tencent LINGO Lab

ZHINENGYIDIAN

Sogou

AISPEECH

LIULISHUO

China Mobile

Alibaba Group

GTCOM

Platinum Sponsors

Microsoft

Baidu

Leyan Tech

Laiye

Huawei

GRID SUM

LENOVO

AITC

Unisound

XIAOMI

CVTE

ByteDance

WISERS

Golden Sponsors

NiuParser

SoftBank

Genelife

Contents – Part I

Knowledge Graph/IE

Machine Learning for NLP

Machine Translation

NLP Applications

Contents – Part II

Text Mining

Short Papers

Conversational Bot/QA/IR

Question Answering for Technical Customer Support

Yang Li[1]([✉]), Qingliang Miao[1], Ji Geng[1], Christoph Alt[2],
Robert Schwarzenberg[2], Leonhard Hennig[2], Changjian Hu[1], and Feiyu Xu[1]

[1] Lenovo, Building H, No. 6, West Shangdi Road, Haidian District Beijing, China
{liyang54,miaoql1,hucj1,fxu}@lenovo.com, jgeng@uestc.edu.cn
[2] DFKI, Alt-Moabit 91c, 10559 Berlin, Germany
{christoph.alt,robert.schwarzenberg,leonhard.hennig}@dfki.de

Abstract. Human agents in technical customer support provide users
with instructional answers to solve a task. Developing a technical support
question answering (QA) system is challenging due to the broad variety
of user intents. Moreover, user questions are noisy (for example, spelling
mistakes), redundant and have various natural language expresses, which
are challenges for QA system to match user queries to corresponding
standard QA pair. In this work, we combine question intent categories
classification and semantic matching model to filter and select correct
answers from a back-end knowledge base. Using a real world user chatlog dataset with 60 intent categories, we observe that while supervised
models, perform well on the individual classification tasks. For semantic matching, we add muti-info (answer and product information) into
standard question and emphasize context information of user query (captured by GRU) into our model. Experiment results indicate that neural
multi-perspective sentence similarity networks outperform baseline models. The precision of semantic matching model is 85%.

Keywords: Question and Answer · Answer selection
Semantic matching

1 Introduction

"My Wi-Fi is not working anymore!!!" – most mobile device users probably
have faced this or similar issues in the past. Solving such questions is the task
of technical customer support agents (CSAs). For frequent questions and user
intents, for which solutions often exist in the form of user guides and question-
answering knowledge base (QA-KB), this is a repetitive and time consuming
process. For example, our QA-KB contains 242 unique standard QA pairs, and
these QA pairs are assigned to 60 intents. QA system would significantly reduce
the time CSAs have to invest in solving common intents, which they could then
focus on more complex or previously unseen customer problems.

© Springer Nature Switzerland AG 2018
M. Zhang et al. (Eds.): NLPCC 2018, LNAI 11108, pp. 3–15, 2018.
https://doi.org/10.1007/978-3-319-99495-6_1

With the work described in this paper, we aim to automatize this task of matching instructional answers from a QA-KB to user queries. Our work combines two models, the first is question intent categories classification which for getting candidate QA pairs from QA-KB, the second is semantic matching which focuses on selecting the correct result from candidate QA pairs. Intent categories classification faces a big challenge due to the broad variety of intents in QA-KB.

> **User query:**
> Hi, I am a new moto user. I have a moto z plus. My English is bad! How do I add a google account? Can you tell me how to do it?
> **Standard question:**
> Add or remove accounts.
> **Answer:**
> You can use multiple Google Accounts and Microsoft Exchange Active Sync accounts on your device. Add account: You may need to obtain details from system administrator about the service to which the account connects. For example, you may need to know the account's domain or server address.1. Touch Apps > Settings2. Touch Accounts3. Select the type of account, e.g. Exchange, Email, Google

Fig. 1. An example of actual business data and its corresponding QA entry.

Figure 1 shows an example of real world user query and its corresponding QA entry in QA-KB. The first field is user query, the second field is standard question and the last one is answer. Table 1 lists statistic information of the data set. Through careful analysis, we obtain following characteristics. First, user queries are usually noisy (for example, spelling mistakes) and contain background or redundant information besides true intent. Second, standard questions are typically short, concise and often realized as title-like phrases, e.g. "Add or remove accounts" because this format is easy for user and CSAs to read. In addition, we can see that user query is 5 times longer than standard question so it's difficult to match these two contents. Third, answer is quite long (the average length is 78.2) and contains some information that related to user query. Due to the characteristics of the data, we add answer information into standard question. Specially, we use Gated Recurrent Unit (GRU) [1] to get context information of user query and learn long-term dependencies before multi-perspective CNN [2].

Table 1. The statistics of data.

	Num	Average length(word level)
User query	6808	31.9
Standard question	242	5.2
Answer	242	78.2

2   Related Work

Existing work on QA systems in the customer service domain has focused on answering Ubuntu OS-related questions [3], insurance domain [4] and customer relationship management [5]. Both studies show that it is in principle possible to handle longer dialogs in an unsupervised fashion and answer complex questions with the help of a noisy training set and an unstructured knowledge source. Lowe et al. [3] use a large corpus of support dialogs in the operating system domain to train an end-to-end dialog system for answering customer questions. Their results indicate that end-to-end trained systems can achieve good performance but perform poorly on dialogs that require specific domain knowledge which the model possibly never observed. In contrast, in our work we adopt a classical classification approach followed by semantically matching a user question to a set of results from a QA-KB, in order to cope with the limited amount of training data.

Most previous work on semantic matching has focused on handcrafted features. Due to the variety of word choices and inherent ambiguities in natural languages, bag-of-word approaches with simple surface-form word matching tend to poor prediction precision [6]. As a result, researchers put more emphasis on exploiting syntactic and semantic structure which are more complex and time consuming. Representative examples include methods based on deeper semantic analysis [7] and quasi-synchronous grammars [8] that match the dependency parse trees of the two sentences. Instead of focusing on the high-level semantic representation, Yih et al. turn their attention to improve the shallow semantic component, lexical semantics [9].

As development of neural network, recent work has moved away from handcrafted features and towards modeling with distributed representations and neural network architectures. Hu et al. propose two general CNN architectures ARC-I and ARC-II for matching two general sentences, and ARC-II consider the interaction between input two sentences [10]. Liu et al. propose a dual attentive neural network framework(DANN) to embed question topics and user network structures for answer selection. DANN first learns the representation of questions and answers by CNN. Then DANN learns interactions of questions and answers which is guided via user network structures and semantic matching of question topics with double attention [11]. He et al. propose a model for comparing sentences that uses a multiplicity of perspectives. They first use a CNN model to extract features at multiple levels of granularity and then use multiple similarity metrics to measure sentence similarity [2]. Feng et al. create and release an insurance domain QA corpus. The paper demonstrate 13 proposed neural network model architectures for selecting the matched answer [4]. Gaurav et al. use character n-gram embedding instead of word embedding and noisy pretraining for the task of question paraphrase identification [12]. Wu et al. propose a multi-turn sequential matching network SMN which matches two sentences in the context on multiple granularity, and distills important matching information from each pair with convolution and pooling operations. And then, a recurrent neural network (RNN) model is used to model sentence relationships [13]. These models either calculate the similarity of users surface form question (Qu) and standard query (Q_i) in Q-A KB or calculate the similarity of Qu and answer (A_i). Our work comprehensive use Qu, Q_i and A_i. Besides we use GRU to get context information of Qu and learns long-term dependencies before CNN.

3   The Proposed Approach

3.1   Problem Formalization

The goal of our approach is to identify the Q_iA_i from n candidate QA pairs $\{(Q_1, A_1), \ldots, (Q_n, A_n)\}$ of Q-A KB that best matches Qu. Q_i is a concise and representative question such as "Connect to a Wifi network" which prototypically stands for other questions that can be answered by A_i. Figure 1 shows the example of an user query and the corresponding QA pair of QA-KB.

We hypothesize the QA pair that shares the most semantic similarity with the Qu. Following a common information retrieval approach, we use a pairwise scoring function S(Q_iA_i, Qu) to sort all candidate of the question expressed by user. Our method has two main steps, the first is intent category classification to select relevant candidate QA pairs from QA-KB. The second is semantic matching to select the best matching one from candidate QA pairs.

3.2   Intent Category Classification

Determining the correct intent category significantly reduces the number of candidate QA pairs. The dataset contains 60 intents such as "Wifi", "Screen Unlock", "Google Account", etc.

The question intent category classifier estimates the probability $p(I|Qu)$ where I denotes the intent. Our baseline approaches are Gradient Boosted Decision Trees and a linear SVM. For feature extraction, the Qu is tokenized, followed by stop-word removal and transformation into a bag-of-words representation. The classifiers use tfidf weighted unigram and bigram features. We also implement a bidirectional LSTM model [15]. In this model, each $w_i \in$ Qu is represented by an embedding $e_i \in R^d$ that we obtain from a set of pretrained distributed word representations E = $[e_1, \ldots, e_W]$. The BiLSTM output is passed to a fully-connected layer followed by a ReLU non-linearity and softmax normalization, s.t. $p(I|Qu)$ is computed as follows

$$SM(ReLU(FC(BiLSTM(E))))(Qu) \tag{1}$$

3.3   Semantic Matching

In this section, we present innovative solutions that incorporate multi-info and context information of user question into multi-perspective CNN to fulfill question paraphrase identification. The architecture of our neural network is shown in Fig. 2. The work has two same subnetworks that processing Qu and Q_iA_i

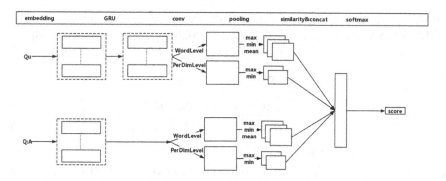

Fig. 2. Multi-perspective sentence similarity network with GRU.

in parallel after getting context by GRU. The following layer extracts features at multiple levels of granularity and uses multiple types of pooling. After that, sentence representations are compared with several granularities using multiple similarity metrics such as cosine similarity and L2 euclidean distance that are distilled into a matching vector followed by a linear projection and softmax normalization.

Multi-info. To the data, Qu is quite long, Q_i is short and contains less information. Besides, the A_i is quite long and contains some information that related to Qu. In this work, we concat Q_i and A_i of QA-KB then to compute $S(Q_iA_i,$ Qu$)$. User queries are always concerned with a specific product but some related standard questions for different products may be the same in the QA-KB. As you can see the example "moto z plus" in Fig. 1 which is a mobile name. Due to we do not consider the influence of different mobile, we directly replace these mobiles by the same word "Mobile". We use Product-KB and CRF algorithm to recognize the mobile in Qu. The ontology of Product-KB are constructed by senior businessmen and front-line customer service staff. Pink part of Fig. 2 indicates the structure of the Product-KB. In Product-KB, every mobile has its surface names which are mined from huge chat log. Most surface mobile name of Qu can be recognized by Product-KB.

Knowledge base hardly contains all mobiles and their corresponding surface names so we use CRF to recognize the mobile as a supplement. Features of mobile recognition are char level ngrams and word level ngrams. Maximum char level ngrams is 6 and word level ngrams is 3.

Context Multi-perspective CNN. After getting the multi-info, the input of our network are Qu and Q_iA_i. Both of them need to transfer all letters to lowercase. Given an user query Qu and a response candidate Q_iA_i, the model looks up an embedding table and represents Qu and Q_iA_i as Qu$=[e_{u,1},e_{u,2},...,e_{u,L}]$ and $Q_iA_i=[e_{s,1},e_{s,2},...,e_{s,L}]$ respectively, where $e_{u,j}$ and $e_{s,j} \in R^d$ are the embeddings of the j-th word of Qu and Q_iA_i respectively. L is the max length of the two sequences. Before feed into Multi-Perspective CNN, we first employ a GRU to transform Qu to hidden vectors conM_{Qu}. Suppose that $\text{conM}_{Qu} = [h_{u,1}, h_{u,1}, \ldots, h_{u,L}]$ are the hidden vectors of Qu, then $h_{u,i}$ is defined by

$$z_i = \sigma(W_z e_{u,i} + U_z h_{u,i-1}) \tag{2}$$

$$r_i = \sigma(W_r e_{u,i} + U_r h_{u,i-1}) \tag{3}$$

$$\overline{h}_{u,i} = tanh(W_h e_{u,i} + U_h(r_i \odot h_{u,i-1})) \tag{4}$$

$$h_{u,i} = z_i \odot \overline{h}_{u,i} + (1 - z_i) \odot h_{u,i-1} \tag{5}$$

where $h_{u,0} = 0$, z_i and r_i are an update gate and a reset gate respectively, $\sigma(.)$ is a sigmoid function, and W_z, W_r, W_h, U_z, U_r, U_h are parameters. The model only gets context information of Qu and learns long-term dependencies by GRU because $Q_i A_i$ is not a sequential sentence. conM_{Qu} and $Q_i A_i$ are then processed by the same CNN subnetworks. This work applies to multi-perspective convolutional filters: word level filters and embedding level filters. Word level filters operate over sliding windows while considering the full dimensionality of the word embeddings, like typical temporal convolutional filters. The embedding level filters focus on information at a finer granularity and operate over sliding windows of each dimension of the word embeddings. Embedding level filters can find and extract information from individual dimensions, while word level filters can discover broader patterns of contextual information. We use both kinds of filters allow more information to be extracted for richer sentence modeling.

For each output vector of a convolutional filter, the model converts it to a scalar via a pooling layer. Pooling helps a convolutional model retain the most prominent and prevalent features, which is helpful for robustness across examples. One widely adopted pooling layer is max pooling, which applies a max operation over the input vector and returns the maximum value. In addition to max pooling, the model uses two other types of pooling, min and mean, to extract different aspects of the filter matches.

Multi-similarity. After the sentence models produce representations for Qu and $Q_i A_i$ then to calculate the similarity of their representations. One straight forward way to compare them is to flatten their representations into two vectors, then use standard metrics like cosine similarity. However, this may not be optimal because different regions of the flattened sentence representations are from different underlying sources. Flattening might discard useful compositional information for computing similarity. We therefore perform structured comparisons over particular regions of the sentence representations.

The model uses rules to identify local regions whose underlying components are related. These rules consider whether the local regions are: (1) from the same filter type; (2) from the convolutional filter with the same window size; (3) from the same pooling type; (4) from the same specific filter of the underlying convolution filter type. Then we use same algorithms as MPCNN [2] to calculate similarity matching vector. MPCNN use two algorithms by three similarity metrics(Cosine distance, L2 Euclidean distance, Manhattan distance) to compare local regions. The first algorithm works on the output of holistic filters only, while the other uses the outputs of both the holistic and per-dimension filters.

4   Experiments and Discussion

We evaluate our approaches for question intent category classification, as well as semantic matching using real world user chatlog data. Next, we will introduce the dataset, QA-KB and experiment results separately.

4.1   Data Set

The dataset mainly consists of user and agent conversation records, in which user question and technical answer are stated. Each conversation record includes the full text of each utterance, chat starting and ending time, user and agent ids, and optionally a product id and an intent category assigned by the customer service agent. From 80216 user and agent conversation records, we extract 6808 user questions and annotated with a gold standard QA pair, an intent category and a product id. The distribution over the top 30 intent categories (out of 60) is shown in Fig. 4.

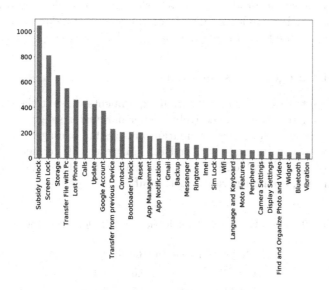

Fig. 3. Distribution of intent categories (top 30) for user question.

4.2   QA-KB and Product-KB

The KB module stores the answers of question and its relevant product. A diagram capturing the simplified structure of the KB is depicted in Fig. 4. The

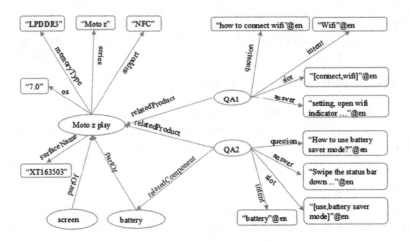

Fig. 4. Structure of the product-KB and QA-KB.

left part is the hardware and software parameters of mobile product, such as operating system, memory type, supported features, components and surface names. The right part shows QA pairs that include standard question name, corresponding answers, products, relevant slot, and an intent category. In the current version, KB includes 20 mobile products, 242 standard questions. KB totally includes more than 150000 triples.

4.3   Intent Category Classification

For question intent category classification experiments we split the dataset into 80/20 train and test sets, respectively. Hyper-parameter selection is done on the training set via 5-fold cross validation and results averaged over multiple runs are reported on the test set. For BiLSTM we use 300 dimensional GloVe word embeddings [14]. Table 2 shows the evaluation results on the dataset. The baselines perform, even outperforming the BiLSTM model.

Table 2. Intent Category Classification Results for User Question.

Model	Precision	Recall	F1
GBDT	0.67	0.68	0.67
BiLSTM	0.68	0.70	0.69
SVM	0.74	0.76	0.75

From the detailed per category results (SVM) in Table 3 we find that some categories (e.g. "Google Account and Transfer from previous Device") achieve a

Table 3. Intent category classification results for user question, Top 10 categories.

Model	Precision	Recall	F1
Subsidy unlock	0.83	0.93	0.88
Screen lock	0.84	0.92	0.88
Storage	0.79	0.85	0.82
Transfer file w. PC	0.77	0.91	0.83
Lost phone	0.87	0.91	0.89
Calls	0.79	0.87	0.83
Google account	0.64	0.72	0.68
Update	0.77	0.92	0.84
Bootloader unlock	0.93	0.67	0.78
Transfer p. device	0.67	0.74	0.70

disproportional lower performance. For example, "Google Account" is often confused with "Reset as a Google account" is generally a main topic when trying to reset a device (e.g., "Android smartphone"). It is also noteworthy that "Subsidy Unlock", "Bootloader Unlock" and "Screen Lock" are frequently confused. This is best illustrated by the example "Hi i need pin for unlock red to my moto g", which has the true category "Subsidy Unlock" but is categorized as "Screen Lock". Without knowledge about the mobile phones & contracts domain it is very difficult to understand that the customer is referring to a "pin" (subsidy unlock code) for "red" (mobile service provider) and not the actual PIN code for unlocking the phone. This example also symbolizes a common problem in customer support, where users unfamiliar with the domain are not able to describe their information need in the domain-specific terminology.

4.4   Semantic Matching

For semantic matching we evaluate TFIDF and WMD as unsupervised baselines for obtaining the most semantically similar QA pair to a given Qu. Supervised approaches include the sequential matching network (SMN), a multi-perspective CNN (MPCNN) with and without a GRU layer for user question encoding.

TFIDF and WDM. Our first baseline(TFIDF) use a tfidf weighted bag-of-words representation of Q_iA_i and Qu to estimate the semantic relatedness by cosine similarity $\cos(Q_iA_i, Qu)$.

The second baseline(WDM) leverages the semantic information of distributed word representations [16]. To this end, we replace the tokens in Q_iA_i and Qu with their respective embeddings and then compute the word mover distance [17] between the embeddings.

SMN and MPCNN. In addition to the unsupervised method we also use SMN [13] and MPCNN [2], which treats semantic matching as a classification task. The SMN first represents Q_iA_i and Qu by their respective sequence of word embeddings Ei and Ek before encoding both separately with a recurrent network, GRU [1] in this case. A word-word similarity matrix Mw and a sequence similarity matrixMs is constructed from Ei and Ek, and important matching information is distilled into a matching vector vm via a convolutional layer followed by max-pooling. vm is further projected using a fully connected layer followed by a softmax.

The MPCNN [2] first represents Q_iA_i and Qu, the following layer extracts features at multiple levels of granularity and uses multiple types of pooling. Afterwards, sentence representations are compared with several granularities using multiple similarity metrics such as cosine similarity and L2 euclidean distance. The results are distilled into a matching vector followed by a linear projection and softmax normalization.

Model Result. The description of MPCNN_GRU model is showed in Chap. 3.4. For all models except TF-IDF, we use 300 dimensional GloVe word embeddings [14]. To obtain negative samples, for each Qu, we randomly select 5 standard queries with the same intent and 5 standard queries with different intents. To alleviate the impact of unbalanced training data, we oversample positive samples. As the standard questions Q_i of most QA pairs (Q_i, A_i) are usually less then 10 tokens, we also evaluate the impact on model performance when adding the answer A_i as additional context (up to 500 characters) to Q_i. For the experimentation we randomly split the dataset 80/20 into train and test set and repeat the experiment 5 times. Hyperparameter selection is done on 10% of the training set and results are reported on the test set.

Table 4 shows the precision of each model on the semantic matching task. We see that the MPCNN and MPCNN_GRU outperform the unsupervised baseline approaches, with a 43% error reduction achieved with the MPCNN_GRU model. Intuitively it makes sense to provide the models with additional context that can be used to learn a better representation of semantic similarity. Adding a GRU to the MPCNN to encode contextual information and long-range dependencies in the user query does not really improve performance. The SMN's precision and

Table 4. Semantic matching results for user question.

Model	Without answer	With answer	Average response time
TF-IDF	0.62	0.60	null
WMD	0.60	0.58	null
SMN	0.62	0.68	200ms
MPCNN	0.72	0.84	50ms
MPCNN_GRU	0.72	0.85	55ms

recall scores are much lower than those of the MPCNN models, and only slightly higher than those of the unsupervised approaches.

Beside the precision of each semantic matching model, we also conduct experiments to evaluate the efficiency of each models. The machine configuration information in our experiment is 2 i7 CPUs with 14 cores, a memory of 125G and a disk of 930.4G. The last column of Table 4 shows the results. From Table 4, we can see MPCNN is faster than other two models. When adding GRU, the average response time increases 5ms. SMN model is slowest, because the neural network structure of SMN is more complicated than MPCNN_GRU and MPCNN. The experiment results indicate MPCNN_GRU and MPCNN is capable for real time system.

4.5   The Importance of Intent Classification for Semantic Matching

Question intent categories classification is an important step to narrow down answer candidates. In this section, we compare models with a baseline to highlight the effectiveness of intent categories classification. The baseline uses the same model as MPCNN and MPCNN_GRU, without intent categories classification so the model directly matching with all QA pairs (262) in QA-KB. Table 5 indicates that the precision of semantic matching with intent outperforms baseline models.

Table 5. Semantic matching results on baseline for User Question.

	Without intent	With intent
MPCNN	0.63	0.84
MPCNN_GRU	0.65	0.85

5   Conclusion

In this paper we presented a approach for question answering in the complex and little-explored domain of technical customer support. Our approach incorporates intent classification and semantic matching to select an answer from knowledge base. Question intent classification for a dataset with 60 intent categories and model performs reasonably well on the individual classification tasks. In semantic matching, we incorporate multi-info and context information into multi-perspective CNN to fulfill question paraphrase identification. The precision of semantic matching is 85%. Our approach outperforms baseline models. For future research, we plan to train an end-to-end model jointly add more QA pairs into QA-KB to solve more problems of customers.

References

1. Chung, J., Gulcehre, C., Cho, K.H.: Empirical evaluation of gated recurrent neural networks on sequence modeling. arXiv preprint arXiv:1412.3555 (2014)
2. He, H., Gimpel, K., Lin, J.: Multi-perspective sentence similarity modeling with convolutional neural networks. In: 20th International Proceedings on Empirical Methods in Natural Language Processing, pp. 1576–1586. ACL, Stroudsburg (2015)
3. Lowe, R.T., Pow, N., Serban, I.V.: Training end-to-end dialogue systems with the Ubuntu dialogue corpus. Dialogue Discourse **8**(1), 31–65 (2017)
4. Feng, M., Xiang, B., Glass, M.R.: Applying deep learning to answer selection: a study and an open task. In: 3rd International Proceedings on Automatic Speech Recognition and Understanding (ASRU), pp. 813–820. IEEE, Piscataway (2015)
5. Li, X., Li, L., Gao, J.: Recurrent reinforcement learning: a hybrid approach. Computer Science (2015)
6. Bilotti, M.W., Ogilvie, P., Callan, J.: Structured retrieval for question answering. In: 30th International Proceedings on SIGIR Conference on Research and Development in Information Retrieval, pp. 351–358. ACM, New York (2007)
7. Shen, D., Lapata, M.: Using semantic roles to improve question answering. In: 12th International Proceedings on Joint Conference on Empirical Methods in Natural Language Processing and Computational Natural Language Learning, pp. 12–21. ACL, Stroudsburg (2007)
8. Wang, M., Smith, N.A., Mitamura, T.: What is the Jeopardy model? A quasi-synchronous grammar for QA. In: 12th International Proceedings on Joint Conference on Empirical Methods in Natural Language Processing and Computational Natural Language Learning, pp. 22–32. ACL, Stroudsburg (2007)
9. Yih, W., Chang, M.W., Meek, C.: Question answering using enhanced lexical semantic models. In: 51st International Proceedings on Association for Computational Linguistics, pp. 1744–1753. ACL, Stroudsburg (2013)
10. Hu, B., Lu, Z., Li, H.: Convolutional neural network architectures for matching natural language sentences. In: 23rd International Proceedings on Neural Information Processing Systems, pp. 2042–2050. Springer, Berlin (2014)
11. Liu, Z., Li, M., Bai, T., Yan, R., Zhang, Y.: A dual attentive neural network framework with community metadata for answer selection. In: Huang, X., Jiang, J., Zhao, D., Feng, Y., Hong, Y. (eds.) NLPCC 2017. LNCS (LNAI), vol. 10619, pp. 88–100. Springer, Cham (2018). https://doi.org/10.1007/978-3-319-73618-1_8
12. Tomar, G.S., Duque, T.: Neural paraphrase identification of questions with noisy pretraining. In: 22th International Proceedings on Empirical Methods in Natural Language Processing, pp. 142–147. ACL, Stroudsburg (2017)
13. Wu, Y., Wu, W., Xing, C.: Sequential matching network: a new architecture for multi-turn response selection in retrieval-based Chatbots. In: 55th Annual Meeting of the Association for Computational Linguistics, pp. 496–505. ACL, Stroudsburg (2017)
14. Pennington, J., Socher, R., Manning, C.: Glove: global vectors for word representation. In: 19th International Proceedings on Empirical Methods in Natural Language Processing, pp. 1532–1543. ACL, Stroudsburg (2014)
15. Hochreiter, S., Schmidhuber, J.: Long short-term memory. Neural Comput. **9**(8), 1735–1780 (1997)

16. Mikolov, T., Sutskever, I., Chen, K.: Distributed representations of words and phrases and their compositionality. In: 9th International Proceedings on Advances in Neural Information Processing System, pp. 3111–3119. MIT Press, Massachusetts (2013)
17. Kusner, M., Sun, Y., Kolkin, N.: From word embeddings to document distances. In: 32nd International Proceedings on International Conference on Machine Learning, pp. 957–966. ACM, New York (2015)

Perception and Production of Mandarin Monosyllabic Tones by Amdo Tibetan College Students

Zhenye Gan[1,2(✉)], Jiafang Han[1,2], and Hongwu Yang[1,2]

[1] College of Physics and Electronic Engineering,
Northwest Normal University, Lanzhou 73000, China
ganzy@nwnu.edu.cn
[2] Engineering Research Center of Gansu Province for Intelligent Information
Technology and Application, Lanzhou 73000, China

Abstract. The purpose of the work is to research the error patterns of production and perception of Mandarin monosyllabic tone by college students from Amdo Tibetan agricultural and pastoral areas, and make the analysis of the causes of acoustics in both errors. We do the work through the two experiments of perception and production of tone. We use the methods of combining the speech engineering and experimental phonetics. Results show that the error rate of tone perception is highly correlated [r = 0.92] with that of tone production. The level of Mandarin in Amdo Tibetan agricultural area is higher than that in pastoral area both in terms of tone perception and production. The hierarchy of difficulty for the four grades in agricultural area is as follows: sophomore > freshman > junior > senior, pastoral area is as follows: freshman > sophomore > junior > senior. The hierarchy of difficulty for the four tones both in agricultural and pastoral areas is as follows: Tone 2 > Tone 3 > Tone 1 > Tone 4. Tone 2 and 3 are most likely to be confused. There is no obvious tone shape bias of the four tones, but the tone domain is narrow and the location of the tone domain is lower than standard Mandarin both in agriculture and pastoral areas.

Keywords: Amdo tibetan college students · Agriculture and pastoral areas
Tonal production and perception

1 Introduction

Amdo Tibetan is different from Mandarin. The most significant difference is that Mandarin is a tonal language while Amdo Tibetan is not. Previous studies have found that the background of native language influenced the learning of the tone of the second language [1, 2]. Therefore, students in the Amdo Tibetan area have difficulty in the learning of Mandarin tone. Liu [3] had pointed out that tone was an important factor,

Foundation item: National Natural Science Foundation of China (Grant No. 61262055, 11664036).

© Springer Nature Switzerland AG 2018

M. Zhang et al. (Eds.): NLPCC 2018, LNAI 11108, pp. 16–26, 2018.
https://doi.org/10.1007/978-3-319-99495-6_2

which directly affected the level of Mandarin. Tone should be the focus of research in the process of learning Mandarin.

In recent years, studies of Mandarin tone have focused on the learning situation of international students [4–8]. There were relatively few studies on the learning of Mandarin tone by Tibetans, especially the Amdo Tibetan. Literatures [9–11] conducted related research on the analysis of Mandarin tone of Amdo Tibetans to summarize the characteristics of Mandarin monosyllables tone errors by Amdo Tibetans. Most of the studies focused on the research of tone pronunciation by the experimental phonetics and linguistic methods. And the selection of experimental personnel was limited to a certain area. However, learning a language was as important as hearing and pronunciation. Only byhearing correctly can you have a correct pronunciation [12]. For most of the college students from the Amdo Tibetan area, they used Tibetan as their primary language both in primary and middle school, supplemented by Mandarin. When they came to the University of Han nationality and entered into the whole Mandarin environment suddenly, which makes it difficult for them to communicate and learn in life. Therefore, it is necessary to start research from the university to get a more comprehensive law.□

This paper was based on the needs of the Mandarin learning by college students from Amdo Tibetan area. According to the different level of Mandarin in the internal of Amdo Tibetan area and the sudden change of the Mandarin environment of college students from the Amdo Tibetan area, the college students from Amdo Tibetan agriculture and pastoral areas were taken as research subjects, the research subjects were in the four grades of freshman to senior. Agricultural area refers to areas where crops are the main source of life while pastoral area refers to areas where livestock and cattle are the main sources of life. In this paper, we selected suitable experimental subjects through field investigations to research the tone perception and production of Mandarin by college students from Amdo Tibetan area. We used the methods of combining the speech engineering and experimental phonetics. The main works of the paper were as follows. Firstly, we make analysis of the perception bias of the Mandarin tone by college students from Amdo pastoral and agriculture areas. Secondly, we make analysis of the error patterns of production of Mandarin tone by college students from Amdo pastoral and agriculture areas. Thirdly, we make the analysis of the causes of acoustics in both two errors.

2 Perception of Mandarin Tone

2.1 Experimental Corpus

Literature [13] once pointed out that the tone acquisition of Central Asian students had more difficulties in continuous speech than that in a single word. Therefore, the corpus used in this experiment consists of 50 long sentences in Mandarin. Each long sentence consists of 4 short sentences. Each short sentence has 5 monosyllables. The tone of last syllables of each short sentence is Tone 1, Tone 2, Tone 3, and Tone 4. The experimental corpus is 200 monosyllables and distributed evenly among four tones. The corpus is covering all monosyllable structures basically.

2.2 Experimental Subjects

The experimental corpus of standard Mandarin was produced by a postgraduate, a woman of Han nationality. The subjects were college students from Amdo Tibetan area and none of them had difficulty in hearing and speaking.

2.3 Experimental Process

Step1, Standard Mandarin was recorded by the above-mentioned female, which was recorded with 16 kHz and 16 bit in a sound proof studio. We denoise and segmented the speech signals into perception speech signals.
Step2, we designed a questionnaire according to the perception speech signals.
Step3, the subjects were asked to listen to the speech signals and write out the most probable tone marks of the monosyllabic they had heard. The tone of the monosyllables would be marked as follows: T1, T2, T3 and T4, which represent the four tones.

A total of 185 questionnaires were distributed and we finally obtained134 valid data. We removed some results such as an outgoing result that is not in line with the general rule. The valid data included 60 in pastoral area and 74 in agriculture area. The error rate of each tone was separately counted. All data were tested for normal distribution and then analyzed.

2.4 Experimental Results

As shown in Figs. 1 and 2, it can be seen that the perceptual error rate of college students in pastoral area is higher than that in agriculture area. Tones 2 and 3 have higher error rates in the four tones. Tone 4 is best judged, which is related to the reason that the tone of Amdo Tibetan had only one habit tone.

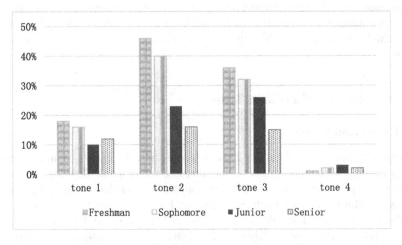

Fig. 1. The result of tone perception in pastoral area

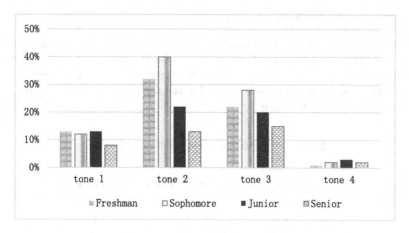

Fig. 2. The result of tone perception in agriculture area

It can be concluded that the hierarchy of difficulty for the four tones both in agriculture and pastoral areas is as follows: Tone 2 > Tone 3 > Tone 1 > Tone 4. The hierarchy of difficulty for the four grades in pastoral area is as follows: freshman > sophomore > junior > senior, agriculture area is as follows: sophomore > freshman > junior > senior. We find that the perception proficiency of freshman is better than sophomore in agriculture area which has stayed in school for one year. This is more relevant to that freshman has more chances to contact Chinese than sophomore in agriculture area. The progress of society is affecting their Mandarin proficiency. It will be further analyzed by using the follow experiment of tone production.

3   Production of Mandarin Tone

The experiment of tone production includes two parts: the tone annotation and the production of tone pronunciation experiment. We will first make tonal annotation experiment to select appropriate speech corpora. And then make further analysis of the error pattern of production in tone pronunciation.

Tone annotation is that several listeners judge the correct tone and the type of tone by subjects, and then decide the label of the tone. The tone of the monosyllables would be labeled as follows: T1, T2, T3 and T4, which represent the four tones.

3.1   Experimental Corpus

To make the result of the comparison between the tone perception and production correctly, the corpus used in this experiment is chosen from the perceptual corpus. It consists of 40 long sentences in Mandarin. There are a total of 160 monosyllables and distributed evenly among the four tones.

3.2 Experimental Subjects

Experimental subjects included 16 speakers of Amdo Tibetan. None had difficulty in hearing and speaking. The detailed information of the 16 speakers of Amdo Tibetan is as showed in the following Table 1. Through a large number of questionnaires, 16 speakers whose native language was Amdo Tibetan were selected. They had the same background in learning Mandarin.

Table 1. Experimental subjects information table

Area	Grade				
	Freshman	Sophomore	Junior	Senior	Total
Agriculture	2	2	2	2	8
Pastoral	2	2	2	2	8
Total	4	4	4	4	16

3.3 Experimental Process

In this experiment, 6 listeners of Han nationality were invited and didn't know the original correct tone. Each subject had 160 speech data. 6 listeners listened to the data of 16 speakers and judged the tone of the last word of each short sentence, and then made a mandatory tone judgment and labeled it as T1, T2, T3 andT4.

3.4 Experimental Results

The following Table 2 shows the result of comparing the results of the annotation in pastoral area with standard Mandarin. Table 3 shows the result of comparing agriculture areas with the standard mandarin. T1, T2, T3 and T4 represent the four tones.

Table 2. Results of annotation of Pastoral area

Correct tone	Tone annotation			
	Compare annotation result with correct tone			
	T 1	T 2	T 3	T 4
T 1	**75.0%**	7.5%	5.0%	5.0%
T 2	10.0%	**55.0%**	22.5%	5.0%
T 3	7.5%	27.5%	**65.0%**	5.0%
T 4	7.5%	10.0%	7.5%	**85.0%**

Comparing the results of the two groups, the accuracy of the Tone 4 is the highest both in the agriculture and pastoral areas and the lowest accuracy is Tone 2. We can also see the obvious differences from the comparison between the agriculture area and the pastoral area. The pronunciation accuracy of the four tones of the agriculture area is higher than that of the pastoral area. The overall accuracy rate is higher than 50%. From

Table 3. Results of annotation of agriculture area

Correct tone	Tone annotation			
	Compare annotation result with correct tone			
	T 1	T 2	T 3	T 4
T 1	**80.0%**	7.5%	2.5%	5.0%
T 2	10.0%	**67.5%**	17.5%	2.5%
T 3	7.5%	20.0%	**70.0%**	2.5%
T 4	2.5%	5.0%	10.0%	**90.0%**

Tables 2 and 3, we can also find that Tone 2 and 3 are most easily confused. It is probably because of the similar phonetic properties in Tone 2 and Tone 3, which both have a rising portion.

Later we will make further error analysis of the correct pronunciation signals in the production of Mandarin tone experiment.

3.5 Correlation Between Perception and Production

A comparison of perception and production is shown in Table 4. The error rate of tone perception is highly correlated [r = 0.92] with that of tone production. The formula of r is as follows (1):

$$r(x, y) = \frac{\text{Cov}(xy)}{\sqrt{D(X)D(y)}} \tag{1}$$

The formula of r is the cross-correlation function of the two sequences. Where x and y are the correct rates of the four tones in pastoral and agricultural areas (x: 75%, 55%, 65%, 85%; y: 80%, 67.5%, 70%, 90%).

Table 4. Overall error rate of perception and production

Experimental type	Tone			
	Error rate			
	T 1	T 2	T 3	T 4
Production	22.5%	28.75%	32.5%	12.5%
Perception	13%	29%	24%	2%

The tonal production error rate ranges of 12% to 33%, the perception error rates ranges of 2% to 29%. The level of tone perception is greater than that of production.

4   Production of Mandarin Tone Pronunciation

We will do an analysis of the error pattern production of tonal pronunciation with the speech signal of the correct annotation result from the preceding annotation experiment. As shown in Fig. 3, the experiment of production of Mandarin tone pronunciation is roughly divided into the following sections.

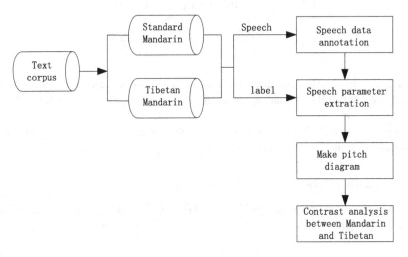

Fig. 3. The process of production of Mandarin Tone pronunciation

4.1   Experimental Corpus

A total of 2350 monosyllables were correctly pronounced according to the results of Tables 2 and 3. We classified the 2350 monosyllabic counterparts into 2 groups according the area of the subjects. According to the grade, speech data of 2 groups were divided into four groups separately. There were a total of 8 groups of speech data recorded by Amdo Tibetan, and with another group of standard Mandarin speech data recorded by native Mandarin speakers. There are a total of 9 groups of speech data.

4.2   Experimental Process

Step 1, the data of each group was divided into Tone 1, Tone 2, Tone 3, and Tone 4, which was used as the experimental speech signal of tone production.

Step2, we labeled the experimental speech signals by Visual Speech software, and then normalized the duration of syllables by straight algorithm to extract and modify the pitch [14].

Step 3, a program was written to realize the extraction of the maximum, minimum, and T value of the pitch data of each syllable obtained. The detailed approach was to divide the pitch data of each syllable into 10 equal parts by the duration. Each of which takes their average value to obtain the pitch data of the 10 average points that we need.

We used the formula proposed by [15] to normalize to convert the pitch data to T values, and then made pitch contours with T value. The formula of normalized processing was as follows (2):

$$T = 5 * \frac{\log_x - \log_L}{\log_H - \log_L} \tag{2}$$

Where H and L are the highest and lowest f0 of a group of all people's regulatory domains, and X is any given point of a pitch contour. The output of T is a value between 0 to 5.

The data obtained in the preceding step were statistically analyzed to obtain tone curves of 8 groups Mandarin monosyllables by Amdo Tibetan students. We made comparison between the tone curves of the speech data of 8 groups and compared the speech data of 8 groups with the standard Mandarin.

4.3   Experimental Results

As showed in Figs. 4 and 5, this is comparison graph of the monosyllable tone curves between standard Mandarin monosyllables and Mandarin by students from Amdo pastoral and agriculture areas. We make compares from the following aspects. Firstly, we make analyze between the four grades that freshman to senior. It is clear that compared with the standard Mandarin from the two figures, the monosyllable tone shape of the four grades both in Amdo pastoral and agriculture areas has obvious differences both in tone shape and tone domain. Secondly, we make analyze between the four tones that from Tone 1 to Tone 4. Tone 1, Tone 2, and Tone 3 of Mandarin monosyllable by students both in Amdo pastoral and agriculture areas have obvious differences in the tone shape and tone domain. It is still different from Mandarin in tone domain while the tone shape of Tone 4 is similar to standard Mandarin.

The result of tone production by college students from Amdo pastoral area. We compare the difference of the tone shape curves in Fig. 4 between Mandarin monosyllabic by college students from Amdo pastoral area and standard Mandarin.

Tone 1: The tone shape of freshman students from Amdo pastoral area shows a downward trend. The tone value is (23). The tone shape of junior shows a trend of decreasing first and then increasing, and the tone value is (34). The tone shape of sophomore and senior are similar to standard Mandarin. The location of the tone domain of the four grades is lower than standard Mandarin. In general, the Mandarin proficiency of senior and junior are better than sophomore and freshman.

Tone 2: In terms of tone shape, the shape tail of the tone by freshman, sophomore and senior are not rising enough and the middle of the tone shape is decreasing too much. The tone shape of junior is closest to standard Mandarin among the four grades. In terms of tone value, the tone value of sophomore and senior are closer to standard Mandarin than junior and freshman.

Tone 3: In terms of tone shape, the tone shape of the freshman, junior and senior are similar to standard Mandarin. The shape tail of the tone by freshman and senior are not rising enough. The height of the tone shape of junior at the beginning is not sufficient. The tone shape of sophomore is very different from standard Mandarin. The shape of

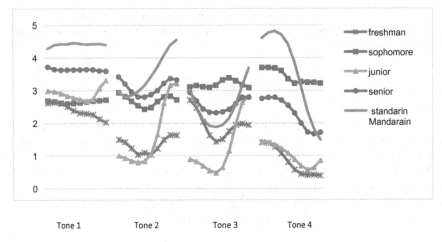

Tone 1   Tone 2   Tone 3   Tone 4

Note: The ordinate is the normalized T value

Fig. 4. Mandarin monosyllabic tone shape curves of students from Amdo pastoral area (Note: The ordinate is the normalized T value)

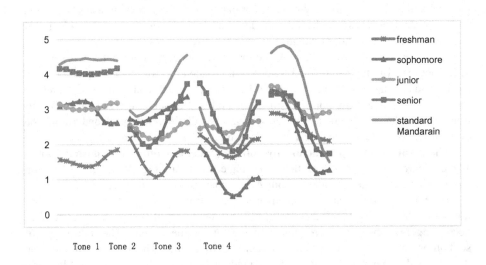

Tone 1   Tone 2   Tone 3   Tone 4

Note: The ordinate is the normalized T value

Fig. 5. Mandarin monosyllabic tone shape curves of students from Amdo agriculture area (Note: The ordinate is the normalized T value)

Tone 3 emphasizes the rising and falling lifting too much and ignores the need to reduce tone in the middle of the Tone 3, which is leading to a very strange shape.

Tone 4: The tone shape of the four grades from pastoral area is similar to standard Mandarin. But the tone domain is narrower than standard Mandarin. In general, the Mandarin proficiency of senior and sophomore are better than junior and freshman.

As above analyzed, the hierarchy of difficulty for the four grades from pastoral area is as follows: freshman > sophomore > junior > senior.

The result of tone production by college students from Amdo agriculture area. We compare the difference of the tone shape curves in Fig. 5 between Mandarin monosyllabic by college students from Amdo agriculture area and standard Mandarin.

Tone 1: The tone shape of freshmen shows a trend of decreasing first and then increasing, the tone value is (12), which is much lower than the standard Mandarin. The tone shape of sophomore shows a trend of decrease, the tone value is (34).The tone shape of junior and senior are similar to standard Mandarin. In general, the Mandarin proficiency of senior and Junior are better than sophomore and freshman.

Tone 2: The tone shape and tone domain of junior and senior are both similar to standard Mandarin. Senior and junior students have a higher level of Mandarin than freshman and sophomore. In general, the tone shape of Tone 2 is similar to the shape of Tone 3. It is probably because of the similar phonetic properties in Tone 2 and Tone 3, which both have a rising portion.

Tone 3: In terms of tone shape, the middle of the tone shape of junior is not decreasing enough. The tail of the tonal shape by freshman, sophomore is not rising enough. The tone shape of senior students is very close to standard Mandarin. In terms of tone value, the difference between sophomore and standard Mandarin is relatively big.

Tone 4: The tone shape of the four grades is both similar to standard Mandarin. The tone domain is a little narrower than the standard Mandarin. In general, the level of Tone 4 in the agriculture area is higher than the pastoral area.

In summary, it is evident that the hierarchy of difficulty for the four grades in pastoral area is as follows: freshman > sophomore > junior > senior, agriculture area is as follows: sophomore > freshman > junior > senior. We find that the Mandarin proficiency of students in pastoral area was increasing as the increasing of grade level, the poorest pronunciation of Mandarin tones by sophomore in agriculture area is due to the fact that the society is continuously improving. The freshman in agriculture area compared to pastoral area has more chances to contact Mandarin.

We compared Fig. 4 with Fig. 5, the pronunciation of college students in agriculture area is better than that in the pastoral area. It can be concluded that the hierarchy of difficulty for the four tones both in agriculture and pastoral areas is as follows: Tone 2 > Tone 3 > Tone 1 > Tone 4. In terms of tone domain, the agriculture area is obviously better than the pastoral area. The tone domain of the pastoral and agriculture area is narrow, and the location of the tone domain is much lower than standard Mandarin.

5 Conclusion

In this paper, the college students from the Amdo rural and pastoral areas were selected for experiment subjects. Two experiments were conducted on the error patterns of Mandarin tone perception and production by college students from Amdo Tibetan area.

From the above researches, it can be found that tone is a major difficulty for Amdo Tibetan students. And the confusion of Tone 2 and Tone 3 is a common error for students from Amdo Tibetan area. It is probably because of the similar phonetic properties in Tone 2 and Tone 3, which both have a rising portion. The error rate of tone perception is highly correlated [r = 0.92] with that of tone production. Most of them can imitate the tone shape of the four tones, but the tone domain is difficult to imitate. It is related to that Amdo Tibetan is a non-tonal language. The Amdo Tibetan has only one habit tone, which makes it difficult for them to grasp the tone domain of the four tones correctly. In addition, there is a big difference between college students from the Amdo agriculture and pastoral areas on the Mandarin proficiency. The reason of the difference is that the economic in agriculture area is higher than that in pastoral area, and there are fewer people speak Mandarin in pastoral area than that in agriculture area. In addition to the negative transfer of their native language, the different economic development in the internal of the Amdo Tibetan area has a great influence on Mandarin learning.

References

1. Ding, H.: Perception and production of Mandarin disyllabic tones by German learners. Speech Prosody **184**, 378–381 (2012)
2. Weltens, B.: Language attrition in progress. Language **66**(3), 37–49 (1986)
3. Yali, L.: Perception and production of Mandarin tones of pupils from Xinjiang ethnic areas. J. Tsinghua Univ. **53**(6), 823–827 (2013)
4. Gao, M.: Perception and production of mandarin tones by swedish students. Iran. J. Radiol. **29**(4), 533–538 (2010)
5. Hoffmann, C.: Disyllabic Mandarin lexical tone perception by native Dutch speakers: a case of adult perceptual asymmetry. J. Acoust. Soc. Am. **134**(5), 4230 (2013)
6. Zhang, J.: Influences of vowels on the perception of nasal codas in Mandarin for Japanese and Chinese natives. J. Tsinghua Univ. **57**(2), 164–169 (2017)
7. Lijuan, G.: Tone production in Mandarin Chinese by American students. In:Chan, K.M., Kang, H. (eds.) Conference 2008, NACCL, Ohio, vol. 9999, pp. 123–138 (2008)
8. Ding, H.: An investigation of tone perception and production in German learners of Mandarin. Arch. Acoust. **36**(3), 509–518 (2011)
9. Haifeng, D.: A Study of single Chinese characters of Tibetan learners of Amdo Tibetan. J. Lang. Lit. Stud. **14**, 17–18 (2011)
10. Yan, L.: Acoustic research on the monosyllabic pitch pattern of Amdo-Mandarin Inter language and Tibetan Amdo Dialect. Hanzangyu Xuebao **1**, 51–63 (2015)
11. Wu Yong, F.: A study of International Error on Putonghua Inter language of Tibetan language Anduo Dialect Region. J. Qinghai Nationalities Univ. (Soc. Sci.) **38**(02), 135–140 (2012)
12. Yali, L.: The train methods of perception of Mandarin tones of pupils in Xinjiang ethnic areas. Tech. Acoust. **33**(s2), 328–330 (2014)
13. Wenbo, C.: Analysis of Chinese phonetic teaching for students in Central Asia. Lang. Transl. **3**, 73–76 (2009)
14. Irino, T.: Speech segregation using an auditory vocoder with event- synchronous enhancements. IEEE Trans. Speech Audio Process. **14**(6), 2212–2221 (2006)
15. Feng, S.: Exploration of Experimental Phonology. Beijing Peking University Press, Beijing (2009)

Effective Character-Augmented Word Embedding for Machine Reading Comprehension

Zhuosheng Zhang[1,2], Yafang Huang[1,2], Pengfei Zhu[1,2,3], and Hai Zhao[1,2(✉)]

[1] Department of Computer Science and Engineering,
Shanghai Jiao Tong University, Shanghai, China
{zhangzs,huangyafang}@sjtu.edu.cn
[2] Key Laboratory of Shanghai Education Commission for Intelligent Interaction and
Cognitive Engineering, Shanghai Jiao Tong University, Shanghai 200240, China
10152510190@stu.ecnu.edu.cn, zhaohai@cs.sjtu.edu.cn
[3] School of Computer Science and Software Engineering,
East China Normal University, Shanghai, China

Abstract. Machine reading comprehension is a task to model relationship between passage and query. In terms of deep learning framework, most of state-of-the-art models simply concatenate word and character level representations, which has been shown suboptimal for the concerned task. In this paper, we empirically explore different integration strategies of word and character embeddings and propose a character-augmented reader which attends character-level representation to augment word embedding with a short list to improve word representations, especially for rare words. Experimental results show that the proposed approach helps the baseline model significantly outperform state-of-the-art baselines on various public benchmarks.

Keywords: Question answering · Reading comprehension
Character-augmented embedding

1 Introduction

Machine reading comprehension (MRC) is a challenging task which requires computers to read and understand documents to answer corresponding questions, it is indispensable for advanced context-sensitive dialogue and interactive systems [12,34,36]. There are two main kinds of MRC, user-query types [13,24] and cloze-style [7,10,11]. The major difference lies in that the answers for the former are usually a span of texts while the answers for the latter are words or phrases.

This paper was partially supported by National Key Research and Development Program of China (No. 2017YFB0304100), National Natural Science Foundation of China (No. 61672343 and No. 61733011), Key Project of National Society Science Foundation of China (No. 15-ZDA041), The Art and Science Interdisciplinary Funds of Shanghai Jiao Tong University (No. 14JCRZ04).

© Springer Nature Switzerland AG 2018
M. Zhang et al. (Eds.): NLPCC 2018, LNAI 11108, pp. 27–39, 2018.
https://doi.org/10.1007/978-3-319-99495-6_3

Most of recent proposed deep learning models focus on sentence or paragraph level attention mechanism [5,8,14,25,30] instead of word representations. As the fundamental part in natural language processing tasks, word representation could seriously influence downstream MRC models (readers). Words could be represented as vectors using word-level or character-level embedding. For word embeddings, each word is mapped into low dimensional dense vectors directly from a lookup table. Character embeddings are usually obtained by applying neural networks on the character sequence of each word and the hidden states are used to form the representation. Intuitively, word-level representation is good at capturing wider context and dependencies between words but it could be hard to represent rare words or unknown words. In contrast, character embedding is more expressive to model sub-word morphologies, which facilitates dealing with rare words.

Table 1. A cloze-style reading comprehension example.

Passage	1 早上，青蛙、小白兔、刺猬和大蚂蚁高高兴兴过桥去赶集。 2 不料，中午下了一场大暴雨，哗啦啦的河水把桥冲走了。 3 天快黑了，小白兔、刺猬和大蚂蚁都不会游泳。 4 过不了河，急得哭了。 5 这时，青蛙想，我可不能把朋友丢下，自己过河回家呀。 6 他一面劝大家不要着急，一面动脑筋。 7 嘀，有了！ 8 他说："我有个朋友住在这儿，我去找他想想办法。 9 青蛙找到了他的朋友_____，请求他说："大家过不了河了，请帮个忙吧！ 10 鼹鼠说："可以，请把大家领到我家里来吧。 11 鼹鼠把大家带到一个洞口，打开了电筒，让小白兔、刺猬、大蚂蚁和青蛙跟着他，"大家别害怕，一直朝前走。 12 走呀走呀，只听见上面"哗啦哗啦"的声音，象唱歌。 13 走着走着，突然，大家看见了天空，天上的月亮真亮呀。 14 小白兔回头一瞧，高兴极了："哈，咱们过了河啦！ 15 嗬，真了不起。 16 原来，鼹鼠在河底挖了一条很长的地道，从这头到那头。 17 青蛙、小白兔、刺猬和大蚂蚁是多么感激鼹鼠啊！ 18 第二天，青蛙、小白兔、刺猬和大蚂蚁带来很多很多同伴，杠着木头，抬着石头，要求鼹鼠让他们来把地道挖大些，修成河底大"桥"。 19 不久，他们就把鼹鼠家的地道，挖成了河底的一条大隧道，大家可以从河底过河，还能通车，真有劲哩！	1 In the morning, the frog, the little white rabbit, the hedgehog and the big ant happily crossed the bridge for the market. 2 Unexpectedly, a heavy rain fell at noon, and the water swept away the bridge. 3 It was going dark. The little white rabbit, hedgehog and big ant cannot swim. 4 Unable to cross the river, they were about to cry. 5 At that time, the frog made his mind that he could not leave his friend behind and went home alone. 6 Letting his friends take it easy, he thought and thought. 7 Well, there you go! 8 He said, "I have a friend who lives here, and I'll go and find him for help." 9 The frog found his friend _____ and told him, "We cannot get across the river. Please give us a hand!" 10 The mole said, "That's fine, please bring them to my house." 11 The mole took everyone to a hole, turned on the flashlight and asked the little white rabbit, the hedgehog, the big ant and the frog to follow him, saying, "Don't be afraid, just go ahead." 12 They walked along, hearing the "walla-walla" sound, just like a song. 13 All of a sudden, everyone saw the sky, and the moon was really bright. 14 The little white rabbit looked back and rejoiced: "ha, the river crossed!". 15 "Oh, really great." 16 Originally, the mole dug a very long tunnel under the river, from one end to the other. 17 How grateful the frog, the little white rabbit, the hedgehog and the big ant felt to the mole! 18 The next day, the frog, the little white rabbit, the hedgehog, and the big ant with a lot of his fellows, took woods and stones. They asked the mole to dig tunnels bigger, and build a great bridge under the river. 19 It was not long before they dug a big tunnel under the river, and they could pass the river from the bottom of the river, and it could be open to traffic. It is amazing!
Query	青蛙找到了他的朋友_____，请求他说："大家过不了河了，请帮个忙吧！"	The frog found his friend _____ and told him, "We cannot get across the river. Please give us a hand!"
Answer	鼹鼠	the mole

As shown in Table 1, the passages in MRC are quite long and diverse which makes it hard to record all the words in the model vocabulary. As a result, reading comprehension systems suffer from out-of-vocabulary (OOV) word issues, especially when the ground-truth answers tend to include rare words or named entities (NE) in cloze-style MRC tasks.

To form a fine-grained embedding, there have been a few hybrid methods that jointly learn the word and character representations [15, 19, 32]. However, the passages in machine reading dataset are content-rich and contain massive words and characters, using fine-grained features, such as named entity recognition and part-of-speech (POS) tags will need too high computational cost in return, meanwhile the efficiency of readers is crucial in practice.

In this paper, we verify the effectiveness of various simple yet effective character-augmented word embedding (CAW) strategies and propose a CAW Reader. We survey different CAW strategies to integrate word-level and character-level embedding for a fine-grained word representation. To ensure adequate training of OOV and low-frequency words, we employ a short list mechanism. Our evaluation will be performed on three public Chinese reading comprehension datasets and one English benchmark dataset for showing our method is effective in multi-lingual case.

2   Related Work

Machine reading comprehension has been witnessed rapid progress in recent years [8, 22, 26–29, 31, 33, 35]. Thanks to various released MRC datasets, we can evaluate MRC models in different languages. This work focuses on cloze-style ones since the answers are single words or phrases instead of text spans, which could be error-prone when they turn out to be rare or OOV words that are not recorded in the model vocabulary.

Recent advances for MRC could be mainly attributed to attention mechanisms, including query-to-passage attention [7, 14], attention-over-attention [5] and self attention [30]. Different varieties and combinations have been proposed for further improvements [8, 25]. However, the fundamental part, word representation, which proves to be quite important in this paper, has not aroused much interest. To integrate the advantages of both word-level and character-level embeddings, some researchers studied joint models for richer representation learning where the common combination method is the concatenation. Seo et al. [25] concatenated the character and word embedding and then fed the joint representation to a two-layer Highway Network. FG reader in [32] used a fine-grained gating mechanism to dynamically combine word-level and character-level representations based on word property. However, this method is computationally complex and requires extra labels such as NE and POS tags.

Not only for machine reading comprehension tasks, character embedding has also benefited other natural language process tasks, such as word segmentation [2, 3], machine translation [18, 19], tagging [1, 9, 17] and language modeling [21, 23]. Notablely, Cai et al. [3] presented a greedy neural word segmenter where high-frequency word embeddings are attached to character embedding via average pooling while low-frequency words are represented as character embedding. Experiments show this mechanism helps achieve state-of-the-art word segmentation performance, which partially inspires our reader design.

3   Model

In this section, we will introduce our model architecture, which is consisted of a fundamental word representation module and a gated attention learning module.

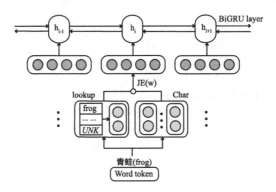

Fig. 1. Overview of the word representation module.

3.1   Word Representation Module

Figure 1 illustrates our word representation module. The input token sequence is first encoded into embeddings. In the context of machine reading comprehension tasks, word only representation generalizes poorly due to the severe word sparsity, especially for rare words. We adopt two methods to augment word representations, namely, a short list filtering and character enhancement.

Actually, if all the words in the dataset are used to build the vocabulary, the OOV words from the test set will not be well dealt with for inadequate training. To handle this issue, we keep a short list L for specific words. If word w is in L, the immediate word embedding $\mathbf{e}_w$ is indexed from word lookup table $M^w \in \mathbb{R}^{d \times s}$ where s denotes the size (recorded words) of lookup table and d denotes the embedding dimension. Otherwise, it will be represented as the randomly initialized default word (denoted by a specific mark UNK). Note that only the word embedding of the OOV words will be replaced by the vectors of UNK (denoted by $\mathbf{e}_u$) while their character embedding $\mathbf{e}_c$ will still be processed using the original word. In this way, the OOV words could be tuned sufficiently with expressive meaning after training.

In our experiments, the short list is determined according to the word frequency. Concretely, we sort the vocabulary according to the word frequency from high to low. A frequency filter ratio γ is set to filter out the low-frequency words (rare words) from the lookup table. For example, $\gamma = 0.9$ means the least frequent 10% words are replaced with the default UNK notation.

Character-level embeddings have been widely used in lots of natural language processing tasks and verified for the OOV and rare word representations. Thus, we consider employing neural networks to compose word representations from smaller units, i.e., character embedding [15,21], which results in a hybrid mechanism for word representation with a better fine-grained consideration. For a given word w, a joint embedding (JE) is to straightforwardly integrate word embedding $\mathbf{e}_w$ and character embedding $\mathbf{e}_c$.

$$JE(w) = \mathbf{e}_w \circ \mathbf{e}_c$$

where $\circ$ denotes the joint operation. Specifically, we investigate concatenation (*concat*), element-wise summation (*sum*) and element-wise multiplication (*mul*). Thus, each passage P and query Q is represented as $\mathbb{R}^{d \times k}$ matrix where d denotes the dimension of word embedding and k is the number of words in the input.

Finally by combining the short list mechanism and character enhancement, $JE(w)$ can be rewritten as

$$JE(w) = \begin{cases} \mathbf{e}_w \circ \mathbf{e}_c \text{ if } w \in L \\ \mathbf{e}_u \circ \mathbf{e}_c \text{ otherwise} \end{cases}$$

The character embedding $\mathbf{e}_c$ can be learned by two kinds of networks, recurrent neural network (RNN) or convolutional neural network (CNN)[1].

RNN Based Embedding. The character embedding $\mathbf{e}_c$ is generated by taking the final outputs of a bidirectional gated recurrent unit (GRU) [4] applied to the vectors from a lookup table of characters in both forward and backward directions. Characters $w = \{x_1, x_2, \ldots, x_l\}$ of each word are vectorized and successively fed to forward GRU and backward GRU to obtain the internal features. The output for each input is the concatenation of the two vectors from both directions: $\overleftrightarrow{h_t} = \overrightarrow{h_t} \parallel \overleftarrow{h_t}$ where h_t denotes the hidden states.

Then, the output of BiGRUs is passed to a fully connected layer to obtain the a fixed-size vector for each word and we have $\mathbf{e}_c = W \overleftrightarrow{h_t} + b$.

CNN Based Embedding character sequence $w = \{x_1, x_2, \ldots, x_l\}$ is embedded into vectors M using a lookup table, which is taken as the inputs to the CNN, and whose size is the input channel size of the CNN. Let W_j denote the Filter matrices of width l, the substring vectors will be transformed to sequences $c_j (j \in [1, l])$:

$$c_j = [\ldots; \tanh(W_j \cdot M_{[i:i+l-1]} + b_j); \ldots]$$

---

[1] Empirical study shows the character embeddings obtained from these two networks perform comparatively. To focus on the performance of character embedding, we introduce the networks only for reproduction. Our reported results are based on RNN based character embeddings.

where $[i : i+l-1]$ indexes the convolution window. A *one-max-pooling* operation is adopted after convolution $s_j = \mathbf{max}(c_j)$. The character embedding is obtained through concatenating all the mappings for those l filters.

$$\mathbf{e}_c = [s_1 \oplus \cdots \oplus s_j \oplus \cdots \oplus s_l]$$

3.2 Attention Learning Module

To obtain the predicted answer, we first apply recurrent neural networks to encode the passage and query. Concretely, we use BiGRUs to get contextual representations of forward and backward directions for each word in the passage and query and we have G_p and G_q, respectively.

Then we calculate the gated attention following [8] to obtain the probability distribution of each word in the passage. For each word p_i in G_p, we apply soft attention to form a word-specific representation of the query $q_i \in G_q$, and then multiply the query representation with the passage word representation.

$$\alpha_i = softmax(G_q^\top p_i)$$
$$\beta_i = G_q \alpha_i$$
$$x_i = p_i \odot \beta_i$$

where $\odot$ denotes the element-wise product to model the interactions between p_i and q_i. The passage contextual representation $\tilde{G}_p = \{x_1, x_2, \ldots, x_k\}$ is weighted by query representation.

Inspired by [8], multi-layered attentive network tends to focus on different aspects in the query and could help combine distinct pieces of information to answer the query, we use K intermediate layers which stacks end to end to learn the attentive representations. At each layer, the passage contextual representation $\tilde{G}_p$ is updated through above attention learning. Let q_k denote the k-th intermediate output of query contextual representation and G_P represent the full output of passage contextual representation $\tilde{G}_p$. The probability of each word $w \in C$ in the passage as being the answer is predicted using a softmax layer over the inner-product between q_k and G_P.

$$r = softmax((q_k)^\top G_P)$$

where vector p denotes the probability distribution over all the words in the passage. Note that each word may occur several times in the passage. Thus, the probabilities of each candidate word occurring in different positions of the passage are added together for final prediction.

$$P(w|p, q) \propto \sum_{i \in I(w,p)} r_i$$

where $I(w, p)$ denotes the set of positions that a particular word w occurs in the passage p. The training objective is to maximize $\log P(A|p, q)$ where A is the correct answer.

Finally, the candidate word with the highest probability will be chosen as the predicted answer. Unlike recent work employing complex attention mechanisms, our attention mechanism is much more simple with comparable performance so that we can focus on the effectiveness of our embedding strategies.

Table 2. Data statistics of PD, CFT and CMRC-2017.

	PD			CFT	CMRC-2017		
	Train	Valid	Test	Human	Train	Valid	Test
# Query	870,710	3,000	3,000	1,953	354,295	2,000	3,000
Avg # words in docs	379	425	410	153	324	321	307
Avg # words in query	38	38	41	20	27	19	23
# Vocabulary	248,160				94,352		

4   Evaluation

4.1   Dataset and Settings

Based on three Chinese MRC datasets, namely People's Daily (PD), Children Fairy Tales (CFT) [7] and CMRC-2017 [6], we verify the effectiveness of our model through a series of experiments[2]. Every dataset contains three parts, *Passage*, *Query* and *Answer*. The *Passage* is a story formed by multiple sentences, and the *Query* is one sentence selected by human or machine, of which one word is replaced by a placeholder, and the *Answer* is exactly the original word to be filled in. The data statistics is shown in Table 2. The difference between the three Chinese datasets and the current cloze-style English MRC datasets including Daily Mail, CBT and CNN [10] is that the former does not provide candidate answers. For the sake of simplicity, words from the whole passage are considered as candidates.

Besides, for the test of generalization ability in multi-lingual case, we use the Children's Book Test (CBT) dataset [11]. We only consider cases of which the answer is either a NE or common noun (CN). These two subsets are more challenging because the answers may be rare words.

For fare comparisons, we use the same model setting in this paper. We randomly initialize the $100d$ character embeddings with the uniformed distribution in the interval $[-0:05, 0:05]$. We use word2vec [20] toolkit to pre-train $200d$ word embeddings on *Wikipedia* corpus[3], and randomly initialize the OOV words. For both the word and character representation, the GRU hidden units are 128. For

---

[2] In the test set of CMRC-2017 and human evaluation test set (Test-human) of CFT, questions are further processed by human and the pattern of them may not be in accordance with the auto-generated questions, so it may be harder for machine to answer.

[3] https://dumps.wikimedia.org/.

optimization, we use stochastic gradient descent with ADAM updates [16]. The initial learning rate is 0.001, and after the second epoch, it is halved every epoch. The batch size is 64. To stabilize GRU training, we use gradient clipping with a threshold of 10. Throughout all experiments, we use three attention layers.

Table 3. Accuracy on PD and CFT datasets. All the results except ours are from [7].

Model	Strategy	PD		CFT
		Valid	Test	Test-human
AS Reader	-	64.1	67.2	33.1
GA Reader	-	64.1	65.2	35.7
CAS Reader	-	65.2	68.1	35.0
CAW Reader	concat	64.2	65.3	37.2
	sum	65.0	68.1	38.7
	mul	**69.4**	**70.5**	**39.7**

4.2   Results

PD & CFT. Table 3 shows the results on PD and CFT datasets. With improvements of 2.4% on PD and 4.7% on CFT datasets respectively, our CAW Reader model significantly outperforms the CAS Reader in all types of testing. Since the CFT dataset contains no training set, we use PD training set to train the corresponding model. It is harder for machine to answer because the test set of CFT dataset is further processed by human experts, and the pattern quite differs from PD dataset. We can learn from the results that our model works effectively for out-of-domain learning, although PD and CFT datasets belong to quite different domains.

CMRC-2017. Table 4 shows the results[4]. Our CAW Reader (mul) not only obtains 7.27% improvements compared with the baseline Attention Sum Reader (AS Reader) on the test set, but also outperforms all other single models. The best result on the valid set is from WHU, but their result on test set is lower than ours by 1.97%, indicating our model has a satisfactory generalization ability.

We also compare different CAW strategies for word and character embeddings. We can see from the results that the CAW Reader (mul) significantly outperforms all the other three cases, word embedding only, concatenation and summation, and especially obtains 8.37% gains over the first one. This reveals

---

[4] Note that the test set of CMRC-2017 and human evaluation test set (Test-human) of CFT are harder for the machine to answer because the questions are further processed manually and may not be in accordance with the pattern of auto-generated questions.

Table 4. Accuracy on CMRC-2017 dataset. Results marked with † are from the latest official CMRC Leaderboard (http://www.hfl-tek.com/cmrc2017/leaderboard. html). The best results are in bold face. WE is short for word embedding.

Model	CMRC-2017	
	Valid	Test
Random Guess †	1.65	1.67
Top Frequency †	14.85	14.07
AS Reader †	69.75	71.23
GA Reader	72.90	74.10
SJTU BCMI-NLP †	76.15	77.73
6ESTATES PTE LTD †	75.85	74.73
Xinktech †	77.15	77.53
Ludong University †	74.75	75.07
ECNU †	77.95	77.40
WHU †	**78.20**	76.53
CAW Reader (WE only)	69.70	70.13
CAW Reader (concat)	71.55	72.03
CAW Reader (sum)	72.90	74.07
CAW Reader (mul)	77.95	**78.50**

that compared with concatenation and sum operation, the element-wise multiplication might be more informative, because it introduces a similar mechanism to endow character-aware *attention* over the word embedding. On the other hand, too high dimension caused by concatenation operation may lead to serious over-fitting issues[5], and sum operation is too simple to prevent from detailed information losing.

CBT. The results on CBT are shown in Table 5. Our model outperforms most of the previous public works. Compared with GA Reader with word and character embedding concatenation, i.e., the original model of our CAW Reader, our model with the character augmented word embedding has 2.4% gains on the CBT-NE test set. FG Reader adopts neural gates to combine word-level and character-level representations and adds extra features including NE, POS and word frequency, but our model also achieves comparable performance with it. This results on both languages show that our CAW Reader is not limited to dealing with Chinese but also for other languages.

---

[5] For the best concat and mul model, the training/validation accuracies are 97.66%/71.55, 96.88%/77.95%, respectively.

Table 5. Accuracy on CBT dataset. Results marked with ‡ are of previously published works [7,8,32].

Model	CBT-NE		CBT-CN	
	Valid	Test	Valid	Test
Human ‡	-	81.6	-	81.6
LSTMs ‡	51.2	41.8	62.6	56.0
MemNets ‡	70.4	66.6	64.2	63.0
AS Reader ‡	73.8	68.6	68.8	63.4
Iterative Attentive Reader ‡	75.2	68.2	72.1	69.2
EpiReader ‡	75.3	69.7	71.5	67.4
AoA Reader ‡	77.8	72.0	72.2	69.4
NSE ‡	**78.2**	**73.2**	**74.3**	**71.9**
GA Reader ‡	74.9	69.0	69.0	63.9
GA word char concat ‡	76.8	72.5	73.1	69.6
GA scalar gate ‡	78.1	72.6	72.4	69.1
GA fine-grained gate ‡	78.9	74.6	72.3	70.8
FG Reader ‡	**79.1**	**75.0**	**75.3**	**72.0**
CAW Reader	78.4	74.9	74.8	71.5

5   Analysis

We conduct quantitative study to investigate how the short list influence the model performance on the filter ratio from $[0.1, 0.2, \ldots, 1]$. Figure 2 shows the results on the CMRC-2017 dataset. Our CAW reader achieves the best accuracy when $\gamma = 0.9$. It indicates that it is not optimal to build the vocabulary among the whole training set, and we can reduce the frequency filter ratio properly to

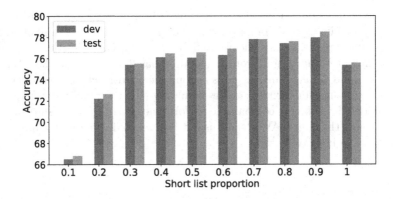

Fig. 2. Quantitative study on the influence of the short list.

promote the accuracy. In fact, training the model on the whole vocabulary may lead to over-fitting problems. Besides, improper initialization of the rare words may also bias the whole word representations. As a result, without a proper OOV representation mechanism, it is hard for a model to deal with OOV words from test sets precisely.

6 Conclusion

This paper surveys multiple embedding enhancement strategies and proposes an effective embedding architecture by attending character representations to word embedding with a short list to enhance the simple baseline for the reading comprehension task. Our evaluations show that the intensified embeddings can help our model achieve state-of-the-art performance on multiple large-scale benchmark datasets. Different from most existing works that focus on either complex attention architectures or manual features, our model is more simple but effective. Though this paper is limited to the empirical verification on MRC tasks, we believe that the improved word representation may also benefit other tasks as well.

References

1. Bai, H., Zhao, H.: Deep enhanced representation for implicit discourse relation recognition. In: Proceedings of the 27th International Conference on Computational Linguistics (COLING 2018) (2018)
2. Cai, D., Zhao, H.: Neural word segmentation learning for Chinese. In: Proceedings of the 54th Annual Meeting of the Association for Computational Linguistics (ACL 2016), pp. 409–420 (2016)
3. Cai, D., Zhao, H., Zhang, Z., Xin, Y., Wu, Y., Huang, F.: Fast and accurate neural word segmentation for Chinese. In: Proceedings of the 55th Annual Meeting of the Association for Computational Linguistics (ACL 2017), pp. 608–615 (2017)
4. Cho, K., Merrienboer, B.V., Gulcehre, C., Bahdanau, D., Bougares, F., Schwenk, H., Bengio, Y.: Learning phrase representations using rnn encoder-decoder for statistical machine translation. In: Proceedings of the 2014 Conference on Empirical Methods in Natural Language Processing (EMNLP 2014), pp. 1724–1734 (2014)
5. Cui, Y., Chen, Z., Wei, S., Wang, S., Liu, T., Hu, G.: Attention-over-attention neural networks for reading comprehension. In: Proceedings of the 55th Annual Meeting of the Association for Computational Linguistics (ACL 2017), pp. 1832–1846 (2017)
6. Cui, Y., Liu, T., Chen, Z., Ma, W., Wang, S., Hu, G.: Dataset for the first evaluation on chinese machine reading comprehension. In: Calzolari (Conference Chair), N., Choukri, K., Cieri, C., Declerck, T., Goggi, S., Hasida, K., Isahara, H., Maegaard, B., Mariani, J., Mazo, H., Moreno, A., Odijk, J., Piperidis, S., Tokunaga, T. (eds.) Proceedings of the Eleventh International Conference on Language Resources and Evaluation (LREC 2018). European Language Resources Association (ELRA) (2018)

7. Cui, Y., Liu, T., Chen, Z., Wang, S., Hu, G.: Consensus attention-based neural networks for Chinese reading comprehension. In: Proceedings of the 26th International Conference on Computational Linguistics (COLING 2016), pp. 1777–1786 (2016)
8. Dhingra, B., Liu, H., Yang, Z., Cohen, W.W., Salakhutdinov, R.: Gated-attention readers for text comprehension. In: Proceedings of the 55th Annual Meeting of the Association for Computational Linguistics (ACL 2017), pp. 1832–1846 (2017)
9. He, S., Li, Z., Zhao, H., Bai, H., Liu, G.: Syntax for semantic role labeling, to be, or not to be. In: Proceedings of the 56th Annual Meeting of the Association for Computational Linguistics (ACL 2018) (2018)
10. Hermann, K.M., Kocisky, T., Grefenstette, E., Espeholt, L., Kay, W., Suleyman, M., Blunsom, P.: Teaching machines to read and comprehend. In: Advances in Neural Information Processing Systems (NIPS 2015), pp. 1693–1701 (2015)
11. Hill, F., Bordes, A., Chopra, S., Weston, J.: The goldilocks principle: reading children's books with explicit memory representations. arXiv preprint arXiv:1511.02301 (2015)
12. Huang, Y., Li, Z., Zhang, Z., Zhao, H.: Moon IME: neural-based Chinese Pinyin aided input method with customizable association. In: Proceedings of the 56th Annual Meeting of the Association for Computational Linguistics (ACL 2018), System Demonstration (2018)
13. Joshi, M., Choi, E., Weld, D.S., Zettlemoyer, L.: TriviaQA: a large scale distantly supervised challenge dataset for reading comprehension. In: ACL, pp. 1601–1611 (2017)
14. Kadlec, R., Schmid, M., Bajgar, O., Kleindienst, J.: Text understanding with the attention sum reader network. In: Proceedings of the 54th Annual Meeting of the Association for Computational Linguistics (ACL 2016), pp. 908–918 (2016)
15. Kim, Y., Jernite, Y., Sontag, D., Rush, A.M.: Character-aware neural language models. In: Proceedings of the Thirtieth AAAI Conference on Artificial Intelligence (AAAI 2016), pp. 2741–2749 (2016)
16. Kingma, D., Ba, J.: Adam: a method for stochastic optimization. arXiv preprint arXiv:1412.6980 (2014)
17. Lample, G., Ballesteros, M., Subramanian, S., Kawakami, K., Dyer, C.: Neural architectures for named entity recognition. arXiv preprint arXiv:1603.01360 (2016)
18. Ling, W., Trancoso, I., Dyer, C., Black, A.W.: Character-based neural machine translation. arXiv preprint arXiv:1511.04586 (2015)
19. Luong, M.T., Manning, C.D.: Achieving open vocabulary neural machine translation with hybrid word-character models. arXiv preprint arXiv:1604.00788 (2016)
20. Mikolov, T., Chen, K., Corrado, G., Dean, J.: Efficient estimation of word representations in vector space. arXiv preprint arXiv:1301.3781 (2013)
21. Miyamoto, Y., Cho, K.: Gated word-character recurrent language model. In: Proceedings of the 2016 Conference on Empirical Methods in Natural Language Processing (EMNLP 2016), pp. 1992–1997 (2016)
22. Munkhdalai, T., Yu, H.: Reasoning with memory augmented neural networks for language comprehension. In: Proceedings of the International Conference on Learning Representations (ICLR 2017) (2017)
23. Peters, M.E., Neumann, M., Iyyer, M., Gardner, M., Clark, C., Lee, K., Zettlemoyer, L.: Deep contextualized word representations. In: Conference of the North American Chapter of the Association for Computational Linguistics: Human Language Technologies (NAACL 2018) (2018)

24. Rajpurkar, P., Zhang, J., Lopyrev, K., Liang, P.: SQuAD: 100,000+ questions for machine comprehension of text. In: Proceedings of the 2016 Conference on Empirical Methods in Natural Language Processing (EMNLP 2016), pp. 2383–2392 (2016)
25. Seo, M., Kembhavi, A., Farhadi, A., Hajishirzi, H.: Bidirectional attention flow for machine comprehension. In: Proceedings of the International Conference on Learning Representations (ICLR 2017) (2017)
26. Sordoni, A., Bachman, P., Trischler, A., Bengio, Y.: Iterative alternating neural attention for machine reading. arXiv preprint arXiv:1606.02245 (2016)
27. Trischler, A., Ye, Z., Yuan, X., Suleman, K.: Natural language comprehension with the EpiReader. In: Proceedings of the 2016 Conference on Empirical Methods in Natural Language Processing (EMNLP 2016), pp. 128–137 (2016)
28. Wang, B., Liu, K., Zhao, J.: Conditional generative adversarial networks for commonsense machine comprehension. In: Proceedings of the Twenty-Sixth International Joint Conference on Artificial Intelligence (IJCAI 2017), pp. 4123–4129 (2017)
29. Wang, S., Jiang, J.: Machine comprehension using Match-LSTM and answer pointer. In: Proceedings of the International Conference on Learning Representations (ICLR 2016) (2016)
30. Wang, W., Yang, N., Wei, F., Chang, B., Zhou, M.: Gated self-matching networks for reading comprehension and question answering. In: Proceedings of the 55th Annual Meeting of the Association for Computational Linguistics (ACL 2017), pp. 189–198 (2017)
31. Wang, Y., Liu, K., Liu, J., He, W., Lyu, Y., Wu, H., Li, S., Wang, H.: Multi-passage machine reading comprehension with cross-passage answer verification. In: Proceedings of the 56th Annual Meeting of the Association for Computational Linguistics (ACL 2018) (2018)
32. Yang, Z., Dhingra, B., Yuan, Y., Hu, J., Cohen, W.W., Salakhutdinov, R.: Words or characters? Fine-grained gating for reading comprehension. In: Proceedings of the International Conference on Learning Representations (ICLR 2017) (2017)
33. Zhang, Z., Huang, Y., Zhao, H.: Subword-augmented embedding for cloze reading comprehension. In: Proceedings of the 27th International Conference on Computational Linguistics (COLING 2018) (2018)
34. Zhang, Z., Li, J., Zhu, P., Zhao, H.: Modeling multi-turn conversation with deep utterance aggregation. In: Proceedings of the 27th International Conference on Computational Linguistics (COLING 2018) (2018)
35. Zhang, Z., Zhao, H.: One-shot learning for question-answering in Gaokao history challenge. In: Proceedings of the 27th International Conference on Computational Linguistics (COLING 2018) (2018)
36. Zhu, P., Zhang, Z., Li, J., Huang, Y., Zhao, H.: Lingke: A fine-grained multi-turn chatbot for customer service. In: Proceedings of the 27th International Conference on Computational Linguistics (COLING 2018), System Demonstrations (2018)

Mongolian Grapheme to Phoneme Conversion by Using Hybrid Approach

Zhinan Liu[1], Feilong Bao[1(✉)], Guanglai Gao[1], and Suburi[2]

[1] College of Computer Science,
Inner Mongolia University, Huhhot 010021, China
lzn_bung@163.com, {csfeilong, csggl}@imu.edu.cn
[2] Inner Mongolia Public Security Department, Huhhot 010021, China
sunbuer@163.com

Abstract. Grapheme to phoneme (G2P) conversion is the assignment of converting word to its pronunciation. It has important applications in text-to-speech (TTS), speech recognition and sounds-like queries in textual databases. In this paper, we present the first application of sequence-to-sequence (Seq2Seq) Long Short-Term Memory (LSTM) model with the attention mechanism for Mongolian G2P conversion. Furthermore, we propose a novel hybrid approach of combining rules with Seq2Seq LSTM model for Mongolian G2P conversion, and implement the Mongolian G2P conversion system. The experimental results show that: Adopting Seq2Seq LSTM model can obtain better performance than traditional methods of Mongolian G2P conversion, and the hybrid approach further improves G2P conversion performance. The word error rate (WER) relatively reduces by 10.8% and the phoneme error rate (PER) approximately reduces by 1.6% through comparing with the Mongolian G2P conversion method being used based on the joint-sequence models, which completely meets the practical requirements of Mongolian G2P conversion.

Keywords: Mongolian · Grapheme-to-phoneme · Sequence-to-sequence
LSTM

1 Introduction

Grapheme-to-phoneme conversion (G2P) refers to the task of converting a word from the orthographic form (sequence of letters/characters/graphemes) to its pronunciation (a sequence of phonemes). It has a wide range of applications in speech synthesis [1–3], automatic speech recognition (ASR) [4–6] and speech retrieval [7, 8].

One of the challenges in G2P conversion is that the pronunciation of any grapheme depends on a variety of factors including its context and the etymology of the word. Another complication is that output phone sequence can be either shorter than or longer than the input grapheme sequence. Typical approaches to G2P involve using rule-based methods and joint-sequence models. While rule-based methods are effective to handle new words, they have some limitations: designing the rules is hard and requires specific linguistic skills, and it is extremely difficult to capture all rules for natural languages. To overcome the above limitations, another called statistics-based method are proposed,

© Springer Nature Switzerland AG 2018
M. Zhang et al. (Eds.): NLPCC 2018, LNAI 11108, pp. 40–50, 2018.
https://doi.org/10.1007/978-3-319-99495-6_4

in which joint-sequence models are well performing and popular. In joint-sequence models, the alignment is provided via some external aligner [9–11]. However, since the alignment is a latent variable—a means to an end rather than the end itself, it is worthy to consider whether we can do away with such explicit alignment.

In recent years, some work on the G2P problem has used neural network-based approaches. Specifically, long short-term memory (LSTM) networks have recently been explored [12]. LSTMs (and, more generally, recurrent neural networks) can model varying contexts ("memory") and have been successful for a number of sequence prediction tasks. When used in a sequence-to-sequence (Seq2Seq) model, as in [13], which includes an encoder RNN and a decoder RNN, the encoder RNN encoder input sequence token by token, they in principle require no explicit alignments between the input (grapheme sequence) and output (phoneme sequence), as the model is trained in an end-to-end fashion. Bahdanau et al. [14] proposed a related model with an attention mechanism for translation that makes the model better, and Toshniwal et al. [15] introduce an attention mechanism and improve performance of G2P conversion.

For Mongolian G2P conversion, Bao et al. [16] proposed a rule-based method and the method based on joint-sequence model, where the latter method showed better performance than the former method. However, performance of current Mongolian G2P conversion is inferior to other languages. In this paper, we first introduce a Seq2Seq LSTM model with attention mechanism, which proved is useful in other sequence prediction tasks. We obtain better performance than traditional methods for Mongolian G2P conversion. Seq2Seq LSTM model is generative language model, conditioned on an input sequence, the model using an attention mechanism over the encoder LSTM states will not overfit and generalize much better than the plain model without attention mechanism. Taking account of the shortcomings of statistics-based method that can't exactly decode all words in the dictionary and Mongolian characteristics is the majority of Out Of Vocabulary (OOV) words with suffixes connected to the stem using Narrow Non-Break Space (NNBS), we proposed a novel hybrid approach combining of rules with Seq2Seq LSTM model to covert Mongolian word, and we obtain better performance for Mongolian G2P conversion.

In the next section, we will discuss traditional methods for Mongolian G2P conversion. In the remainder we will focus on Seq2Seq LSTM model with an attention mechanism for Mongolian G2P conversion. We will lay the theoretical foundations and undertake a detailed exposition of this model in Sect. 3, and then we will introduce the hybrid approach in Sect. 4. Section 5 presents experimental results demonstrating the better performance of the proposed method, and analyze the consequences of the method. Finally, in Sect. 6 we conclude this paper and look forward to the future of Mongolian G2P conversion technology.

2   Related Work

The Mongolian G2P conversion that was firstly considered in the context of Mongolian text-to-speech (TTS) applications. In this section, we will summarize two traditional approaches to Mongolian G2P conversion.

The one approach is rule-based Mongolian G2P conversion. The written form and spoken form of Mongolian are not one-to-one, and the vowels and consonants may increase, fall off and change. Through in-depth study of Mongolian pronunciation rules, three rules, vowels pronunciation variation rule, consonant binding rule and vowel-harmony rule, are employed in Mongolian G2P conversion. Firstly, Mongolian word is converted by using vowels pronunciation variation rule, and then conversion is followed according to the consonant binding rule, finally, the vowel-harmony rule is used. The rule-based method overcomes the limitations of simple dictionary look-up. However, this method consists of two drawbacks: firstly, designing the rules is hard and requires specific linguistic skills. Mongolian frequently show irregularities, which need to be captured by exceptional rules or exceptional lists. Secondly, the interdependence between rules can be quite complex, so rule designers have to cross-check if the outcome of applying the rules is correct in all cases. This makes development and maintenance of rule systems very tedious in practice. Moreover, a rule-based G2P system is still likely to make mistakes when presented with an exceptional word, not considered by the rule designer.

Another Mongolian G2P conversion approach is based on joint-sequence model. The model needs to find a joint vocabulary of graphemes and phonemes (named graphone) by aligning letters and phonemes, and uses graphone sequence to generate the orthographic form and pronunciation of a Mongolian word. The probability of a graphone sequence is

$$p(C = c_1 \ldots c_T) = \prod_{t=1}^{T} p(c_t | c_1 \ldots c_{t-1}) \qquad (1)$$

Where each c is a graphone unit. The conditional probability $p(c_t | c_1 \ldots c_{t-1})$ is estimated using an n-gram language model.

To date, this model has produced the better performance on common benchmark datasets. Sequitur G2P is a good established G2P conversion tool using joint-sequence n-gram modelling so that it is very convenient to perform an experiment. In the next section, we will introduce Seq2Seq LSTM model with attention mechanism.

3 Seq2Seq LSTM Model with an Attention Mechanism

Neural Seq2Seq model has recently shown promising results in several tasks, especially translation [17, 18]. Because the G2P problem is in fact largely analogous to the translation problem, with a many-to-many mapping between subsequences of input labels and subsequences of output labels and with potentially long-range dependencies, so this model is also frequently used in G2P conversion [13, 15, 19]. We first apply the Seq2Seq LSTM model with attention mechanism to Mongolian G2P conversion, here, we describe in detail the Seq2Seq LSTM model used in Mongolian G2P conversion.

The Seq2Seq LSTM model follow the LSTM Encoder-Decoder framework [20], the encoder reads the input Mongolian letters sequence, a sequence of vectors $x = (x_1, \ldots, x_2)$, the LSTM computes the $h_1, \ldots, h_T$ (h_t is control state at timestep t) and $m_1, \ldots, m_T$ (m_t is memory state at timestep t) as follows.

$$i_t = sigm(W_1 x_t + W_2 h_{t-1}) \tag{2}$$

$$i_t' = tanh(W_3 x_t + W_4 h_{t-1}) \tag{3}$$

$$f_t = sigm(W_5 x_t + W_6 h_{t-1}) \tag{4}$$

$$o_t = sigm(W_7 x_t + W_8 h_{t-1}) \tag{5}$$

$$m_t = m_{t-1} \odot f_t + i_t \odot i_t' \tag{6}$$

$$h_t = m_t \odot o_t \tag{7}$$

Where the operator $\odot$ represents element-wise multiplication, the matrices $W_1, \ldots, W_8$ and the vector h_0 are the parameters of the model, and all the nonlinearities are computed element-wise. The above equations are merged as:

$$h_t = f(x_t, h_{t-1}) \tag{8}$$

In above equation, f represents an LSTM.

The decoder is another LSTM to produce the output sequence (phonemes sequence y $(y_1, \ldots, y_{T_B})$) and trained to predict the next phoneme y_t given the attention vector c_t and all the previously predicted phonemes sequence $\{y_1, \ldots, y_{t-1}\}$, each conditional probability is modeled as

$$p(y_t | y_1, \ldots, y_{t-1}, x) = g(y_{t-1}, s_t, c_t) \tag{9}$$

Where g is a nonlinear, potentially multi-layered, function that outputs the probability of y_t, and s_t represents the hidden state of the LSTM at timestep t, the attention vector c_t [21] concatenating with s_t became the new hidden state to predict y_t. To computed the attention vector c_t at each output time t over the input Mongolian letters sequence $x(x_1, \ldots, x_{T_A})$ as following:

$$u_i^t = v^T tanh\left(W_1' h_i + W_2' s_t\right) \tag{10}$$

$$a_i^t = softmax(u_i^t) \tag{11}$$

$$c_t = \sum_{i=1}^{T_A} a_i^t h_i \tag{12}$$

The vector v and matrices W_1', W_2' are learnable parameters of the model. The vector u_i^t has length T_A and its $i - th$ item contains a score of how much attention should be put on the $i - th$ hidden encoder state h_i. These scores are normalized by softmax to create the attention mask a^t over encoder hidden decoder.

4   Hybrid Approach to Mongolian G2P Conversion

Taking account of the shortcomings of statistics-based method that can't exactly decode all words in the dictionary and Mongolian characteristics is the majority of Out Of Vocabulary (OOV) words with suffixes connected to the stem using Narrow Non-Break Space (NNBS), we proposed a novel hybrid approach of combining rules with Seq2Seq LSTM model to covert Mongolian word.

4.1   Rules

The rules include two parts. The first part overcomes the disadvantages of the method based on Seq2Seq LSTM model that this method can't ensure that all words in the dictionary are exactly decoded. In rules, for those words in the dictionary, their accurate phonemes sequences can be got by dictionary look-up. Because Mongolian is Agglutinative Language, the majority of Mongolian word with suffixes connected to the stem using Narrow Non-Break Space (NNBS), following work in [22], we also called those NNBS suffixes. The NNBS suffixes refer to case suffixes, reflexive suffixes and partly plural suffixes. They are used very flexible that each stem can add several NNBS suffixes to change Mongolian word form. The another part is to handle Mongolian word with NNBS suffixes, the pronunciation of Mongolian word with NNBS suffixes follow two rules, one rule called NNBS suffixes' rules is that NNBS suffixes' pronunciation depends on the form of stem's phoneme sequence, which is different due to varying form of stem's phonemes sequence. Another rule named stem's rules is that stem's phonemes sequence can be changed according to NNBS suffixes' phonemes sequence. We define four forms of stem's phonemes sequence for NNBS suffixes' rules as following:

- Form 1: The word-stem is a positive word and stem's phonemes sequence ends with a vowel, there are two cases. The first case is that stem's phoneme sequence ending with a vowel in the set {al, vl, ael, vi, vae, va, av}, then check whether Il or I exists in the stem's phoneme sequence. The second case is similar to the first case except that stem's phoneme sequence ending with a vowel in the set {wl, oel, wi, w}.
- Form 2: The word-stem is a negative word and stem's phonemes sequence ends with a vowel, there are two cases. The first case is that stem's phoneme sequence ending with a vowel in the set {el, ul, El, ui, ue, Yl, e, u}, then check whether il or i exists in the stem's phoneme sequence. The second case is similar to the first case except that stem's phoneme sequence ending with a vowel in the set {ol, Ol, o}.
- Form 3: The word-stem is a positive word and stem's phonemes sequence ends with a consonant, whether the first vowel searched from back to front exists in the set {a, v, Y, ae, as1, as2, vi, al, vl, ael, va, vae} or in the set {w, oe, wi, wl, oel, ws}, if Il or I is encountered, the next vowel should be searched forward from Il or I.
- Form 4: The word-stem is a negative word and stem's phonemes sequence ends with a consonant, whether the first vowel searched from back to front exists in the set {e, u, es, ui, El, el, ul, Yl, ue} or in the set {o, os, ol, Ol}, if il or i is encountered, the next vowel should be searched forward from il or i.

We list parts NNBS suffixes and their different phoneme sequence corresponding to varying form of the sequence of the stem in Table 1. The stem's rules include two parts, one part named as stem_rule1 is to determine whether NNBS suffixes' phonemes sequence starts with a long vowel, another part is to judge that whether stem's phonemes follows the stem_rule2. The stem's rules are showed as following in Table 2.

Table 1. Parts NNBS suffixes and their different phoneme sequence corresponding to varying form of the sequence of stem.

Mongolian	Latin	Form 1	Form 2	Form 3	Form 4
᠊ᠶᠢᠨ	-yin	g il l	g il l	il l	Il l
᠊ᠳᠦ	-dv	d	d	asl d	ws d
᠊ᠪᠠᠷ	-bar	g ar r	g wr r	ar r	wr r
᠊ᠢᠶᠡᠷ	-iyer	al r	el r	ol r	wl r
᠊ᠲᠡᠢ	-tei	t El	t Ol	t ael	t oel
᠊ᠠᠬᠠ	-aqa	al s	wl s	el s	ol s
᠊ᠠᠬᠠ ᠪᠠᠨ	-aqa-ban	g al s al n	g wl s wl n	al s al n	wl s wl n
᠊ᠶᠢᠨ ᠢᠶᠡᠨ	-yin-iyen	g il n h el n	g il n h ol n	il n h el n	il n h ol n

Table 2. The stem's rules. C1, C2, V1, C3 and V, C1, V1, C2 are the last four phonemes of the stem's phonemes sequence, C1, C2 and C3 are the consonants, V and V1 are the vowels, A represents the phonemes sequence of the NNBS suffixes of Mongolian word.

Stem's phonemes	Stem_rule1	Stem_rule2	Word's Phonemes
—$C_1 C_2 V_1 C_3$	A stars with a long vowel	V1 belongs to the set {as1, as2, es, ws, os}, C1 and C2 form a composite consonant.	—$C_1 C_2 C_3 + A$
—$V C_1 V_1 C_2$	A stars with a long vowel	V1 don't belong to the set {j, q, x, y, I, i}.	—$V C_1 C_2 + A$
Others	—	—	— + A

4.2   Combining Rules with Seq2Seq LSTM Model

Combining rules with Seq2Seq LSTM model for Mongolian G2P conversion are shown in Fig. 1. The procedures mainly comprise the following parts. Firstly, we transliterate all Mongolian words and its phonemes sequence in the dictionary to the Latin form, and transliterate input Mongolian word to the Latin form, if input Mongolian word exists in the dictionary, we can directly get word's phonemes sequence through the dictionary look-up. Secondly, Mongolian word does not exist in dictionary, if Mongolian word is with NNBS suffixes, we make use of combining rules with Seq2Seq LSTM model to handle this word. Instead of with NNBS suffixes, we decode Mongolian word by using Seq2Seq LSTM model. Finally, Mongolian word's phonemes sequence can be generated.

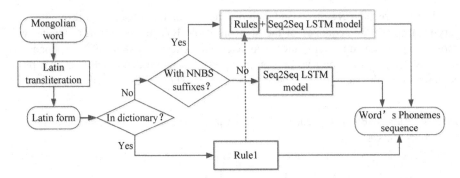

Fig. 1. The structure of combination rules and Seq2Seq LSTM model. The blue boxes represent that using rules to handle word, and Rule1 included in Rules is dictionary look-up. The green boxes represent that using Seq2Seq LSTM model to decode word, and the yellow box represents using hybrid approach. (Color figure online)

For Mongolian word with NNBS suffixes, we handle this word by using the hybrid method (see in Fig. 2). We firstly segment Mongolian word to stem and NNBS suffixes. If stem exists in the hash table, we can get stem's phonemes sequence from hash table, instead of this situation, we decode stem by using Seq2Seq LSTM model. Depending on stem's phonemes sequence and NNBS suffixes' rules, we can get NNBS suffixes' phonemes sequence, and then modify stem's phonemes sequence through NNBS suffixes' phonemes sequence and stem's rules. Mongolian word's phonemes

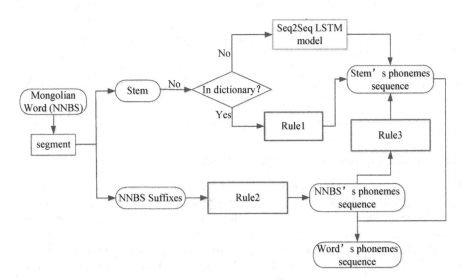

Fig. 2. The procedure of using the proposed method to handle Mongolian word with NNBS suffixes. The blue boxes represent rules included Rule1 (hash table look-up), Rule2 (NNBS suffixes' rules), and Rule3 (stem's rules). The green boxes represent that using Seq2Seq LSTM model to decode stem. (Color figure online)

sequence can be produced by jointing stem's phonemes sequence modified with NNBS suffixes' phonemes sequence.

5 Experiments

5.1 Data Set

This paper uses the Mongolian Orthography dictionary as an experimental dataset, Mongolian word with NNBS suffixes accounts for 15 percent of this dictionary. The dataset consisted of a training set of 33483 pairs of word and its phonemes sequence, a validation set of 1920 pairs of word and its phonemes and a test set of 3940 pairs of word and its phonemes. The evaluation criteria of the model are the word error rate (WER) and the phoneme error rate (PER).

$$WER = 1 - \frac{NW_{correct}}{NW_{wtotal}} \tag{13}$$

$$PER = 1 - \frac{NP_{ins} + NP_{del} + NP_{sub}}{NP_{ptotal}} \tag{14}$$

Where $NW_{correct}$ represents number of correctly decoded Mongolian words, NW_{wtotal} is the number of Mongolian words, NP_{ptotal} is the total number of phonemes corresponding to all the Mongolian words converted, NP_{ins}, NP_{del} and NP_{sub} is the quality of insertion errors, deletion errors and substitute errors of total phonemes, respectively.

5.2 Setting and Result

The baseline systems used in this paper are a rule-based G2P conversion system and a G2P conversion system based on the joint-sequence model. The performance of the rule-based G2P conversion system is that WER is 32.3% and PER is 7.6%, apparently, its result is not good. For Sequitur G2P, we tune the model order (n-gram) on the development set used to adjust the discount parameters of joint-sequence model. We found that the experimental result is better when the order is between 6 and 10, however, WER and PER are difficult to lower when the order is more than 10. The Table 3 shows the experimental result of Sequitur G2P.

We use the same dataset to train Seq2Seq LSTM model. We choose the width of the network' LSTM layers from the set {64,128,256,512,1024,2048}, number of layers from {1,2,3,4}, and choose stochastic gradient descent (SGD) as optimization method for network training, and learning rate is 0.5. In our experiment, we found that WER and PER are increasing as the width of LSTM layers and the number of layers increasing. We take better results (in Table 4) out of experimental results.

Comparing Table 3 with Table 4, we find that the performance of Mongolian G2P conversion based on Seq2Seq LSTM model is better than based on the joint-sequence model. We take the Seq2Seq LSTM model (1024x1) whose performance is best in

Table 3. Sequitur G2P's experimental result, where we test the joint-sequence model by decoding the test set and the train set. Best result for single model in bold.

Model	The test set		The train set	
	WER	PER	WER	PER
Model order 6	16.4%	3.4%	4.1%	0.8%
Model order 7	16.4%	3.4%	3.4%	0.7%
Model order 8	16.3%	3.6%	3.3%	0.7%
Model order 9	**16.3%**	**3.2%**	**2.9%**	**0.5%**
Model order 10	16.3%	3.2%	2.9%	0.5%

Table 4. The results of different Seq2Seq LSTM models tested, best result in bold

Model	The test set		The train set	
	WER	PER	WER	PER
LSTM model(512 × 1)	8.7%	2.4%	2.3%	0.6%
LSTM model(1024 × 1)	**8.0%**	**2.2%**	**1.0%**	**0.3%**
LSTM model(128 × 2)	9.7%	2.6%	3.5%	0.9%
LSTM model(256 × 2)	8.8%	2.4%	2.4%	0.6%

Table 4 to combine with rules for Mongolian G2P conversion. We firstly randomly take the same amount of the test dataset of pairs of Mongolian words and their phoneme sequences, and write them in a dictionary for look-up. We get the experimental result to compare with above methods (see in Table 5).

Table 5. The comparing result of testing the best joint-sequence model (order 9), the best seq2seq LSTM model (1024 × 1) and the hybrid method.

Method	The test set		The train set	
	WER	PER	WER	PER
Model order 9	16.3%	3.2%	2.9%	0.5%
LSTM model(1024 × 1)	8.0%	2.2%	1.0%	0.3%
LSTM model(1024 × 1) + rules	**5.5%**	**1.6%**	**0.7%**	**0.2%**

We can see from Table 5, the performance using Seq2Seq model (1024 × 1) is better than using the best joint-sequence model, the WER and the PER reduce by 8.3% and 1.0%, respectively. Although using the same Seq2Seq LSTM model (1024 × 1), the method based on combing rules with Seq2Seq LSTM model performs better, it's WER and PER is 5.5% and 1.6%, respectively. There are two reasons for performance improvement after combining rules. Firstly, if Mongolian word exists in dictionary, we can get exact word's phonemes sequence through rules (dictionary look-up), this approach is apparently more accurate than Seq2Seq LSTM model. Secondly, Mongolian words with NNBS suffixes are ordinary, because of characteristics of stem's pronunciation and NNBS suffixes' pronunciation, it is difficult to exactly cope with this situation by only using Seq2Seq LSTM model, combing rules (stem's rules and NNBS suffixes' rules) can get more accurate phonemes sequence.

6   Conclusion

In this paper, we present the first application of Seq2Seq LSTM model with attention mechanism for Mongolian G2P conversion, the experimental results show that the Mongolian G2P conversion based on Seq2Seq model can get better performance than the previous methods. We continuously adjusted the parameters of the model. We obtain a best Seq2Seq LSTM model (1024x1), and we use the best model to combine with rules for Mongolian G2P conversion, and experimental results became better than the method only using Seq2Seq model. This method proposed is of profound significance to Mongolian G2P conversion, meantime, it is greatly beneficial to the study of Mongolian speech synthesis, speech retrieval and speech recognition. When we go further, and try model fusion for Mongolian G2P conversion, we assume that model fusion may make significant advances.

Acknowledgement. This research was supported by the China national natural science foundation (No. 61563040, No. 61773224) and Inner Mongolian nature science foundation (No. 2016ZD06).

References

1. Hojo, N., Ijima, Y., Mizuno, H.: An investigation of DNN-based speech synthesis using speaker codes. In: INTERSPEECH, pp. 2278–2282 (2016)
2. Merritt, T., Clark, R., Wu, Z.: Deep neural network-guided unit selection synthesis. In: IEEE International Conference on Acoustics, Speech and Signal Processing, pp. 5145–5149. IEEE (2016)
3. Liu, R., Bao, F., Gao, G., Wang, Y.: Mongolian text-to-speech system based on deep neural network. In: Tao, J., Zheng, T.F., Bao, C., Wang, D., Li, Y. (eds.) NCMMSC 2017. CCIS, vol. 807, pp. 99–108. Springer, Singapore (2018). https://doi.org/10.1007/978-981-10-8111-8_10
4. Graves, A., Mohamed, A., Hinton, G.: Speech recognition with deep recurrent neural networks. In: IEEE International Conference on Acoustics, Speech and Signal Processing, pp. 6645–6649. IEEE (2013)
5. Wang, Y., Bao, F., Zhang, H., Gao, G.: Research on Mongolian speech recognition based on FSMN. In: Huang, X., Jiang, J., Zhao, D., Feng, Y., Hong, Y. (eds.) NLPCC 2017. LNCS (LNAI), vol. 10619, pp. 243–254. Springer, Cham (2018). https://doi.org/10.1007/978-3-319-73618-1_21
6. Zhang, H., Bao, F., Gao, G.: Mongolian speech recognition based on deep neural networks. In: Sun, M., Liu, Z., Zhang, M., Liu, Y. (eds.) CCL 2015. LNCS (LNAI), vol. 9427, pp. 180–188. Springer, Cham (2015). https://doi.org/10.1007/978-3-319-25816-4_15
7. Bao, F., Gao, G., Bao, Y.: The research on Mongolian spoken term detection based on confusion network. Commun. Comput. Inf. Sci. **321**(1), 606–612 (2012)

8. Lu, M., Bao, F., Gao, G.: Language model for Mongolian polyphone proofreading. In: Sun, M., Wang, X., Chang, B., Xiong, D. (eds.) CCL/NLP-NABD -2017. LNCS (LNAI), vol. 10565, pp. 461–471. Springer, Cham (2017). https://doi.org/10.1007/978-3-319-69005-6_38

9. Bisani, M., Ney, H.: Joint-sequence models for grapheme-to-phoneme conversion. Speech Commun. **50**(5), 434–451 (2008)

10. Chen, S.F.: Conditional and joint models for grapheme-to-phoneme conversion. In: European Conference on Speech Communication and Technology, INTERSPEECH 2003, Geneva, Switzerland, DBLP (2003)

11. Jiampojamarn, S., Kondrak, G., Sherif, T.: Applying many-to-many alignments and hidden markov models to letter-to-phoneme conversion. In: Proceedings of Human Language Technology Conference of the North American Chapter of the Association of Computational Linguistics, Proceedings, USA, pp. 372–379 (2008)

12. Rao, K., Peng, F., Sak, H.: Grapheme-to-phoneme conversion using long short-term memory recurrent neural networks. In: IEEE International Conference on Acoustics, Speech and Signal Processing, pp. 4225–4229. IEEE (2015)

13. Yao, K., Zweig, G.: Sequence-to-sequence neural net models for grapheme-to-phoneme conversion. Computer Science (2015)

14. Bahdanau, D., Cho, K., Bengio, Y.: Neural machine translation by jointly learning to align and translate. Computer Science (2014)

15. Toshniwal, S., Livescu, K.: Jointly learning to align and convert graphemes to phonemes with neural attention models. In: Spoken Language Technology Workshop. IEEE (2017)

16. Bao, F., Gao, G.: Research on grapheme to phoneme conversion for Mongolian. Appl. Res. Comput. **30**(6), 1696–1700 (2013)

17. Luong, M.T., Sutskever, I., Le, Q.V.: Addressing the rare word problem in neural machine translation. Bull. Univ. Agric. Sci. Vet. Med. Cluj-Napoca Vet. Med. **27**(2), 82–86 (2014)

18. Jean, S., Cho, K., Memisevic, R.: On using very large target vocabulary for neural machine translation. Computer Science (2014)

19. Milde, B., Schmidt, C., Köhler, J.: Multitask sequence-to-sequence models for grapheme-to-phoneme conversion. In: INTERSPEECH, pp. 2536–2540 (2017)

20. Sutskever, I., Vinyals, O., Le, Q.V.: Sequence to sequence learning with neural networks. In: NIPS, pp. 3104–3112 (2014)

21. Vinyals, O., Kaiser, L., Koo, T.: Grammar as a foreign language. Eprint Arxiv, pp. 2773–2781 (2014)

22. Wang, W., Bao, F., Gao, G.: Mongolian named entity recognition system with rich features. In: The 26th International Conference on Computational Linguistics, pp. 505–512. Proceedings of the Conference, Japan (2016)

From Plots to Endings: A Reinforced Pointer Generator for Story Ending Generation

Yan Zhao[1], Lu Liu[1], Chunhua Liu[1], Ruoyao Yang[1], and Dong Yu[1,2(✉)]

[1] Beijing Language and Culture University, Beijing, China
zhaoyan.nlp@gmail.com, luliu.nlp@gmail.com, chunhualiu596@gmail.com,
xmffaf@163.com, yudong_blcu@126.com
[2] Beijing Advanced Innovation for Language Resources of BLCU, Beijing, China

Abstract. We introduce a new task named Story Ending Generation (SEG), which aims at generating a coherent story ending from a
sequence of story plot. We propose a framework consisting of a Generator and a Reward Manager for this task. The Generator follows
the pointer-generator network with coverage mechanism to deal with
out-of-vocabulary (OOV) and repetitive words. Moreover, a mixed loss
method is introduced to enable the Generator to produce story endings
of high semantic relevance with story plots. In the Reward Manager, the
reward is computed to fine-tune the Generator with policy-gradient reinforcement learning (PGRL). We conduct experiments on the recently-
introduced ROCStories Corpus. We evaluate our model in both automatic evaluation and human evaluation. Experimental results show that
our model exceeds the sequence-to-sequence baseline model by 15.75%
and 13.57% in terms of CIDEr and consistency score respectively.

Keywords: Story Ending Generation · Pointer-generator
Policy gradient

1 Introduction

Story generation is an extremely challenging task in the field of NLP. It has
a long-standing tradition and many different systems have been proposed in
order to solve the task. These systems are usually built on techniques such
as planning [12,21] and case-based reasoning [5,25], which rely on a fictional
world including characters, objects, places, and actions. The whole system is
very complicated and difficult to construct.

We define a subtask of story generation named Story Ending Generation
(SEG), which aims at generating a coherent story ending according to a sequence
of story plot. A coherent ending should have a high correlation with the plot
in terms of semantic relevance, consistency and readability. Humans can easily
provide a logical ending according to a series of events in the story plot. The core

© Springer Nature Switzerland AG 2018
M. Zhang et al. (Eds.): NLPCC 2018, LNAI 11108, pp. 51–63, 2018.
https://doi.org/10.1007/978-3-319-99495-6_5

objective of this task is to simulate the mode of people thinking to generate story endings, which has the significant application value in many artificial intelligence fields.

SEG can be considered as a Natural Language Generation (NLG) problem. Most studies on NLG aim at generating a target sequence that is semantically and lexically matched with the corresponding source sequence. Encoder-decoder framework for sequence-to-sequence learning [26] has been widely used in NLG tasks, such as machine translation [3] and text summarization [4,15,22]. Different from the above NLG tasks, SEG pays more attention to the consistency between story plots and endings. From different stories, we have observed that some OOV words in the plot, such as entities, may also appear in the ending. However, traditional sequence-to-sequence models replace OOV words by the special UNK token, which makes it unable to make correct predictions for these words. Moreover, encoder-decoder framework is of inability to avoid generating repetitive words. Another two limitations of encoder-decoder framework are exposure bias [19] and objective mismatch [18], resulting from Maximum Likelihood Estimation (MLE) loss. To overcome these limitations, some methods [6,10,20,23,27] have been explored.

In this paper, we propose a new framework to solve the SEG problem. The framework consists of a Generator and a Reward Manager. The Generator follows a pointer-generator network to produce story endings. The Reward Manager is used for calculating the reward to fine tune the Generator through PGRL. With a stable and healthy environment that the Generator provides, PGRL can take effect to enable the generated story endings much more sensible. The key contributions of our model are as follows:

- We apply copy and coverage mechanism [23] to traditional sequence-to-sequence model as the Generator to handle OOV and repetitive words, improving the accuracy and fluency of generated story endings.
- We add a semantic relevance loss to the original MLE loss as a new objective function to encourage the high semantic relevance between story plots and generated endings.
- We define a Reward Manager to fine tune the Generator through PGRL. In the Reward Manager, we attempt to use different evaluation metrics as reward functions to simulate the process of people writing a story.

We conduct experiments on the recently-introduced ROCStories Corpus [14]. We utilize both automatic evaluation and human evaluation to evaluate our model. There are word-overlap and embedding metrics in the automatic evaluation [24]. In the human evaluation, we evaluate generated endings in terms of consistency and readability, which reflect the logical coherence and fluency of those endings. Experimental results demonstrate that our model outperforms previous basic neural generation models in both automatic evaluation and human evaluation. Better performance in consistency indicates that our model has strong capability to produce reasonable sentences.

2 Related Work

2.1 Encoder-Decoder Framework

Encoder-decoder framework, which uses neural networks as encoder and decoder, was first proposed in machine translation [3, 26] and has been widely used in NLG tasks. The encoder reads and encodes a source sentence into a fixed-length vector, then the decoder outputs a new sequence from the encoded vector. Attention mechanism [2] extends the basic encoder-decoder framework by assigning different weights to input words when generating each target word. [4, 15, 22] apply attention-based encoder-decoder model to text summarization.

2.2 Copy and Coverage Mechanisms

The encoder-decoder framework is unable to deal with OOV words. In most NLP systems, there usually exists a predefined vocabulary, which only contains top-K most frequent words in the training corpus. All other words are called OOV and replaced by the special UNK token. This makes neural networks difficult to learn a good representation for OOV words and some important information would be lost. To tackle this problem, [7, 28] introduce pointer mechanism to predict the output words directly from the input sequence. [6] incorporate copy mechanism into sequence-to-sequence models and propose CopyNet to naturally combine generating and copying. Other extensions of copy mechanism appear successively, such as [13]. Another problem of the encoder-decoder framework is repetitive words in the generated sequence. Accordingly coverage model in [27] maintains a coverage vector for keeping track of the attention history to adjust future attention. A hybrid pointer-generator network introduced by [23] combines copy and coverage mechanism to solve the above problems.

2.3 Reinforcement Learning for NLG

The encoder-decoder framework is typically trained by maximizing the log-likelihood of the next word given the previous ground-truth input words, resulting in exposure bias [19] and objective mismatch [18] problems. Exposure bias refers to the input distribution discrepancy between training and testing time, which makes generation brittle as error accumulate. Objective mismatch refers to using MLE at training time while using discrete and non-differentiable NLP metrics such as BLEU at test time. Recently, it has been shown that both the two problems can be addressed by incorporating RL in captioning tasks. Specifically, [19] propose the MIXER algorithm to directly optimize the sequence-based test metrics. [10] improve the MIXER algorithm and uses a policy gradient method. [20] present a new optimization approach called self-critical sequence training (SCST). Similar to the above methods, [16, 18, 29] explore different reward functions for video captioning. Researchers also make attempts on other NLG tasks such as dialogue generation [9], sentence simplification [31] and abstract summarization [17], obtaining satisfying performances with RL.

Although many approaches for NLG have been proposed, SEG is still a challenging yet interesting task and worth trying.

3   Models

Figure 1 gives the overview of our model. It contains a Generator and a Reward Manager. The Generator follows the pointer-generator network with coverage mechanism to address the issues of OOV words and repetition. A mixed loss method is exploited in the Generator for improving semantic relevance between story plots and generated endings. The Reward Manager is utilized to produce the reward for PGRL. The reward can be calculated by evaluation metrics or other models in the Reward Manager. Then it is passed back to the Generator for updating parameters. Following sections give more detailed descriptions of our models.

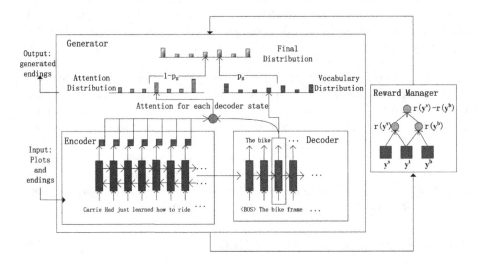

Fig. 1. Overview of our model

3.1   Attention-Based Encoder-Decoder Model

Our attention-based encoder-decoder baseline model is similar to the framework in [2]. Given a sequence of plot words of length T_e, we feed the word embeddings into a single-layer bidirectional LSTM to compute a sequence of encoder hidden states $h_i^e = \{h_1^e, h_2^e, ..., h_{T_e}^e\}$. At each decoding step t, a single LSTM decoder takes the previous word embedding and context vector c_{t-1}, which is calculated by attention mechanism, as inputs to produce decoder hidden state h_t^d.

We concatenate the context vector c_t and decoder hidden state h_t^d to predict the probability distribution P_v over all the words in the vocabulary:

$$P_v = softmax(W_1(W_2[h_t^d, c_t] + b_2) + b_1) \tag{1}$$

where W_1, W_2, b_1, b_2 are all learnable parameters. [a, b] means the concatenation of a and b.

MLE is usually used as the training objective for sequence-to-sequence tasks. We denote $y_t^* = \{y_1^*, y_2^*, ..., y_{T_d}^*\}$ as the ground truth output ending. The cross entropy loss function is defined as:

$$L_{mle} = -\sum_{t=1}^{T_d} \log P_v(y_t^*) \tag{2}$$

3.2   Pointer-Generator Network with Coverage Mechanism

From the dataset, we find that some words in the story plot will also appear in the ending. It makes sense that the story ending usually describes the final states of some entities, which are related to the events in the story plot. Thus we follow the hybrid pointer-generator network in [23] to copy words from the source plot text via pointing [28], in this way we can handle some OOV words. In this model, we accomplish the attention-based encoder-decoder model in the same way as Sect. 3.1. Additionally, we choose top-k words to build a vocabulary and calculate a generation probability p_g to weight the probability of generating words from the vocabulary.

$$p_g = sigmoid(W_{c_t} c_t + W_{h_t} h_t + W_{y_t} y_t + b_p) \tag{3}$$

where c_t, h_t^d, y_t represent the context vector, the decoder hidden state and the decoder input at each decoding step t respectively. W_{c_t}, W_{h_t}, W_{y_t} are all weight parameters and b_p is a bias.

Furthermore, the attention distributions of duplicate words are merged as $P_{att}(w_t)$. We compute the weighted sum of vocabulary distribution $P_v(w_t)$ and $P_{att}(w_t)$ as the final distribution:

$$P_{fin}(w_t) = p_g P_v(w_t) + (1 - p_g)P_{att}(w_t) \tag{4}$$

The loss function is the same as that in attention-based encoder-decoder model, with $P_v(w_t)$ in Eq. (2) changed to $P_{fin}(w_t)$.

To avoid repetition, we also apply coverage mechanism [27] to track and control coverage of the source plot text. We utilize the sum of attention distributions over all previous decoder steps as the coverage vector s^t. Then the coverage vector is added into the calculation of attention score e_i^t to avoid generating repetitive words:

$$e_i^t = v^T tanh(W_1^{att} h_i^e + W_2^{att} h_t^d + W_3^{att} s_i^t) \tag{5}$$

where W_1^{att}, W_2^{att}, W_3^{att}, and v^T are learnable parameters.

Moreover, a coverage loss is defined and added to the loss function to penalize repeatedly attending to the same locations:

$$L_{poi} = -\sum_{t=1}^{T_d}[\log P_{fin}(w_t) + \beta \sum_{i=1}^{T_e} min(\alpha_i^t, s_i^t)] \tag{6}$$

where β is a hyperparameter and $min(a, b)$ means the minimum of a and b.

3.3 Mixed Loss Method

Pointer-generator network has the capacity of generating grammatically and lexically accurate story endings. These story endings are usually of low semantic relevance with plots, which fails to meet our requirements of satisfying story endings. To overcome this weakness, we add a semantic similarity loss to the original loss as the new objective function.

There are some different ways to obtain the semantic vectors, such as the average pooling of all word embeddings or max pooling of the RNN hidden outputs. Intuitively, the bidirectional LSTM encoder can fully integrate the context information from two directions. Therefore the last hidden output of the encoder $h_{T_e}^e$ is qualified to represent the semantic vector of the story plot. Similar to [11], we select $h_{T_e}^e$ as the plot semantic vector v_{plot}, and the last hidden output of decoder subtracting last hidden output of the encoder as the semantic vector of the generated ending v_{gen}:

$$v_{plot} = h_{T_e}^e \tag{7}$$

$$v_{gen} = h_{T_d}^d - h_{T_e}^e \tag{8}$$

Semantic Relevance: Cosine similarity is typically used to measure the matching affinity between two vectors. With the plot semantic vector v_{plot} and the generated semantic vector v_{gen}, the semantic relevance is calculated as:

$$S_{sem} = cos(v_{plot}, v_{gen}) = \frac{v_{plot} \cdot v_{gen}}{\|v_{plot}\| \|v_{gen}\|} \tag{9}$$

Mixed Loss: Our objective is maximizing the semantic relevance between story plots and generated endings. As a result, we combine the similarity score S_{sem} with the original loss as a mixed loss:

$$L_{mix} = -S_{sem} + L_{poi} \tag{10}$$

The mixed loss method encourages our model to generate story endings of high semantic relevance with plots. In addition, it makes the Generator more stable for applying RL algorithm.

3.4 Policy-Gradient Reinforcement Learning

The Generator can generate syntactically and semantically correct sentences with the above two methods. However, models trained with MLE still suffer from exposure bias [19] and objective mismatch [18] problems. A well-known policy-gradient reinforcement learning algorithm [30] can directly optimize the non-differentiable evaluation metrics such as BLEU, ROUGE and CIDEr. It has good performance on several sequence generation tasks [17,20].

In order to solve the problems, we cast the SEG task to the reinforcement learning framework. An *agent* interacting with the external environment in reinforcement learning can be analogous to our generator taking words of the story

---

Algorithm 1. The reinforcement learning algorithm for training the Generator $G_{\theta'}$

Input: ROCstories $\{(x, y)\}$;
Output: Generator $G_{\theta'}$;
1 Initialize G_θ with random weights θ;
2 Pre-train G_θ using MLE on dataset $\{(x, y)\}$;
3 Initialize $G_{\theta'} = G_\theta$;
4 **for** *each epoch* **do**
5   Generate an ending $y^b = (y_1^b, \ldots, y_T^b)$ according to $G_{\theta'}$ given x;
6   Sample an ending $y^s = (y_1^s, \ldots, y_T^s)$ from the probability distribution $P(y_t^s)$;
7   Compute reward $r(y^b)$ and $r(y^s)$ defined in the Reward Manager;
8   Compute L_{rl} using Eq.(11);
9   Compute L_{total} using Eq.(12);
10   Back-propagate to compute $\nabla_{\theta'} \mathcal{L}_{total}(\theta')$;
11   Update Generator $G_{\theta'}$ using ADAM optimizer with learning rate lr
12 **end**
13 **return** $G_{\theta'}$

---

plot as inputs and then producing outputs. The parameters of the agent define a *policy*, which results in the agent picking an *action*. In our SEG task, an action refers to generating a sequence as story ending. After taking an action, the agent computes the *reward* of this action and updates its internal *state*.

Particularly, we use the SCST approach [20] to fine-tune the Generator. This approach designs a loss function, which is formulated as:

$$L_{rl} = (r(y^b) - r(y^s)) \sum_{t=1}^{T} \log P(y_t^s) \qquad (11)$$

where $y^s = (y_1^s, ..., y_T^s)$ is a sequence sampled from the probability distribution $P(y_t^s)$ at each decoding time step t. y^b is the baseline sequence obtained by greedy search from the current model. $r(y)$ means the reward for the sequence y, computed by the evaluation metrics. Intuitively, the loss function L_{rl} enlarges the log-probability of the sampled sequence y^s if it obtains a higher reward than the baseline sequence y^b. In the Reward Manager, we try several different metrics as reward functions and find that BLEU-4 produces better results than others.

To ensure the readability and fluency of the generated story endings, we also define a blended loss function, which is a weighted combination of the mixed loss in Sect. 3.3 and the reinforcement learning loss:

$$L_{total} = \mu L_{rl} + (1 - \mu) L_{mix} \qquad (12)$$

where μ is a hyper-parameter controlling the ratio of L_{rl} and L_{mix}. This loss function can make a trade-off between the RL loss and mixed loss in Sect. 3.3.

The whole reinforcement learning algorithm for training the Generator is summarized as Algorithm 1.

4 Experiments

4.1 Dataset

ROCStories Corpus is a publicly available collection of short stories released by [14]. There are 98161 stories in training set and 1871 stories in both validation set and test set. A complete story in the corpus consists of five sentences, in which the first four and last one are viewed as the plot and ending respectively. The corpus captures a variety of causal and temporal commonsense relations between everyday events. We choose it for our SEG task because of its great performance in quantity and quality.

4.2 Experimental Setting

In this paper, we choose attention-based sequence-to-sequence model (Seq2Seq) as our baseline. Additionally, we utilize pointer-generator network with coverage mechanism (PGN) to deal with OOV words and avoid repetition. Then we train pointer-generator network with mixed loss method (PGN+Sem_L) and PGRL algorithm (PGN+RL) respectively. Finally, we integrate the entire model with both mixed loss method and PGRL algorithm (PGN+Sem_L+RL).

We implement all these models with Tensorflow [1]. In all the models, the LSTM hidden units, embedding dimension, batch size, dropout rate and beam size in beam search decoding are set to 256, 512, 64, 0.5 and 4 respectively. We use ADAM [8] optimizer with an initial learning rate of 0.001 when pre-training the generator and 5×10^{-5} when running RL training. The weight coefficient of coverage loss β is set to 1. The ratio μ between RL loss and mixed loss is 0.95. Through counting all the words in the training set, we obtain the vocab size 38920 (including extra special tokens UNK, PAD and BOS). The size of vocabulary is 15000 when training the pointer-generator network. The coverage mechanism is used after 10-epoch training of single pointer-generator network. We evaluate the model every 100 global steps and adopt early stopping on the validation set.

4.3 Evaluation Metrics

For SEG, a story may have different kinds of appropriate endings for the same plot. It is unwise to evaluate the generated endings from a single aspect. Therefore we apply automatic evaluation and human evaluation in our experiments.

Automatic Evaluation: We use the evaluation package nlg-eval[1] [24], which is a publicly available tool supporting various unsupervised automated metrics for NLG. It considers not only word-overlap metrics such as BLEU, METEOR, CIDEr and ROUGE, but also embedding-based metrics including SkipThoughts Cosine Similarity (STCS), Embedding Average Cosine Similarity (EACS), Vector Extrema Cosine Similarity (VECS), and Greedy Matching Score (GMS).

---

[1] https://github.com/Maluuba/nlg-eval.

Human Evaluation: We randomly select 100 stories from test set and define two criteria to implement human evaluation. Consistency refers to the logical coherence and accordance between story plots and endings. Readability measures the quality of endings in grammar and fluency. Five human assessors are asked to rate the endings on a scale of 0 to 5.

4.4 Automatic Evaluation

Results on Word-Overlap Metrics. Results on word-overlap metrics are shown in Table 1. Obviously, PGN+sem_L+RL achieves the best result among all the models. This indicates that our methods are effective on producing accurate story endings.

Table 1. Results on word-overlap metrics.

Models	BLEU-1	BLEU-2	BLEU-3	BLEU-4	METEOR	ROUGE-L	CIDEr
Seq2Seq	26.17	10.54	5.29	3.03	10.80	26.84	47.48
PGN	28.07	11.39	5.53	3.02	10.87	27.80	51.09
PGN+Sem_L	28.21	11.56	5.81	3.33	11.08	28.15	53.21
PGN+RL	28.05	11.50	5.69	3.17	10.83	27.46	49.83
PGN+Sem_L+RL	**28.51**	**11.92**	**6.16**	**3.53**	**11.10**	**28.52**	**54.96**

From the results, we have some other observations. PGN surpasses the Seq2Seq baseline model, especially in BLEU-1 (+1.9) and CIDEr (+3.61). This behaviour suggests that copy and coverage mechanisms can effectively handle OOV and repetitive words so as to improve scores of word-overlap metrics. Compared with PGN, the results of PGN+Sem_L have an increase in all the word-overlap metrics. This improvement benefits from our mixed loss method based on semantic relevance. More interestingly, PGN+RL performs poorly while PGN+Sem_L+RL obtains an improvement. We attribute this to an insufficiency of applying RL directly into PGN. Results on PGN+Sem_L+RL prove that mixed loss method shows its effectiveness and it motivates RL to take effect.

Results on Embedding Based Metrics. We compute cosine similarities between generated endings and plots. For comparison, the cosine similarity between target endings and plots is provided as the ground-truth reference. Evaluation results are illustrated in Table 2.

Embedding-based metrics tend to acquire more semantics than word-overlap metrics. From Table 2, all the models are likely to generate endings with less discrepancy in terms of the embedding based metric. It can also be observed that scores of all models surpass that of the ground-truth reference. This indicates that nearly every model can generate endings which have higher cosine similarity scores with the plot. But it cannot just measure these endings by calculating these scores.

Table 2. Results on embedding based metrics.

Models	STCS-p	EACS-p	VECS-p	GMS-p
Ground truth	66.94	87.03	45.64	70.75
Seq2Seq	67.94	89.98	46.37	73.23
PGN	68.15	89.20	48.96	74.64
PGN+Sem_L	67.90	90.02	48.60	74.61
PGN+RL	68.07	89.97	49.48	74.90
PGN+Sem_L+RL	67.84	89.50	48.44	74.45

Table 3. Human evaluation results.

Models	Consistency	Readability
Ground truth	4.33	4.83
Seq2Seq	2.80	4.33
PGN	2.95	4.38
PGN+Sem_L	3.00	**4.43**
PGN+RL	2.92	4.36
PGN+Sem_L+RL	**3.18**	4.41

4.5   Human Evaluation

Table 3 presents human evaluation results. Apparently, PGN+Sem_L+RL and PGN+Sem_L achieve the best results in terms of consistency and readability respectively. The readability score of PGN+sem_L+RL is good enough, with the difference of 0.02 compared to the best result (PGN+Sem_L). We can also observe that readability scores of all the models are basically equivalent. It manifests that all the models have the ability to generate grammatically and lexically correct endings. Therefore, we only analyze the consistency scores as follows.

The consistency score of PGN increases by 5.37% compared with Seq2Seq. This is attributed to the copy and coverage mechanism discouraging OOV and repetitive words. The score of PGN+Sem_L is 1.69% higher than PGN. With mixed loss method, the semantic relevance between story plots and endings is improved, leading to better performance in consistency. PGN+RL gets a lower score than PGN. This indicates that PGN is not prepared for incorporating RL, and RL alone can not directly promote PGN. In contrast, the score of PGN+Sem_L+RL is 6% higher than PGN+Sem_L. We can conclude that PGN with mixed loss method rather than simple PGN is more capable of stimulating RL to take effect.

In order to demonstrate the generative capability of different models, we present some endings generated by different models in Table 4. Compared with other models, the endings generated by PGN+Sem_L+RL are not only fluent

Table 4. Examples of plots, target endings and generated endings of all models

Model	Example-1	Example-2
Plot	Juanita realizes that she needs warmer clothing to get through winter. She looks for a jacket but at first everything she finds is expensive. Finally she finds a jacket she can afford. She buys the jacket and feels much better	My dad took me to a baseball game when I was little. He spent that night teaching me all about the sport. He showed me every position and what everything meant. He introduced me to one of my favorite games ever
Target	She is happy	Now, playing or seeing baseball on TV reminds me of my father
Seq2Seq	Juanita is happy that she is happy that she is happy	I was so happy that he was so happy
PGN	Juanita is happy that she needs through winter clothing	I was so excited to have a good time
PGN+Sem_L	Juanita is happy to have warmer clothing to winter	I was so happy to have a good time
PGN+RL	Juanita is happy that she has done through winter	My dad told me I had a great time
PGN+Sem_L+RL	Juanita is happy that she has **a new warmer clothing**	I am going to **play with** my dad

but also contain new information (words that are bold). Thus, we conclude that our model reaches its full potential under the joint of mixed loss method and RL.

5 Conclusion

In this work we propose a framework consisting of a Generator and a Reward Manager to solve the SEG problem. Following the pointer-generator network with coverage mechanism, the Generator is capable of handling OOV and repetitive words. A mixed loss method is also introduced to encourage the Generator to produce story endings of high semantic relevance with story plots. The Reward Manager can fine tune the Generator through policy-gradient reinforcement learning, promoting the effectiveness of the Generator. Experimental results on ROCStories Corpus demonstrate that our model has good performance in both automatic evaluation and human evaluation.

Acknowledgements. This work is funded by Beijing Advanced Innovation for Language Resources of BLCU, the Fundamental Research Funds for the Central Universities in BLCU (No. 17PT05).

References

1. Abadi, M., et al.: TensorFlow: a system for large-scale machine learning. CoRR abs/1605.08695 (2016). http://arxiv.org/abs/1605.08695
2. Bahdanau, D., Cho, K., Bengio, Y.: Neural machine translation by jointly learning to align and translate. arXiv preprint arXiv:1409.0473 (2014)
3. Cho, K., et al.: Learning phrase representations using RNN encoder-decoder for statistical machine translation. arXiv preprint arXiv:1406.1078 (2014)
4. Chopra, S., Auli, M., Rush, A.M.: Abstractive sentence summarization with attentive recurrent neural networks. In: Proceedings of the 2016 Conference of the North American Chapter of the Association for Computational Linguistics: Human Language Technologies, pp. 93–98 (2016)
5. Gervs, P., Daz-agudo, B., Peinado, F., Hervs, R.: Story plot generation based on CBR. J. Knowl. Based Syst. **18**, 2–3 (2005)
6. Gu, J., Lu, Z., Li, H., Li, V.O.: Incorporating copying mechanism in sequence-to-sequence learning. arXiv preprint arXiv:1603.06393 (2016)
7. Gulcehre, C., Ahn, S., Nallapati, R., Zhou, B., Bengio, Y.: Pointing the unknown words. arXiv preprint arXiv:1603.08148 (2016)
8. Kingma, D.P., Ba, J.: Adam: a method for stochastic optimization. arXiv preprint arXiv:1412.6980 (2014)
9. Li, J., Monroe, W., Ritter, A., Galley, M., Gao, J., Jurafsky, D.: Deep reinforcement learning for dialogue generation. arXiv preprint arXiv:1606.01541 (2016)
10. Liu, S., Zhu, Z., Ye, N., Guadarrama, S., Murphy, K.: Improved image captioning via policy gradient optimization of spider. arXiv preprint arXiv:1612.00370 (2016)
11. Ma, S., Sun, X.: A semantic relevance based neural network for text summarization and text simplification. arXiv preprint arXiv:1710.02318 (2017)
12. Meehan, J.R.: The metanovel: writing stories by computer. Ph.D. thesis, New Haven, CT, USA (1976). aAI7713224
13. Miao, Y., Blunsom, P.: Language as a latent variable: discrete generative models for sentence compression. arXiv preprint arXiv:1609.07317 (2016)
14. Mostafazadeh, N., et al.: A corpus and evaluation framework for deeper understanding of commonsense stories. arXiv preprint arXiv:1604.01696 (2016)
15. Nallapati, R., Zhou, B., Gulcehre, C., Xiang, B., et al.: Abstractive text summarization using sequence-to-sequence RNNs and beyond. arXiv preprint arXiv:1602.06023 (2016)
16. Pasunuru, R., Bansal, M.: Reinforced video captioning with entailment rewards. arXiv preprint arXiv:1708.02300 (2017)
17. Paulus, R., Xiong, C., Socher, R.: A deep reinforced model for abstractive summarization. arXiv preprint arXiv:1705.04304 (2017)
18. Phan, S., Henter, G.E., Miyao, Y., Satoh, S.: Consensus-based sequence training for video captioning. arXiv preprint arXiv:1712.09532 (2017)
19. Ranzato, M., Chopra, S., Auli, M., Zaremba, W.: Sequence level training with recurrent neural networks. arXiv preprint arXiv:1511.06732 (2015)
20. Rennie, S.J., Marcheret, E., Mroueh, Y., Ross, J., Goel, V.: Self-critical sequence training for image captioning. arXiv preprint arXiv:1612.00563 (2016)
21. Riedl, M.O., Young, R.M.: Narrative planning: balancing plot and character. CoRR abs/1401.3841 (2014). http://arxiv.org/abs/1401.3841
22. Rush, A.M., Chopra, S., Weston, J.: A neural attention model for abstractive sentence summarization. arXiv preprint arXiv:1509.00685 (2015)

23. See, A., Liu, P.J., Manning, C.D.: Get to the point: summarization with pointer-generator networks. arXiv preprint arXiv:1704.04368 (2017)
24. Sharma, S., Asri, L.E., Schulz, H., Zumer, J.: Relevance of unsupervised metrics in task-oriented dialogue for evaluating natural language generation. arXiv preprint arXiv:1706.09799 (2017)
25. Stede, M.: Scott R. Turner, the creative process. A computer model of storytelling and creativity. Hillsdale, NJ: Lawrence Erlbaum, 1994. ISBN 0-8058-1576-7, £49.95. 298 pp. Nat. Lang. Eng. **2**(3), 277–285 (1996). http://dl.acm.org/citation.cfm?id=974680.974687
26. Sutskever, I., Vinyals, O., Le, Q.V.: Sequence to sequence learning with neural networks. In: Advances in Neural Information Processing Systems, pp. 3104–3112 (2014)
27. Tu, Z., Lu, Z., Liu, Y., Liu, X., Li, H.: Modeling coverage for neural machine translation. arXiv preprint arXiv:1601.04811 (2016)
28. Vinyals, O., Fortunato, M., Jaitly, N.: Pointer networks. In: Advances in Neural Information Processing Systems, pp. 2692–2700 (2015)
29. Wang, X., Chen, W., Wu, J., Wang, Y.F., Wang, W.Y.: Video captioning via hierarchical reinforcement learning. arXiv preprint arXiv:1711.11135 (2017)
30. Williams, R.J.: Simple statistical gradient-following algorithms for connectionist reinforcement learning. In: Sutton, R.S. (ed.) Reinforcement Learning. SECS, vol. 173, pp. 5–32. Springer, Boston (1992). https://doi.org/10.1007/978-1-4615-3618-5_2
31. Zhang, X., Lapata, M.: Sentence simplification with deep reinforcement learning. arXiv preprint arXiv:1703.10931 (2017)

A3Net: Adversarial-and-Attention Network for Machine Reading Comprehension

Jiuniu Wang[1,2], Xingyu Fu[1], Guangluan Xu[1(✉)], Yirong Wu[1,2], Ziyan Chen[1], Yang Wei[1], and Li Jin[1]

[1] Key Laboratory of Technology in Geo-spatial Information
Processing and Application System, Institute of Electronics, CAS, Beijing, China
gluanxu@mail.ie.ac.cn
[2] School of Electronic, Electrical and Communication Engineering,
University of Chinese Academy of Sciences, Beijing, China
wangjiuniu16@mails.ucas.ac.cn

Abstract. In this paper, we introduce Adversarial-and-attention Network (A3Net) for Machine Reading Comprehension. This model extends existing approaches from two perspectives. First, adversarial training is applied to several target variables within the model, rather than only to the inputs or embeddings. We control the norm of adversarial perturbations according to the norm of original target variables, so that we can jointly add perturbations to several target variables during training. As an effective regularization method, adversarial training improves robustness and generalization of our model. Second, we propose a multi-layer attention network utilizing three kinds of high-efficiency attention mechanisms. Multi-layer attention conducts interaction between question and passage within each layer, which contributes to reasonable representation and understanding of the model. Combining these two contributions, we enhance the diversity of dataset and the information extracting ability of the model at the same time. Meanwhile, we construct A3Net for the WebQA dataset. Results show that our model outperforms the state-of-the-art models (improving Fuzzy Score from 73.50% to 77.0%).

Keywords: Machine Reading Comprehension · Adversarial training
Multi-layer attention

1 Introduction

Machine reading comprehension (MRC) aims to teach machines to better read and comprehend, and answer questions posed on the passages that they have seen [5]. In this paper, we propose a novel model named Adversarial-and-attention Network (A3Net) for MRC.

© Springer Nature Switzerland AG 2018
M. Zhang et al. (Eds.): NLPCC 2018, LNAI 11108, pp. 64–75, 2018.
https://doi.org/10.1007/978-3-319-99495-6_6

The understanding of neural network is shallow, and it is easy to be disturbed by adversarial examples [7]. So we adopt Adversarial training(AT) [3] as a regularization method to improve our model's generality and robustness. Previous works apply adversarial perturbations mainly on input signals [3] or word embeddings [11], acting as a method to enhance data. While we blend these perturbations into different model layers, especially where question-passage interaction takes place.

The state-of-the-art models have been proved to be effective in English MRC datasets such as CNN/DailyMail [5] and SQuAD [12], such as Match-LSTM [18], BIADF [13], SAN [10], and ReasoNet [14]. They all use attention mechanism and pointer network [17] to predict the span answer. However, these models tend to apply attention function on the limited layer. Thus they would ignore some deep-seated information. To solve this problem, we adopt multi-layer attention to extract information at each level. In the low-level, attention weight is highly affected by the similarity of word embedding and lexical structure(e.g. *affix, part of speech*, etc.), which contains syntactic information. While in the high-level, attention weight reflects the abstract concept correlation between passage and question, which contains semantic information.

To sum up, our contributions can be summarized as follows:

- We blend adversarial training to each layer of our model. Not only can adversarial training enhance the information representation ability, but also improve extraction ability of the whole model.
- We apply multi-layer attention to each layer of our model. In this way, our model can make efficient interactions between questions and passages, so as to find which passage span is needed.
- We propose a novel neural network named A3Net for Machine Reading Comprehension, which gains the best result on the WebQA dataset.

2   Related Work

Adversarial Training. Szegedy et al. [15] found that deep neural network might make mistakes when adding small worst-case perturbations to input. This kind of inputs is called adversarial examples. Many models cannot defend the attack of adversarial examples, including widely used state-of-the-art neural networks such as CNN and RNN. In recent years, there are several methods for regularizing the parameters and features of a deep neural network during training. For example, by randomly dropping units, dropout is widely used as a simple way to prevent neural networks from overfitting.

Adversarial training(AT) [3] is a kind of regularizing learning algorithms. It was first proposed to fine tune the task of image classification. By adding perturbations to input signals during training, neural network gains the ability to tolerant the effect of adversarial example. Miyato et al. [11] first adopt AT to text classification. They add perturbations to word embedding and obtain similar benefits like that in image classification. Following this idea, Wu et al. [20] utilizes AT to Relation Extraction and improves the precision. In order

Table 1. An outline of attention mechanism used in state-of-the-art architectures.

Model	Syntactic attention	Semantic attention	Self-match attention
DrQA [2]	$\checkmark$		
FastQA [19]	$\checkmark$		
Match-LSTM [18]		$\checkmark$	
BIDAF [13]		$\checkmark$	
R-Net [4]		$\checkmark$	$\checkmark$
SAN [10]			$\checkmark$
FusionNet [6]	$\checkmark$	$\checkmark$	$\checkmark$

to improve the model stability and generality, we adopt adversarial training to several target variables within our model.

Attention Mechanism. Attention mechanism has demonstrated success in a wide range of tasks. Bahdanau et al. [1] first propose attention mechanism and apply it to neural machine translation. And then, it is widely used in MRC tasks. Attention near embedding module aims to attend the embedding from the question to the passage [2]. Attention after context encoding extracts the high-level representation in the question to augment the context. Self-match attention [16] takes place before answer module. It dynamically refines the passage representation by looking over the whole passage and aggregating evidence relevant to the current passage word and question.

As shown in Table 1, three different types of attention mechanisms are widely used in state-of-the-art architectures. DrQA [2] simply use a bilinear term to compute the attention weights, so as to get word level question-aware passage representation. FastQA [19] combines feature into the computation of word embedding attention weights. Match-LSTM [18] applies LSTM to gain more context information, and concatenate attentions from two directions. A less memory attention mechanism is introduced in BIDAF [13] to generate bi-directional attention flow. R-Net [4] extends self-match attention to refine information over context. SAN [10] adopts self-match attention and uses stochastic prediction dropout to predict answer during training. Huang et al. [6] summarizes previous research and proposes the fully-aware attention to fuse different representations over the whole model. Different from the above models, we utilize three kinds of method to calculate attention weights, which helps our model to interchange information frequently, so as to select the appropriate words in passages.

3   Proposed Model

In this paper, we use span extraction method to predict the answer to a specific question $Q = \{q_1, q_2, .., q_J\}$ based on the related passage $P = \{p_1, p_2, ..., p_T\}$.

As depicted in Fig. 1, our model can be decomposed into four layers: Embedding Layer, Representation Layer, Understanding Layer and Pointer Layer. And

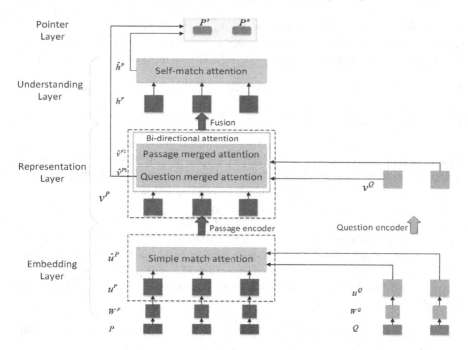

Fig. 1. The overall structure of our model. The dotted box represents concatenate operation.

three attention mechanisms are applied to different layers. Simple match attention is adopted in Embedding Layer, extracting syntactic information between the question and passage. While in Representation Layer, bi-directional attention raises the representation ability by linking and fusing semantic information from the question words and passage words. In Understanding Layer, we adopt self-match attention to refine overall understanding.

3.1 Embedding Layer

Input Vectors. We use randomly initialized character embeddings to represent text. Firstly, each word in P and Q is represented as several character indexes. Afterwards, each character is mapped to a high-density vector space(w^P and w^Q). In order to get a fixed-size vector for each word, 1D max pooling is used to merge character vectors into word vector(u^P and u^Q). Simple match attention is then applied to match word level information, which can be calculated as follows:

$$\hat{u}^P = SimAtt(u^P, u^Q) \tag{1}$$

where $SimAtt(\cdot)$ denotes the function of simple match attention.

Simple Match Attention. Given two sets of vector $v^A = \{v_1^A, v_2^A, ..., v_N^A\}$ and $v^B = \{v_1^B, v_2^B, ..., v_M^B\}$, we can synthesize information in v^B for each vector in v^A. Firstly we get the attention weight α_{ij} between i-th word of A and j-th word of B by $\alpha_{ij} = softmax(\exp(< v_i^A, v_j^B >))$, where $<>$ represents inner product. Then calculate the sum for every vector in v^B weighted by α_{ij} to get the attention representation $\hat{v}_i^A = \sum_j \alpha_{ij} v_j^B$. This attention variable $\hat{v}^A$ can be denoted as $\hat{v}^A = \{\hat{v}_1^A, \hat{v}_2^A, ..., \hat{v}_N^A\} = SimAtt(v^A, v^B)$.

3.2 Representation Layer

To better extract semantic information, we utilize RNN encoders to produce high-level representation $v_1^Q, ..., v_J^Q$ and $v_1^P, ..., v_T^P$ for all words in the question and passage respectively. The encoders are made up of bi-directional Simple Recurrent Unit (SRU) [8], which can be denoted as follows:

$$v_t^P = BiSRU(v_{t-1}^P, [u_t^P; \hat{u}_t^P]), v_j^Q = BiSRU(v_{j-1}^Q, u_j^Q) \tag{2}$$

Bi-directional attention. is applied in this layer to combine the semantic information between questions and passages. Similar to the attention flow layer in BIDAF, we compute question merged attention $\hat{v}^{P1}$ and passage merged attention $\hat{v}^{P2}$ with bi-directional attention. The similarity matrix is computed by $S_{ij} = \beta(v_i^P, v_j^Q)$, we choose

$$\beta(v_i^P, v_j^Q) = W_{(S)}^T [v_i^P; v_j^Q; v_i^P \cdot v_j^Q] \tag{3}$$

where $W_{(S)}$ is trainable parameters, $\cdot$ is element-wise multiplication, [;] is vector concatenation across row.

Question merged attention signifies which question words are most relevant to each passage words. Question merged attention weight (the i-th word in the passage to a certain word in question) is computed by $a_{i:} = softmax(S_{i:}) \in R^J$. Subsequently, each attended question merged vector is denoted as $\hat{v}_i^{P1} = \sum_j a_{ij} v_j^Q$. *Passage merged attention* signifies which context words have the closest similarity to one of the question words and hence critical for answering the question. The attended passage-merged vector is $\tilde{v}^{P2} = \sum_i b_i v_i^P$, where $b = softmax(max_{col}(S))$ and $b \in R^T$, the maximum function $max_{col}()$ is performed across the column. Then $\tilde{v}^{P2}$ is tiled T times to $\hat{v}^{P2} \in R^{2d \times T}$, where d is the length of hidden vectors.

3.3 Understanding Layer

The above bi-directional attention representation $\hat{v}_i^{P1}$ and $\hat{v}^{P2}$ is concatenated with word representation v^p to generate the attention representation $\hat{v}^P$.

$$\hat{v}^P = [\hat{v}^{P1}; \hat{v}^{P2}; v^P] \tag{4}$$

Then we use a bi-directional SRU as a Fusion to fuse information, which can be represented as $h_t^P = BiSRU(h_{t-1}^{P-1}, \hat{v}_t^P)$.

In order to take more attention over the whole passage, we apply self-match attention in Understanding Layer. Note that the computing function is the same as simple match attention, but its two inputs are both h^P

$$\hat{h}^P = SimAtt(h^P, h^P) \tag{5}$$

3.4 Pointer Layer

Pointer network is a sequence-to-sequence model proposed by Vinyals et al. [17] In Pointer Layer, we adopt pointer network to calculate the possibility of being the start or end position for every word in the passage. Instead of using a bilinear function, we take a linear function (which is proved to be simple and effective) to get the probability of start position P^s and end position P^e

$$P^s = softmax(W_{Ps}[\hat{h}_i^P; \hat{v}_i^{P1}]) \tag{6}$$

Training. During training, we minimize the cross entropy of the golden span start and end $L(\theta) = \frac{1}{N} \sum_{k}^{N} (\log(P_{i_k^s}^s) + \log(P_{i_k^e}^e))$, where i_k^s, i_k^e are the predicted answer span for the k-th instance.

Prediction. We predict the answer span to be i_k^s, i_k^e with the maximum $P_{i^s}^s + P_{i^e}^e$ under the constraint $0 \leq i^e - i^s \leq 10$.

3.5 Adversarial Training

Adversarial training applies worst case perturbations on target variables. As is shown in Fig. 2, we denote X as the target variable and θ as the parameters of the model. Different from previous works, X can be set as each variable in our model, adversarial training adds adversarial cost function $L_{adv}(X; \theta)$ to the original cost function. The equation of $L_{adv}(X; \theta)$ is described as follows:

$$L_{adv}(X; \theta) = L(X + r_{adv}; \theta), r_{adv} = \arg \max_{||r|| \leq \varepsilon} L(X + r; \hat{\theta}) \tag{7}$$

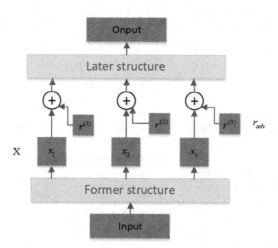

Fig. 2. The computation graph of adversarial training. X denotes target variable, r_{adv} denotes adversarial perturbation. The Input of the model is mapped into target variable X by Former Structure. And then Later Structure generates the Output based on the target variable X combined with adversarial perturbation r_{adv}.

where r is a perturbation on the target variable and $\hat{\theta}$ is a fixed copy to the current parameters. When optimizing parameters, the gradients should not propagate through r_{adv}. One problem is that we cannot get the exact value of r_{adv} simply following Eq. (7), since the computation is intractable. Goodfellow et al. [11] approximate the value of r_{adv} by linearizing $L(X; \hat{\theta})$ near X

$$r_{adv} = \varepsilon ||X|| \frac{g}{||g||}, g = \nabla_X L(X|\hat{\theta}) \tag{8}$$

where $||\cdot||$ denotes the norm of variable $\cdot$, and ε is an intensity constant to adjust the relative norm between $||r_{adv}||$ and $||X||$. So the norm of r_{adv} is decided by $||X||$, and it could be different during each training sample and training step.

4   Experiments

In this section, we evaluate our model on the WebQA dataset. Outperforming the baseline model in the original paper (Fussy F1 73.50%), we obtain 77.01% with multi-layer attention and adversarial training.

4.1   Dataset and Evaluation Metrics

WebQA [9] is a large scale real-world Chinese QA dataset. Table 2 gives an example from WebQA dataset. Its questions are from user queries in search engines and its passages are from web pages. Different from SQuAD, question-passage pairs in WebQA are matched more weakly. We use annotated evidence(shown in Table 3) to train and evaluate our model. There is an annotated golden answer to each question. So we can measure model performance by comparing predicted answers with golden answers. It can be evaluated by precision (P), recall (Q) and F1-measure (F1):

$$P = \frac{|C|}{|A|}, R = \frac{|C|}{|Q|}, F1 = \frac{2PR}{P + R} \tag{9}$$

Table 2. An example from WebQA.

Question:
What kind of color can absorb all the seven colors of the sunshine? (哪种颜色的物体能把太阳的七种颜色全部吸收？)
Passage:
Black absorbs light of various colors (黑色能够吸收各种颜色的光)
Answer: Black(黑色)

Table 3. Statistics of WebQA dataset.

Dataset	Question		Annotated evidence	
	#	word#	#	word#
Train	36,145	374,500	140,897	10,757,652
Validation	3,018	36,666	5,412	233,911
Test	3,024	36,815	5,445	234,258

where $|C|$ is the number of correctly answered questions, $|A|$ is the number of produced answers, and $|Q|$ is the number of all questions.

The same answer in WebQA may have different surface forms, such as "Beijing" v.s. "Beijing city". So we use two ways to count correctly answered questions, which are referred to as "strict" and "fuzzy". Strict matching means the predicted answer is identical to the standard answer; Fuzzy matching means the predicted answer is a synonym of the standard answer.

4.2   Model Details

In our model, we use randomly initialized 64-dimensional character embedding and hidden vector length d is set to 100 for all layers. We utilize 4-layer Passage encoder and Question encoder. And Fusion SRU is set to 2-layer. We also apply dropout between layers, with a dropout rate of 0.2. The model is optimized using AdaDelta with a minibatch size of 64 and an initial learning rate of 0.1. Hyper-parameter ε is selected on the WebQA validation dataset.

4.3   Main Results

The evaluation results are shown in Table 4. Different models are evaluated on the WebQA test dataset, including baseline models(LSTM+softmax and LSTM+CRF), BIDAF and A3Net. A3Net(without AT) denotes our base model which does not apply adversarial training(AT); A3Net(random noise) denotes control experiment which replaces adversarial perturbations with random Gaussian noise with a scaled norm. Baseline models utilize sequence label method to mark the answer, while others adopt pointer network to extract the answer. Sequence label method can mark several answers for one question, leading high recall(R) but low precision(P). So we adopt pointer network to generate one answer for each question. In this condition, evaluation metrics(P, R, F1) are equal. Thus we can use *Score* to evaluate our model. Besides, Fuzzy evaluation is closer to real life, so we mainly focus on *Fuzzy Score*.

Based on single layer attention and pointer network, BIDAF obtains the obvious promotion (Fuzzy F1 74.43%). Benefit from multi-layer attention, A3Net

Table 4. Evaluation results on the test dataset of WebQA.

Model	Strict score			Fuzzy Score		
	P(%)	R(%)	F1(%)	P(%)	R(%)	F1(%)
LSTM+softmax	59.38	68.77	63.73	63.58	73.63	68.24
LSTM+CRF	63.72	**76.09**	69.36	67.53	**80.63**	73.50
BIDAF	70.04	70.04	70.04	74.43	74.43	74.43
A3Net(without AT)	71.03	71.03	71.03	75.46	75.46	75.46
A3Net(random noise)	71.28	71.28	71.28	75.89	75.89	75.89
A3Net	**72.51**	72.51	**72.51**	**77.01**	77.01	**77.01**

(without AT) gains 0.97 point promotion in Fuzzy F1 compared to BIDAF, which indicates that multi-layer attention is useful. Our model would get another 1.12 point promotion in Fuzzy F1 after jointly adopting adversarial training on target variable w^P and $\hat{v}^P$.

A common misconception is that perturbation in adversarial training is equivalent to random noise. In actually, noise is a far weaker regularization than adversarial perturbations. An average noise vector is approximately orthogonal to the cost gradient in high dimensional input spaces. While adversarial perturbations are explicitly chosen to consistently increase the cost. To demonstrate the superiority of adversarial training over the addition of noise, we include control experiments which replaced adversarial perturbations with random perturbations from a Gaussian distribution. We use random noise to replace worst case perturbations on each target variable, which only lead slightly improvement. It indicates that AT can actually improve the robustness and generalization of our model.

4.4 Ablation on Base Model Structure

Next, we investigate the ablation study on the structure of our base model. From Table 5 (*A3Net base model* is same with *A3Net (without AT)* in Table 4), we can tell that both Strict Score and Fuzzy Score would drop when we omit any attention. It indicates that each attention layer in A3Net base model is essential.

4.5 Adversarial Training on Different Target Variables

We evaluate the predicted result when we apply adversarial training on different target variables. As is shown in Table 6, applying adversarial training on each target variable can improve Fuzzy Score as well as Strict Score in different degree. It indicates that adversarial training can work as a regularizing method not just for word embeddings, but also for many other variables in our model. Note that the Score is improved significantly when applying AT on embedding variable w^P and attention variable $\hat{v}^P$. It reveals that adversarial training can improve representing ability for both inputs and non-input variables. Finally, we obtain the best result when applying AT on both w^P and $\hat{v}^P$.

Table 5. Comparison of different configurations of base model. The symbols in this table is corresponding with Fig. 1.

Model	Strict score (%)	Fuzzy Score (%)
A3Net base model (without $\hat{u}^P$)	70.57	74.93
A3Net base model (without $\hat{v}^{P1}$)	70.77	75.18
A3Net base model (without $\hat{v}^{P1}$ and $\hat{v}^{P2}$)	70.63	74.56
A3Net base model (without $\hat{h}^P$)	70.70	75.23
A3Net base model	**71.03**	**75.46**

Table 6. Comparison of adversarial training results on different target variables. The symbols in this table is corresponding with Fig. 1

Target variable	Strict score	Fuzzy Score	Target variable	Strict score	Fuzzy Score
None (base model)	71.03	75.46	$\hat{v}^{P1}$	71.85	76.28
w^P	71.95	76.62	$\hat{h}^P$	71.56	76.42
u^P	72.06	76.39	$\hat{v}^P$	72.28	76.81
$\hat{u}^P$	71.32	75.92	w^P and $\hat{v}^P$	**72.51**	**77.01**

We also evaluate adversarial training on two target variables (w^P and $\hat{v}^P$) under different intensity constant ε. As shown in Fig. 3, we repeat experiment 3 times for each target variable on each constant ε, and get the average Fuzzy Score and its std. error. For AT on attention variable $\hat{v}^P$, we obtain the best performance when ε is 0.5×10^{-4}; While for AT on character embedding variable w^P, we obtain the best performance when ε is 2×10^{-4}. It indicates we needs larger adversarial perturbation for low-level variable. While comparable smaller intensity benefits its training for high-level variable. We can explain this phenomenon in two different views. Firstly, w^P and $\hat{v}^P$ are in different concept levels. w^P contains syntactic meaning, and represents as character embedding vectors. Most of the vectors can still hold original meaning under small perturbation, because most points in embedding space have no real meanings. But $\hat{v}^P$ contains semantic meaning. Any perturbation on it would change its meaning, thus our model is sensitive to the perturbation on $\hat{v}^P$. Secondly, w^P and $\hat{v}^P$ are in different layers of our model. $\hat{v}^P$ is closer to Pointer Layer, which could have more influence on the output of the model and computation of cost function.

4.6 Effective of Adversarial Training

Figure 4(a) shows the Fuzzy Score on the test dataset and Fig. 4(b) shows the loss value on the training dataset of A3Net in different configurations. The meaning

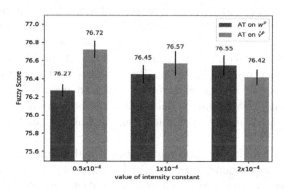

Fig. 3. Effect of intensity constant when AT on target variable w^P and $\hat{v}^P$.

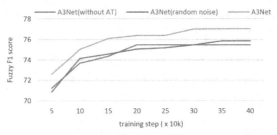

(a) Fuzzy Score (test) under different training step.

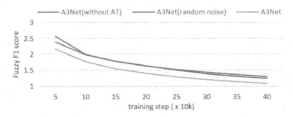

(b) Loss value (train) under different training step.

Fig. 4. Fuzzy Score (test) and Loss value (train) under different training step.

of *without AT* and *random noise* are the same with that in Table 4. The data curves of the base model and the random noise model are close to each other in both two figures. It indicates that random noise has limited effect on our model. Within each training step, the Fuzzy Score of our final model is the highest, and its loss value is the lowest. It demonstrates that adversarial training can lead to better performance with less training step.

5   Conclusions

This paper proposes a novel model called Adversarial-and-attention Network (A3Net), which includes adversarial training and multi-layer attention.

Adversarial training works as a regularization method. It can be applied to almost every variable in the model. We blend adversarial training into each layer of the model by controlling the relative intensity of norm between adversarial perturbations and original variables. Results show that applying adversarial perturbations on some high-level variables can lead even better performance than that on input signals. Our model obtains the best performance by jointly applying adversarial training to character embedding and high-level attention representation.

We use simple match attention and bi-directional attention to enhance the interaction between questions and passages. Simple match attention on Embedding Layer refines syntactic information. In addition, bi-directional attention on Representation Layer refines semantic information. Furthermore, self-much attention is used on Understanding Layer to refine the overall information among the whole passages. Experiments on the WebQA dataset show that our model outperforms the state-of-the-art models.

References

1. Bahdanau, D., Cho, K., Bengio, Y.: Neural machine translation by jointly learning to align and translate. In: Proceedings of ICLR (2015)
2. Chen, D., Fisch, A., Weston, J., Bordes, A.: Reading Wikipedia to answer open-domain questions. In: Proceedings of ACL, pp. 1870–1879 (2017)
3. Goodfellow, I.J., Shlens, J., Szegedy, C.: Explaining and harnessing adversarial examples. In: Proceedings of ICLR (2015)
4. Natural Language Computing Group: R-net: machine reading comprehension with self-matching networks (2017)
5. Hermann, K.M., et al.: Teaching machines to read and comprehend. In: Proceedings of NIPS, pp. 1693–1701 (2015)
6. Huang, H.Y., Zhu, C., Shen, Y., Chen, W.: FusionNet: fusing via fully-aware attention with application to machine comprehension. In: Proceedings of ICLR (2018)
7. Jia, R., Liang, P.: Adversarial examples for evaluating reading comprehension systems. In: Proceedings of EMNLP, pp. 2021–2031 (2017)
8. Lei, T., Zhang, Y.: Training RNNs as fast as CNNs. arXiv preprint arXiv:1709.02755 (2017)
9. Li, P., et al.: Dataset and neural recurrent sequence labeling model for open-domain factoid question answering. arXiv preprint arXiv:1607.06275 (2016)
10. Liu, X., Shen, Y., Duh, K., Gao, J.: Stochastic answer networks for machine reading comprehension. In: Proceedings of NAACL (2018)
11. Miyato, T., Dai, A.M., Goodfellow, I.: Adversarial training methods for semi-supervised text classification. In: Proceedings of ICLR (2017)
12. Rajpurkar, P., Zhang, J., Lopyrev, K., Liang, P.: Squad: 100,000+ questions for machine comprehension of text. In: Proceedings of EMNLP, pp. 2383–2392 (2016)
13. Seo, M., Kembhavi, A., Farhadi, A., Hajishirzi, H.: Bidirectional attention flow for machine comprehension. In: Proceedings of ICLR (2017)
14. Shen, Y., Huang, P.S., Gao, J., Chen, W.: ReasoNet: Learning to stop reading in machine comprehension. In: Proceedings of SIGKDD, pp. 1047–1055. ACM (2017)
15. Szegedy, C., et al.: Intriguing properties of neural networks. In: Proceedings of ICLR (2014)
16. Tan, C., Wei, F., Yang, N., Du, B., Lv, W., Zhou, M.: S-net: from answer extraction to answer generation for machine reading comprehension. In: Proceedings of AAAI (2018)
17. Vinyals, O., Fortunato, M., Jaitly, N.: Pointer networks. In: Proceedings of NIPS, pp. 2692–2700 (2015)
18. Wang, S., Jiang, J.: Machine comprehension using Match-LSTM and answer pointer. In: Proceedings of ICLR (2017)
19. Weissenborn, D., Wiese, G., Seiffe, L.: Making neural QA as simple as possible but not simpler. In: Proceedings of CoNLL, pp. 271–280 (2017)
20. Wu, Y., Bamman, D., Russell, S.: Adversarial training for relation extraction. In: Proceedings of EMNLP, pp. 1778–1783 (2017)

When Less Is More: Using Less Context Information to Generate Better Utterances in Group Conversations

Haisong Zhang[1], Zhangming Chan[2], Yan Song[1], Dongyan Zhao[2], and Rui Yan[2(✉)]

[1] Tencent AI Lab, Beijing, China
{hansonzhang,clksong}@tencent.com
[2] Institute of Computer Science and Technology, Peking University, Beijing, China
{chanzhangming,zhaody,ruiyan}@pku.edu.cn

Abstract. Previous research on dialogue systems generally focuses on the conversation between two participants. Yet, group conversations which involve more than two participants within one session bring up a more complicated situation. The scenario is real such as meetings or online chatting rooms. Learning to converse in groups is challenging due to different interaction patterns among users when they exchange messages with each other. Group conversations are structure-aware while the structure results from different interactions among different users. In this paper, we have an interesting observation that fewer contexts can lead to better performance by tackling the structure of group conversations. We conduct experiments on the public Ubuntu Multi-Party Conversation Corpus and the experiment results demonstrate that our model outperforms baselines.

Keywords: Group conversations · Context modeling
Dialogue system

1 Introduction

Dialogue systems such as chatbots and virtual assistants have been attracting great attention nowadays [17,18,21,22,27,28]. To launch dialogue systems with moderate intelligence, the first priority for computers is to learn how to converse by imitating human-to-human conversations. Researchers have paid great efforts on learning to converse between two participants, either single-turn [5,14,16,31] or multi-turn [12,25,29,33]. The research is valuable but is still quite simple in reality: two-party conversations do not cover all possible conversation scenarios.

A more general scenario is that conversations may have more than two interlocutors conversing with each other [9,32], known as "group conversations". In real-world scenarios, group conversations are rather common, such as dialogues

---

H. Zhang and Z. Chan—contribute equally.

© Springer Nature Switzerland AG 2018
M. Zhang et al. (Eds.): NLPCC 2018, LNAI 11108, pp. 76–84, 2018.
https://doi.org/10.1007/978-3-319-99495-6_7

in online chatting rooms, discussions in forums, and debates, etc. Learning for group conversations is of great importance, and is more complicated than two-party conversations which requires extra work such as understanding the relations among utterances and users throughout the conversation.

Table 1. An example of group conversations in the IRC dataset. The conversation involves multiple participants and lasts for multiple turns.

User	Utterance
User 1	"i'm on 15.10"
User 2	@User 1 "have you tried using... "
User 1	@User 2 "nope. but this might... "
User 3	"i read on the internets..."
User 2	"yeah. i'm thinking ..."

For example, in Ubuntu Internet Relay Chat channel (IRC), one initiates a discussion about an Ubuntu technical issue as illustrated in Table 1: multiple users interact with each other as a group conversation. Different pieces of information are organized into a structure-aware conversation session due to different responding relation among users. Some utterances are closely related because they are along the same discussion *thread*, while others are not. We characterize such an insight into the structure formulation for group conversations.

Group conversations are naturally multi-party and multi-turn dialogues. Compared with two-party conversations in Fig. 1(a), a unique issue for group conversations is to incorporate multiple threads of interactions (which indicate "structures") into the conversations formulation, as illustrated in Fig. 1(b).

Learning to generate utterances in group conversations is challenging: we need a uniform framework to model the structure-aware conversation sessions, which is non-trivial. To this end, we propose a tree-structured conversation model to formulate group conversations. The branches of the tree characterize different interaction threads among the users. We learn to encode the group conversation along the tree branches by splitting the tree as sequences (Fig. 1(c)). Given the learned representations, the model generates the next utterance in group conversations. Due to the tree-structured frame of group conversations, we will not use all utterances across turns to generate the target utterance: only the utterances along the target tree branch will be used. With **fewer** contexts used, we obtain even **better** generation results. It is interesting to see that "less" is "more".

To sum up, our contributions in this paper are:

- We are the first to investigate structure-aware formulation for group conversations. The model organizes the utterance flows in the conversation into a tree-based frame, which is designed especially for the group conversation scenario.

- To the best of our knowledge, we are also the first to investigate the task of generation-based conversation in groups. With less context information used by ruling out utterances from irrelevant branches, we generate better results.

Experiments are conducted on the Ubuntu Multi-party Conversation Corpus, a public dataset for group conversation studies. The experimental result shows that our model outperforms the state-of-the-art baselines on all metrics.

2   Background Knowledge

Dialog systems can be categorized as generation-based and retrieval-based approaches. Generation-based methods produce responses with natural language generators, which are learned from conversation data [5,14,17,23]; while retrieval-based ones retrieve responses by ranking and selecting existing candidates from a massive data repository [24,25,27,30]. Researchers also investigate how to ensemble generation-based ones and retrieval-based ones together [11,15]. In this paper, we focus on the generation-based conversational model.

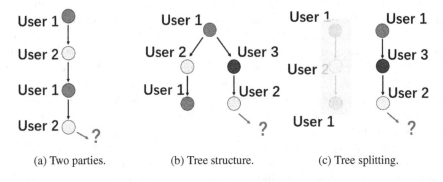

(a) Two parties.   (b) Tree structure.   (c) Tree splitting.

Fig. 1. We illustrate the difference of conversations between two participants in (a) and the group conversation of Table 1 in (b). Group conversations are structure-aware and formulated as trees (b) and we split the tree into sequences (c). "Irrelevant" utterances on other sequences are not used for generation (shaded in the figure).

Early work in dialog systems focuses on single-turn conversations. It outputs a response given only one utterance as the input [5,14,20]. However, a normal conversation lasts for several turns. Conversation models without context information is insufficient [22]. Context representation learning is proposed to solve the problem. Interested readers may refer to a recent survey paper for more details [26].

Multi-turn dialogue systems take in the current message and all previous utterances as contexts and output a response which is appropriate given the entire conversation session. Recently, methods are proposed to capture long-span dependencies in contexts by concatenation [18,27], latent variable-based models

[8,13] or hierarchical approaches [12,25]. In this paper, we target at multi-turn conversations.

Most previous studies focus on two-party conversations, while the group conversation is a more general case of multi-turn conversations which involve more than two participants. It is more difficult to understand group conversations. Ouchi et al. [9] proposed to select the responses associated with addressees in multi-party conversations, which is basically a retrieval-based model by matching. Zhang et al. [32] extended the work by introducing an interactive neural network modeling, which is also for retrieval-based matching.

None of the work focuses on generation-based group conversations. More importantly, none of the work incorporates structure-aware formulation for group conversation models.

3 Group Conversation Model

In this section, we introduce our model for group conversations. First, we need to organize the conversation session according to the responding structures among users. We propose to construct the conversation context into a tree structure. With the established tree structure, we encode information for generation. Finally, we generate the target utterance in the decoding process.

Our problem formulation is straightforward. Given a group conversation with T utterances, we denote $X = \{X_i\}_{i=1}^{T}$. Each X_i is an utterance sentence. The goal is to generate the next utterance $Y = (y_1, y_2, \ldots, y_m)$ by estimating the probability $p(Y|X) = \prod p(y_t|y_{<t}, X)$.

Structure information is vital for group conversations. With different responding relationships, we construct different tree structures and accordingly, encode different information for the group conversations. Since we model the conversations as *trees*, we add the utterances with direct responding relationships onto the same branch of the tree. In this way, different responding relationships lead to different tree-branch structures.

Tree-based Formulation. Given the responding relationship among users, it is straightforward to establish the tree. If an utterance X_i is responding to the utterance X_j where $i > j$, we add an edge between X_i and X_j. X_j is the parent node of X_i while X_i is the child node of X_j. Generally, each utterance responds to one utterance but multiple subsequent utterances can respond to the same utterance. Suppose X_i and X_k both respond to X_j where $i > j$ and $k > j$. In this case, X_i and X_k are sibling nodes at the same level with a common parent node X_j.

As illustrated in Table 1, sometimes an utterance is addressed to a particular user *explicitly*. In this situation, we establish the tree without any ambiguity. In other cases, an utterance is not explicitly addressed to any user. To make the model practical, we introduce an assumption that if not explicitly designated, the utterance is addressing to the most recent utterance in the context. It is a simple assumption but holds in majority circumstances. For the group conversation in Table 1, we establish the tree in Fig. 1(b).

Splitting. Given a tree-structured conversation session, there are multiple sequences with shared nodes. We split the multiple sequences into separate sequences by duplicating the shared nodes, which is shown in Fig. 1(c). In this way, the conversation is represented by multiple sequences. Sequences have unique advantages over trees in batch learning. We identify which sequence the target utterance will be addressed to, and learn the embeddings of utterances along this sequence. We decode the target utterance based on the learned representation. Utterances from other sequences (i.e., other branches) will not be used for context information encoding and decoding.

Hierarchical Encoding. Our model is based on the encoder-decoder framework using the sequence-to-sequence model [19]. We implement with gated recurrent units (GRU) [3]. The encoder converts a sequence of embedding inputs $(\mathbf{x}_1, \mathbf{x}_2, \ldots, \mathbf{x}_n)$ to hidden representations $(\mathbf{h}_1, \mathbf{h}_2, \ldots, \mathbf{h}_n)$ by:

$$\mathbf{h}_t = \mathrm{GRU}(\mathbf{h}_{t-1}, \mathbf{x}_t) \tag{1}$$

Our model is established based on the hierarchical representations [6,12]. A hierarchical model draws on the intuition that just as the integration of words creates the overall meaning of an utterance, and furthermore the integration of multiple utterances creates the overall meaning of several utterances. To be more specific, we first obtain representation vectors at the utterance level by putting one layer of a recurrent neural network with GRU units on top of its containing words. The vector output at the ending time-step is used to represent the entire utterance sentence.

To build the representation for multiple utterances along a branch, another layer of GRU is placed on top of all utterances, computing representations sequentially for each time step. Representation computed at the final time step is used to represent the long sequence (i.e., the tree branch). Thus one GRU operates at the word-level, leading to the acquisition of utterance-level representations that are used as inputs into the second GRU that acquires the overall representations.

Decoding. After we obtain the encoded information, the decoder takes as input a context vector $\mathbf{c}_t$ and the embedding of a previously decoded word $\mathbf{y}_{t-1}$ to update its state st using another GRU:·

$$\mathbf{s}_t = \mathrm{GRU}(\mathbf{s}_{t-1}, [\mathbf{c}_t; \mathbf{y}_{t-1}]) \tag{2}$$

$[\mathbf{c}_t; \mathbf{y}_{t-1}]$ is the concatenation of the two vectors, serving as the input to GRU units. The context vector $\mathbf{c}_t$ is designed to dynamically attend on important information of the encoding sequence during the decoding process [1]. Once the state vector $\mathbf{s}_t$ is obtained, the decoder generates a token by sampling from the output probability distribution $\mathbf{o}_t$ computed from the decoder's state $\mathbf{s}_t$:

$$\begin{aligned} y_t \sim \mathbf{o}_t &= p(y_t | y_1, y_2, \ldots, y_{t-1}, \mathbf{c}_t) \\ &= \mathrm{softmax}(\mathbf{W}_o \mathbf{s}_t) \end{aligned} \tag{3}$$

Table 2. Experimental results of different models based on automatic evaluations.

Model	BLEU-1	BLEU-2	BLEU-3	BLEU-4	METEOR
NRM	9.85	3.04	1.38	0.67	3.98
C-Seq2Seq	10.45	4.13	2.08	1.02	3.43
HRED	11.23	4.40	2.45	1.42	4.38
Our method	**11.73**	**6.06**	**4.28**	**3.29**	**4.86**

4 Experiments

Data. We run experiments using the public Ubuntu Corpus[1] [8] for training and testing. The original data comes from the logs of Ubuntu IRC chat room where users discuss technical problems related to Ubuntu. The corpus consists of huge amount of records including over 7 million response utterances and 100 million words. We organize the dataset as tree-structured samples of 380k conversation sessions.

To be more specific, we take the last utterance in each session as the target to be generated and other utterances as inputs. We randomly divide the corpus into train-dev-test sets: 5,000 sessions for validation, 5,000 sessions for testing and the rest for training. We report results on the test set.

Baselines. We evaluate our model against a series of baselines. We include the context-insensitive baseline and context-aware methods (either non-hierarchical or hierarchical).

- *NRM.* Shang et al. [14] proposed the single-turn conversational model without contexts incorporated, namely neural responding machine (NRM). For NRM, only the penultimate utterance is used to generate the last utterance. It is performed using the Seq2Seq model with attention.
- *Context-Seq2Seq.* The context-sensitive seq2seq means that given a session, we use the last utterance as the target and all other utterances as the inputs. We concatenate all input utterances into a long utterance [18]. The concatenated contexts do not distinguish word or sentence hierarchies.
- *HRED.* The Hierarchical Recurrent Encoder-Decoder (HRED) model is a strong context-aware baseline which consists both word-level encoders and sentence-level encoders [12]. In this way, context utterances are encoded in two hierarchies as the training data.

None of these models takes the structure in group conversations into account. Our model incorporates structures into the hierarchical context-aware conversational model, where indicates a new insight.

Evaluation Metrics. We use the evaluation package released by [2] to evaluate our model and baselines. The package includes BLEU-1 to 4 [10] and

---

[1] http://dataset.cs.mcgill.ca/ubuntu-corpus-1.0/.

METEOR [4]. All these metrics evaluate the word overlap between the generated utterances and the ground truth targets. Still, note that these evaluation metrics have plenty room for improvement as to dialogue evaluations [7].

Results and Analysis. Table 2 shows the evaluation results. We observe that the performance is improved incrementally. From NRM to C-Seq2Seq, the improvement may be ascribed that context information is important for conversations with more than one turn. It concurs with many previous studies [25,27]. Hierarchical information depicted by fine-grained representation learning is also demonstrated to be useful [12,22]. None of the baselines formulates structure information, while our method utilizes a tree-based frame. Our method outperforms baselines in all metrics: structure-aware information is shown to be effective in group conversations.

Note that in our method, after splitting the tree into multiple sequences, we actually discard part of the context utterances during the encoding process. It is surprising that the model achieves even better results. We understand that within a group conversation, only the relevant information is useful to generate the target utterance. Irrelevant utterances on the other branches of the tree (i.e., other sequences) might be the noises for generation. It is interesting to see that "less becomes more" in group conversations.

5 Conclusion

In this paper, we proposed a tree-based model frame for structure-aware group conversations. According to different responding relations, we organize the group conversation as a tree with different branches involving multiple conversation threads. We split the established tree into multiple sequences, and we only use the target sequence to generate the next utterance. This method is quite simple but rather effective. We have performance improvement in terms of automatic evaluations, which indicate less context information results in better generations in group conversations. In other words, "less" is "more".

Acknowledgments. We would like to thank the anonymous reviewers for their constructive comments. This work was supported by the National Key Research and Development Program of China (No. 2017YFC0804001), the National Science Foundation of China (NSFC No. 61672058). Rui Yan was sponsored by CCF-Tencent Open Research Fund and Microsoft Research Asia (MSRA) Collaborative Research Program.

References

1. Bahdanau, D., Cho, K., Bengio, Y.: Neural machine translation by jointly learning to align and translate. In: ICLR (2015)
2. Chen, X., Fang, H., Lin, T.Y., Vedantam, R., Gupta, S., Dollár, P., Zitnick, C.L.: Microsoft coco captions: data collection and evaluation server. CoRR abs/1504.00325 (2015)

3. Chung, J., Gulcehre, C., Cho, K., Bengio, Y.: Empirical evaluation of gated recurrent neural networks on sequence modeling. In: ICLR (2015)
4. Denkowski, M.J., Lavie, A.: Meteor universal: language specific translation evaluation for any target language. In: WMT@ACL (2014)
5. Li, J., Galley, M., Brockett, C., Spithourakis, G., Gao, J., Dolan, B.: A persona-based neural conversation model. In: Proceedings of the 54th Annual Meeting of the Association for Computational Linguistics (Volume 1: Long Papers), vol. 1, pp. 994–1003 (2016)
6. Li, J., Luong, T., Jurafsky, D.: A hierarchical neural autoencoder for paragraphs and documents. In: Proceedings of the 53rd Annual Meeting of the Association for Computational Linguistics and the 7th International Joint Conference on Natural Language Processing (Volume 1: Long Papers), vol. 1, pp. 1106–1115 (2015)
7. Liu, C.W., Lowe, R., et al.: How not to evaluate your dialogue system: an empirical study of unsupervised evaluation metrics for dialogue response generation. In: EMNLP 2016, pp. 2122–2132 (2016)
8. Lowe, R.J., Pow, N., Serban, I., Pineau, J.: The ubuntu dialogue corpus: a large dataset for research in unstructured multi-turn dialogue systems. In: SIGDIAL Conference (2015)
9. Ouchi, H., Tsuboi, Y.: Addressee and response selection for multi-party conversation. In: EMNLP, pp. 2133–2143 (2016)
10. Papineni, K., Roucos, S.E., Ward, T., Zhu, W.J.: BLEU: a method for automatic evaluation of machine translation. In: ACL (2002)
11. Qiu, M., Li, F.L., Wang, S., Gao, X., Chen, Y., Zhao, W., Chen, H., Huang, J., Chu, W.: AliMe chat: a sequence to sequence and rerank based chatbot engine. In: ACL 2017, pp. 498–503 (2017)
12. Serban, I.V., Sordoni, A., Bengio, Y., Courville, A.C., Pineau, J.: Building end-to-end dialogue systems using generative hierarchical neural network models. In: AAAI, pp. 3776–3784 (2016)
13. Serban, I.V., Sordoni, A., Lowe, R., Charlin, L., Pineau, J., Courville, A.C., Bengio, Y.: A hierarchical latent variable encoder-decoder model for generating dialogues. In: AAAI 2017, pp. 3295–3301 (2017)
14. Shang, L., Lu, Z., Li, H.: Neural responding machine for short-text conversation. In: Proceedings of the 53rd Annual Meeting of the Association for Computational Linguistics and the 7th International Joint Conference on Natural Language Processing (Volume 1: Long Papers), vol. 1, pp. 1577–1586 (2015)
15. Song, Y., Li, C.T., Nie, J.Y., Zhang, M., Zhao, D., Yan, R.: An ensemble of retrieval-based and generation-based human-computer conversation systems. In: IJCAI 2018 (2018)
16. Song, Y., Tian, Z., Zhao, D., Zhang, M., Yan, R.: Diversifying neural conversation model with maximal marginal relevance. In: IJCNLP 2017, pp. 169–174 (2017)
17. Song, Y., Yan, R., Feng, Y., Zhang, Y., Zhao, D., Zhang, M.: Towards a neural conversation model with diversity net using determinantal point processes. In: AAAI 2018, pp. 5932–5939 (2018)
18. Sordoni, A., Galley, M., Auli, M., Brockett, C., Ji, Y., Mitchell, M., Nie, J.Y., Gao, J., Dolan, B.: A neural network approach to context-sensitive generation of conversational responses. In: Proceedings of the 2015 Conference of the North American Chapter of the Association for Computational Linguistics: Human Language Technologies, pp. 196–205 (2015)
19. Sutskever, I., Vinyals, O., Le, Q.V.: Sequence to sequence learning with neural networks. In: NIPS (2014)

20. Tao, C., Gao, S., Shang, M., Wu, W., Zhao, D., Yan, R.: Get the point of my utterance! learning towards effective responses with multi-head attention mechanism. In: IJCAI 2018 (2018)
21. Tao, C., Mou, L., Zhao, D., Yan, R.: RUBER: an unsupervised method for automatic evaluation of open-domain dialog systems. In: AAAI 2018, pp. 722–729 (2018)
22. Tian, Z., Yan, R., Mou, L., Song, Y., Feng, Y., Zhao, D.: How to make context more useful? An empirical study on context-aware neural conversational models. In: Annual Meeting of the Association for Computational Linguistics, pp. 231–236 (2017)
23. Vinyals, O., Le, Q.: A neural conversational model. arXiv preprint arXiv:1506.05869 (2015)
24. Wang, H., Lu, Z., Li, H., Chen, E.: A dataset for research on short-text conversations. In: EMNLP, pp. 935–945 (2013)
25. Wu, Y., Wu, W., Xing, C., Zhou, M., Li, Z.: Sequential matching network: A new architecture for multi-turn response selection in retrieval-based chatbots. In: Proceedings of the 55th Annual Meeting of the Association for Computational Linguistics (Volume 1: Long Papers), vol. 1, pp. 496–505 (2017)
26. Yan, R.: "Chitty-Chitty-Chat Bot": Deep learning for conversational AI. In: IJCAI 2018 (2018)
27. Yan, R., Song, Y., Wu, H.: Learning to respond with deep neural networks for retrieval-based human-computer conversation system. In: Proceedings of the 39th International ACM SIGIR Conference on Research and Development in Information Retrieval, pp. 55–64. ACM (2016)
28. Yan, R., Song, Y., Zhou, X., Wu, H.: Shall i be your chat companion? Towards an online human-computer conversation system. In: CIKM 2016, pp. 649–658 (2016)
29. Yan, R., Zhao, D.: Coupled context modeling for deep chit-chat: towards conversations between human and computer. In: Proceedings of the 24th ACM SIGKDD International Conference on Knowledge Discovery and Data Mining (2018)
30. Yan, R., Zhao, D., E., W.: Joint learning of response ranking and next utterance suggestion in human-computer conversation system. In: Proceedings of the 40th International ACM SIGIR Conference on Research and Development in Information Retrieval, SIGIR 2017, pp. 685–694 (2017)
31. Yao, L., Zhang, Y., Feng, Y., Zhao, D., Yan, R.: Towards implicit content-introducing for generative short-text conversation systems. In: EMNLP 2017, pp. 2190–2199 (2017)
32. Zhang, R., Lee, H., Polymenakos, L., Radev, D.: Addressee and response selection in multi-party conversations with speaker interaction RNNs. In: AAAI (2018)
33. Zhou, X., Dong, D., Wu, H., Zhao, S., Yu, D., Tian, H., Liu, X., Yan, R.: Multi-view response selection for human-computer conversation. In: EMNLP 2016, pp. 372–381 (2016)

I Know There Is No Answer: Modeling Answer Validation for Machine Reading Comprehension

Chuanqi Tan[1(✉)], Furu Wei[2], Qingyu Zhou[3], Nan Yang[2], Weifeng Lv[1], and Ming Zhou[2]

[1] Beihang University, Beijing, China
tanchuanqi@nlsde.buaa.edu.cn, lwf@buaa.edu.cn
[2] Microsoft Research Asia, Beijing, China
{fuwei,nanya,mingzhou}@microsoft.com
[3] Harbin Institute of Technology, Harbin, China
qyzhgm@gmail.com

Abstract. Existing works on machine reading comprehension mostly focus on extracting text spans from passages with the assumption that the passage must contain the answer to the question. This assumption usually cannot be satisfied in real-life applications. In this paper, we study the reading comprehension task in which whether the given passage contains the answer is not specified in advance. The system needs to correctly refuse to give an answer when a passage does not contain the answer. We develop several baselines including the answer extraction based method and the passage triggering based method to address this task. Furthermore, we propose an answer validation model that first extracts the answer and then validates whether it is correct. To evaluate these methods, we build a dataset SQuAD-T based on the SQuAD dataset, which consists of questions in the SQuAD dataset and includes relevant passages that may not contain the answer. We report results on this dataset and provides comparisons and analysis of the different models.

Keywords: Machine reading comprehension · Answer validation

1 Introduction

Machine reading comprehension, which attempts to enable machines to answer questions after reading a passage, has attracted much attention from both research and industry communities in recent years. The release of large-scale manually created datasets such as SQuAD [12] and TriviaQA [5] has brought great improvement for model training and testing of machine learning algorithms on the related research area. However, most existing reading comprehension datasets assume that there exists at least one correct answer in the passage set. Current models therefore only focus on extracting text spans from passages to

© Springer Nature Switzerland AG 2018
M. Zhang et al. (Eds.): NLPCC 2018, LNAI 11108, pp. 85–97, 2018.
https://doi.org/10.1007/978-3-319-99495-6_8

answer the question, but do not determine whether an answer even exists in the passage for the question. Although the assumption simplifies the problem, it is unrealistic for real-life applications. Modern systems usually rely on an independent component to pre-select relevant passages, which cannot guarantee that the candidate passage contains the answer.

In this paper, we study the reading comprehension task in which whether the given passage contains the answer is not specified in advance[1]. For the question whose passage contains the answer, the system needs to extract the correct text span to answer the question. For the question whose passage does not contain the answer, the system needs to correctly refuse to give the answer. We develop several baseline methods following previous work on answer extraction [12] and answer triggering [21]. We implement the answer extraction model [19] to predict the answer. We then use the probability of the answer to judge whether it is correct. In addition, we propose two methods to improve the answer extraction model by considering that there may be no answer. The first is to add a no-answer option with a padding position for the passage that does not contain the answer and supervise the model to predict this padding position when there is no answer. The second is to control the probability of the answer by modifying the objective function for the passage that does not contain the answer. Second, we develop the passage triggering based method, which first determines whether the passage contains the answer then extracts the answer only in the triggered passage. Finally, we propose the answer validation method, which first extracts the answer in the passage then validates whether it is correct.

To test the above methods, we build a new dataset SQuAD-T based on the SQuAD dataset. For each question in the SQuAD dataset, we use Lucene[2], an off-the-shelf tool, to retrieve the top relevant passage from the whole SQuAD passage set. If the top passage is the original corresponding passage in the SQuAD dataset that contains the answer, we treat the question and passage pair as a positive example. Otherwise, we treat the question and the top-ranked passage that does not contain the answer as a negative example. Table 1 shows two examples in the SQuAD-T dataset. In the first example, the passage contains the correct answer "Denver Broncos" (underlined). In the second example, the passage does not contain the answer. We use precision, recall and F_1 scores for the positive examples and overall accuracy for all data to evaluate this task.

Experiments show that both the answer extraction model with the no-answer option and the modified objective function improve the results of the answer extraction model. Our answer validation model achieves the best F_1 score and overall accuracy on the SQuAD-T test set. Further analysis indicates that our proposed answer validation model performs better in refusing to give the answers when passages do not contain the answers without performance degradation when passages contain the answers.

---

[1] We notice Rajpurkar et al. also address this problem [11] when this paper is under review.

[2] http://lucene.apache.org.

Table 1. Examples in the SQuAD-T dataset. The first example contains the answer "Denver Broncos" (underlined). The second example does not contain the answer to the question.

Question: Which NFL team represented the AFC at Super Bowl 50?   **Passage:** The American Football Conference (AFC) champion   <u>Denver Broncos</u> defeated the National Football Conference (NFC)   champion Carolina Panthers 24-10 to earn their third Super Bowl title
Question: Where did Super Bowl 50 take place?   **Passage:** In addition to the Vince Lombardi Trophy that all Super   Bowl champions receive, the winner of Super Bowl 50 will also receive   a large, 18-karat gold-plated "50"

2 Related Work

Previous methods achieve promising results on the SQuAD dataset for reading comprehension. Since the passage must contain the answer to the question in the SQuAD dataset, state-of-the-art methods usually answer the question by predicting the start and end positions of the answer in the passage [4,13,18,20]. Unlike the SQuAD dataset that only has one passage for a question, the TriviaQA dataset [5] and the MS-MARCO dataset [9] contain multiple paragraphs or passages for a question. However, since the datasets still guarantee that it must contain the answer, state-of-the-art methods do not discriminate which passage contains the answer, but concatenate all passages to predict one answer [3,15,18].

Yang et al. [21] propose an answer triggering task with the WikiQA dataset. It aims to detect whether there is at least one correct answer in the set of candidate sentences for the question, and selects one of the correct answer sentences from the candidate sentence set if yes. Several feature-based methods [21] and deep learning methods [6,22] are proposed for this task. Chen et al. [1] tackle the problem of open-domain question answering, which combines the document retrieval (finding the relevant articles) with machine comprehension of text (identifying the answer spans from those articles). It only evaluates the coverage of the retrieval result and the accuracy of the final answer, but does not address the problem of the retrieved document not containing the answer.

3 Approach

Previous reading comprehension tasks usually aim to extract text spans from the passage to answer the question. In this work, the task is advanced that whether the given passage contains the answer is not specified. For the question whose passage contains the answer, the system needs to correctly extract the answer. Otherwise, the system needs to refuse to answer the question that there is no answer in the passage.

To solve this problem, we develop three categories of methods. First, we implement an answer extraction model and propose two methods to improve it for the passage that may not contain the answer. Second, we develop the passage triggering based method, which first judges whether the passage contains the answer then extracts the answer only in the triggered passage. Finally, we propose an answer validation model, which first extracts the answer then validates whether it is correct.

3.1   Answer Extraction Based Method

In this work, we implement the answer extraction model following match-LSTM [17] and R-Net [19], which have shown the effectiveness in many reading comprehension tasks.

Consider a question $Q = \{w_t^Q\}_{t=1}^m$ and a passage $P = \{w_t^P\}_{t=1}^n$, we first convert the words to their respective word-level embeddings and character-level embeddings. The character-level embeddings are generated by taking the final hidden states of a bi-directional GRU [2] applied to embeddings of characters in the token. We then use a bi-directional GRU to produce new representation $u_1^Q, \ldots, u_m^Q$ and $u_1^P, \ldots, u_n^P$ of all words in the question and passage respectively:

$$u_t^Q = \text{BiGRU}_Q(u_{t-1}^Q, [e_t^Q, char_t^Q]), u_t^P = \text{BiGRU}_P(u_{t-1}^P, [e_t^P, char_t^P]) \quad (1a)$$

Given question and passage representations $\{u_t^Q\}_{t=1}^m$ and $\{u_t^P\}_{t=1}^n$, [17] introduce match-LSTM, which combines the passage representation u_j^P with the passage-aware question representation c_t^Q to aggregate the question information to words in the passage, where $c_t^Q = att(u^Q, [u_t^P, v_{t-1}^P])$ is an attention-pooling vector of the whole question u^Q. [19] propose adding a gate to the input ($[u_t^P, c_t^Q]$) of GRU to determine the the of passage parts.

$$s_j^t = \text{v}^T\tanh(W_u^Q u_j^Q + W_u^P u_t^P + W_v^P v_{t-1}^P) \quad (2a)$$

$$a_i^t = \exp(s_i^t)/\Sigma_{j=1}^m\exp(s_j^t) \quad (2b)$$

$$c_t^Q = \Sigma_{i=1}^m a_i^t u_i^Q \quad (2c)$$

$$g_t = \text{sigmoid}(W_g[u_t^P, c_t^Q]) \quad (2d)$$

$$[u_t^P, c_t^Q]^* = g_t \odot [u_t^P, c_t^Q] \quad (2e)$$

$$v_t^P = \text{GRU}(v_{t-1}^P, [u_t^P, c_t^Q]^*) \quad (2f)$$

We then obtain the question-aware passage representation v_t^P for all positions in the passage.

Following previous methods used on the SQuAD, we use pointer networks [16] to predict the start and end positions of the answer. Given the passage representation $\{v_t^P\}_{t=1}^n$, the attention mechanism is utilized as a pointer to select the start position (p^1) and end position (p^2) from the passage, which can be formulated as follows:

$$s_j^t = v^T \tanh(W_h^P v_j^P + W_h^a h_{t-1}^a) \tag{3a}$$

$$a_i^t = \exp(s_i^t)/\Sigma_{j=1}^n \exp(s_j^t) \tag{3b}$$

$$p^t = \text{argmax}(a_1^t, \ldots, a_n^t) \tag{3c}$$

Here h_{t-1}^a represents the last hidden state of the answer recurrent network (pointer network). The input of the answer recurrent network is the attention-pooling vector based on current predicted probability a^t:

$$c_t = \Sigma_{i=1}^n a_i^t v_i^P, h_t^a = \text{GRU}(h_{t-1}^a, c_t) \tag{4a}$$

When predicting the start position, h_{t-1}^a represents the initial hidden state of the answer recurrent network. We utilize the question vector r^Q as the initial state of the answer recurrent network. $r^Q = att(u^Q, v_r^Q)$ is an attention-pooling vector of the question based on the parameter v_r^Q:

$$s_j = v^T \tanh(W_u^Q u_j^Q + W_v^Q v_r^Q) \tag{5a}$$

$$a_i = \exp(s_i)/\Sigma_{j=1}^m \exp(s_j) \tag{5b}$$

$$r^Q = \Sigma_{i=1}^m a_i u_i^Q \tag{5c}$$

The objective function is to minimize the following cross entropy:

$$\mathcal{L} = -\Sigma_{t=1}^2 \Sigma_{i=1}^n [y_i^t \log a_i^t + (1 - y_i^t) \log(1 - a_i^t)] \tag{6a}$$

where $y_i^t \in \{0, 1\}$ denotes a label. $y_i^t = 1$ means i is a correct position, otherwise $y_i^t = 0$.

This model is trained on the positive examples in the SQuAD-T dataset. When predicting the answer, the answer extraction model outputs two probabilities at the start and end positions, respectively. We multiply them for the probability of each text span to select the answer. If the probability of the answer is higher than a threshold pre-selected on the development set, we output it as the final answer, otherwise we refuse to answer this question.

Answer Extraction with No-Answer Option

The answer extraction model has two issues. First, we can only train it with positive examples in which the passage contains the answer. Second, the score is relative since the probability of the answer is normalized in each passage. To handle these issues, we propose improving the answer extraction model with a no-answer option. Levy et al. [8] propose adding a trainable bias to the confidence score p^t to allow the model to signal that there is no answer in the relation extraction task. Similarly, we add a padding position for the passage and supervise the model to predict this position when the passage does not contain the answer. In addition to the prediction strategy in the answer extraction model, we refuse to give an answer when the model predicts the padding position.

Answer Extraction with Modified Objective Function

We develop another strategy to improve the answer extraction model by modifying the objective function. For the positive example, we use the original objective function in the answer extraction, for which the probability is set to 1 for correct start and end positions, otherwise it is 0. For the negative example, we modify the objective function as follows:

$$\mathcal{L} = -\Sigma_{t=1}^{2}\Sigma_{i=1}^{n}[y_i^t \log a_i^t + (1 - y_i^t)\log(1 - a_i^t)] \tag{7a}$$

where $y_i^t = \frac{1}{n}$ for all positions.

3.2   Passage Triggering Based Method

Unlike the answer extraction based methods that extract and judge the answer in one model, the passage triggering based method divides this task into two steps. We first apply a passage triggering model to determine whether the passage contains the answer. We then apply the answer extraction model only on the triggered passage for the answer.

For passage triggering, we follow the above-mentioned matching strategy to obtain the question-aware passage representation $\{v_j^P\}_{j=1}^n$ in Eq. 2 and the question representation r^Q in Eq. 5. We apply an attention pooling to aggregate the matching information to a fix length vector.

$$s_j = v^T \tanh(W_v^P v_j^P + W_v^Q r^Q) \tag{8a}$$
$$a_i = \exp(s_i)/\Sigma_{j=1}^n \exp(s_j) \tag{8b}$$
$$r^P = \Sigma_{i=1}^n a_i v_i^P \tag{8c}$$

We then feed r^P to a multi-layers perceptron for the decision. The objective function is to minimize the following cross entropy:

$$\mathcal{L} = -\sum_{i=1}^{N}[y_i \log p_i + (1 - y_i)\log(1 - p_i)] \tag{9a}$$

where p_i is the probability that the passage contains the answer. y_i denotes a label, $y_i = 1$ means the passage contains the answer, otherwise it is 0.

When predicting the answer, we first judge whether the passage contains the answer by comparing the probability with a pre-selected threshold on the development set. For the triggered passage, we then apply the extraction model for the answer.

3.3   Answer Validation Based Method

There is an issue posed by answer information not being considered in the passage triggering based method. To this end, we propose the answer validation model, which first extracts an answer then validates whether it is correct.

We first apply the answer extraction model to obtain the answer span. Next, we incorporate the answer information into the encoding part by adding additional features f_t^s and f_t^e, to indicate the start and end positions of the extracted answer span. $f_t^s = 1$ and $f_t^e = 1$ mean the position t is the start and end of the answer span, respectively.

$$u_t^P = \text{BiGRU}_P(u_{t-1}^P, [e_t^P, char_t^P, f_t^s, f_t^e]) \tag{10a}$$

Unlike the answer extraction that predicts the answer on the passage side, answer validation needs to judge whether the question is well answered. Therefore, we reverse the direction of all above-mentioned equations to aggregate the passage information with the question. Specifically, we reverse Eq. 2 to obtain the passage-aware question representations,

$$s_j^t = v^{\text{T}}\tanh(W_u^P u_j^P + W_u^Q u_t^Q + W_v^Q v_{t-1}^Q) \tag{11a}$$

$$a_i^t = \exp(s_i^t)/\Sigma_{j=1}^n \exp(s_j^t) \tag{11b}$$

$$c_t^P = \Sigma_{i=1}^n a_i^t u_i^P \tag{11c}$$

$$g_t = \text{sigmoid}(W_g[u_t^Q, c_t^P]) \tag{11d}$$

$$[u_t^Q, c_t^P]^* = g_t \odot [u_t^Q, c_t^P] \tag{11e}$$

$$v_t^Q = \text{GRU}(v_{t-1}^Q, [u_t^Q, c_t^P]^*) \tag{11f}$$

Based on the Eq. 11, we obtain the v_t^Q for each position of questions. We then make the decision by judging whether each question word is well answered by the passage and answer with three steps. First, we measure the passage-independent importance of question words. We hold that the importance of each word in the question should not vary no matter what the passage and answer are. For example, the interrogative and name entity are usually more important than the conjunction and stopwords. Therefore, we apply the gate mechanism to select the important information, which is produced by the original representation of each question word.

$$g_t = \text{sigmoid}(W_g u_t^Q), v_t^Q * = g_t \odot v_t^Q \tag{12a}$$

Next, we obtain the matching score of each question word by a multi-layers perceptron,

$$s_t^Q \propto \exp(W_2(\tanh(W_1 v_t^Q *))) \tag{13a}$$

Finally, we combine the matching score of question words adaptively. We apply the attention mechanism on the matching vector $v_t^Q *$ based on the learned parameter v_s to obtain the weight of each question, and then apply it to weighted-sum the score s_t^Q for the final score s.

$$s_j^t = v^{\text{T}}\tanh(W_v v_s + W_u^Q v_t^Q *) \tag{14a}$$

$$a_i^t = \exp(s_i^t)/\Sigma_{j=1}^m \exp(s_j^t) \tag{14b}$$

$$s = \Sigma_{i=1}^m a_i^t s_t^Q \tag{14c}$$

As both the score and the weight of each question word are normalized, we treat the final score s as the probability that the answer is correct.

$$\mathcal{L} = -\sum_{i=1}^{N}[y_i \log s + (1 - y_i) \log(1 - s)] \tag{15a}$$

where y_i denotes a label, $y_i = 1$ means the answer is correct, otherwise 0.

When predicting the answer, we compare s with a threshold pre-selected on the development set to determine whether to answer the question with the extracted answer.

3.4   Implementation Details

For all above-mentioned models, we use 300-dimensional uncased pre-trained *GloVe* embeddings [10][3] without update during training. We use zero vectors to represent all out-of-vocabulary words. Hidden vector length is set to 150 for all layers. We apply dropout [14] between layers, with a dropout rate of 0.2. The model is optimized using Adam [7] with default learning rate of 0.002.

4   Experiments

To evaluate methods in this task, we build a new dataset SQuAD-T based on the SQuAD dataset and propose using F-measure on the positive examples and over-all accuracy for all data for evaluation. We report results of all above-mentioned models. Experimental results show that our answer validation model achieves the best F_1 score and accuracy on the SQuAD-T dataset. In addition, we provide detailed comparisons and analysis of all methods.

4.1   Dataset Construction

In real-life application (i.e. search engine), given a question (or query), it usually first retrieves the relevant passage then discriminates whether there is an answer. In this work, we simulate this process to build the SQuAD-T dataset based on the SQuAD dataset. Specifically, we use Lucene to index all passages in the SQuAD dataset. Then for each question in the SQuAD dataset, we obtain one relevant passage by searching the question with Lucene using its default ranker based on the vector space model and TF-IDF[4]. We observe that only 65.67% of questions whose most related passages are still original corresponding passages in the SQuAD dataset. We then treat these question and passage pairs as the positive examples in which the passages contain the answer. For other questions, we select the top-ranked passage that does not contain the answer as

---

[3] http://nlp.stanford.edu/data/glove.6B.zip.
[4] Details can be found in https://lucene.apache.org/core/2_9_4/api/core/org/apache/lucene/search/Similarity.html.

the negative example. As the author of SQuAD only publishes the training set and the development set, we split the 10,570 instances in the development set to 5,285 for development and 5,285 for test. The statistics of the SQuAD-T dataset are shown in Table 2.[5]

Table 2. Statistics of the SQuAD-T dataset.

	Train	Dev	Test
Question	86,830	5,285	5,285
Positive	57,024	3,468	3,468
Negative	29,806	1,817	1,817

4.2 Evaluation Metrics

Previous work adopts Exact Match and F_1[6] to evaluate the performance of the reading comprehension model [12]. These metrics are to evaluate the extracted answer in the case that the passage must contain the answer, and hence are not suitable for data in which there is no answer. In this work, we propose using precision, recall and F_1 scores at the question level to evaluate this task. A question is treated as a positive case only if it contains the correct answer in the corresponding passage. Given the question set Q, Q^+ is the set where there is an answer in the passage to answer the question, otherwise Q^-. T^+ is the set where the model gives an answer, otherwise T^-. A^+ is the set where the given answer is correct, otherwise A^-. We define the F-measure as follows:

$$Precision = \frac{|A^+|}{|T^+|}, Recall = \frac{|A^+|}{|Q^+|}, F_1 = \frac{2 \times Precision \times Recall}{Precision + Recall} \quad (16a)$$

In addition, we define the overall accuracy as follows:

$$Acc = \frac{|A^+| + |T^- \cap Q^-|}{|Q|} \quad (17a)$$

4.3 Main Result

We show the result in terms of precision, recall, and F_1 in Table 3, and illustrate the Precision-Recall curves on the development and test set in Fig. 1, respectively. The answer extraction model only achieves 68.59 and 67.50 in terms of F_1 on the development set and test set, respectively. Since the probability of the answer is normalized on each passage, the score is relative and therefore

---

[5] We release the dataset in https://github.com/chuanqi1992/SQuAD-T.

[6] Here the F_1 score is calculated at the token level between the true answer and the predicted answer.

Table 3. Results in terms of precision, recall, and F_1 on the SQuAD-T development and test set. *Significant improvement over the baseline method of the answer extraction (underlined) (t-test, p < 0.05).

Method	Development set			Test set		
	Prec	Rec	F_1	Prec	Rec	F_1
Answer extraction	66.70	70.59	<u>68.59</u>	66.35	68.69	<u>67.50</u>
Answer extraction with no-answer option	74.48	67.50	70.82	74.63	67.44	70.86
Answer extraction with modified objective function	74.63	67.68	70.98	73.09	66.35	69.55
Passage triggering then extraction	59.65	74.65	66.32	58.27	73.33	64.94
Answer validation	69.73	73.41	**71.53***	68.74	73.93	**71.24***

performs worse for judging whether it is a real answer. In addition, this model achieves 78.374 and 77.105 in terms of Exact Match on the positive examples in the development and test set, respectively, which is used to provide the extraction result for the passage triggering based model and answer validation model. It determines the max recall of these related models shown in Fig. 1.

Improving the answer extraction model by adding the no-answer option and modifying the objective function greatly improves the result. The passage triggering based method only achieves 66.32 and 64.94 on the development set and test set, respectively. We observe that the answer validation model obviously outperforms the passage triggering based method since it incorporates the answer information. Our answer validation model outperforms all other baselines and achieves best the F_1 score with 71.53 and 71.24 in the development set and test set, respectively.

We show the overall accuracy on the SQuAD-T development and test set in Table 4. The answer extraction model with the no-answer option and modified objective function consistently improve the result of the answer extraction. Our answer validation model achieves the best overall accuracy of 74.60 on the SQuAD-T test set.

Table 4. Results in terms of overall accuracy on the SQuAD-T development and test set.

Method	Dev	Test
Answer extraction	65.26	64.34
Answer extraction with no-answer option	**74.65**	74.48
Answer extraction with modified objective function	74.14	73.08
Passage triggering then extraction	63.28	62.44
Answer validation	73.98	**74.60**

4.4   Model Analysis

Figure 1 shows the precision-recall curves on the development set and test set, respectively. We observe that the answer extraction based method achieves better precision when the recall is relatively low. With the increase of the recall, the precision of the answer extraction model obviously decreases, which indicates that the score of the answer extraction is not suitable for judging whether it is a real answer. Improving the answer extraction model by the no-answer option or modified objective function can partly relieve this issue. However, we observe that the max recall of these two improved methods is much lower than the answer extraction model in Fig. 1, which indicates that training these two models with negative examples leads to worse extraction precision on the positive examples. Therefore, we argue that the answer extraction model should only be trained on the positive example for better extraction precision. The answer validation model achieves better performance than the passage triggering based method. Our answer validation model almost maintains stable precision with the increase of the recall, which leads to the best F_1 score on the SQuAD-T dataset.

Table 5 shows the detailed result distribution of all methods. We observe that the answer extraction model with no-answer option and modified objective function achieve the lower ratio in Q^-T^+ and higher ratio in Q^-T^-, which shows the effectiveness of refusing to give an answer when there is no answer. However, these two methods sacrifice the precision as they have much lower ratios in $Q^+T^+A^+$. In addition, they also have higher ratios in Q^+T^-. Therefore, if we calculate the F-measure of negative examples, the result of answer extraction with no-answer option and modified function are 80.49 and 78.96, respectively,

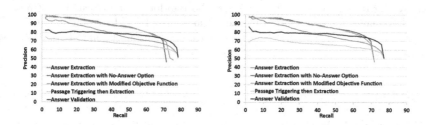

Fig. 1. Precision-Recall curves on the development and test set.

Table 5. The result distribution on the SQuAD-T test set. The values are percentages in corresponding categories.

Method	$Q^+T^+A^+$	$Q^+T^+A^-$	Q^+T^-	Q^-T^+	Q^-T^-
Answer extraction	45.07	7.72	12.83	15.14	19.24
+ No-answer option	44.26	10.88	10.49	**4.16**	**30.22**
+ Modified objective function	43.54	11.18	10.90	**4.84**	**29.54**
Passage triggering then extraction	48.12	14.40	**3.10**	20.06	14.32
Answer validation	**48.51**	13.77	**3.33**	8.29	26.09

which is still lower than 81.79 for the answer validation model. The passage triggering based method achieves a worse result with the highest ratio in Q^-T^+ because it does not consider the answer information when making the decision. The answer validation model achieves the better results in $Q^+T^+A^+$. Meanwhile, it also achieves relative lower ratio in Q^-T^+ and relative higher ratio in Q^-T^- compared with other methods using the answer extraction model to extract the answer, which indicates that it performs better in refusing to give the answers when passages do not contain the answers without performance degradation when passages contain the answers. Our answer validation model therefore achieves the best result in terms of F_1 and overall accuracy in the SQuAD-T test set.

5   Conclusion and Future Work

In this paper, we study the machine reading comprehension task in which whether the passage contains the answer is not specified. Therefore the system needs to correctly refuse to give an answer when a passage does not contain the answer. We develop several baseline methods including the answer extraction based method, the passage triggering based method, and propose the answer validation method for this task. Experiments show that our proposed answer validation model outperforms all other baseline methods on the SQuAD-T test set. We notice that Rajpurkar et al. build a dataset SQuAD2.0 in which questions are written by humans. We will test our methods on this benchmark dataset in the future.

Acknowledgments. We greatly thank Hangbo Bao for helpful discussions. Chuanqi Tan and Weifeng Lv are supported by the National Key R&D Program of China (No. 2017YFB1400200) and National Natural Science Foundation of China (No. 61421003 and 71501003).

References

1. Chen, D., Fisch, A., Weston, J., Bordes, A.: Reading Wikipedia to answer open-domain questions. In: ACL (2017)
2. Cho, K., van Merrienboer, B., Gulcehre, C., Bahdanau, D., Bougares, F., Schwenk, H., Bengio, Y.: Learning phrase representations using RNN encoder-decoder for statistical machine translation. In: EMNLP, pp. 1724–1734. Association for Computational Linguistics (2014)
3. Clark, C., Gardner, M.: Simple and effective multi-paragraph reading comprehension. arXiv preprint arXiv:1710.10723 (2017)
4. Huang, H.Y., Zhu, C., Shen, Y., Chen, W.: FusioNnet: Fusing via fully-aware attention with application to machine comprehension. In: ICLR (2018)
5. Joshi, M., Choi, E., Weld, D., Zettlemoyer, L.: Triviaqa: A large scale distantly supervised challenge dataset for reading comprehension. In: ACL, pp. 1601–1611. Association for Computational Linguistics (2017)

6. Jurczyk, T., Zhai, M., Choi, J.D.: SelQA: a new benchmark for selection-based question answering. In: 2016 IEEE 28th International Conference on Tools with Artificial Intelligence (ICTAI), pp. 820–827. IEEE (2016)
7. Kingma, D., Ba, J.: Adam: a method for stochastic optimization. arXiv preprint arXiv:1412.6980 (2014)
8. Levy, O., Seo, M., Choi, E., Zettlemoyer, L.: Zero-shot relation extraction via reading comprehension. In: CoNLL, pp. 333–342. Association for Computational Linguistics (2017)
9. Nguyen, T., Rosenberg, M., Song, X., Gao, J., Tiwary, S., Majumder, R., Deng, L.: MS MARCO: a human generated machine reading comprehension dataset. arXiv preprint arXiv:1611.09268 (2016)
10. Pennington, J., Socher, R., Manning, C.D.: Glove: global vectors for word representation. In: EMNLP, pp. 1532–1543 (2014)
11. Rajpurkar, P., Jia, R., Liang, P.: Know what you don't know: unanswerable questions for squad. In: ACL (2018)
12. Rajpurkar, P., Zhang, J., Lopyrev, K., Liang, P.: Squad: 100,000+ questions for machine comprehension of text. In: EMNLP (2016)
13. Seo, M., Kembhavi, A., Farhadi, A., Hajishirzi, H.: Bidirectional attention flow for machine comprehension. In: International Conference on Learning Representations (2017)
14. Srivastava, N., Hinton, G.E., Krizhevsky, A., Sutskever, I., Salakhutdinov, R.: Dropout: a simple way to prevent neural networks from overfitting. J. Mach. Learn. Res. **15**, 1929–1958 (2014)
15. Tan, C., Wei, F., Yang, N., Du, B., Lv, W., Zhou, M.: S-Net: from answer extraction to answer generation for machine reading comprehension. AAAI (2018)
16. Vinyals, O., Fortunato, M., Jaitly, N.: Pointer networks. In: Cortes, C., Lawrence, N.D., Lee, D.D., Sugiyama, M., Garnett, R. (eds.) Advances in Neural Information Processing Systems, vol. 28. Curran Associates, Inc. (2015)
17. Wang, S., Jiang, J.: Learning natural language inference with LSTM. In: The 2016 Conference of the North American Chapter of the Association for Computational Linguistics: Human Language Technologies, NAACL HLT 2016 (2016)
18. Wang, S., Jiang, J.: Machine comprehension using match-LSTM and answer pointer. In: International Conference on Learning Representations (2017)
19. Wang, W., Yang, N., Wei, F., Chang, B., Zhou, M.: Gated self-matching networks for reading comprehension and question answering. In: Proceedings of the 55th ACL, pp. 189–198. Association for Computational Linguistics (2017)
20. Xiong, C., Zhong, V., Socher, R.: Dynamic coattention networks for question answering. arXiv preprint arXiv:1611.01604 (2016)
21. Yang, Y., Yih, W.t., Meek, C.: WikiQA: a challenge dataset for open-domain question answering. In: Proceedings of EMNLP, pp. 2013–2018. Citeseer (2015)
22. Zhao, J., Su, Y., Guan, Z., Sun, H.: An end-to-end deep framework for answer triggering with a novel group-level objective. In: EMNLP, pp. 1276–1282. Association for Computational Linguistics (2017)

Learning to Converse Emotionally Like Humans: A Conditional Variational Approach

Rui Zhang and Zhenyu Wang[✉]

Department of Software Engineering, South China University of Technology,
Guangzhou, People's Republic of China
z.rui16@mail.scut.edu.cn, wangzy@scut.edu.cn

Abstract. Emotional intelligence is one of the key parts of human intelligence. Exploring how to endow conversation models with emotional intelligence is a recent research hotspot. Although several emotional conversation approaches have been introduced, none of these methods were able to decide an appropriate emotion category for the response. We propose a new neural conversation model which is able to produce reasonable emotion interaction and generate emotional expressions. Experiments show that our proposed approaches can generate appropriate emotion and yield significant improvements over the baseline methods in emotional conversation.

Keywords: Emotion selection · Emotional conversation

1 Introduction

The ability of a computer to converse in a natural and coherent manner with humans has long been held as one of the primary objectives of artificial intelligence, yet conventional dialog systems continue to face challenges in emotion understanding and expression.

In the past, the research on emotional response generation focused on the domain-specific, task-oriented dialogue systems. These methods are mainly based on some hand-crafted emotion inference rules to choose a reasonable strategy, and retrieve a suitable pre-designed template for response generation.

However, these approaches are not suitable for open-domain chatterbot, since there are large amount of topics and more complex emotion states in chitchat conversation. For instance, if the user says "my cat died yesterday", it is reasonable to generate response like "so sorry to hear that" to express sadness, also it is appropriate to generate response like "bad things always happen, I hope you will be happy soon" to comfort the user. Although some neural network based methods for open-domain emotional conversation [3,17,18,20] have been proposed recently, yet these approaches mainly focused in generating emotional

© Springer Nature Switzerland AG 2018
M. Zhang et al. (Eds.): NLPCC 2018, LNAI 11108, pp. 98–109, 2018.
https://doi.org/10.1007/978-3-319-99495-6_9

expressions and none of them provide a mechanism to determine which emotion category is appropriate for response generation.

This paper presents two kinds of emotion sensitive conditional variational autoencoder (EsCVAE) structure for determining a reasonable emotion category for response generation. Just the same as children learn to converse emotionally through imitation, the principle idea is to model the emotion interaction patterns from large-scale dialogue corpus as some kind of distribution, and sample from this emotion interaction distribution to decide what kind of emotion should be expressed when generating responses.

As far as we know, this is the first work to determine an appropriate emotion category for current large-scale conversation generation model. To sum up, our contributions are as follows:

- We propose the EsCVAE-I and EsCVAE-II model to learn to automatically specify a reasonable emotion category for response generation. Experiments show that our proposed approaches yield significant improvements over the baseline methods.
- We show that there are some frequent emotion interaction patterns in humans dialogue (e.g. happiness-like, angry-disgust), and our models are able to learn such frequent patterns and apply it to emotional conversation generation.

2 Related Work

2.1 Conversation Generation

In recent years, there is a surge of research interest in dialogue system. Due to the development of deep neural network, learning a response generation model within a machine translation (MT) framework from large-scale social conversation corpus becomes possible. Following the principle idea of sequence-to-sequence architecture, recurrent network based models [11,16] and VAE based models [2,10] were successively proposed. The basic idea of VAE is to firstly encode the input x into a probability distribution z, and then to apply a decoder network to reconstruct the original input x using samples from z.

To better control the generative process, the conditional variational autoencoder (CVAE) [4,12] is recently introduced to generate diverse texts conditioned on certain attributes c. The conditional distribution in the CVAE is defined as $p(x, z|c) = p(x|z, c)p(z|c)$. By approximating $p(x|z, c)$ and $p(z|c)$ using deep neural network (parameterized by θ), the generative process of x is: (1) sample a latent variable z from the prior network $p_\theta(z|c)$, and (2) generate x through the response decoder $p_\theta(x|z, c)$. As proposed in [15], CVAE can be trained by maximizing the variational lower bound of the conditional log likelihood. By assuming the z follows Gaussian distribution and introducing a recognition network $q_\phi(z|x, c)$ to approximate the true posterior distribution $p(z|x, c)$, the variational lower bound can be written as:

$$\mathcal{L}(\theta, \phi; x, c) = -KL(q_\phi(z|x, c)||p_\theta(z|c)) + \mathbf{E}_{q_\phi(z|x,c)}[\log p_\theta(x|z, c)] \quad (1)$$
$$\leq \log p(x|c)$$

Further, Zhao et al. [19] introduced Knowledge-Guided CVAE (kgCVAE) to get benefits from linguistic cues. They assumed that in kgCVAE the generation of x also depends on the linguistic cues y, and have proved that the kgCVAE based dialogue model can more easily control the model's output by cooperating with the linguistic cues.

2.2   Emotional Intelligence

Emotional intelligence is one of the key parts of human intelligence. Exploring the influence of emotional intelligence on human-computer interaction has a long history. Experiments show that dialogue systems with emotional intelligence lead to less breakdowns in dialogue [6], and enhance users' satisfaction [9].

In the early studies, a few of emotion modeling approaches were introduced to construct emotional dialogue systems. Polzin et al. [8] proposed a pioneer work in emotional human-computer conversation, which is capable of employing different discourse strategies based on users' affection states. Andre et al. [1] integrates social theory of politeness with cognitive theory of emotions to endow dialogue systems with emotional intelligence. Skowron [13,14] quantize users' emotion via affective profile and respond to users' utterances at content- and affect- related levels.

Recently, neural network based methods for emotional text generation have been investigated. Zhou et al. [20] proposed Emotional Chatting Machine (ECM) to generate emotional response by adopting emotion category embedding, internal emotional memory and external memory. Ghosh et al. [3] introduced Affect-LM which is able to generate emotionally colored conversational text in five specific affect categories with varying affect strengths.

Unfortunately, none of these neural network based approaches provides a mechanism to determine which emotion category is appropriate for emotional response generation. These works, mainly focused in generating emotional expressions, are either driven by pre-defined rules, or in need of specifying target emotion manually. Thus, in this paper we introduce two kinds of EsCVAE model to address this problem.

3   Proposed Models

3.1   Problem Definition

Our models aim to learn the inner relationship of emotional interaction, and to automatically specify a reasonable emotion category for response generation given an input utterance. In practice, however, it is hard to evaluate the appropriateness if we only predict the emotional category. Therefore, we reformulate the task as below:

Given a post utterance $\mathbf{u}_p = (w_{p_1}, w_{p_2}, \cdots, w_{p_m})$ and its emotion category $\mathbf{e}_p$, the goal is to predict a reasonable emotion category $\mathbf{e}_r$ for response generation, and generate a response utterance $\mathbf{u}_r = (w_{r_1}, w_{r_2}, \cdots, w_{r_n})$ which is coherent with $\mathbf{e}_r$, where w_{p_k} and w_{r_k} are the k-th words in the post utterance and the response utterance, respectively. The emotions are divided into six categories {*Anger, Disgust, Happiness, Like, Sadness, Other*}.

3.2   EsCVAE-I: Conditioned on Emotions only

Since most rule-based strategies are triggered according to the emotion type of the input text, we firstly consider a variational autoencoder architecture conditioned on emotional information.

Figure 1(a) delineates an overview of our EsCVAE-I model. An emotional category embedding network is adopted to represent the emotion category of the utterance by a real-value, low dimensional vector, since an emotion category provides a high-level abstraction of an expression of the emotion. We randomly initialize the vector of each emotion categories, and then learn the vectors of emotion category through training. Thus, the emotion categories of the post utterance and the response utterance are represented by emotion embedding vectors e_p and e_r, respectively. A bidirectional GRU network is adopted as the response encoder to encode the response utterance $\mathbf{u}_r$ into a fixed-sized vector u_r, by concatenating the last hidden states of the forward and backward RNN. The post encoder is another GRU network that encodes the post utterance $\mathbf{u}_p$ into a vector u_p.

To capture the inner relationship of emotional interaction, a conditional variational autoencoder architecture is applied in the proposed model. In EsCVAE-I model we consider the post emotion e_p as the condition c and response emotion e_r as the linguistic feature y. The target utterance x is simply the response

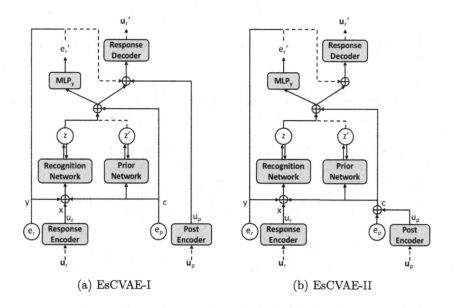

(a) EsCVAE-I   (b) EsCVAE-II

Fig. 1. Illustrations of our proposed models.

utterance u_r. Assuming latent variable z follows isotropic Gaussian distribution, with the recognition network $q_\phi(z|x,c,y) \sim \mathcal{N}(\mu, \sigma^2 \mathbf{I})$ and the prior network $p_\theta(z|c) \sim \mathcal{N}(\mu', \sigma'^2 \mathbf{I})$, we have:

$$\begin{bmatrix} \mu \\ \log(\sigma^2) \end{bmatrix} = W_r \begin{bmatrix} x \\ c \\ y \end{bmatrix} + b_r = W_r \begin{bmatrix} u_r \\ e_p \\ e_r \end{bmatrix} + b_r \tag{2}$$

$$\begin{bmatrix} \mu' \\ \log(\sigma'^2) \end{bmatrix} = \mathrm{MLP}_p(c) = W_p e_p + b_p \tag{3}$$

In the training stage, we obtain samples of z from $\mathcal{N}(z; \mu, \sigma^2 \mathbf{I})$ predicted by the recognition network. The response decoder is a GRU network with initial state $s_0 = W_i[z, c, e_r, u_p] + b_i$, which then predicts the words in x sequentially. An MLP is adopted to predict the response emotion category $e_r' = \mathrm{MLP}_y(z, c)$ based on z and c. While testing, samples of z is obtained from $\mathcal{N}(z; \mu', \sigma'^2 \mathbf{I})$ predicted by the prior network. And the initial state of the response decoder is calculated as: $s_0 = W_i[z, c, e_r', u_p] + b_i$, where e_r' is the predicted response emotion.

The proposed model is trained by minimizing the reconstruction loss while maximizing the variational lower bound. The reconstruction loss is calculated based on the cross entropy error. Following [19] the variational lower bound can be calculated as:

$$\begin{aligned} \mathcal{L}(\theta, \phi; x, c, y) = &-KL(q_\phi(z|x,c,y)||p_\theta(z|c)) \\ &+ \mathbf{E}_{q_\phi(z|x,c,y)}[\log p(x|z,c,y)] \\ &+ \mathbf{E}_{q_\phi(z|x,c,y)}[\log p(y|z,c)] \end{aligned} \tag{4}$$

3.3   EsCVAE-II: Sensitive to both Content-Level and Emotion-Level Information

A natural extension of the previous approach is a model that is also sensitive to the content information, since emotion interactions in dialogue are not only related to the emotional state of the talkers, but also closely related to the topic of the conversation.

In order to get benefits from these features, we propose the EsCVAE-II model. In this model we consider both the content-level and emotion-level information of the post utterance as condition c, by concatenating the utterance and emotion vectors: $c = [u_p, e_p]$. Following the same assumption, the recognition network $q_\phi(z|x,c,y)$ and the prior network $p_\theta(z|c)$ are calculated as:

$$\begin{bmatrix} \mu \\ \log(\sigma^2) \end{bmatrix} = W_r \begin{bmatrix} x \\ c \\ y \end{bmatrix} + b_r = W_r \begin{bmatrix} u_r \\ [u_p, e_p] \\ e_r \end{bmatrix} + b_r \tag{5}$$

$$\begin{bmatrix} \mu' \\ \log(\sigma'^2) \end{bmatrix} = \mathrm{MLP}_p(c) = W_p[u_p, e_p] + b_p \tag{6}$$

Then we obtain samples of z either from $\mathcal{N}(z; \mu, \sigma^2 \mathbf{I})$ predicted by the recognition network (while training) or $\mathcal{N}(z; \mu', \sigma'^2 \mathbf{I})$ predicted by the prior network (while testing). Since the content information of post utterance u_p has been contained in c, the initial state of the response decoder is therefore changed to $s_0 = W_i[z, c, e_r] + b_i$. Details of the EsCVAE-II model are shown in Fig. 1(b).

4   Experiment Setup

4.1   Datasets

The NLPCC Corpus We use the Emotional Conversation Dataset of NLPCC 2017, which consists of 1,119,207 post-response pairs, to evaluate our approaches. The dataset is automatically annotated by a six-way Bi-LSTM classifier which is reported to reach an accuracy of 64%[1].

The STC-2 Corpus We also use the Short Text Conversation Corpus (STC-2) dataset, which consists of 4,433,949 post-response pairs collected from Weibo. We apply the same Bi-LSTM classifier to annotate this corpus.

4.2   Model Details

We implement the encoders as 2-layer GRU and the response generator as single-layer GRU RNNs, with input and hidden dimension of 400 and maximum utterance length of 25. The dimensions of word embedding and emotion embedding are also set to 400. We use a KL term weight linearly annealing from 0 to 1 during training, to avoid vanishingly small KL term in the VAE module as introduced in [2]. Both the prior network and the recognition network consist of 200 hidden neurons.

In order to generate a better response, we adopt the learning to start (LTS) technique [21], which use an additional network layer to predict the first word instead of using a "GO" symbol as the first word. In addition, beam-search is also adopted with beam size of 5.

4.3   Evaluation Results

Since automatically evaluating an open-domain generative dialog model is still an open research challenge [5], we provide the following metrics to measure our models.

Quantitative Analysis The following metrics are proposed to automatically evaluate our models. At the content level, perplexity and BLEU are adopted. At the emotion level, we introduce emotion accuracy and EIP difference degree as the evaluating metrics. We randomly sample 5000 posts for test. Details of quantitative analysis results are reported in Table 1.

---

[1] http://tcci.ccf.org.cn/conference/2017/dldoc/taskgline04.pdf.

Table 1. Performance of each model on automatic measures. Note that our BLEU scores has been normalized to [0,1].

Corpora	Model	Perplexity	BLEU	Acc.	Diff.
NLPCC corpus	Seq2Seq	101.0	0.105	-	1.195
	CVAE	47.2	0.090	-	0.171
	EsCVAE-I	46.1	0.114	0.675	0.085
	EsCVAE-II	**44.3**	**0.139**	**0.690**	**0.072**
STC-2 corpus	Seq2Seq	88.4	0.156	-	0.289
	CVAE	20.7	0.224	-	0.048
	EsCVAE-I	20.9	0.211	0.621	0.042
	EsCVAE-II	**19.0**	**0.230**	**0.627**	**0.035**

BLEU is a popular metric which measures the geometric mean of modified n-gram precision with a length penalty [7]. We use BLEU-3 as lexical similarity metrics and normalize the score to [0,1]. Following [16] we also employ perplexity as an evaluation metric at the context level.

We quantitatively measure emotion accuracy to evaluate the EsCVAE-I and EsCVAE-II, which is defined as the agreement between the expected emotion category e'_r (generated by our models) and emotion category of the corresponding generated response (predicted by the Bi-LSTM classifier mentioned in Sect. 4.1).

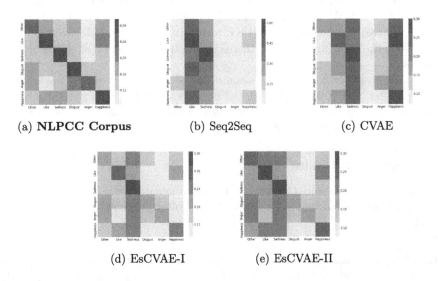

(a) **NLPCC Corpus**   (b) Seq2Seq   (c) CVAE

(d) EsCVAE-I   (e) EsCVAE-II

Fig. 2. Visualization of emotion interaction on NLPCC Corpus. The emotion categories on X/Y-axis (from left/top to right/bottom) are: *Other, Like, Sadness, Disgust, Anger, Happiness*.

To evaluate the imitative ability of the emotional dialogue model, we adopt the EIP difference degree as the evaluating metric. Emotion interaction pattern (EIP) is defined as the pair of emotion categories of a post and its response. The value of an EIP is the conditional probability $P(e_r|e_p) = P(e_r, e_p)/P(e_p)$. We define the EIP difference degree as the agreement between the EIP distribution of the original corpus and the EIP distribution of the generated results, calculated as:

$$\text{Difference degree} = \sum_r \sum_p |P_o(e_r|e_p) - P_g(e_r|e_p)|^2 \tag{7}$$

where $P_o(e_r|e_p)$ stands for the emotion interaction distribution of the generated results, while $P_g(e_r|e_p)$ is the distribution of the original corpora. In other words, a lower difference degree indicates that the model has a better ability to imitate emotion interaction.

As shown in Table 1, we have the following observations: (1) CVAE-based seq2seq model performs much better than vanilla seq2seq model, since vanilla seq2seq model tends to generate "safe" and dull responses, which lead to higher difference degree and higher perplexity, and (2) thanks to the external linguistic feature (response emotion), both EsCVAE models perform better on difference degree metric. Figures 2 and 3 visualize the EIP distribution by heat maps, where the color darkness indicates the strength of an interaction. We found that both EsCVAE models perform better on emotion interaction imitating.

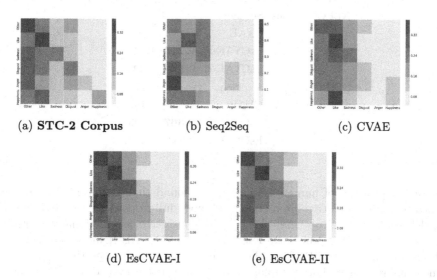

(a) **STC-2 Corpus**   (b) Seq2Seq   (c) CVAE

(d) **EsCVAE-I**   (e) **EsCVAE-II**

Fig. 3. Visualization of emotion interaction on STC-2 Corpus.

Human Evaluation In order to better evaluate the proposed model at both content- and emotion- level, we also analyze the generated results by human evaluation. We recruit 3 annotators for human evaluation experiments. Annotators are asked to score the generated responses in terms of emotional accuracy and naturalness. Naturalness are annotated according to the following criteria:

```
IF (Coherence and Fluency)
    IF (Emotion Appropriateness)
        SCORE 2
    ELSE
        SCORE 1
ELSE
    SCORE 0
```

We extract 300 posts from the test set, and for each model we generate 3 responses for each post. The results of human evaluation for the quality of response are shown in Table 2. We calculate the Fleiss' kappa as the statistical measure of inter-rater consistency, the average score of Fleiss' kappa for naturalness is 0.471 and for emotion accuracy is 0.744. As can be seen, the EsCVAE-I model gets the best performance in emotion accuracy metric, while the EsCVAE-II model achieves better score in naturalness metric.

Table 2. Human evaluation in terms of emotion accuracy and naturalness.

Corpora	Model	Acc.	Naturalness
NLPCC corpus	Seq2Seq	-	1.080
	CVAE	-	1.163
	EsCVAE-I	**0.283**	1.176
	EsCVAE-II	0.250	**1.217**
STC-2 corpus	Seq2Seq	-	1.170
	CVAE	-	1.187
	EsCVAE-I	**0.367**	1.193
	EsCVAE-II	0.353	**1.250**

In the progress of human evaluation, we found that vanilla seq2seq model failed to capture emotion interaction, and most dull responses generated by this model, such as "Me too" and "Haha", are considered as emotionally inappropriate. On the other hand, the EsCVAE-I and EsCVAE-II model are capable of learning frequent EIPs from the training corpus and generating reasonable emotional responses. For example, when the post utterance is in a negative emotional state (anger, disgust, ...), the EsCVAE based models tend to predict the negative emotional category, and vice versa (as shown in Table 3).

In addition, the emotion accuracy in human evaluation is much lower than in automatically evaluation, which indicates that it is still hard to assess a dialogue model at emotion level, since this metric is complicate and subjective.

Table 3. Sample responses generated by each model (original Chinese and English translation).

Post	Model		Response(Original)	Response(Translated)
我老了，已经不会再写诗了。能做的就是让你在我的怀里躲风。 I am old and I can no longer write poetry. What I can do is hide you in my arms. **(Sadness)**	Seq2Seq		我想知道你是怎么做到的。	I want to know how you did it.
	CVAE	Like	好可爱啊	So cute.
		Happiness	哈哈，诗人，原来是这样	Haha, the poet, it turned out to be this way.
	EsCVAE-1	Sadness	你真的很伤感啊!	You are really sad!
		Sadness	你看不见，你永远不懂	You will never understand what you cannot see.
	EsCVAE-2	Like	我也喜欢你的书……	I like your book too.
		Disgust	你的梦还没醒呢?	Didn't you wake up from your dreams?
将来的你，一定会感谢现在拼命努力的你。 In the future, you will certainly thank you for your hard work. **(Like)**	Seq2Seq		我现在也是这样的。	I am like this now.
	CVAE	Other	你知道将来的我吗?	Do you know me in the future?
		Happiness	是的，这让我们在一起!	Yes, this allows us to be together!
	EsCVAE-1	Like	我的青春也不能虚度啊。	My youth cannot be devalued.
		Other	将来的你，你会后悔。	In the future, you will regret it.
	EsCVAE-2	Like	嗯嗯嗯，绝对的。	Um, absolutely.
		Like	坚持就会有收获的。	As long as you persist, there will be a harvest.

5 Conclusion

In this paper, we propose two kinds of emotion sensitive conditional variational autoencoder (EsCVAE) structure for emotional conversation, to model the inner relationship of emotional interaction in human dialogue and generate emotional responses. To our best knowledge, this is the first work to generate an appropriate emotion category for current large-scale conversation generation model. Automatic and manual evaluation results show that EsCVAE based models can predict a reasonable emotion category for response generation by learning emotion interaction pattern from the training corpus.

We leave the exploration of EsCVAEs with attention mechanism for future work. Additional accuracy improvements might be also achieved by extended features (e.g. topics, dialog-act). At the same time, we will improve our model by considering polite rules and persona model to avoid generating offensive responses. We also plan to investigate the applicability of our model for task-oriented conversation.

Acknowledgements. This work is supported by the Science and Technology Program of Guangzhou, China(No. 201802010025), the Fundamental Research Funds for the Central Universities(No. 2017BQ024), the Natural Science Foundation of Guangdong Province(No. 2017A030310428) and the University Innovation and Entrepreneurship Education Fund Project of Guangzhou(No. 2019PT103). The authors also thank the editors and reviewers for their constructive editing and reviewing, respectively.

References

1. André, E., Rehm, M., Minker, W., Bühler, D.: Endowing spoken language dialogue systems with emotional intelligence. In: André, E., Dybkjær, L., Minker, W., Heisterkamp, P. (eds.) ADS 2004. LNCS (LNAI), vol. 3068, pp. 178–187. Springer, Heidelberg (2004). https://doi.org/10.1007/978-3-540-24842-2_17
2. Bowman, S.R., Vilnis, L., Vinyals, O., Dai, A., Jozefowicz, R., Bengio, S.: Generating sentences from a continuous space. In: Proceedings of The 20th SIGNLL Conference on Computational Natural Language Learning, pp. 10–21 (2016)
3. Ghosh, S., Chollet, M., Laksana, E., Morency, L.P., Scherer, S.: Affect-LM: a neural language model for customizable affective text generation. In: Proceedings of the 55th Annual Meeting of the Association for Computational Linguistics (Volume 1: Long Papers), vol. 1, pp. 634–642 (2017)
4. Hu, Z., Yang, Z., Liang, X., Salakhutdinov, R., Xing, E.P.: Toward controlled generation of text. In: International Conference on Machine Learning, pp. 1587–1596 (2017)
5. Liu, C.W., Lowe, R., Serban, I., Noseworthy, M., Charlin, L., Pineau, J.: How not to evaluate your dialogue system: An empirical study of unsupervised evaluation metrics for dialogue response generation. In: Proceedings of the 2016 Conference on Empirical Methods in Natural Language Processing, pp. 2122–2132 (2016)
6. Martinovski, B., Traum, D.: Breakdown in human-machine interaction: the error is the clue. In: Proceedings of the ISCA Tutorial and Research Workshop on Error Handling in Dialogue Systems, pp. 11–16 (2003)
7. Papineni, K., Roukos, S., Ward, T., Zhu, W.J.: BLEU: a method for automatic evaluation of machine translation. In: Proceedings of the 40th Annual Meeting on Association for Computational Linguistics, pp. 311–318. Association for Computational Linguistics (2002)
8. Polzin, T.S., Waibel, A.: Emotion-sensitive human-computer interfaces. In: ISCA Tutorial and Research Workshop (ITRW) on Speech and Emotion (2000)
9. Prendinger, H., Mori, J., Ishizuka, M.: Using human physiology to evaluate subtle expressivity of a virtual quizmaster in a mathematical game. Int. J. Hum. Comput. Stud. **62**(2), 231–245 (2005)
10. Semeniuta, S., Severyn, A., Barth, E.: A hybrid convolutional variational autoencoder for text generation. In: Proceedings of the 2017 Conference on Empirical Methods in Natural Language Processing, pp. 627–637 (2017)
11. Shang, L., Lu, Z., Li, H.: Neural responding machine for short-text conversation. In: Proceedings of the 53rd Annual Meeting of the Association for Computational Linguistics and the 7th International Joint Conference on Natural Language Processing (Volume 1: Long Papers), vol. 1, pp. 1577–1586 (2015)
12. Shen, X., Su, H., Li, Y., Li, W., Niu, S., Zhao, Y., Aizawa, A., Long, G.: A conditional variational framework for dialog generation. In: Proceedings of the 55th Annual Meeting of the Association for Computational Linguistics (Volume 2: Short Papers), vol. 2, pp. 504–509 (2017)

13. Skowron, M.: Affect listeners: acquisition of affective states by means of conversational systems. In: Esposito, A., Campbell, N., Vogel, C., Hussain, A., Nijholt, A. (eds.) Development of Multimodal Interfaces: Active Listening and Synchrony. LNCS, vol. 5967, pp. 169–181. Springer, Heidelberg (2010). https://doi.org/10.1007/978-3-642-12397-9_14

14. Skowron, M., Rank, S., Theunis, M., Sienkiewicz, J.: The good, the bad and the neutral: affective profile in dialog system-user communication. In: D'Mello, S., Graesser, A., Schuller, B., Martin, J.-C. (eds.) ACII 2011. LNCS, vol. 6974, pp. 337–346. Springer, Heidelberg (2011). https://doi.org/10.1007/978-3-642-24600-5_37

15. Sohn, K., Lee, H., Yan, X.: Learning structured output representation using deep conditional generative models. In: Advances in Neural Information Processing Systems, pp. 3483–3491 (2015)

16. Vinyals, O., Le, Q.: A neural conversational model. arXiv preprint arXiv:1506.05869 (2015)

17. Yuan, J., Zhao, H., Zhao, Y., Cong, D., Qin, B., Liu, T.: Babbling - The HIT-SCIR system for emotional conversation generation. In: Huang, X., Jiang, J., Zhao, D., Feng, Y., Hong, Y. (eds.) NLPCC 2017. LNCS (LNAI), vol. 10619, pp. 632–641. Springer, Cham (2018). https://doi.org/10.1007/978-3-319-73618-1_53

18. Zhang, R., Wang, Z., Mai, D.: Building emotional conversation systems using multi-task Seq2Seq learning. In: Huang, X., Jiang, J., Zhao, D., Feng, Y., Hong, Y. (eds.) NLPCC 2017. LNCS (LNAI), vol. 10619, pp. 612–621. Springer, Cham (2018). https://doi.org/10.1007/978-3-319-73618-1_51

19. Zhao, T., Zhao, R., Eskenazi, M.: Learning discourse-level diversity for neural dialog models using conditional variational autoencoders. In: Proceedings of the 55th Annual Meeting of the Association for Computational Linguistics (Volume 1: Long Papers), vol. 1, pp. 654–664 (2017)

20. Zhou, H., Huang, M., Zhang, T., Zhu, X., Liu, B.: Emotional chatting machine: emotional conversation generation with internal and external memory. arXiv preprint arXiv:1704.01074 (2017)

21. Zhu, Q., Zhang, W., Zhou, L., Liu, T.: Learning to start for sequence to sequence architecture. arXiv preprint arXiv:1608.05554 (2016)

Response Selection of Multi-turn Conversation with Deep Neural Networks

Yunli Wang[1], Zhao Yan[2], Zhoujun Li[1(✉)], and Wenhan Chao[1]

[1] Beihang University, Beijing, China
{wangyunli,lizj,chaowenhan}@buaa.edu.cn
[2] Tencent, Beijing, China
zhaoyan@tencent.com

Abstract. This paper describes our method for sub-task 2 of Task 5: multi-turn conversation retrieval, in NLPCC2018. Given a context and some candidate responses, the task is to choose the most reasonable response for the context. It can be regarded as a matching problem. To address this task, we propose a deep neural model named RCMN which focus on modeling relevance consistency of conversations. In addition, we adopt one existing deep learning model which is advanced for multi-turn response selection. And we propose an ensemble strategy for the two models. Experiments show that RCMN has good performance, and ensemble of two models makes good improvement. The official results show that our solution takes 2nd place. We open the source of our code on GitHub, so that other researchers can reproduce easily.

Keywords: Multi-turn conversation · Response selection
Relevance consistency

1 Introduction

The task 5 of NLPCC 2018 focus on how to utilize context to conduct multi-turn human-computer conversations. It contains two sub-tasks: response generation and response retrieval. We signed up the response retrieval task, which is to select the most reasonable response for context from some given candidates. The data set is real multi-turn human-to-human conversations in Chinese, and it is in open domain.

The retrieval task can be regarded as a matching problem, which is to give a matching score between context and response. The challenges of this task are how to make full use of context and response: (1) identify important information in context for response, (2) model the relationship between context and response, (3) model the relationship between utterances in context.

Sometimes the relevance intensity of utterances in different conversations may be quite different, especially for open domain conversations. Contexts with different relevance intensity often have different requirement of relevance between the context and a proper response. So, we think the last one of those challenges

© Springer Nature Switzerland AG 2018
M. Zhang et al. (Eds.): NLPCC 2018, LNAI 11108, pp. 110–119, 2018.
https://doi.org/10.1007/978-3-319-99495-6_10

is the most important for this task. To tackle these challenges, we propose a model named RCMN using the self-matching information in context to add a local relevance threshold for matching. In addition, we also adopt an existing model named SMN [11]. To the best of our knowledge, SMN is state-of-the-art model for multi-turn conversations matching. SMN focus on the relevance between response and each utterances of context and considers the matching information in both word level and sentence level. We finally ensemble the two models to predict. In Sect. 4, we will introduce our method in detail. Experiments in Sect. 5 show that RCMN is a comparable model to SMN. And because the two models have different significant diversity, model ensemble has achieved good improvement. Experiments details will be introduced in Sect. 5.

2   Related Work

As mentioned, this task can be regarded as a matching problem. Matching tasks can be divided into single turn conversation matching task and multi-turn conversation matching task. And the task is closer to multi-turn conversation matching.

For single turn text matching task, there are several notable works: Huang et al. (2013) propose a neural network structure which use word hashing strategy and full-connection feed-forward neural network for text matching [3]. This is an early application of neural networks to text matching. Hu et al. (2014) adopt convolutional neural network for the representation of query and response, propose two neural network structures: ARC-I and ARC-II [2]. Wan et al. (2016) adopt bidirectional LSTM for query and responses representation respectively and explore three different interaction functions for modeling matching signals between query and response [9]. Yang et al. (2016) propose a value shared weight strategy and a question attention network for text matching [13]. Wan et al. (2016) propose 2D-GRU for accumulating matching information in word interaction tensor [10]. Xiong et al. (2017) adopt kernel pooling for dealing with interaction tensor [12].

Recently, researchers begin to pay attention to multi-turn conversations matching. Lowe et al. (2015) concatenated the utterances of context and then treated multi-turn matching as single turn matching [6]. Zhou et al. (2016) propose a multi-view model including an utterance view and a word view to improve multi-turn response selection [14]. Wu et al. (2017) match a response with each utter-ance at first and accumulate matching information instead of sentences by a GRU, thus useful information for matching can be sufficiently retained [11].

3   Problem Formalization

In this task, the training data consists of raw multi-turn conversations. In the testing data, there are 10 candidates for each dialogue session. Among candidates for each session, only one reply is the ground truth while other candidates are randomly sampled from the data sets. We can abstract the retrieval problem as follow:

Assume that we have a data set $D = \{c_i, r_i, y_i\}_{i=1}^N$. c_i represents the context of conversation. Each c_i consists of k_i utterances: $c_i = \{u_1, u_2, ..., u_k\}_i$. r_i represents a candidate response of context. y_i represents whether r_i is ground truth for c_i. $y_i = 1$ means r_i is a proper response for c_i, otherwise $y_i = 0$. Thus, our goal is to learning a model M, for each pair of c_i and r_i, $M(c_i, r_i)$ given a matching degree between them. When $y_i = 1$, $M(c_i, r_i)$ is expected to output a value that as close to 1 as possible, and the situation is opposite when $y_i = 0$. Therefore, we can transform this matching task into a classification task during the training process. And In the prediction process, for each c_i, we use M to measure the matching degree for c_i and all the candidate responses of c_i, then we choose the r_j in candidate responses set $\{r_1, r_2, ..., r_t\}$ with the highest matching degree as the correct response.

Note that testing data has no y_i for sample i. And the training data consists of raw conversation so we should transform it into the form of D. We will introduce how to process the training data in Sect. 5.

4   System Description

As mentioned, our system adopted two deep neural networks and ensemble them for predicting. In this Section, we will first introduce RCMN we proposed in detail. And then we will introduce SMN briefly. Finally, we will introduce how we ensemble the two models.

4.1   Relevance Consistency Matching Network

As mentioned, sometimes the relevance intensity of utterances in different conversations may be quite different. Assume that there are two conversations, one has strong relevance between each two utterances of context denoted as c_1, the other has weak relevance denoted as c_2, the proper responses for c_1 and c_2 are r_1 and r_2 respectively. If there is a model that not consider the relevance intensity in context, the model is likely to predict that $M(c_1, r_1)$ is close to 1 but $M(c_2, r_2)$ is close to 0, but the right output is both 1. To solve this problem in RCMN, we consider to use the self-matching information in context. So, we first use RNN [8] to get sentence level representation of utterances in context and response. Then we let each two utterances in context and response do an interaction, then we can get an interaction tensor. Hence, the interaction tensor contains information of context relevance intensity and relevance between response and each utterance in context. Considering the outstanding performance of convolutional neural network in image processing and pattern recognition, we employ CNN [5] for extracting local features in interaction tensor into high level representation of matching features which is named final matching vector. To transfer the final matching vector into matching score, we adopt a Multilayered perception which can do a nonlinear mapping. The architecture of RCMN is shown in Fig. 1.

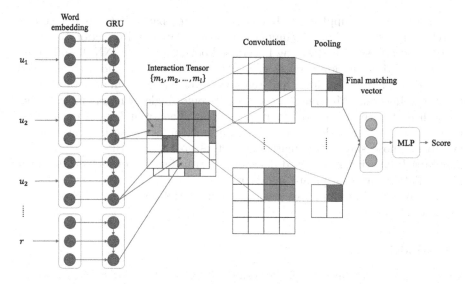

Fig. 1. The architecture of RCMN.

RCMN consists of 4 parts: Sentence RNN Layer, Global Interaction Layer, Convolution and Pooling Layer, Output Layer.

Sentence RNN Layer transfer utterances and response with word embedding representation into a sentence level representation. Each utterance and response will be encoded into one vector. We use the last state of GRU [1] as output. Then the Global Interaction Layer use these vectors to generate interactions of each (u, r) and (u, u) pair. We denote Sentence RNN Layers output as:

$$S = [u_1, u_2, ..., u_k, r] \tag{1}$$

In this layer, we use a series of matrixes denoted as $[w_1, w_2, ..., w_l]$ as interaction weight. For w_i, we can get an interaction matrix m_i, m_i is computed as follow:

$$m_i = S^T w_i S \tag{2}$$

After Global Interaction Layer, we will get l interaction matrixes which can also be called interaction tensor. The interaction tensor is the input of Convolution and Pooling Layer which contains a convolution operation and a max pooling operation. Each m_i will be a channel of convolution. The output of Convolution and Pooling Layer will be concatenated into one vector named final matching vector. Finally, Output Layer which is a Multilayer perceptron use the final matching vector to calculate the final matching score between context and response.

4.2   Sequential Matching Network

For better modeling the relationship of response and each utterance in context, we adopt SMN which consider the matching information in both word level and sentence level as a supplement. It consists of three parts: Utterance-Response Matching Layer, Matching Accumulation Layer, Matching Prediction Layer.

Utterance-Response Matching Layer used both CNN and RNN. It concerned the matching information between each utterance of context and response. And it considered both word level and sentence level matching information. After Utterance-Response Matching Layer, we got some single turn matching vectors for each utterance of context and response pair. Then Matching Accumulation Layer used GRU to model the matching information of the entire conversation, by letting those single turn matching vectors go through this GRU. Finally, Matching Prediction Layer used dense layers transfers the output of Matching Accumulation Layer into matching degree.

Due to space limitation, we will not discuss SMN further more. See [11] for more details.

4.3   Model Ensemble

We use the weighted average of the results of each model as an ensemble result. To get a better ensemble result, weight of each model should be positively related to models performance. We denote the weight of M_i using $train_d$ to train as $w_{m_i}^d$, the data set d as $d = \{train_d, val_d, test_d\}$, the precision of M_i on val_d when using $train_d$ for training as $p_{m_i}^d$. For convenience, we set the $w_{m_i}^d$ as follow:

$$w_{m_i}^d = \frac{p_{m_i}^d}{\sum_j^L (\sum_d^H p_{m_j}^d)} \tag{3}$$

where L means the number of models, H means the number of validation data sets. In this paper, L is equal to 2. We denote the result of M_i on sample q when use $train_d$ for training as $r_{m_i}^d$. Then the ensemble result ER of q can be formalized as follow:

$$ER_q = \frac{\sum_{i=1}^L (\sum_{d=1}^H w_{m_i}^d \cdot r_{m_i}^d)}{L \cdot H} \tag{4}$$

5   Experiments

Our code is available at https://github.com/jimth001/NLPCC2018_Multi_Turn_Response_Selection. It is implemented using python 3.6. Main external packages we used are TensorFlow[1], thulac[2], word2vec[3], NumPy[4].

5.1   Data Sets and Metrics

We used the data sets provided by NLPCC 2018. Table 1 gives the statistics.

---

[1] https://tensorflow.google.cn/

[2] http://thulac.thunlp.org/

[3] https://pypi.org/project/word2vec/

[4] http://www.numpy.org/

Table 1. Statistics of origin data sets.

	Train	Test
Number of contexts	5000K	10K
Positive responses per context	1	1
Negative responses per context	-	9
Average turns of context	3.10	3.10
Average length of utterance	10.18	10.77

Participants are required submit the index of the ground truth they inferred for each conversation. The evaluation metric is precision of retrieved results. We give a formalization description as follows:

$$percision = \frac{\sum_i^N I(\arg\min_j(M(c_i, r_{i_j})) = \arg\min_k(y_{i_k}))}{|D|} \tag{5}$$

$$r_{i_j} \in \{r_1, r_2, ..., r_t\}_i \tag{6}$$

$$y_{i_k} \in \{y_1, y_2, ..., y_t\}_i \tag{7}$$

where $\{r_1, r_2, ..., r_t\}_i$ are candidate responses of c_i, $\{y_1, y_2, ..., y_t\}_i$ are the labels for $\{r_1, r_2, ..., r_t\}_i$. $I(\cdot)$ is an indicator function.

5.2   Experiments on Training Data

In the training and validation process, we shuffle the training data provided by organizer and partition it into 8 folds. We choose one-fold as validation set, one-fold as testing set, and others as training set. And we use random sampling to select 9 negative responses for each conversation in validation and testing set. We denote the validation set, training set and testing set as val_1, $train_1$ and $test_1$ respectively. For further experiments, we use the same method to generate val_2, $train_2$ and $test_2$. Table 2 gives the statistics. We use validation set to select hyper parameters, and use testing set to test the effect of our models.

Table 2. Statistics of data sets divided from training set.

	$train_1$	val_1	$test_1$	$train_2$	val_2	$test_2$
Number of contexts	3750K	625K	625K	3750K	625K	625K
Positive responses per context	1	1	1	1	1	1
Negative responses per context	-	9	9	-	9	9
Average turns of context	3.10	3.09	3.10	3.10	3.10	3.09
Average length of utterance	10.59	10.61	10.57	10.59	10.58	10.59

Because the training data only has raw conversations, which means that there is no (c, r, y) pairs in the data. So, we manually construct these pairs using the following strategy: For each conversation in training data, take the last turn as r, other as c, and the y is 1. In the training process, for each c, random sampling 2 sentences in the whole data set as r, and the corresponding y is 0.

Our preprocessing of text is simple. We selected thulac as our tokenizer and removed stop words using default stop words list in thulac. We adopted word2vec [7] to generate word embedding. We set main parameters of thulac as follows: user_dict is None, T2S is True, seg_only is True, filt is True. And main parameters of word2vec are set as follows: size is 200, window is 8, sample is 1e-5, cbow is 0, min_count is 6. Parameters not mentioned are set as default value. Only $train_1$ is used for pre-train word embedding, and all of our experiments is based on it. The word embedding is not trainable in both models.

Finally, we set some same parameters for both two models: the maximum context length is 10, the utterance length is 50. We padded zeros if context length and utterance length is less than 10 and 50 respectively. If context length and utterance length is more than 10 and 50, we chose the last 10 and 50 respectively. The parameters were updated by stochastic gradient descent with Adam algorithm [4] on a single 1080Ti GPU. The initial learning rate is 0.001, and the parameters of Adam, $\beta 1$ and $\beta 2$ are 0.9 and 0.999 respectively. We adopted cross entropy loss for training.

We finally set SMN parameters following [11]: the dimensionality of the hidden states of GRU in Utterance-Response Matching Layer is 200, window size of convolution and pooling is $(3, 3)$, the dimensionality of the hidden states of GRU in Matching Accumulation Layer is 50.

We finally set parameters of RCMN as follows: the dimensionality of the hidden states of GRU is 200, the shape of interaction weight is $(200, 2, 200)$, window size of convolution and pooling is $(3, 3)$, the number of filters is 8, the dimensionality of the hidden layer of MLP is 50.

The experiments result based on training data released by organizer is shown in Table 3.

5.3   Experiment Result on Testing Data and Analysis

We used $train_1$ and $train_2$ to train RCMN and SMN. And ensemble these models using the method introduced in Sect. 4. The official ranking results on test data sets are summarized in Table 4.

We conducted ablation experiments on official test data to examine the usefulness of models and training data sets. Experiment result is shown in Table 5. We can see that each model and data set contribute to the final result. The contribution of $train_1$ is more than the contribution of $train_2$. The contribution of RCMN is more than the contribution of SMN. Result of single model on official test data is shown in Table 6.

Experiments shown in Tables 3 and 5 show that RCMN is a comparable model to state-of-the-art model for multi-turn conversation response selection. Table 3 also shows that RCMN and SMN are insensitive to data not seen on this data set, because that the performance has not significant improvement when training data and validation data have overlap. For example, $train_1$ and val_1

Table 3. Experiments result on training data.

Training set	Model name	Testing set	Precision
$train_1$	SMN	val_1	60.10
$train_2$	SMN	val_1	58.80
$train_1$	RCMN	val_1	59.00
$train_2$	RCMN	val_1	59.56
$train_1$	RCMN+SMN	val_1	61.91
$train_2$	RCMN+SMN	val_1	62.58
$train_1$ and $train_2$	RCMN+SMN	val_1	63.48
$train_1$	SMN	val_2	60.64
$train_2$	SMN	val_2	58.96
$train_1$	RCMN	val_2	59.81
$train_2$	RCMN	val_2	58.91
$train_1$	RCMN+SMN	val_2	62.66
$train_2$	RCMN+SMN	val_2	62.27
$train_1$ and $train_2$	RCMN+SMN	val_2	63.64
$train_1$	SMN	$test_1$	60.17
$train_2$	SMN	$test_1$	58.81
$train_1$	RCMN	$test_1$	59.08
$train_2$	RCMN	$test_1$	59.67
$train_1$	RCMN+SMN	$test_1$	62.02
$train_2$	RCMN+SMN	$test_1$	62.69
$train_1$ and $train_2$	RCMN+SMN	$test_1$	63.55
$train_1$	SMN	$test_2$	60.69
$train_2$	SMN	$test_2$	58.86
$train_1$	RCMN	$test_2$	59.82
$train_2$	RCMN	$test_2$	58.92
$train_1$	RCMN+SMN	$test_2$	62.72
$train_2$	RCMN+SMN	$test_2$	62.23
$train_1$ and $train_2$	RCMN+SMN	$test_2$	63.69

Table 4. Official results on test data.

System name	Precision
ECNU	62.61
wyl_buaa	59.03
Yiwise-DS	26.68
laiye_rocket	18.13
ELCU_NLP	10.54

Table 5. Evaluation results of model ablation.

Models and data sets	Precision	Loss of performance
All	59.03	-
– SMN	57.51	1.52
– SMN on $train_1$	58.84	0.19
– SMN on $train_2$	58.86	0.17
– RCMN	56.41	2.61
– RCMN on $train_1$	58.22	0.81
– RCMN on $train_2$	58.68	0.35
– $train_1$	57.44	1.59
– $train_2$	58.33	0.70

Table 6. Single model on official test data.

Training set	Model name	Precision
$train_1$	SMN	56.11
$train_2$	SMN	53.24
$train_1$	RCMN	55.92
$train_2$	RCMN	55.58

have no overlap but $train_1$ and val_2 have. Model Ablation shows that RCMN slightly outperform SMN on official testing data. We think that RCMN does not obviously outperform SMN on NLPCC 2018 data set because RCMN does not consider the matching information in word level and the relevance intensity of conversations in the data set is not quite different. Table 6 shows that SMN using $train_2$ for training is weaker than RCMN using $train_1$ or $train_2$ for training on official testing data. This is probably because $train_2$ is more different on relevance intensity of conversations from official testing data. If so, this is also a proof of that RCMN can adapt to different conversations having different relevance intensity. Ensemble result shows that model ensemble brings good improvement. We think its because of the significant diversity of the two models.

6   Conclusion

In this paper, we proposed RCMN which considered self-matching information in context for response selection. RCMN focus on better modeling the relevance consistency of conversations but lacks in considering word level matching information. We also employed an existing model named SMN. Experiments in Sect. 5 show that RCMN is a comparable model to SMN, which is regarded as state-of-the-art model for multi-turn conversation response selection. We also proposed an ensemble strategy for the two models. And experiments show that ensemble of models makes good improvement. We have given some reasonable analysis for experiments result in Sect. 5. The official results show that our solution takes 2nd place. In feature work, we will pay more attention to capture more matching information in single turn matching and to model relevance intensity more effectively.

Acknowledgments. This work was supported in part by the Natural Science Foundation of China (Grand Nos. U61672081, 1636211, 61370126), and Beijing Advanced Innovation Center for Imaging Technology (No. BAICIT-2016001) and National Key R&D Program of China (No. 2016QY04W0802).

References

1. Chung, J., Gulcehre, C., Cho, K.H., Bengio, Y.: Empirical evaluation of gated recurrent neural networks on sequence modeling. Eprint Arxiv (2014)
2. Hu, B., Lu, Z., Li, H., Chen, Q.: Convolutional neural network architectures for matching natural language sentences. Adv. Neural Inf. Process. Syst. **3**, 2042–2050 (2015)
3. Huang, P.S., He, X., Gao, J., Deng, L., Acero, A., Heck, L.: Learning deep structured semantic models for web search using clickthrough data. In: ACM International Conference on Conference on Information & Knowledge Management, pp. 2333–2338 (2013)
4. Kingma, D.P., Ba, J.: Adam: a method for stochastic optimization. Computer Science (2014)
5. Lecun, Y., et al.: Backpropagation applied to handwritten zip code recognition. Neural Comput. **1**(4), 541–551 (2014)
6. Lowe, R., Pow, N., Serban, I., Pineau, J.: The ubuntu dialogue corpus: a large dataset for research in unstructured multi-turn dialogue systems. Computer Science (2015)
7. Mikolov, T., Sutskever, I., Chen, K., Corrado, G., Dean, J.: Distributed representations of words and phrases and their compositionality. Adv. Neural Inf. Process. Syst. **26**, 3111–3119 (2013)
8. Rumerlhar, D.E.: Learning representation by back-propagating errors. Nature **323**(3), 533–536 (1986)
9. Wan, S., Lan, Y., Guo, J., Xu, J., Pang, L., Cheng, X.: A deep architecture for semantic matching with multiple positional sentence representations, pp. 2835–2841 (2015)
10. Wan, S., Lan, Y., Xu, J., Guo, J., Pang, L., Cheng, X.: Match-SRNN: modeling the recursive matching structure with spatial RNN. Comput. Graph. **28**(5), 731–745 (2016)
11. Wu, Y., et al.: Sequential matching network: a new architecture for multi-turn response selection in retrieval-based chatbots. In: Meeting of the Association for Computational Linguistics, pp. 496–505 (2017)
12. Xiong, C., Dai, Z., Callan, J., Power, R., Power, R.: End-to-end neural ad-hoc ranking with kernel pooling, pp. 55–64 (2017)
13. Yang, L., Ai, Q., Guo, J., Croft, W.B.: aNMM: ranking short answer texts with attention-based neural matching model. In: ACM International on Conference on Information and Knowledge Management, pp. 287–296 (2016)
14. Zhou, X., et al.: Multi-view response selection for human-computer conversation. In: Conference on Empirical Methods in Natural Language Processing, pp. 372–381 (2016)

Learning Dialogue History for Spoken Language Understanding

Xiaodong Zhang, Dehong Ma, and Houfeng Wang[(✉)]

Institute of Computational Linguistics, Peking University, Beijing 100871, China
{zxdcs,madehong,wanghf}@pku.edu.cn

Abstract. In task-oriented dialogue systems, spoken language understanding (SLU) aims to convert users' queries expressed by natural language to structured representations. SLU usually consists of two parts, namely intent identification and slot filling. Although many methods have been proposed for SLU, these methods generally process each utterance individually, which loses context information in dialogues. In this paper, we propose a hierarchical LSTM based model for SLU. The dialogue history is memorized by a turn-level LSTM and it is used to assist the prediction of intent and slot tags. Consequently, the understanding of the current turn is dependent on the preceding turns. We conduct experiments on the NLPCC 2018 Shared Task 4 dataset. The results demonstrate that the dialogue history is effective for SLU and our model outperforms all baselines.

Keywords: Spoken language understanding · Dialogue history
Hierarchical LSTM

1 Introduction

In recent years, task-oriented dialogue systems have a rapid development and widespread application, e.g., voice assistant in mobiles and intelligent customer service. Spoken language understanding (SLU) plays an import role in task-oriented dialogue systems. An utterance of a user is often first transcribed to text by an automatic speech recognizer (ASR). The SLU then interprets the meaning of the utterance and convert the unstructured text to structured representations. The result of SLU is passed to dialogue management module to update dialogue state and make dialogue policy. Therefore, the performance of SLU is critical to a task-oriented dialogue system.

SLU usually consists of two parts, namely intent identification and slot filling. Intent identification can be viewed as an utterance classification problem, and slot filling can be viewed as a sequence labeling problem. Take the utterance in Table 1 as an example. The intent of the utterance is *navigation*. There are two slots in the utterance, i.e., the origin *the Forbidden City* and the destination *Peking University*. With the IOB (Inside, Outside, Beginning) annotation

© Springer Nature Switzerland AG 2018
M. Zhang et al. (Eds.): NLPCC 2018, LNAI 11108, pp. 120–132, 2018.
https://doi.org/10.1007/978-3-319-99495-6_11

Table 1. An example utterance for SLU.

Utterance	Go	to	Peking	University	from	the	Forbidden	City
Slots	O	O	B-dest	I-dest	O	B-orig	I-orig	I-orig
Intent	Navigation							

method, the slot labels for each word are listed in Table 1. The category of intent and slot is usually defined by domain experts.

For intent identification, lots of classifiers have been used by previous work, including support vector machines (SVM) [5], Adaboost [22], and convolutional neural networks (CNN) [24]. For slot filling, the popular methods include maximum entropy Markov models (MEMM) [12], conditional random fields (CRF) [17], and recurrent neural networks (RNN) [13]. In consideration of intent identification and slot filling are correlative, recent works have focused on the joint model of the two tasks [4,10,26].

Previous work mainly processes each utterance individually, which does not make use of dialogue context. In task-oriented dialogue systems, dialogue history is important to the understanding of the current utterance. As is shown in Table 2, the third turns in Dialogue 1 and 2 are both "Cancel". If only looking at this turn, it is impossible to identify whether the intent is to *cancel navigation* or *cancel phone call*. The same problem occurs in slot filling. If only looking at the second turns in Dialogue 3 and 4, it is hard to tell whether "Washington" is a name of a place or a person. Fortunately, with the preceding turns, it is possible to make the correct predictions. Therefore, how to model and use dialogue history for SLU is worth of research.

Table 2. Some sample dialogues.

ID	Dialogue
1	A: Go to Peking University
	B: Planning route to Peking University
	A: Cancel
2	A: Call Tom
	B: Calling Tom for you
	A: Cancel
3	A: Where are you going?
	B: Washington
4	A: Who would you like to call?
	B: Washington

For modeling dialogue history, Serban et al. propose a hierarchical recurrent encoder-decoder (HRED) model, which uses a high-level context RNN to keep track of past utterances [19]. However, HRED is only designed for non-task-oriented conversational dialogue systems. Following HRED, we propose a hierarchical long short-term memory based model for SLU (HLSTM-SLU). In HLSTM-SLU, a low-level LSTM is used to learn inner-turn representations. Max-pooling is used to obtain a fixed-length vector of a turn. A high-level LSTM reads the vectors iteratively and memorize useful dialogue history. The contextual representation of a turn is used in two aspects. On one hand, it is utilized to make prediction for the intent of the current turn. One the other hand, it is concatenated with low-level representation to predict the slot label for each word. Consequently, the dialogue history is available for both intent identification and slot filling and contributes to the two tasks.

The rest of the paper is organized as follows. The related work is surveyed in Sect. 2. In Sect. 3, we introduce the proposed HLSTM-SLU model. Section 4 discusses the experimental setup and results on the NLPCC dataset. The conclusion is given in Sect. 5.

2   Related Work

The study of SLU surged in the 1990s with the Air Travel Information System (ATIS) project [16]. The early systems are basically rule-based [23]. Developers are required to write quantities of syntactic or semantic grammars. Due to the informality of spoken language and ASR errors, rule-based methods have run into the bottleneck and data-driven statistical approaches have become the mainstream.

For intent identification, word n-grams are typically used as features with generic entities, such as dates, locations. Many classifiers have been used by prior work, such as SVM [5] and Adaboost [22]. Some researchers tried to use syntactic information. Hakkani et al. presented an approach populating heterogeneous features from syntactic and semantic graphs of utterance for call classification [6]. Tur et al. proposed a dependency parsing based sentence simplification approach that extracts a set of keywords and uses those in addition to entire utterances for completing SLU tasks [22]. For slot filling, Moschitti et al. employed syntactic features via syntactic tree kernels with SVM [15]. Many works used CRF because it is a well-performed model for sequence labelling [17]. Jeong and Lee proposed triangular CRF, which coupled an additional random variable for intent on top of a standard CRF [8]. Mairesse et al. presented an efficient technique that learned discriminative semantic concept classifiers whose output was used to recursively construct a semantic tree, resulting in both slot and intent labels [11].

Recently, various deep learning models have been explored in SLU. The initial try is deep belief networks (DBN), which have been used in call routing classification [18] and slot filling [3]. Tur et al. used deep convex networks (DCN) for domain classification and produced higher accuracy than a boosting-based classifier [21]. RNN has shown excellent performance on slot filling and outperformed CRF [13,14]. Yao et al. improved RNN by using transition features and the sequence-level optimization criterion of CRF to explicitly model dependencies of output labels [25]. Many joint models are proposed to leverage the correlation of the two tasks. Xu and Sarikaya improved the triangular CRF model by using convolutional neural networks (CNN) to extract features automatically [24]. Zhang and Wang used representations learned by Grated Recurrent Unit (GRU) to predict slot labels and used a vector obtained by max-pooling of these representations to predict intent labels [26]. Liu and Lane proposed a conditional RNN model that can be used to jointly perform SLU and language modeling [10].

To model dialogue history, Serban et al. propose a HRED model, which uses a high-level context RNN to keep track of past utterances [19]. Furthermore, Serban et al. extended HRED by introducing Variational Autoencoder (VAE) and proposed VHRED model [20]. VHRED can generate more meaningful and diverse responses. HRED and VHRED are both used in conversational dialogue systems, while our propose HLSTM-SLU is designed for task-oriented dialogue systems.

3   The Proposed Method

In this section, we describe the HLSTM-SLU model in detail. Due to the dataset used in the paper is a Chinese dataset, we use Chinese dialogues as the input to introduce our model. The model can be applied to English dialogues with minor modifications. We consider a dialogue $D = \{T_1, \cdots, T_m\}$ as a sequence of m turns. Each T_i is a sequence of n Chinese characters, i.e., $T_i = \{c_{i,1}, \cdots, c_{i,n}\}$. The outputs of the model for T_i are intent label $\hat{y}_i^T$ and slot labels $\hat{y}_i^c = \{\hat{y}_{i,1}^c, \cdots, \hat{y}_{i,n}^c\}$. An overview of the model is illustrated in Fig. 1.

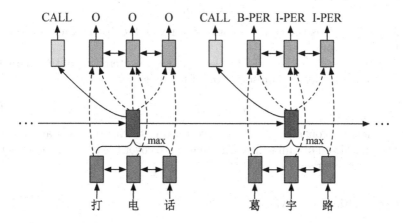

Fig. 1. The structure of HLSTM-SLU. (Color figure online)

3.1 Inputs

HLSTM-SLU processes Chinese text at the character level. The input of the model contains two parts, i.e., characters and additional features. The jth character in the ith turn $c_{i,j}$ is mapped to a low-dimensional dense vector $e_{i,j} \in \mathbb{R}^{d_e}$ via a lookup table, where d_e is the dimension of character embeddings. The additional feature of $c_{i,j}$ is a 0–1 sparse vector $a_{i,j} \in \mathbb{R}^{d_a}$, where d_a is the total number of features. We use two kinds of features, namely part-of-speech and domain lexicons. Domain lexicons contain collected words for several domains, e.g., singers, songs, and so on.

Table 3 lists feature vectors of an example utterance " 小苹果 " (little apple). Feature "B-A" and "I-A" denote the beginning and following characters of adjectives respectively. Similarly, "B-N" and "I-N" are the beginning and following characters of nouns, and "B-song" and "I-song" are the matching result of the song lexicon. The omitted features are all 0.

Table 3. An example of additional feature vectors.

		B-A	I-A	B-N	I-N		B-song	I-song	
小	...	1	0	0	0	...	1	0	...
苹	...	0	0	1	0	...	0	1	...
果	...	0	0	0	1	...	0	1	...

The sparse feature vector is convert to a low-dimensional dense vector via a linear transformation. Formally, the dense feature vector $\overline{a}_{i,j}$ for $a_{i,j}$ is computed by

$$\overline{a}_{i,j} = W_a a_{i,j} \tag{1}$$

where $W_a \in \mathbb{R}^{d_{\overline{a}} \times d_a}$ is the transformation matrix and $d_{\overline{a}}$ is the dimension of the dense feature vector.

The input $x_{i,j} \in \mathbb{R}^{d_e + d_{\overline{a}}}$ for LSTM layer is the concatenation of the character embedding and the dense representation of features, i.e.,

$$x_{i,j} = [c_{i,j}; \overline{a}_{i,j}] \tag{2}$$

where $[\cdot; \cdot]$ denotes concatenation of two vectors. With this representation as the input, the LSTM can obtain lexical information to cover the shortage of sequence modeling at the character level.

3.2 Hierarchical LSTM

RNN is a family of neural network that can process variable-length sequences. It uses a recurrent hidden state to take into account the influence of past states. Concretely, it takes a sequence of vectors $X = \{x_1, \cdots, x_n\}$ as input and outputs a sequence $H = \{h_1, \cdots, h_n\}$ that is the representation of the sequence at each time step.

It was hard to for the vanilla RNN to capture long-term dependencies because the gradients tend to vanish or explode [1]. Some more sophisticated activation functions with gating units were designed, among which the most representative variant is long short-term memory (LSTM) [7]. LSTM uses several gates to control the proportion of the input and the proportion of the previous state to forget. Formally, the computation of LSTM is as follows:

$$i_t = \sigma(W_i x_t + U_i h_{t-1} + b_i) \tag{3}$$

$$f_t = \sigma(W_f x_t + U_f h_{t-1} + b_f) \tag{4}$$

$$o_t = \sigma(W_o x_t + U_o h_{t-1} + b_o) \tag{5}$$

$$\hat{c}_t = tanh(W_c x_t + U_c h_{t-1} + b_c) \tag{6}$$

$$c_t = f_t \odot c_{t-1} + i_t \odot \hat{c}_t \tag{7}$$

$$h_t = o_t \odot tanh(c_t) \tag{8}$$

where i_t, f_t, o_t are input gate, forget gate and output gate respectively, σ is a sigmoid function, W_i, W_f, W_o, W_c, U_i, U_f, U_o, U_c are weight matrices, b_i, b_f, b_c are biases, and x_t is the input at the time step t.

A bidirectional LSTM consists of a forward and a backward LSTM. The forward LSTM reads the input sequence as it is ordered and calculates forward hidden states ($\overrightarrow{h}_1, ..., \overrightarrow{h}_n$). The backward LSTM reads the sequence in the reserve order and calculates backward hidden states ($\overleftarrow{h}_1, ..., \overleftarrow{h}_n$). The bidirectional hidden state is obtained by concatenating the forward and backward hidden state, i.e.,

$$h_t = [\overrightarrow{h}_t; \overleftarrow{h}_t] \tag{9}$$

A dialogue can be considered as a sequence at two levels, i.e., a sequence of characters for each turn and a sequence of turns. Intuitively, HLSTM-SLU models a dialogue with two LSTMs: one at the character level and one at the turn level. The character-level LSTM is bidirectional. The hidden states $h_{i,j}^c$ for the jth character in the ith turn is calculated by

$$h_{i,j}^c = BLSTM(x_{i,j}, h_{i,j-1}^c, h_{i,j+1}^c) \tag{10}$$

where $BLSTM(x_t, h_{t-1}, h_{t+1})$ is an abbreviation for Eq. (9), and $x_{i,j}$ is as described in Eq. (2). The character-level LSTM is illustrated by the red part in Fig. 1.

Then, a fixed-length representation for the ith turn is obtained by the max-pooling of all hidden states in the turn, i.e.,

$$r_i^T = \max_{k=1}^{n} h_{i,k}^c \tag{11}$$

where the max function is an element-wise function, and n is the number of characters in the turn.

The turn-level LSTM is unidirectional. This is because only preceding turns are available in an online system. Therefore, the representation of the ith turn with dialogue history r_i^D is calculated by

$$r_i^D = LSTM(r_i^T, r_{i-1}^D) \tag{12}$$

where $LSTM(x_t, h_{t-1})$ is an abbreviation for Eqs. (3)–(8). The turn-level LSTM is illustrated by the blue part in Fig. 1. In this way, r_i^D can learn important information from preceding turns, which is useful for making predictions at the current turn.

3.3   Intent and Slot Tagger

The r_i^D is used for predicting intent labels, which is illustrated by the yellow part in Fig. 1. Instead of making individual prediction for each turn, we select the best label sequence of all turns with a CRF, which has demonstrated good performance on many sequence labeling tasks [2,9]. The CRF layer is not illustrated in Fig. 1 for simplification. Concretely, for a dialogue $D = \{T_1, \cdots, T_m\}$, the unnormalized score for intent tags at T_i is calculated by a fully-connected network with r_i^D as input, i.e.,

$$s_i^T = W_s^T r_i^D + b_s^T \tag{13}$$

where W_s^T is a weight matrix, b_s^T is a bias vector, s_i^T is the score vector and $s_{i,k}^T$ is the score of the kth intent tag for the ith turn.

For a sequence of tags $y^T = \{y_1^T, \cdots, y_m^T\}$, the score of the sequence is

$$S(y^T) = \sum_{i=1}^{m} s_{i,y_i^T}^T + \sum_{i=1}^{m-1} A_{y_i^T, y_{i+1}^T} \tag{14}$$

where A is a matrix of transition scores for intent tags and $A_{j,k}$ represents the transition score from the tag j to tag k.

The probability for the sequence y^T is computed by a softmax over all possible tag sequences, i.e.,

$$p(y^T) = \frac{e^{S(y^T)}}{\sum_{\tilde{y}^T \in Y^T} e^{S(\tilde{y}^T)}} \tag{15}$$

where Y^T is all possible intent tag sequences.

During training, the objective is to maximize the log-probability of the correct tag sequence. During decoding, the predicted intent tag sequence is the one with the maximum score:

$$\hat{y}^T = \underset{\tilde{y}^T \in Y^T}{\operatorname{argmax}} S(\tilde{y}^T) \tag{16}$$

Although all possible sequences are enumerated in Eqs. (15) and (16), they can be computed efficiently using dynamic programming.

As for slot filling, to make use of the dialogue history, the model rereads a turn with the contextual turn representation, which is illustrated by the green part in Fig. 1. Concretely, for a turn T_i, the hidden states of the character-level LSTM

$h_{i,j}^c$ and the turn representation with dialogue history r_i^D, another bidirectional LSTM is used to learn the character representation with dialogue history $h_{i,j}^d$. Formally, we have

$$h_{i,j}^d = BLSTM(x_{i,j}^d, h_{i,j-1}^d, h_{i,j+1}^d) \qquad (17)$$

$$x_{i,j}^d = [h_{i,j}^c; r_i^D] \qquad (18)$$

We also use the CRF layer for predict slot tags for each turn. The calculation is similar to Eqs. (13)–(16) and we describe it briefly. For each turn T_i, we have

$$s_{i,j}^c = W_s^c h_{i,j}^d + b_s^c \qquad (19)$$

$$S(y_i^c) = \sum_{j=1}^{n} s_{i,j,y_{i,j}^c}^c + \sum_{i=1}^{n-1} B_{y_{i,j}^c, y_{i,j+1}^c} \qquad (20)$$

$$p(y_i^c) = \frac{e^{S(y_i^c)}}{\sum_{\tilde{y}_i^c \in Y_i^c} e^{S(\tilde{y}_i^c)}} \qquad (21)$$

$$\hat{y}_i^c = \underset{\tilde{y}_i^c \in Y_i^c}{\operatorname{argmax}} S(\tilde{y}_i^c) \qquad (22)$$

The loss function of the model is the sum of negative log-probability of the correct tag sequence for both intent and slot, i.e.,

$$L = -\frac{1}{m}\left(\log(p(y^T)) + \sum_{i=1}^{m} \log(p(y_i^c))\right) \qquad (23)$$

4   Experiments

4.1   Dataset

We use NLPCC 2018 Shared Task 4 dataset[1] in our experiment. The dataset is a sample of the real query log from a Chinese commercial task-oriented dialog system. It includes three domains, namely *music, navigation* and *phone call,* and an additional domain *OTHERS*, which is the data not covered by the three domains. The dataset contains 4705 dialogues with 21352 turns in the training set and 1177 dialogues with 5350 turns in the testing set. The statistics of the intents and slots in the dataset is listed in Table 4.

We use two metrics given by the task organizer. The first metric is macro F1 of intents and the second one is the precision of both intents and slots. The detailed equations can be found at the guideline[2].

---

[1] http://tcci.ccf.org.cn/conference/2018/taskdata.php.
[2] http://tcci.ccf.org.cn/conference/2018/dldoc/taskgline04.pdf.

Table 4. The number of intents and slots in the NLPCC dataset.

Intent	Train	Test	Slot	Train	Test
music.play	6425	1641	Song	3941	983
			Singer	1745	473
			Theme	191	45
			Style	69	26
			Age	48	15
			Toplist	44	16
			Emotion	38	2
			Language	29	2
			Instrument	14	7
			Scene	7	0
music.pause	300	75	-		
music.prev	5	4	-		
music.next	132	34	-		
navigation.navigation	3961	1039	Destination	3805	1014
			custom_destination	132	17
			Origin	14	9
navigation.open	245	56	-		
navigation.start_navigation	33	4	-		
navigation.cancel_navigation	835	206	-		
phone_call.make_a_phone_call	2796	674	phone_num	1100	256
			contact_name	779	224
phone_call.cancel	22	18	-		
OTHERS	6598	1599	-		
Total	21352	5350		11956	3089

4.2 Experimental Setup

We compare our model with the following baselines:

- **BLSTM** A bidirectional LSTM model with only character embeddings as input.
- **BLSTM-FT:** A bidirectional LSTM model with character embeddings and additional features as input.
- **BLSTM-CRF:** Compared to BLSTM-FT, it adds a CRF layer for decoding slot tag sequences.
- **HLSTM-Intent:** The difference between HLSTM-Intent and the proposed HLSTM-SLU is that it only uses dialogue history for intent identification, but not for slot filling.

We only use the official provided data for training without using additional data. We shuffle the data randomly at dialogue granularity and set apart

10% dialogues as development set. All hyper-parameters are tuned on the development set. Jieba[3] is used for part-of-speech tagging. The neural networks are implemented using TensorFlow. The dimension of character embeddings and dense feature vector are both 100. The character embeddings are pretrained by GloVe[4] with a large unlabeled Chinese corpus. The dimension of character-level LSTM is 300 (each direction is 150), and the dimension of turn-level LSTM is 300. Models are trained using Adam optimization method using the learning rate set to 0.001.

As shown in Table 4, the data is imbalanced for different intents, which does harm to the performance of the model. We solve the problem from two aspects. First, we use over sampling method. The dialogues that contains the scarce intents are copied for many times. Second, we write some simple rules to identify some intents.

To correct ASR errors, we perform error corrections for song slot using the song lexicon in the dataset. If a song predicted by the model is not found in the lexicon, we try to find a song from the lexicon that has the same pinyin with the predicted song. If still not found, we try to find a song of which the Levenshtein distance with the predicted song is equal to 1. If multiple candidates exist, we randomly select one.

4.3 Experimental Results

The results are demonstrated in Table 5. We can see that the BLSTM preforms worst on both metrics. With additional features, the BLSTM-FT outperforms BLSTM by a large margin. This is because the part-of-speech and lexicon matching information is useful for predicting slot tags. The BLSTM-CRF improves the results further. The reason is that the CRF layer model the transition of tags explicitly, which reduces the situation of incomplete or redundant slot values and illegal tag sequences. HLSTM-Intent uses dialogue history for intent identification, which improves the score of intent significantly. Our HLSTM-SLU model uses dialogue history not only for intent identification but also for slot filling and obtains the best performance on the two metrics. The results support our argument that the dialogue history is important for the two tasks in SLU.

4.4 Case Study

We give the result of three models for an example dialogue to show the difference of the models intuitively. The dialogue is selected from the testing set. It contains two turns, as follows.

- **Turn 1:** 蓝牙打电话 (Make a call with Bluetooth.)
- **Turn 2:** 张解 (Zhang Jie.)

---

[3] https://github.com/fxsjy/jieba.
[4] https://nlp.stanford.edu/projects/glove/.

Table 5. Results of HLSTM-SLU and baselines.

Method	F1 for Intent	Precision for intent & Slot
BLSTM	88.56	83.38
BLSTM-FT	89.24	86.95
BLSTM-CRF	89.31	88.62
HLSTM-Intent	93.61	90.21
HLSTM-SLU	94.19	90.84

The results of three models are demonstrated in Table 6. For BLSTM-CRF, the Turn 2 is misclassified as *OTHERS*. It is because the BLSTM-CRF process each turn individually. The obvious clue of making a phone call in Turn 1 is not utilized when processing Turn 2. The HLSTM-Intent uses dialogue history for intent prediction and therefore it predicts the intent of Turn 2 correctly. However, it does not identify the contact name. The HLSTM-SLU uses dialogue history for both tasks and the prediction is totally correct.

Table 6. The comparative result with baselines on an example dialogue.

Model	Dialogue	Intent	Slot
BLSTM-CRF	Turn 1	phone_call.make_a_phone_call	-
	Turn 2	OTHERS	-
HLSTM-Intent	Turn 1	phone_call.make_a_phone_call	-
	Turn 2	phone_call.make_a_phone_call	-
HLSTM-SLU	Turn 1	phone_call.make_a_phone_call	-
	Turn 2	phone_call.make_a_phone_call	contact_name = 张解

5   Conclusion

Dialogue history provides import information for SLU in dialogues. In this paper, we propose a HLSTM-SLU model to represent dialogue history and use it for SLU. The HLSTM-SLU uses a character-level LSTM to learn inner-turn representation. Then, the dialogue history is memorized by a turn-level LSTM and it is used to assist the prediction of intent and slot tags. The experimental results demonstrate that the dialogue history is effective for SLU and our model outperforms all baselines. In future work, we will test HLSTM-SLU on English datasets to investigate the generalization of the model.

Acknowledgments. Our work is supported by National Natural Science Foundation of China (No. 61433015).

References

1. Bengio, Y., Simard, P., Frasconi, P.: Learning long-term dependencies with gradient descent is difficult. IEEE Trans. Neural Netw. **5**(2), 157–166 (1994)
2. Collobert, R., Weston, J., Bottou, L., Karlen, M., Kavukcuoglu, K., Kuksa, P.: Natural language processing (almost) from scratch. J. Mach. Learn. Res. **12**(Aug), 2493–2537 (2011)
3. Deoras, A., Sarikaya, R.: Deep belief network based semantic taggers for spoken language understanding. In: INTERSPEECH, pp. 2713–2717 (2013)
4. Guo, D., Tur, G., Yih, W.t., Zweig, G.: Joint semantic utterance classification and slot filling with recursive neural networks. In: SLT, pp. 554–559. IEEE (2014)
5. Haffner, P., Tur, G., Wright, J.H.: Optimizing SVMs for complex call classification. In: ICASSP, vol. 1, pp. 632–635. IEEE (2003)
6. Hakkani-Tür, D., Tur, G., Chotimongkol, A.: Using syntactic and semantic graphs for call classification. In: Proceedings of the ACL Workshop on Feature Engineering for Machine Learning in Natural Language Processing (2005)
7. Hochreiter, S., Schmidhuber, J.: Long short-term memory. Neural Comput. **9**(8), 1735–1780 (1997)
8. Jeong, M., Lee, G.G.: Triangular-chain conditional random fields. IEEE Trans. Audio Speech Lang. Process. **16**(7), 1287–1302 (2008)
9. Lample, G., Ballesteros, M., Subramanian, S., Kawakami, K., Dyer, C.: Neural architectures for named entity recognition. In: NAACL-HLT, pp. 260–270 (2016)
10. Liu, B., Lane, I.: Joint online spoken language understanding and language modeling with recurrent neural networks. In: SIGDIAL, p. 22 (2016)
11. Mairesse, F., et al.: Spoken language understanding from unaligned data using discriminative classification models. In: ICASSP, pp. 4749–4752. IEEE (2009)
12. McCallum, A., Freitag, D., Pereira, F.C.: Maximum entropy Markov models for information extraction and segmentation. In: ICML, vol. 17, pp. 591–598 (2000)
13. Mensil, G., et al.: Using recurrent neural networks for slot filling in spoken language understanding. IEEE/ACM Trans. Audio Speech Lang. Process. **23**(3), 530–539 (2015)
14. Mesnil, G., He, X., Deng, L., Bengio, Y.: Investigation of recurrent-neural-network architectures and learning methods for spoken language understanding. In: INTERSPEECH, pp. 3771–3775 (2013)
15. Moschitti, A., Riccardi, G., Raymond, C.: Spoken language understanding with Kernels for syntactic/semantic structures. In: ASRU, pp. 183–188. IEEE (2007)
16. Price, P.J.: Evaluation of spoken language systems: the ATIS domain. In: Speech and Natural Language (1990)
17. Raymond, C., Riccardi, G.: Generative and discriminative algorithms for spoken language understanding. In: Eighth Annual Conference of the International Speech Communication Association (2007)
18. Sarikaya, R., Hinton, G.E., Ramabhadran, B.: Deep belief nets for natural language call-routing. In: ICASSP, pp. 5680–5683. IEEE (2011)
19. Serban, I.V., Sordoni, A., Bengio, Y., Courville, A.C., Pineau, J.: Building end-to-end dialogue systems using generative hierarchical neural network models. AAAI **16**, 3776–3784 (2016)
20. Serban, I.V., et al.: A hierarchical latent variable encoder-decoder model for generating dialogues. In: AAAI, pp. 3295–3301 (2017)
21. Tur, G., Deng, L., Hakkani-Tür, D., He, X.: Towards deeper understanding: deep convex networks for semantic utterance classification. In: ICASSP, pp. 5045–5048. IEEE (2012)

22. Tur, G., Hakkani-Tür, D., Heck, L., Parthasarathy, S.: Sentence simplification for spoken language understanding. In: ICASSP, pp. 5628–5631. IEEE (2011)
23. Ward, W., Issar, S.: Recent improvements in the CMU spoken language understanding system. In: Proceedings of the Workshop on Human Language Technology, pp. 213–216 (1994)
24. Xu, P., Sarikaya, R.: Convolutional neural network based triangular CRF for joint intent detection and slot filling. In: ASRU, pp. 78–83. IEEE (2013)
25. Yao, K., Peng, B., Zweig, G., Yu, D., Li, X., Gao, F.: Recurrent conditional random field for language understanding. In: ICASSP, pp. 4077–4081. IEEE (2014)
26. Zhang, X., Wang, H.: A joint model of intent determination and slot filling for spoken language understanding. In: IJCAI, pp. 2993–2999 (2016)

A Neural Question Generation System Based on Knowledge Base

Hao Wang, Xiaodong Zhang, and Houfeng Wang[(✉)]

MOE Key Lab of Computational Linguistics,
Peking University, Beijing 100871, China
{hhhwang,zxdcs,wanghf}@pku.edu.cn

Abstract. Most of question-answer pairs in question answering task are generated manually, which is inefficient and expensive. However, the existing work on automatic question generation is not good enough to replace manual annotation. This paper presents a system to generate questions from a knowledge base in Chinese. The contribution of our work contains two parts. First we offer a neural generation approach using long short term memory (LSTM). Second, we design a new format of input sequence for the system, which promotes the performance of the model. On the evaluation of KBQG of NLPCC 2018 Shared Task 7, our system achieved 73.73 BLEU, and took the first place in the evaluation.

Keywords: Question answering · Generation · Knowledge base

1 Introduction

Question answering (QA) is a practical and challenging problem of artificial intelligence (AI). QA contains several tasks, like text based QA, knowledge based QA and table based QA. To train a QA system, it is important to get labeled data with high quality. The performance of a QA system is highly depend on the quality of the training QA pairs. However, most of labeled QA pairs, such as [2,15], are labeled manually, which is inefficient and expensive. Because of that, size of the training sets are limited.

Automatic question generation has already been a challenge task for expanding size of data for QA systems. Zhou et al. [18] worked on generating questions from raw text, which has rich information. However, structured knowledge contains less text information, which make it harder to generate questions from them. Although there are some work on question generation with multiple facts [4], questions that only contain one fact are much more common than those with several facts. Therefore, we work on generating questions with only one fact.

Most of previous works, like Rus et al. [12], are based on hand-crafted rules. However, designing a well-defined rules takes much human effort, and it performs terrible on new situations, which means the rules need to be updated

© Springer Nature Switzerland AG 2018
M. Zhang et al. (Eds.): NLPCC 2018, LNAI 11108, pp. 133–142, 2018.
https://doi.org/10.1007/978-3-319-99495-6_12

frequently. Besides, limited to the form of templates, the questions generated by those methods are lack of diversity.

There are also some previous works based on seq2seq model [3], which is widely used by machine translation [6,10], summarization [7,11] and dialog generation [14,17]. Serban et al. [13] works on the SimpleQuestions [2] dataset in English, and release an English question generation dataset based on SimpleQuestions. Liu et al. [9] works on generating Chinese questions by using a template-based manner, but the generated questions is still different from those asked by human. As we can see from Fig. 1, conjunctions and question words in the output of T-seq2seq are generated improperly.

In this paper, we propose a new system to generate questions from a Chinese structured knowledge base. In NLPCC 2018 Shared Task 7, our system won the first place on the KBQG subtask. Our contributions contain three parts: (1) we improve the seq2seq model, which performs better on question generation task; (2) our system has different input with the previous work, which helps the model to generate questions with proper conjunction words; and (3) we evaluate our model by comparing with other different models, and show the contribution of each part in our model.

Fact	Input Triple	T-seq2seq	Ours
#1	张力 ‖‖ 儿子 ‖‖ 张量 Zhang Li ‖‖ Son ‖‖ Zhang Liang	张力的儿子**是什么**？ **What** is Zhang Li's son?	张力的儿子**叫什么**？ **What's the name** of Zhang Li's son?
#2	董学升 ‖‖ 祖籍 ‖‖ 潮州市潮安区彩塘镇 Dong Xuesheng ‖‖ Ancestral Home ‖‖ Caitang Town	董学升的祖籍**是什么**？ **What** is Dong Xuesheng's ancestral home?	董学升的祖籍**在哪**？ **Where** is Dong Xuesheng's ancestral home?
#3	广州富力足球俱乐部 ‖‖ 成立时间 ‖‖ 2011年6月25日 Fuli Football Club ‖‖ Time of foundation ‖‖ June 25th, 2011	广州富力足球俱乐部是**什么时间**成立的？ **What is the foundation time** of Fuli Football Club?	广州富力足球俱乐部是**什么时候**成立的？ **When** was Fuli Football Club founded?

Fig. 1. The questions generated by Template-based seq2seq learning (T-seq2seq) [9] and our system. The questions our model generated are more similar to those asked by human. The differences of generation are shown in black.

2   Task Description

A KB, or knowledge base, is a structured database, which contains some entities and the relationships between those entities. Each relationship connects two entities. For example, *Phoenix Mountain* and *Shanxi Province* are connected by the relation *location*, which means that *Phoenix Mountain* is located in *Shanxi Province*. Those two entities with a relationship formed a triple of KB.

Given a triple $I = (\text{subject}, \text{relationship}, \text{object})$, and the goal of our system is to generate a question $q = \{c_1, c_2, ..., c_n\}$, which can be answered based on the given triple. Besides, to use those generated question to train the QA model, we hope the generation is similar with those asked by human.

The performance of the system is evaluated by comparing the questions asked by human and machine.

3   Structure of Question Generation Model

In this section, we introduce the structure of our generation model. Our model is based on seq2seq model, which is an universal neural model on generating sequences. The model can be divided into an encoder of input triple, and a question decoder.

3.1   Construction of Input Sequence

In each triple I, we have subject, relationship and object. As we know, for most cases, subject and relationship should be shown in the generation. Although object would not be shown directly in the question, it is also important for the system to understand the target of the question. Therefore, the answer object is converted into some special tokens, which are shown on Fig. 2. <date> represents all phrases of dates, and <time> represents the specific time. For example, in the input sequence, "1月1日 12:00" (12:00, Jan.1) will be replaced by "<date> <time>". <number> represents a number is replaced here. Name of books are convert to <book>. <sp_name> represents all other phrases in western characters, numbers and some symbols, which are widely used of name of products and some foreign people. It is easy to find out those phrases by using patterns. When generating input sequence, we convert those phrases into special tokens, and the sequence is named as token_seq.

Token Name	Examples
<date>	2018-01-01, 2018年1月1日
<time>	12:00, 12:00:00, 12点
<number>	100, 100.0, 1.0E+2
<book>	《高等数学》
<sp_name>	000-A01-B02, PKU

Fig. 2. Some examples of special tokens used in our system.

However, it is hard to find Chinese name by using patterns, but it is important for the system to recognize those name. Therefore, we use HanLP toolkit[1], a framework that provides some core processing tools, to find names of people in Chinese. In HanLP, names can be found by role tagging [16]. After converting the

---

[1] https://github.com/hankcs/HanLP.

input sequence to token_seq, we replace words in the object of triple, excluding some conjunctions and adjectives, to their POS tags. The new sequence is called token_pos_seq.

For subjects of triples are used to replace <ent> token of the output sequence, they are not part of the input sequences. Besides, a special token <is> is inserted to split the relationship and the object. Examples of the input sequences are shown in Fig. 3. In Fig. 3, "nr" represents the token is name of a person, and all POS tags which start with "v" represent verbs, and those start with "q" represent quality words.

Original Triple	Name	Input Sequence
恩万科沃·卡努 ‖ 欧冠联赛 ‖ 出场55次，进5球 Nwankwo Kanu ‖ UEFA Champions League ‖ 55 appearance, 5 goals	token_pos_seq	欧冠联赛 <is> vi **<number>** qv vf **<number>** UEFA Champions League <is> vi **<number>** qv vf **<number>**
	token_seq	欧冠联赛 <is> 出场 **<number>** 次，进 **<number>** 球 UEFA Champions League <is> **<number>** appearance, **<number>** goals
棉花枯腐病 ‖ 病原拉丁学名 ‖ macrophomina phaseoli ashby Cotton rot ‖ Pathogenic Latin name ‖ macrophomina phaseoli ashby	token_pos_seq	病原拉丁学名 <is> **<sp_name>** Pathogenic Latin name <is> **<sp_name>**
	token_seq	病原拉丁学名 <is> **<sp_name>** Pathogenic Latin name <is> **<sp_name>**
赵佩茹 ‖ 师承 ‖ 相声老前辈焦寿海先生 Zhao Peiru ‖ Teacher ‖ Mr. Jiao Shouhai, Senior of Xiangsheng	token_pos_seq	师承 <is> nr Teacher <is> nr
	token_seq	师承 <is> 相声老前辈焦寿海先生 Teacher ‖ Mr. Jiao Shouhai, Senior of Xiangsheng

Fig. 3. Some examples of input sequence.

3.2   Encoder

Given a input sequence $X = \{x_1, x_2, ..., x_n\}$, where n is the length of input sequence, and suppose we have a vocabulary V, which contains characters and tokens. Each $x_i \in \mathbb{R}^{|m|}$ is a one-hot vector, which represents a token in the vocabulary. The encoder converts those tokens first into their embeddings $\{e_1, e_2, ..., e_n\}$, by looking up the embedding matrix $E \in \mathbb{R}^{|V| \times m}$, where m is the size of embedding. Then, we use a 3-layer LSTM network to encode the input sequence x. The structure of each LSTM layer is:

$$f_t = \sigma(W_f \cdot [h_{t-1}, e_t] + b_f) \tag{1}$$

$$i_t = \sigma(W_i \cdot [h_{t-1}, e_t] + b_i) \tag{2}$$

$$\tilde{C}_t = \tanh(W_C \cdot [h_{t-1}, e_t] + b_C) \tag{3}$$

$$C_t = f_t * C_{t-1} + i_t * \tilde{C}_t \tag{4}$$

$$o_t = \sigma(W_o \cdot [h_{t-1}, e_t] + b_o) \tag{5}$$

$$h_t = o_t * \tanh(C_t) \tag{6}$$

In Eqs.(1–6), h_t is the hidden state of the LSTM layer at time t, and $e_t, o_t \in \mathbb{R}_n$ are input and output of each layer at step t. $W_f, b_f, W_i, b_i, W_C, b_C, W_o, b_o$ are trainable parameters.

3.3   Decoder

The structure of the decoder is shown in Fig. 4. Our decoder is a neural network that models the conditional probability $p(y|H)$, while $H = \{h_1, h_2, ..., h_n\}$ is the hidden states of encoder. We use an LSTM network with attention and CopyNet [5] as our decoder.

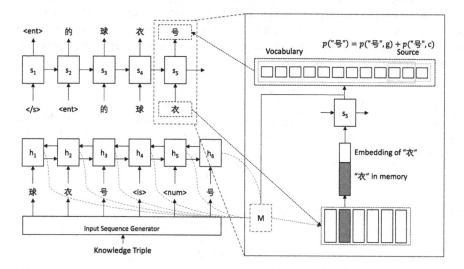

Fig. 4. Structure of the decoder. Hidden states of the encoder are used in decoding.

We have a vocabulary list V, which is same as the vocabulary for encoder, and use <UNK> to represent the OOV (out-of-vocabulary) word. Since there are some words that not contained in V are useful, words in the input sequence X are also included in generation candidates, which makes CopyNet to output some OOV words. To read the states from encoders, we build a memory matrix M, which is initialized by $\{h_1, h_2, ..., h_n\}$, and will be updated while decoding. For each step t, we can get the probability of generating any target word y_t as,

$$p(y_t|\mathbf{s}_t, y_{t-1}, c_t, M) = p(y_t, \mathbf{g}|s_t, y_{t-1}, c_t, M) + p(y_t, \mathbf{c}|s_t, y_{t-1}, c_t, M) \quad (7)$$

where $\mathbf{g}$ represents the generate-mode, and $\mathbf{c}$ represents the copy-mode. The two probability are calculated by

$$p(y_t, \mathbf{g}|\cdot) = \begin{cases} \frac{1}{Z}e^{\psi_g(y_t)}, & y_t \in V \\ 0, & y_t \in X \cap \overline{V}; \\ \frac{1}{Z}e^{\psi_g(<\text{UNK}>)} & y_t \notin X \cup V \end{cases} \quad (8)$$

$$p(y_t, \mathbf{c}|\cdot) = \begin{cases} \frac{1}{Z}\sum_{j:x_j=y_t} e^{\psi_c(y_t)}, & y_t \in X \\ 0 & y_t \notin X \end{cases}. \quad (9)$$

In Eqs. (8–9), ψ_g and ψ_c are score functions for two modes, while Z is the normalization term shared by both two probabilities. The score of copy-mode is

$$\psi_c(y_t) = \tanh(h_j^T W_c) s_t, \, x_j \in X. \tag{10}$$

In Eq. (10), hidden states in M are used to represent the input sequence. Location of each token in the sequence is important for copying. ψ_g is similar with the scorer function in the basic seq2seq model. It is based on a 3-layer LSTM network with attention. Input of the first LSTM layer is embedding of the former output word. Attention of our model is inspired by [10], which gets a great improvement to seq2seq models. That is,

$$\psi_g(y_t) = v_i^T W_s \tanh(W_t[s_t, a_t]) \tag{11}$$

In Eq. (11), v_i is the one-hot indicator vector for v_i, and s_t is the t-th hidden state of the decoder LSTM network. a_t is the attention vector calculated by hidden states:

$$a_t = \sum_{i=1}^{n} \alpha_{t_i} h_i \tag{12}$$

while α_{t_i} is the normalization weight calculated by:

$$\alpha_{t_i} = \frac{e^{g(s_t, h_i)}}{\sum_{j=1}^{n} e^{g(s_t, h_j)}}, \tag{13}$$

where $g(s_t, h_j)$ represents the relevance between the hidden state for decoder and encoder. In our model, the relevance score is measured by

$$g(s_t, h_j) = s_t^T h_j. \tag{14}$$

The input y_{t-1} of decoder is different with the original seq2seq network. It is represent as $[e(y_{t-1}); \zeta(y_{t-1})]^T$, where $e(y_{t-1})$ is the word embedding of y_{t-1}, and $\zeta(y_{t-1})$ is the weighted sum of M, i.e.

$$\zeta(y_{t-1}) = \sum_{\tau=1}^{n} \rho_{t\tau} h_\tau, \tag{15}$$

$$\rho_{t\tau} = \begin{cases} \dfrac{1}{K} p(x_\tau, \mathbf{c} | s_{t-1}, M), & x_\tau = y_{t-1} \\ 0, & \text{otherwise} \end{cases}. \tag{16}$$

In (16), K is the normalization term.

To avoid the improper topic words generated by seq2seq model, we use a template-based method. That is, the system is set to generate questions by replacing the subject entity to a <ent> token. Then, to get the complete question, we convert <ent> back to the subject. Subject entities in all training data and test golds are also replaced to <ent>, and those cases without <ent> are not be used to train the model.

3.4   Train

Seq2seq model with CopyNet and attention is an end-to-end model, which can be trained using back-propagation. We train the model in batches. The objective of training is to minimize the Cross-Entropy (CE) loss:

$$\mathcal{L} = -\sum_{i=1}^{n} \log p(y_i^{ans}),\qquad(17)$$

where y_i^{ans} is the i-th word of the answer. The model can be learned to improve both generate-mode and copy-mode. If the target token can be found in source sequence, the copy-mode is contributed to the model and can be trained, while the copy-mode is discouraged when the token is not in the input.

4   Evaluation

In this section, we first introduce the dataset we use. Then we show the settings of our model. After that, we compare our system with other systems.

4.1   Dataset and Evaluation Metrics

We take our evaluation on the dataset released by knowledge based question generation (KBQG) subtask of NLPCC 2018 Shared Task 7. There are 24,479 answer-question pairs in the training set, and 357 answer-question pairs in the test set. The evaluation of shared task is based on BLEU-4 score with case insensitive. All the questions are generated by human. In the train set, each case has one question. And in the test set, each case is corresponded to three answers. For each case, the BLEU score is determined by the most similar answer with the output of model.

4.2   Settings of the System

The preprocessing on input sequence of our system is described in Sect. 3.1. The input sequences are split by Chinese character, punctuations and special tokens after preprocessing. All the embeddings of tokens are 200-dimensional vectors pre-trained on Chinese Wikipedia documents by word2vec, and all of the vectors are trainable. Encoder and decoder share the same embedding and vocabulary. Based on the training set, we build a vocabulary with 15,435 words, containing special tokens. We use Adam optimizer [8] for optimization, with 1e-3 as the initial learning rate. The batch size of the training stage is 32. All the implementations of our system is based on TensorFlow framework [1]. To evaluate the system during training, we extract the training set to a training set with 20,000 pairs randomly, and rest of the pairs into a validation set. All these configurations are based on the performance of our validation set.

We offer four systems to solve the result. Three of them are single models with different input sequences. One of them is character-by-character raw input, and the other two have token_seq and token_pos_seq as their input respectively. There is also an ensemble model, whose result is constructed by three sub-models using token_seq as input sequence, and another three sub-models using token_pos_seq. Two of them are submitted to the competition, their name are shown in Table 1. All these sub-models are trained separately, and on test cases, the $p(y_t|\cdot)$ is calculated by:

$$p(y_t|\cdot) = \sum_{i=1}^{6} \frac{1}{Z} p_i(p_t|\cdot), \tag{18}$$

while Z is the normalization term of all sub-models.

4.3 Result

The BLEU-4 of the systems is shown in Table 1. To show the contribution of each part of our system, we offer three system whose settings are different with our system. Our ensemble system and single system are all much better than other systems of the task. Compared with raw input, the special designed input sequences lead to 4 BLEU of improvement. The ensemble model combines the advantage of different input sequences and stochastic learning of each sub-models.

Table 1. Comparison between other systems in KBQG task.

System name	BLEU-4
AQG, SWU	49.79
AQG-PAC_greedy_relation_predict, SWU	51.46
AQD-PAC_soft_relation_predict, SWU	51.62
AQG-question_sentence_relation_predict, SWU	53.32
CCNU-319	59.82
LPAI, Soochow Univ	63.67
unique AI group, CCNU	64.38
Ours with raw input (single model)	66.18
Ours with token_pos_seq (single model, ICL-1)	70.21
Ours with token_seq (single model)	70.81
Ours (ensemble model, ICL-2)	**73.73**

4.4 Error Analysis

To find the reason of improper generations, we analysis some cases manually. First, in most cases whose relationships are not in the training set, the generated questions are in the same pattern "<ent> 的 *relationship* 是什么/多少/谁?" (What/ How many/ Who is the *relationship* of <ent>?).

It is better than most earlier models, whose question words in their generation are "是什么" (what) only, but still needs to be improved. Besides, some
of generated questions contain incomplete words. For example, for the triple
"达克鲁斯效力于哪个俱乐?"(Which *Clu* does Da Cruz work for?), "部"(b) of
the word "俱乐部"(Club) is missed. The reason of this fault is that the our model
is character based, and it is hard to for the system to learn word information to
decoder.

5   Conclusion

In this paper, we propose a neural generation system to generate questions from
structured knowledge base. Our system achieves a great performance in the
competition of NLPCC 2018 Shared Task 7. Then, we analysis the contribution
of each feature of our model by different experiments. After that, we evaluate
our system manually, to know the reason of errors in generations.

In future work, it is important to improve the accuracy of generated words
and fluency of sentences. Besides, it is also important to get a dataset with diverse
questions, which can help exploring to improve the diversity of the question
generation systems.

Acknowledgments. Our work is supported by National Natural Science Foundation
of China (No. 61370117).

References

1. Abadi, M., Barham, P., Chen, J., Chen, Z., Davis, A., Dean, J., Devin, M.,
   Ghemawat, S., Irving, G., Isard, M.: TensorFlow: a system for large-scale machine
   learning. OSDI **16**, 265–283 (2016)
2. Bordes, A., Usunier, N., Chopra, S., Weston, J.: Large-scale simple question
   answering with memory networks. CoRR abs/1506.02075 (2015)
3. Cho, K., van Merrienboer, B., Gülçehre, Ç., Bahdanau, D., Bougares, F., Schwenk,
   H., Bengio, Y.: Learning phrase representations using RNN encoder-decoder for
   statistical machine translation. In: Proceedings of the 2014 Conference on Empirical Methods in Natural Language Processing, EMNLP 2014, 25–29 October 2014,
   Doha, Qatar, A meeting of SIGDAT, a Special Interest Group of the ACL, pp.
   1724–1734 (2014)
4. Colin, E., Gardent, C., Mrabet, Y., Narayan, S., Perez-Beltrachini, L.: The
   WebNLG challenge: generating text from DBPedia data. In: INLG 2016 - Proceedings of the Ninth International Natural Language Generation Conference, 5–8
   September 2016, Edinburgh, UK, pp. 163–167 (2016)
5. He, S., Liu, C., Liu, K., Zhao, J.: Generating natural answers by incorporating
   copying and retrieving mechanisms in sequence-to-sequence learning. In: Proceedings of the 55th Annual Meeting of the Association for Computational Linguistics,
   ACL 2017, Vancouver, Canada, 30 July–4 August, Volume 1: Long Papers, pp.
   199–208 (2017)

6. Johnson, M., Schuster, M., Le, Q.V., Krikun, M., Wu, Y., Chen, Z., Thorat, N., Viégas, F.B., Wattenberg, M., Corrado, G., Hughes, M., Dean, J.: Google's multilingual neural machine translation system: enabling zero-shot translation. TACL **5**, 339–351 (2017)

7. Kim, M., Moirangthem, D.S., Lee, M.: Towards abstraction from extraction: multiple timescale gated recurrent unit for summarization. In: Proceedings of the 1st Workshop on Representation Learning for NLP, Rep4NLP@ACL 2016, Berlin, Germany, 11 August 2016, pp. 70–77 (2016)

8. Kingma, D.P., Ba, J.: Adam: a method for stochastic optimization. CoRR abs/1412.6980 (2014)

9. Liu, T., Wei, B., Chang, B., Sui, Z.: Large-scale simple question generation by template-based Seq2seq learning. In: Huang, X., Jiang, J., Zhao, D., Feng, Y., Hong, Y. (eds.) NLPCC 2017. LNCS (LNAI), vol. 10619, pp. 75–87. Springer, Cham (2018). https://doi.org/10.1007/978-3-319-73618-1_7

10. Luong, T., Pham, H., Manning, C.D.: Effective approaches to attention-based neural machine translation. In: Proceedings of the 2015 Conference on Empirical Methods in Natural Language Processing, EMNLP 2015, Lisbon, Portugal, 17–21 September 2015, pp. 1412–1421 (2015)

11. Nallapati, R., Zhou, B., dos Santos, C.N., Gülçehre, Ç., Xiang, B.: Abstractive text summarization using sequence-to-sequence RNNs and beyond. In: Proceedings of the 20th SIGNLL Conference on Computational Natural Language Learning, CoNLL 2016, Berlin, Germany, 11–12 August 2016, pp. 280–290 (2016)

12. Rus, V., Wyse, B., Piwek, P., Lintean, M., Stoyanchev, S., Moldovan, C.: The first question generation shared task evaluation challenge. In: Proceedings of the 6th International Natural Language Generation Conference, pp. 251–257. Association for Computational Linguistics (2010)

13. Serban, I.V., García-Durán, A., Gülçehre, Ç., Ahn, S., Chandar, S., Courville, A.C., Bengio, Y.: Generating factoid questions with recurrent neural networks: the 30m factoid question-answer corpus. In: Proceedings of the 54th Annual Meeting of the Association for Computational Linguistics, ACL 2016, 7–12 August 2016, Berlin, Germany, Volume 1: Long Papers (2016)

14. Song, Y., Yan, R., Li, X., Zhao, D., Zhang, M.: Two are better than one: an ensemble of retrieval- and generation-based dialog systems. CoRR abs/1610.07149 (2016)

15. Yin, W., Schütze, H., Xiang, B., Zhou, B.: ABCNN: attention-based convolutional neural network for modeling sentence pairs. TACL **4**, 259–272 (2016)

16. Zhang, H.P., Liu, Q.: Automatic recognition of chinese personal name based on role tagging. Chin. J. Comput. Chin. Edn. **27**(1), 85–91 (2004)

17. Zhou, H., Huang, M., Zhang, T., Zhu, X., Liu, B.: Emotional chatting machine: emotional conversation generation with internal and external memory. In: Proceedings of the Thirty-Second AAAI Conference on Artificial Intelligence, New Orleans, Louisiana, USA, 2–7 February 2018 (2018)

18. Zhou, Q., Yang, N., Wei, F., Tan, C., Bao, H., Zhou, M.: Neural question generation from text: a preliminary study. In: Huang, X., Jiang, J., Zhao, D., Feng, Y., Hong, Y. (eds.) NLPCC 2017. LNCS (LNAI), vol. 10619, pp. 662–671. Springer, Cham (2018). https://doi.org/10.1007/978-3-319-73618-1_56

Knowledge Graph/IE

ProjR: Embedding Structure Diversity for Knowledge Graph Completion

Wen Zhang, Juan Li, and Huajun Chen[✉]

Zhejiang University, Hangzhou, China
{wenzhang2015,lijuan18,huajunsir}@zju.edu.cn

Abstract. Knowledge graph completion aims to find new true links between entities. In this paper, we consider an approach to embed a knowledge graph into a continuous vector space. Embedding methods, such as TransE, TransR and ProjE, are proposed in recent years and have achieved promising predictive performance. We discuss that a lot of substructures related with different relation properties in knowledge graph should be considered during embedding. We list 8 kinds of substructures and find that none of the existing embedding methods could encode all the substructures at the same time. Considering the structure diversity, we propose that a knowledge graph embedding method should have diverse representations for entities in different relation contexts and different entity positions. And we propose a new embedding method ProjR which combines TransR and ProjE together to achieve diverse representations by defining a unique combination operator for each relation. In ProjR, the input head entity-relation pairs with different relations will go through a different combination process. We conduct experiments with link prediction task on benchmark datasets for knowledge graph completion and the experiment results show that, with diverse representations, ProjR performs better compared with TransR and ProjE. We also analyze the performance of ProjR in the 8 different substructures listed in this paper and the results show that ProjR achieves better performance in most of the substructures.

Keywords: Diversity structures · Knowledge graph embedding
Knowledge graph completion

1 Introduction

Knowledge graphs (KGs) are built to represent knowledge and facts in the world and have become useful resources for many artificial intelligence tasks such as web search and question answering. A knowledge graph could be regarded as a multi-relational directed graph with entities as nodes and relations as labeled edges. An instance of an edge is a fact triple in the form of (h, r, t) and h, r, t denote the head entity, relation, and tail entity respectively. For example, (Steve Jobs, $isFounderOf$, Apple Inc.) represents the fact that Steve Jobs is the

© Springer Nature Switzerland AG 2018
M. Zhang et al. (Eds.): NLPCC 2018, LNAI 11108, pp. 145–157, 2018.
https://doi.org/10.1007/978-3-319-99495-6_13

founder of company Apple. Many huge knowledge graphs have been built automatically or semi-automatically in recent years, such as Yago [19], WordNet [13] and Google Knowledge Graph[1]. Though typical knowledge graphs may contain more than millions of entities and billions of facts, they still suffer from incompleteness.

Much work has focused on knowledge graph completion. Some tried to extract more triples through large text corpora, such as OpenIE [1] and DeepDive [26]. Others tried to get new triples by reasoning based on the existing knowledge graph. Traditional reasoning methods including ontology inference machine, such as Pellet[2], which relies on a well-defined ontology. Knowledge graph embedding is another way to complete knowledge graphs. It tries to embed all the elements in a knowledge graph, including entities and relations, into a continuous vector space while preserving certain properties of the original KG. KG embedding makes it possible to complete the knowledge graph through implicit inference by calculations between the vector representations.

We regard the basic process of most knowledge graph embedding methods as two steps: (1) get combined representation of the given head entity-relation pair through a combination operator $C(h, r)$ and then (2) compare the similarity between candidate tail entities and the combined representation through a similarity operator $S(C(h, r), t)$. The goal is to make the similarity between true candidate entities and combined representation as large as possible and the similarity for false candidate entities as small as possible.

There are a lot of knowledge graph embedding methods proposed in the last few years. One of the simple but effective method is TransE [4] which represents each entity and relation as a vector. The combination operator is defined as an addition and the similarity operator is defined based on distance. ProjE [16] is another knowledge graph embedding method whose basic idea is almost the same with TransE. ProjE also assigns each entity and relation with one vector. The combination operator of ProjE is a global linear layer and the similarity operator is defined as a dot product between the candidate vector and combined representation. ProjE performs better than TransE in link prediction task because of the different definition of the final loss function.

Considering the different properties of relations, improved methods, such as TransH [21], TransR [12] and TransD [9], are proposed based on TransE. Those methods inspired us to consider the structure diversity of knowledge graph as relations of different properties always correspond to different graph structures. We list 8 kinds of graph structures in this paper and find that none of the previous embedding methods can properly encode all of these structures at the same time. Some methods such as TransR can encode N-1, N-N, 1-N-1 structures but not one-relation-circle structures. Some methods, such as ProjE, could encode one-relation-circle structures but not N-1, N-N, 1-N-1 structures.

---

[1] https://www.google.com/intl/en-419/insidesearch/features/search/knowledge.html.

[2] http://pellet.owldl.com/.

The reason why TransR can encode 1-N, N-N, and 1-N-1 structures is that before calculating the similarity, TransR uses a relation specific matrix to project entity vector to relation space and get a new entity representation. In other words, every entity in TransR has one vector representation under each different relation context. The reason why ProjE can encode one-relation-circle structures is that the vector representation for one entity under different entity position (as a head entity or as a tail entity) is different. So the key point to enable embedding methods to encode the diversity of knowledge graph is that every entity should have diverse representations in different relation contexts and entity positions.

In this paper, we propose a new embedding method ProjR which combines TransR and ProjE together. To achieve diverse representation for entities in different relation contexts, we define a unique combination operator for each relation in ProjR. To achieve diverse representations for entities under different entity position, we follow the process of ProjE to project head entity vector through a matrix during combination process and not project tail entity during similarity operator. We conduct link prediction experiments on knowledge graph completion benchmark datasets and evaluate the results in the same way as previous work. The evaluation results show that ProjR achieves better predictive performance compared with both TransR and ProjE, and also some other embedding methods. We also analyze the performance of ProjR with the 8 different substructures listed in Table 6 and the results show that ProjR achieves better performance in most of the substructures compared with ProjE.

The contributions of this paper are as follows:

- We list 8 kinds of graph structures to analyze the structure diversity of knowledge graph and also find examples from real knowledge graph Freebase for each structure.
- We analyze the ability of most related embedding models to encode the structure diversity and find that diverse representations for entities in different relation contexts and entity positions are helpful for encoding the structure diversity in knowledge graph.
- We propose a new knowledge graph embedding method ProjR based on the idea of diverse representations for entities. The experiment results on link prediction tasks show that ProjR achieves better performance than TransR and ProjE. The experiment results of the performance of ProjR on different structures also prove that with diverse representations for entities ProjR could handle the structure diversity of knowledge graph more properly.

2   Related Work

We summarize the related methods in Table 1 with information of the score function and the number of parameters to learn during training. The work most related to ours are TransR [12] and ProjE [16].

TransR: TransR [12] is an extended method of TransE [4]. We call them translation-based methods because their basic assumption is that the relation r

Table 1. A summary for the most related methods. The bold lower letter denotes vector representation and the bold upper letter denotes the matric representation. **I** denotes identity matrix. n_e and n_r are the number of entities and relations in knowledge base, respectively. d denotes the embedding dimension.

Model	Sore function $S(h,r,t)$		#Parameters
	$C(h,r)$	$S(h,r,t)$	
TransE	$\mathbf{h}+\mathbf{r}$	$\|C(h,r)-\mathbf{t}\|_{l_{1/2}}$	$(n_e+n_r)d$
TransH	$(\mathbf{h}-\mathbf{w}_r^{\top}\mathbf{h}\mathbf{w}_r+\mathbf{r})$	$\|C(h,r)-(\mathbf{t}-\mathbf{w}_r^{\top}\mathbf{t}\mathbf{w}_r)\|_2^2$	$(n_e+2n_r)d$
TransR	$\mathbf{M}_r\mathbf{h}+\mathbf{r}$	$\|C(h,r)-\mathbf{M}_r\mathbf{t}\|_2^2$	$(n_e+n_r)d+n_rd^2$
TransD	$(\mathbf{r}_p\mathbf{h}_p^{\top}+\mathbf{I}^{k\times k})\mathbf{h}+\mathbf{r}$	$\|C(h,r)-(\mathbf{r}_p\mathbf{t}_p^{\top}+\mathbf{I}^{d\times d})\mathbf{t}\|_2^2$	$2(n_e+n_r)d$
ProjE	$tanh(\mathbf{D}_1\mathbf{h}+\mathbf{D}_2\mathbf{r}+\mathbf{b}_1)$	$\sigma(C(h,r)\cdot\mathbf{t}+\mathbf{b}_2)$	$(n_e+n_r+5)d$

in triple (h,r,t) could be regarded as a translation from the head entity h to the tail entity t. TransE represents each entity and relation with one vector and make the constraint for a true triple (h,r,t) as $\mathbf{h}+\mathbf{r}\approx\mathbf{t}$. Unstructured [3] is a special case of TransE which sets all relation vectors as zero vectors. Considering the different properties of relations, TransH [21] defines a hyperplane for each relation and projects head and tail entities onto the current relation hyperplane before the calculation of distance. Different from TransE and TransH which represent all elements in the same vector space, TransR represents entities in entity space and represents relations in relation space. And entities h and t should be projected into the current relation space through the relation projection matrix $\mathbf{M}_r$. TransD [9] is an extended method of TransR which defines dynamic projection matrices related with both relations and entities. TransH, TransR and TransD all achieve diverse representations for entities in different relation contexts by different projection strategies but unable to encode one-relation-circle because there is only one representation for one entity with same relation context.

ProjE: ProjE [16] is another knowledge graph embedding method which gets the combined representation of an input head entity-relation pair through a global linear layer. Then projecting all the candidate entity vectors onto the combined vector result which could be regarded as a similarity computation. But different from translation-based methods which optimize a margin-based pairwise ranking loss, ProjE optimizes a cross entropy based ranking loss of a list of candidate entities collectively which makes ProjE more flexible with negative candidate sampling and enhance the ability to handle very large datasets. ProjE also points out that the number of negative samples will affect the embedding results obviously. ProjE is able to encode the one-relation-circle structures.

Other Methods: RESCAL [15] regards the whole knowledge graph as a multi-hot tensor and embeds the knowledge graph based on tensor factorization. NTN [17] is a neural tensor network which represents each relation as a tensor. HOLE [14] employs correlation between different dimensions of entity vectors during training of the vector representations. Some methods also combine other information together with fact triples. RTransE [6] and PTransE [11] employ the path information of 2–3 length over knowledge graph. Jointly [20], DKRL [24], TEKE [22] and SSP [23] combine unstructured entity description texts together with the structured triples during training. The external text information makes those methods more likely to cover the out-of-knowledge-graph entities. TKRL [25] considers the information of entity class. But in this paper, we only focus on the methods that only use triples' information.

3 Our Method

In this section, we first introduce 8 kinds of structures to prove the structure diversity of knowledge graph and analyze the ability of the most related embedding methods to encode these 8 kinds of structures. Then we introduce the new method ProjR.

3.1 Structure Diversity of Knowledge Graph

The complex connections between entity nodes cause the structure diversity of knowledge graph. Relations with different properties are always related to different graph structures. We introduce 8 kinds of substructures in this section.

1-1 relation structure means that one entity links to at most one entity through this relation. **1-N** relation structure means that one entity links to more than one entities through this relation. **N-1** relation structure means that there are more than one entities linking to the same entity through this relation. **N-N** relation structure means that one entity link to more than one entities through this relation and one entity also could be linked to more than one entities through this relation. Those four kinds relation properties are first proposed in [21]. **1-N-1** structure means that there are more than one relation that link one same entity to another same entity. **C1, C2, C3** are special case of one-relation-circle (ORC) substructures which is first proposed in [27]. **C1** means that one

Table 2. This table lists the ability of five most related KG embedding methods to encode the substructures and the number of vector representations for every entity and relation. n_r denotes the number of relations in knowledge graph. In the column of "types of structure", "$\sqrt{}$" means model in current row can encode the substructure in current column and "×" means can't.

Method	# representations		Types of substructure							
	Entity	Relation	1-1	1-N	N-1	N-N	1-N-1	C1	C2	C3
TransE [4]	1	1	$\sqrt{}$	×	×	×	×	×	×	×
TransH [21]	n_r	1	$\sqrt{}$	$\sqrt{}$	$\sqrt{}$	$\sqrt{}$	$\sqrt{}$	×	×	×
TransR [12]	n_r	1	$\sqrt{}$	$\sqrt{}$	$\sqrt{}$	$\sqrt{}$	$\sqrt{}$	×	×	×
TransD [9]	n_r	1	$\sqrt{}$	$\sqrt{}$	$\sqrt{}$	$\sqrt{}$	$\sqrt{}$	×	×	×
ProjE [16]	2	2	$\sqrt{}$	×	×	×	×	$\sqrt{}$	$\sqrt{}$	$\sqrt{}$

entity connects to itself by one relation. **C2** means that two entity connects to each other by the same relation. **C3** means that three entities connect to each other through the same relation and the connections form a circle if ignoring the direction of relations.

In Table 2, we analyze the ability of related embedding models to encode these 8 kinds of structures. We regard one method could encode one kind structure if and only if it is possible to let the similarity score of all the true triple participating in the current structure to be nearly the maximum value. We calculate the vector representation of entities and relations following this rule: we regard any projection operation result of one entity as a representation for it. For example, with triple (h, r, t) in TransH, the head entity h and the tail entity t will be projected onto the relation r specific hyperplane before calculating the distance, which means each entity has a vector representation on every relation hyperplane. So there are n_r representations for each entity in TransH.

Although none of them could encode all the structures, we also could conclude that diverse representations for entities will improve the capability of encoding structure diversity. TransH, TransR and TransD are able to encode 1-N, N-1 and N-N relation structures because they separate the representations for entities in different relation contexts and have n_r kinds of representations for every entity. ProjE is able to encode C1, C2 and C3 because the different representations for one entity in different entity positions enable ProjE to decompose the one-relation-circle structures.

To make embedding model more powerful to encode the structure diversity, we combine TransR and ProjE and propose ProjR based on the key idea of diverse representations for entities in different relation contexts and entity positions.

3.2   ProjR

In ProjR, we define a score function to calculate the probability of an input triple (h, r, t) to be true. And we regard the probability score function as two parts: a combination operator and a similarity operator.

Combination Operator: The input of combination operator is head entity-relation pair (h, r). The head entity embedding is set as $\mathbf{h} \in \mathbb{R}^d$ and the relation embedding is set as $\mathbf{r} \in \mathbb{R}^d$. d is the dimension of embedding vectors.

To achieve diverse representations for entities in different relation contexts, ProjR defines a combination operator $C_r(h)$ for each relation r:

$$C_r(h) = \mathbf{c}_{hr} = tanh(\mathbf{D}_r^e \mathbf{h} + \mathbf{D}_r^r \mathbf{r} + \mathbf{b}_c)$$

$\mathbf{D}_r^e \in \mathbb{R}^{d \times d}$ is a diagonal matrix defined for linear transformation of head entity related with relation r. $\mathbf{D}_r^r \in \mathbb{R}^{d \times d}$ is a diagonal matrix defined for linear transformation of relation r. We choose the diagonal matrix instead of normal matrix in consideration of the balance between diverse representation ability and the number of parameters. $\mathbf{b}_c \in \mathbb{R}^d$ is a global bias vector. $tanh(z) = \frac{e^z - e^{-z}}{e^z + e^{-z}}$ is a nonlinear activate function in which the output value will be constrained to $(-1, 1)$. Each combination operator will project entity with a different diagonal matric and generate a specific representation for each entity.

Similarity Operator: After getting the vector result c_{hr} from combination operator for (h, r), ProjR calculates the similarity between the c_{hr} and the tail entity vector t as the final probability score for triple (h, r, t) to be true. To achieve the diverse representations for entities in different entity positions, we use the tail entity vector directly without any projection. Although defining another projection matrix for the tail entity related with relation r will further improve the ability of ProjR to encode diversity of structures in knowledge graph, it will also increase a lot of parameters.

Considering the convenience for computation, we define the similarity operator as follows:

$$S(h, r, t) = \sigma(t \cdot c_{hr})$$

$t \in \mathbb{R}^d$. We use dot product to simulate the similarity between c_{hr} and t. And $\sigma(z) = \frac{1}{1+e^{-z}}$ is used to constrain the final output to $(0, 1)$ as a probability score.

Loss Function: During training, we define the following learning objective:

$$L = - \sum_{(h,r,t) \in \Delta} log(S(h, r, t))$$
$$- \sum_{(h',r,t') \in \Delta'} log(1 - S(h', r, t')) + \lambda \sum_{p \in P} \|p\|$$

Δ is the set of positive triples in training data. And Δ' is the set of false triples generated for each training triple. The negative triple generation will be introduced in next section. P is the set of parameters to be learned in ProjR. $\lambda \sum_{p \in P} \|p\|$ is a regularization term with the summation of L1 norm of all elements in P. λ is the regularization parameter. The training goal is to minimize loss function L.

4 Experiments

In this paper, we conduct the experiment of link prediction and evaluate the embedding results with the benchmark knowledge graphs WN18 and FB15k which are subsets of WordNet [13] and Freebase [2] respectively (Table 3).

Table 3. Statistic of experiment dataset

Dataset	# Rel	# Ent	# Train	# Valid	# Test
WN18	18	40943	141442	5000	5000
FB15k	1345	14951	483142	50000	59071

4.1 Link Prediction

Link prediction aims to predict the missing entity given one entity and relation such as $(h, r, ?)$ and $(?, r, t)$. $(h, r, ?)$ is tail entity prediction given head entity and relation. $(?, r, t)$ is head entity prediction given tail entity and relation.

Data Prepare: As ProjR always predicts tail entities given head entity-relation pair, we regard head entity prediction $(?, r, t)$ as a tail entity prediction $(t, r^{-1}, ?)$. r^{-1} denotes the reverse relation of r. To get the embedding of r^{-1}, for each triple (h, r, t), we add reverse relation triple (t, r^{-1}, h) into training dataset.

To generate negative triples (h', r', t'), we follow the process of ProjE and randomly select m percent of entities in dataset to replace the tail entity t of (h, r, t) which means there will be $m \times n_e$ negative triples for every training triple. n_e is the number of entities in experiment dataset. $m \in (0, 1)$ is a hyperparameter.

Table 4. Results on WN18 and FB15k for link prediction. The result numbers underlined are the best results among TransR, ProjE, and ProjR. The bold result numbers are the best results among all the methods.

Method	WN18				FB15k			
	Mean Rank		Hit@10(%)		Mean Rank		Hit@10(%)	
	Raw	Filter	Raw	Filter	Raw	Filter	Raw	Filter
Unstructured [3]	315	304	35.3	38.2	1074	979	4.5	6.3
RESCAL [15]	1180	1163	37.2	52.8	828	689	28.4	44.1
SE [5]	1011	985	68.5	80.5	273	162	28.8	39.8
SME(linear) [3]	545	533	65.1	74.1	274	154	30.7	40.8
SME(Bilinear) [3]	526	509	54.7	61.3	284	158	31.3	41.3
LFM [8]	469	456	71.4	81.6	283	164	26.0	33.1
TransE [4]	263	251	75.4	89.2	243	125	34.9	47.1
TransH(unif) [21]	318	303	75.4	86.7	211	84	42.5	58.5
TransH(bern) [21]	401	388	73.0	82.3	212	87	45.7	64.4
CTransR(unif) [12]	243	230	78.9	92.3	233	82	44	66.3
CTransR(bern) [12]	231	218	79.4	92.3	199	75	48.4	70.2
TransD(unif) [9]	242	229	79.2	92.5	211	67	49.4	74.2
TransD(bern) [9]	**224**	**212**	79.6	92.2	**194**	91	**53.4**	77.3
TransR(unif) [12]	<u>232</u>	<u>219</u>	78.3	91.7	226	78	43.8	65.5
TransR(bern) [12]	238	225	79.8	92.0	198	77	48.2	68.7
ProjE_listwise [16][a]	–	–	–	–	214	60	48.1	78.8
ProjR(this paper)	356	345	**<u>82.6</u>**	**<u>95.0</u>**	<u>195</u>	**<u>41</u>**	<u>52.3</u>	**<u>83.3</u>**

[a]The link prediction result of ProjE on FB15k is the latest result provided by author after fixing a bug in the original code. The corresponding parameter setting of the results are: embedding dimension $d = 200$, batchsize $b = 512$, learning rate $r = 0.0005$, negative candidate sampling proportion $m = 0.1$ and max iteration number $iter = 50$.

Training: We use Adaptive Moment Estimation (Adam) [10] as the optimizer during training with the default parameter setting: $\beta_1 = 0.9, \beta_2 = 0.999$, $\epsilon = 1e^{-8}$. We also apply a dropout [18] layer on the top of combination operation to prevent overfitting and the hyperparameter of dropout rate is set to 0.5. Before training, we randomly initialize all the entity and relation vectors from a uniform distribution $U[-\frac{6}{\sqrt{k}}, \frac{6}{\sqrt{k}}]$ as suggested in [7]. The diagonal matrices are initialized with identity diagonal matrix. The bias vector is initialized as zero vector. For both datasets, we set the max training iterations to 100.

Evaluation: We evaluate link prediction following the same protocol of previous work: for every testing triple (h, r, t), we first predict t with input (h, r), then predict h with input (t, r^{-1}). To predict t, we replace t with each entity e in experiment dataset and calculate the similarity score through $S(h, r, e)$. Then rank the scores in ascending order and get the rank of the original right tail entity. The processing of head prediction is the same as tail prediction. Aggregating all the ranks of testing triples, we follow the two metrics used in previous work: *Mean Rank* and *Hit@10*. *Mean Rank* is the averaged rank of all the testing triples. *Hit@10* is the proportion of ranking score of testing triple that is not larger than 10. A good link predictor should achieve lower mean rank and higher hit@10. We also follow the *Filter* and *Raw* settings as previous work. *Filter* setting means filtering the triples in training data when generating negative triples to prevent false negative ones. *Raw* setting means without filtering.

Table 5. Experimental results on FB15k by mapping different patterns (%)

Method	Predict Head(Hit@10)				Predict Tail(Hit@10)			
	1-1	1-N	N-1	N-N	1-1	1-N	N-1	N-N
Unstructured [3]	34.5	2.5	6.1	6.6	34.3	4.2	1.9	6.6
SE [5]	35.6	62.6	17.2	37.5	34.9	14.6	68.3	41.3
SME(linear) [3]	35.1	53.7	19.0	40.3	32.7	14.9	61.6	43.3
SME(Bilinear) [3]	30.9	69.6	19.9	38.6	28.2	13.1	76.0	41.8
TransE [4]	43.7	65.7	18.2	47.2	43.7	19.7	66.7	50.0
TransH(unif) [21]	66.7	81.7	30.2	57.4	63.7	30.1	83.2	60.8
TransH(bern) [21]	66.8	87.6	28.7	64.5	65.5	39.8	83.3	67.2
CTransR(unif) [12]	78.6	77.8	36.4	68.0	77.4	37.8	78.0	70.3
CTransR(bern) [12]	81.5	89.0	34.7	71.2	80.8	38.6	90.1	73.8
TransD(unif) [9]	80.7	85.8	47.1	75.6	80.0	54.5	80.7	77.9
TransD(bern) [9]	86.1	**95.5**	39.8	78.5	85.4	50.6	94.4	81.2
TransR(unif) [12]	76.9	77.9	38.1	66.9	76.2	38.4	76.2	69.1
TransR(bern) [12]	78.8	89.2	34.1	69.2	79.2	37.4	90.4	72.1
ProjE_listwise [16]	61.4	90.1	53.2	**83.3**	61.3	63.5	89.4	**85.5**
ProjR(this paper)	**90.3**	93.3	**64.5**	81.0	**90.7**	**78.3**	**96.6**	85.0

Result: The result of link prediction is in Table 4. We directly copy the results of previous methods from their original paper as WN18 and FB15k are two benchmark datasets. The parameter settings for ProjR's results on WN18 in Table 4 are: embedding dimension $d = 100$, learning rate $r = 0.01$, batchsize $b = 2000$, negative candidate sampling proportion $m = 0.001$. The parameter settings on FB15k are: $d = 200, r = 0.01, b = 4000, m = 0.005$. From Table 4, we could conclude that: (1) As a combination method of TransR and ProjE, ProjR achieves better performance on FB15k than ProjE and TransR. And on WN18, ProjR performs better than TransR on Hit@10. (2) Though on some tasks TransD performs better than ProjR, ProjR has less parameters than TransD. The number of parameters in ProjR is $(n_e + 4n_r + 1)k$ while that in TransD is $2(n_e + n_r)$ and $n_e \gg n_r$.

To analyze how diverse representations improve the link prediction results of different types of relations. We also conduct the experiment of link prediction

Table 6. Hit@10 of ProjE and ProjR on some examples of 8 kinds of substructures. The results are in the form of x/y in which x is the result of ProjE and y is the result of ProjR. The third column is the number of testing triples related with current row relation in testing data.

Type	Example relations	#	Hit@10 (%)	
			Head	Tail
1-1	/influence/peer_relationship/peers	21	66.7/**90.5**	66.7/**85.7**
	/business/employment_tenure/person	43	72.1/**79.1**	67.4/**72.1**
	/tv/tv_program/program_creator	23	95.6/95.6	95.7/95.7
1-N	/film/writer/film	105	92.4/**93.3**	86.7/**87.6**
	/location/country/second_level_divisions	68	**100.0/100.0**	28.4/**77.6**
	/people/cause_of_death/people	98	**87.8**/83.7	77.6/**79.6**
N-1	/music/group_member/membership	22	50.0/**59.1**	68.2/**77.3**
	/people/person/nationality	508	2.2/**13.2**	87.4/**93.1**
	/people/person/education/institution	358	52.2/**73.2**	69.3/**79.1**
N-N	/award/award_winner/awards_won	1045	90.4/**93.8**	90.4/**93.0**
	/tv/tv_genre/programs	105	**96.2**/95.2	87.6/**90.5**
	/music/genre/parent_genre	100	75.0/**83.0**	86.0/**90.0**
1-N-1	/award/award_winner/award	655	63.8/**64.7**	94.4/**94.7**
	/award/award_nominee/award_nominations	1555	**83.0**/82.7	94.0/**96.4**
C1	/education/educational_institution/campuses	60	78.3/**100.0**	80.0/**100.0**
	/education/educational_institution_campus	68	80.9/**100.0**	80.9/**100.0**
	/location/hud_county_place/place	48	20.8/**100.0**	22.9/**100.0**
C2	/people/person/spouse_s	54	33.3/**40.7**	27.8/**35.2**
	/influence/influence_node/influenced	235	**71.5**/63.4	**68.5**/55.7
	/location/location/adjoin_s	284	77.8/**88.7**	75.0/**83.4**
C3	/location/location/contains	608	92.6/**97.2**	**85.7**/84.0
	/location/adjoining_relationship/adjoins	284	77.8/**88.7**	75.0/**83.5**

mapping different structures. In this experiment, we only consider 1-1, 1-N, N-1 and N-N as the other four relation properties could be included in this four types. We choose FB15k as the dataset of this experiment because it contains more relations than WN18. The Hit@10 result of filter setting on FB15k mapping different substructures are showed in Table 5.

The results in Table 5 show that ProjR improves the ability to encode 1-1, 1-N, N-1 and N-N relations with diverse representations for entities. Among all structures, the head prediction of N-1 relations and tail prediction of 1-N relations are the most difficult tasks. And ProjR achieves good improvement on these two tasks. Compared with the second best result listed in Table 5, ProjR achieves 11.3% improvement for head prediction of 1-N relations and 14.8% improvement of tail prediction for N-1 relations.

To understand the ability of ProjR to encode the diversity of knowledge graph more deeply. We select two or three relations for each type of structure. We select relations with the principle that the number of testing triples related to the relation should be larger than 20. We compare the filter Hit@10 result of ProjE and ProjR on FB15k for each selected relation. The results are listed in Table 6. The parameter settings for ProjE and ProjR are same as the parameter settings of the results in Table 4. The results show that ProjR achieves better results in the majority of the relations for each substructure. A huge improvement is achieved on C1 structure.

5   Conclusion and Future Work

In this paper, we list 8 kinds of graph substructures to explore the structure diversity of knowledge graph and propose a new embedding method ProjR to encode structure diversity more properly based on the idea of diverse representations for entities in different relation contexts and different entity positions. In link prediction experiments, ProjR achieves better results compared with the two most related methods, TransR and ProjE. We explore the results from coarse to fine to illustrate how ProjR improves the ability to encode the diversity of structures.

There are some interesting topics that we want to explore in the future: (1) as shown in Table 6, the prediction results for different relations range hugely, which means there are still different structures between those relations. (2) Knowledge graphs are dynamic in the real world and new triples are always added to them. But existing knowledge graph embedding methods can not handle the dynamic property of KG.

Acknowledgement. This work is funded by NSFC 61473260/61673338, and Supported by Alibaba-Zhejiang University Joint Institute of Frontier Technologies.

References

1. Banko, M., Ca-farella, M.J., Soderland, S., Broadhead, M., Etzioni, O.: Open information extraction from the web (2007)
2. Bollacker, K.D., Evans, C., Paritosh, P., Sturge, T., Taylor, J.: Freebase: a collaboratively created graph database for structuring human knowledge. In: Proceedings of SIGMOD, pp. 1247–1250 (2008)
3. Bordes, A., Glorot, X., Weston, J., Bengio, Y.: A semantic matching energy function for learning with multi-relational data - application to word-sense disambiguation. Mach. Learn. **94**(2), 233–259 (2014)
4. Bordes, A., Usunier, N., García-Durán, A., Weston, J., Yakhnenko, O.: Translating embeddings for modeling multi-relational data. In: Proceedings of NIPS, pp. 2787–2795 (2013)
5. Bordes, A., Weston, J., Collobert, R., Bengio, Y.: Learning structured embeddings of knowledge bases. In: Proceedings of AAAI (2011)
6. García-Durán, A., Bordes, A., Usunier, N.: Composing relationships with translations. In: Proceedings of EMNLP, pp. 286–290 (2015)
7. Glorot, X., Bengio, Y.: Understanding the difficulty of training deep feedforward neural networks. In: Proceedings of AISTATS, pp. 249–256 (2010)
8. Jenatton, R., Roux, N.L., Bordes, A., Obozinski, G.: A latent factor model for highly multi-relational data. In: Proceddings of NIPS, pp. 3176–3184 (2012)
9. Ji, G., He, S., Xu, L., Liu, K., Zhao, J.: Knowledge graph embedding via dynamic mapping matrix. In: Proceedings of ACL, pp. 687–696 (2015)
10. Kingma, D.P., Ba, J.: Adam: a method for stochastic optimization. CoRR abs/1412.6980 (2014)
11. Lin, Y., Liu, Z., Luan, H., Sun, M., Rao, S., Liu, S.: Modeling relation paths for representation learning of knowledge bases. In: Proceedings of EMNLP, pp. 705–714 (2015)
12. Lin, Y., Liu, Z., Sun, M., Liu, Y., Zhu, X.: Learning entity and relation embeddings for knowledge graph completion. In: Proceedings of AAAI, pp. 2181–2187 (2015)
13. Miller, G.A.: WordNet: a lexical database for English. Commun. ACM **38**(11), 39–41 (1995)
14. Nickel, M., Rosasco, L., Poggio, T.A.: Holographic embeddings of knowledge graphs. In: Proceedings of AAAI, pp. 1955–1961 (2016)
15. Nickel, M., Tresp, V., Kriegel, H.: A three-way model for collective learning on multi-relational data. In: Proceedings of ICML, pp. 809–816 (2011)
16. Shi, B., Weninger, T.: Proje: Embedding projection for knowledge graph completion. In: Proceedings of AAAI, pp. 1236–1242 (2017)
17. Socher, R., Chen, D., Manning, C.D., Ng, A.Y.: Reasoning with neural tensor networks for knowledge base completion. In: Proceedings of NIPS, pp. 926–934 (2013)
18. Srivastava, N., Hinton, G., Krizhevsky, A., Sutskever, I., Salakhutdinov, R.: Dropout: a simple way to prevent neural networks from overfitting. J. Mach. Learn. Res. **15**, 1929–1958 (2014)
19. Suchanek, F.M., Kasneci, G., Weikum, G.: Yago: a core of semantic knowledge. In: Proceedings of WWW, pp. 697–706 (2007)
20. Wang, Z., Zhang, J., Feng, J., Chen, Z.: Knowledge graph and text jointly embedding. In: Proceedings of EMNLP, pp. 1591–1601 (2014)
21. Wang, Z., Zhang, J., Feng, J., Chen, Z.: Knowledge graph embedding by translating on hyperplanes. In: Proceedings of AAAI, pp. 1112–1119 (2014)

22. Wang, Z., Li, J.: Text-enhanced representation learning for knowledge graph. In: Proceedings of IJCAI, pp. 1293–1299 (2016)
23. Xiao, H., Huang, M., Meng, L., Zhu, X.: SSP: semantic space projection for knowledge graph embedding with text descriptions. In: Proceedings of AAAI, pp. 3104–3110 (2017)
24. Xie, R., Liu, Z., Jia, J., Luan, H., Sun, M.: Representation learning of knowledge graphs with entity descriptions. In: Proceedings of AAAI, pp. 2659–2665 (2016)
25. Xie, R., Liu, Z., Sun, M.: Representation learning of knowledge graphs with hierarchical types. In: Proceedings of IJCAI, pp. 2965–2971 (2016)
26. Zhang, C.: DeepDive: a data management system for automatic knowledge base construction (2015)
27. Zhang, W.: Knowledge graph embedding with diversity of structures. In: Proceedings of WWW Companion, pp. 747–753 (2017)

BiTCNN: A Bi-Channel Tree Convolution Based Neural Network Model for Relation Classification

Feiliang Ren[⊠], Yongcheng Li, Rongsheng Zhao,
Di Zhou, and Zhihui Liu

School of Computer Science and Engineering,
Northeastern University, Shenyang 110819, China
renfeiliang@cse.neu.edu.cn

Abstract. Relation classification is an important task in natural language processing (NLP) fields. State-of-the-art methods are mainly based on deep neural networks. This paper proposes a bi-channel tree convolution based neural network model, *BiTCNN*, which combines syntactic tree features and other lexical level features together in a deeper manner for relation classification. First, each input sentence is parsed into a syntactic tree. Then, this tree is decomposed into two sub-tree sequences with top-down decomposition strategy and bottom-up decomposition strategy. Each sub-tree represents a suitable semantic fragment in the input sentence and is converted into a real-valued vector. Then these vectors are fed into a bi-channel convolutional neural network model and the convolution operations re performed on them. Finally, the outputs of the bi-channel convolution operations are combined together and fed into a series of linear transformation operations to get the final relation classification result. Our method integrates syntactic tree features and convolutional neural network architecture together and elaborates their advantages fully. The proposed method is evaluated on the SemEval 2010 data set. Extensive experiments show that our method achieves better relation classification results compared with other state-of-the-art methods.

Keywords: Relation classification · Syntactic parsing tree · Tree convolution
Convolutional neural networks

1 Introduction

The aim of relation classification is that given a sentence in which two entities are labeled, to select a proper relation type from a predefined set for these entities. For example, given a sentence "*The system as described above has its greatest application in an arrayed <e1> configuration </e1> of antenna <e2> elements </e2>*", a relation classification system aims to identify that there is a "*Component-Whole*" relationship from *e2* to *e1*. Obviously, accurate relation classification results would benefit lots of NLP tasks, such as sentence interpretations, Q&A, knowledge graph construction, ontology learning, and so on. Thus, lots of researchers have devoted to this research field.

© Springer Nature Switzerland AG 2018
M. Zhang et al. (Eds.): NLPCC 2018, LNAI 11108, pp. 158–170, 2018.
https://doi.org/10.1007/978-3-319-99495-6_14

For relation classification, early research mostly focused on features based methods. Usually, these methods firstly select some syntactic and semantic features from the given sentences. Then the selected features are fed into some classification models like support vector machines, maximum entropy, etc. Recently, deep neural network (DNN) based methods have been widely explored in relation classification and have achieved state-of-the-art experimental results. The core of these methods is to embed features into real-valued vectors, and then feed these vectors into DNN architectures. Usually, deep convolutional neural networks (CNN) and deep recurrent neural networks (RNN) are two most widely used architectures for relation classification.

In most recent years, inspired by the broad consensus that syntactic tree structures are of great help and the great success of DNN, more and more research attention is being paid to the methods that integrate syntactic tree features into DNN models. However, most of these existing methods used syntactic tree in a very shallow manner: syntactic tree structure is often taken as an intermediate supporter from which a specific kind of context can be extracted for CNN or RNN models. Obviously, such shallow manner does not make full use of the rich semantic information carried by syntactic tree structures.

It is worth noting that Socher et al. (2013a, b) introduced Compositional Vector Grammar (CVG for short), which used a syntactically untied RNN model to learn a syntactic-semantic compositional vector representation for the category nodes in a syntactic tree. Inspired by their work, we propose a new relation classification method that integrates syntactic tree structures into CNN model with a deeper manner. Specifically, in our method, each input sentence is first parsed into a syntactic tree. Then this tree is decomposed into two sub-tree sequences with bottom-up and top-down decomposition methods respectively. Thirdly, each sub-tree is encoded into a real-valued vector. Fourthly, the two sub-tree vector sequences are fed into a bi-channel CNN model to generate final classification result. Experimental results show that our method achieves better results compared with other baseline methods.

2   Related Work

Generally, there are three widely used DNN architectures for relation classification: CNN, RNN, and their combination.

Zeng et al. (2014) proposed a CNN based approach for relation classification. In their method, sentence level features are learned through a CNN model that takes word embedding features and position embedding features as input. In parallel, lexical level features are extracted from some context windows that are around the labeled entities. Then sentence level features and lexical level features are concatenated into a single vector. This vector is fed into a softmax classifier for relation prediction. Wang et al. (2016) proposed a multi-level attention CNN model for relation classification. In their method, two levels of attentions are used in order to better discern patterns in heterogeneous contexts.

Socher et al. (2012) used RNN for relation classification. In their method, they build recursive sentence representations based on syntactic parsing. Zhang and Wang (2015) investigated a temporal structured RNN with only words as input. They used a bi-directional model with a pooling layer on top. Xu et al. (2015a, b) picked up heterogeneous information along the left and right sub-path of the Shortest Dependent Path (SDP) respectively, leveraging RNN with LSTM. In their method, the SDP retains most relevant information to relation classification, while eliminating irrelevant words in the sentence. And the multichannel LSTM networks allow effective information integration from heterogeneous sources over the dependency paths. Meanwhile, a customized dropout strategy regularizes the neural network to alleviate over-fitting. Besides, there is also other similar work. For example, Hashimoto et al. (2013) explicitly weighted phrases' importance in RNNs. Ebrahimi and Dou (2015) rebuilt an RNN on the dependency path between two labeled entities.

Some researchers combined CNN and RNN for relation classification. For example, Vu et al. (2015) investigated CNN and RNN as well as their combination for relation classification. They proposed extended middle context, a new context representation for CNN architecture. The extended middle context uses all parts of the sentence (the relation arguments, left/right and between of the relation arguments) and pays special attention to the middle part. Meanwhile, they used a connectionist bi-directional RNN model and a ranking loss function is introduced for the RNN model. Finally, CNN and RNN were combined with a simple voting scheme. Cai et al. (2016) proposed a bidirectional neural network BRCNN, which consists of two RCNNs that can learn features along SDP inversely at the same time. Specifically, information of words and dependency relations are used with a two-channel RNN model with LSTM units. The features of dependency units in SDP are extracted by a convolution layer. Liu et al. (2015) used a RNN to model the sub-trees, and a CNN to capture the most important features on the SDP.

3 Our Model

Figure 1 demonstrates the architecture of our method. The network takes sentences as input and extracts syntactic tree features and other useful features. These features are converted into real-valued vector representations and then fed into a bi-channel CNN model for relation type decision. From Fig. 1 we can see that there are six main components in our method: tree decomposition, feature extraction, convolution transformation, max-pooling operation, linear transformation and output.

3.1 Tree Decomposition

Each input sentence will firstly be parsed into a syntactic tree by the Stanford Parser. Then this tree is decomposed into two sub-tree sequences with bottom-up decomposition method and top-down decomposition method. These two kinds of decomposition methods, whose algorithms are shown in Figs. 2 and 3 respectively, complement each other and are expected to generate more meaningful and less ambiguous semantic fragments than words.

For the top-down tree decomposition method, its generated sub-trees don't contain any word information. It is expected to extract the common syntactic sub-tree structures for a specific kind of relationship type, and is also expected to alleviate the over-fitting

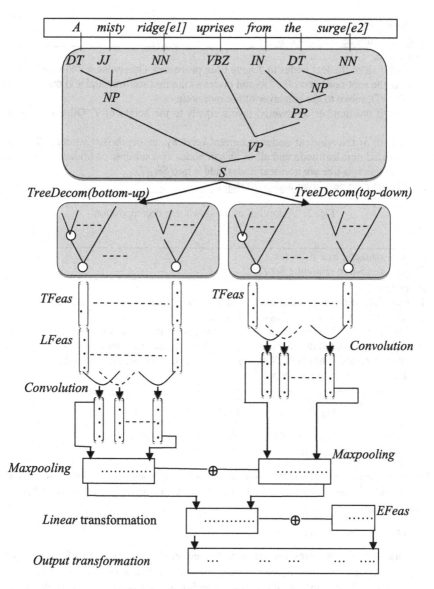

Fig. 1. Architecture of our method

issue by using these abstract sub-tree structures as features. Taking the sentence in Fig. 1 as an example, the sub-tree sequence generated by this method is: (S (NP) (VP)), (NP ((DT) (JJ) (NN))), (VP (VBZ) (PP)), (PP (IN) (NP)), (NP (DT) (NN)).

As for the bottom-up method, it complements with the top-down method. In this method, word information is taken into consideration. Taking the sentence in Fig. 1 as an example, if the hyper parameters in Fig. 3 are set as: h = 3, Δ = 3, and k = 3, the sub-tree sequence generated by this method would be: (NP ((DT A) (JJ misty) (NN ridge))), (VP (VBZ uprises) (PP (IN from) (NP (DT the) (NN surge)))).

Input: a syntactic tree T
Output: a sub-tree sequence $Set_1(T)$

Procedure:
1: Exit if all of the leaf nodes in T have been processed. Otherwise, go to step 2.
2: Take the root node and all of its son nodes as the first sub-tree and add this sub-tree into $Set_1(T)$, move to the son layer of the root node.
3: Exit if the number of covered layers equals to the height of T. Otherwise, go to step 3.
4: Visit all of the non-leaf nodes in current layer by the depth-first strategy. Take an unprocessed non-leaf node and all of its son nodes as a sub-tree candidate, if all of its nodes in this sub-tree are non-leaf nodes, add it into $Set_1(T)$.
5: Move to the next layer and go to step 2.

Fig. 2. Algorithm of top-down tree decomposition

Input: a syntactic tree T
Output: a sub-tree sequence $Set_2(T)$

Procedure:
1: Exit if all of the leaf nodes have been processed. Otherwise, go to step 2.
2: Take the left-most unprocessed leaf node l_i.
3: Move upward h layer from l_i to find its ancestor node a_i and extract the sub-tree t_i that takes a_i as root node. If all the out-degrees in t_i are less than Δ and the height of t_i is less than k, add t_i into $Set_2(T)$, then go to step 2. Otherwise, go to step 4.
4: Let $h = h - 1$ and go to step 3.

Fig. 3. Algorithm of bottom-up tree decomposition

3.2 Feature Extraction

There are three kinds of features used in our model: syntactic tree features, lexical level features and entities level features. In Fig. 1, they are denoted as **TFeas**, **LFeas** and **EFeas** respectively.

- Syntactic Tree Features

Syntactic tree structures can carry more semantic and syntactic information compared with characters, words, or phrases. To make full use of such information, we embed each sub-tree into a d^{tree}-dimensional real-valued vector. Just like a word embedding vector can encode different meaning of the word into a vector, we hope that a tree embedding vectors can also encode as much as possible semantic and syntactic information for a tree structure. In our method, this kind of tree embedding vectors is initialized with the method proposed by Socher et al. (2013a, b). Their method can assign a vector representation for both the nodes in a parsing tree and the whole parsing tree itself. But in our model, we don't care the representations of any inner categories in a parsing tree. Thus, our method assigns a vector representation for the whole tree structure only.

- Lexical Level Features

Lexical level features refer to the features that are related to words. Thus only the sub-trees generated by the bottom-up decomposition method will involve this kind of features. There are two kinds of lexical level features used in our method: word embedding features and position features. The final lexical level features, *LFeas*, are the concatenations of these two kinds of features.

1. Word Embedding Features

Word embedding is a kind of word representation method and is widely used in DNN models. It converts a word into a real-valued vector representation to capture the rich syntactic and semantic features possessed by this word. Generally, a word embedding table is a $d^w*|V|$ real-valued matrix where each column corresponds to a word representation. d^w is the dimension of a embedding vector, and $|V|$ is the vocabulary size.

For a sub-tree t_i that is generated by the bottom-up decomposition method, each of its leaf nodes has a word embedding vector. If there are m leaf nodes in t_i, its word embedding features $wf(t_i)$ would be the arithmetic mean of its m words' embeddings, as shown in:

$$wf(t_i)_j = avg \sum_{k=1}^{m} w_{kj}, \ 1 \leq j \leq d^w \tag{1}$$

2. Position Features

Position features are used to specify which input items are the labeled entities in a sentence or how close an input item is to the labeled entities. They have been proved to be effective for relation classification (Dos et al. (2015); Zeng et al. (2014)). This kind of features maps a distance value into a randomly initiated d^{dst}-dimensional real-valued vector.

For a sub-tree t_i that is generated by the bottom-up decomposition method, each of its leaf nodes has two kinds of position features that are related with e_1 and e_2 respectively. Accordingly, there are two kinds of position features for t_i.

If there are m leaf nodes in t_i, its position features related with e_1, denoted as $pf(e_1)$, would be the arithmetic mean of its m leaf nodes' position features related with e_1. The computation process is shown in formula 2.

$$pf(e_1)_j = avg \sum_{k=1}^{m} p_{kj}, \ 1 \leq j \leq d^{dst} \tag{2}$$

Similarly, the computation process for t_i's position features related with e_2, denoted as $pf(e_2)$, is shown in formula 3.

$$pf(e_2)_j = avg \sum_{k=1}^{m} p_{kj}, \ 1 \leq j \leq d^{dst} \tag{3}$$

Finally, the position features for t_i is the concatenation of $pf(e_1)$ and $pf(e_2)$.

3. Entities Level Features

Previous research shows that words between the two labeled entities could provide more useful cues for relation decision. So an attention scheme is used here to enhance the features extracted from the words that are between the two labeled entities. The enhanced feature is called entities level features and is denoted as **EFeas** in Fig. 1.

We first get the syntactic tree for the text that is from the first labeled entity to the second labeled entity. Then the syntactic tree features and the lexical level features are extracted with the same method introduced previously. Thirdly, these two kinds of features are concatenated to form a new feature vector that is denoted as **EntityF**. The final **EFeas** is generated with formula 4.

$$\textbf{EFeas} = tanh(M_1 * \textbf{EntityF} + b_1) \tag{4}$$

Where $M_1 \in R^{h1 *def}$ is a transformation matrix, h_1 is a hyper parameter that denotes the transformed size, def is the dimension of vector **EntityF**, b_1 is a bias term.

From Fig. 1 we can see that **EFeas** would be concatenated with the linear transformed max-pooling features. To maintain feature consistency, here we use the linear transformed **EntityF** as **EFeas**.

3.3   Convolution Transformation

Convolution transformation is a kind of linear transformation and is expected to extract more abstract features.

For the sub-tree sequence that is generated by the bottom-up method, each of its sub-tree t_i is represented by a vector concatenation of **TFeas(i)** and **LFeas(i)**. The convolution transformation process is written as formula 5.

$$\textbf{CMtrB}_i = M_2 * (\textbf{TFeas}(t_i) \oplus \textbf{LFeas}(t_i)) + b_2 \qquad \forall i \tag{5}$$

For the sub-tree sequence that is generated by the top-down method, each of its sub-tree t_i is represented by **TFeas(i)**. The convolution process is written as formula 6.

$$\textbf{CMtrT}_i = M_3 * \textbf{TFeas}(t_i) + b_3 \qquad \forall i \tag{6}$$

In above formulas, $M_2 \in R^{h2*dbf}$ and $M_3 \in R^{h3*dtf}$ are two transformation matrices, h_2 and h_3 are the sizes of transformed hidden layers, d^{bf} and d^{tf} are the dimensions of sub-tree vectors in the bottom-up and top-down tree sequences respectively. b_2 and b_3 are bias terms. $\oplus$ denotes the concatenation operation.

3.4 Max-Pooling Operation

After convolution transformation, both **CMtrB** and **CMtrT** depend on the length of input sequence. To apply subsequent standard affine operations, max-pooling operation is used to capture the most useful and fixed size local features from the output of convolution transformation. This process is written as formula 7 and 8.

$$p_{bi} = max_n CMtrB(i, n) \qquad 0 \leq i \leq h_2 \tag{7}$$

$$p_{ti} = max_n CMtrT(i, n) \qquad 0 \leq i \leq h3 \tag{8}$$

After max-pooling operation, p_b and p_t will have h_2 and h_3 elements respectively, which are no longer related to the length of input.

3.5 Linear Transformation

After the max-pooling operation, p_b and p_t are concatenated together to form a new vector p. Then p is fed into a linear transformation layer to perform affine transformation. This process is written as formula (9).

$$f = tanh(M_4 * p + b_4) \tag{9}$$

$M_4 \in R^{h4*(h2+h3)}$ is the transformation matrix and h_4 is the size of hidden units in this layer, and b_4 isa bias term.

3.6 Output

After linear transformations, vector f and vector **EFeas** are concatenated together to form a new vector o. Then o is fed into a linear output layer to compute the confidence scores for each possible relationship type. A softmax classifier is further used to get the probability distribution y over all relation labels as formula 10.

$$y = softmax(M_5 * o + b_5) \tag{10}$$

Here $M_5 \in R^{h5*(h1+h4)}$ and h_5 is the number of possible relation types. Softmax is computed with formula 11.

$$y_i = e^{xi} / \sum_m e^{xm} \tag{11}$$

3.7 Dropout Operation

Over-fitting is an issue that cannot be ignored in DNN models. Hinton et al. (2012) proposed dropout method that has been proved to be effective for alleviating over-fitting. This method randomly sets a proportion (called drop rate, a hyper-parameter) of features to zero during training. It is expected to obtain less interdependent network units, thus over-fitting issue is expected to be alleviated. In our method, dropout strategy is taken at feature extraction phase and linear transformation phase. Specially, we take dropout operation on **EntityF**, **TFeas**, **LFeas** in formula $4 \sim 6$ respectively, and on **p** in formula 9. The drop rates for them are denoted as $dp_{1 \sim 4}$ respectively.

3.8   Training Procedure

All the parameters in our method can be denoted as $\theta = (E^w, E^t, E^p, M_1, M_2, M_3, M_4,$ $M_5, b_1, b_2, b_3, b_4, b_5)$, where E^w, E^t and E^p represent the embeddings of word, syntactic tree and position respectively. E^w is initialized by the pre-trained embeddings SENNA (Collobert et al. 2011).Et is initialized with the method introduced by Socher et al. (2013a, b). E^p, transformation matrices, and bias terms are randomly initialized. All the parameters are tuned using the back propagation method. Stochastic gradient descent optimization technique is used for training. Formally, we try to maximize following objective function.

$$J(\theta) = \sum_{i=1}^{N} logy_i \qquad (12)$$

where N is the total number of training samples. During training, each input sentence is considered independently. And each parameter is updated by applying following update rule, in which η is the learning rate.

$$\theta = \theta + \eta * \partial logy / \partial \theta \qquad (13)$$

4   Experiments and Analysis

4.1   Datasets

The SemEval-2010 Task 8 dataset is used to evaluate our method. In this dataset, there are 8000 training sentences and 2717 test sentences. For each sentence, two entities that are expected to be predicted a relation type are labeled. There are 9 relation types whose directions need to be considered and an extra artificial relation "*Other*" which does not need to consider the direction. Thus totally there are 19 relation types in this dataset.

Macro-averaged F1 score (excluding "*Other*"), the official evaluation metric, is used here and the direction is considered. During experiments, all the syntactic trees are generated by the Stanford Parser (Klein and Manning, 2003). We apply a cross-validation procedure on the training data to select suitable hyper-parameters. Finally, the best configurations obtained are: $d^{dst} = 75$, $\eta = 0.001$, $h_{1 \sim 4}$ are 250, 200, 200 and 300 respectively, $dp_{1 \sim 4}$ are all set to 0.5. In Fig. 3, Δ, h and k are set to 3, 3 and 3 respectively. Other parameters, d^w and d^{tree} are 50 and 25 respectively.

4.2 Experimental Results and Analyses

In the first part of our experiment, we evaluate the contributions of different kinds of features and different convolutional channels. To this end, we implement a CNN model that is similar to the one described in Zeng et al. (2014). This CNN model is denoted as *baseline* in which word embedding features and position features are used. Besides, we implement two other CNN models: one is with the bottom-up tree convolution channel, and the other is with the top-down tree convolution channel. Then we investigate how the performance changes when different kinds of features are added. The experimental results are reported in Table 1.

Table 1. Performance of our method with different features

Model	F1
Baseline	82.4
Bottom-up tree convolution(without LFeas)	73.0
Bottom-up tree convolution + *WordEmb feature*	74.6
Bottom-up tree convolution + *Position feature*	74.0
Bottom-up tree convolution + *LFeas*	76.3
Top-down tree convolution	65.5
Our model	84.8

We can see that our model is very effective and it outperforms the baseline system greatly. Also we can see that different convolution channels have different classification performance. Even without **LFeas** features, the bottom-up tree CNN model achieves better performance than the top-down tree CNN model. We think the main reason is that in the bottom-up model, word information is retained and this kind of information would play positive role for relation classification. This can be further proved by the experimental results: the performance improves when different kinds of lexical features are added in the bottom-up tree CNN model.

In the second part of our experiment, we compare our method with several other state-of-the-art DNN based methods. Because the datasets used are the same, we directly copy the experimental results reported in Zeng et al. (2014). The comparison results are shown in Table.

From Table 2 we can see that our method achieves better results compared with other methods. It is also worth noting that our method is the ONLY ONE that does not use any external language-dependent resources like WordNet. This shows the effectiveness of embedding syntactic tree features into CNN architecture for relation classification.

In the third part of our experiment, we compare the classification performance for different types of relationships. The comparison results are reported in Table 3.

From the experimental results we can see that the performance of different types of relationships is very different. Even excluding "*Other*" type, the best performance (for example, "*cause-effect*" and "*entity-destination*") is almost 10% higher than the worst performance (for example, "*product-producer*" and "*content-container*"). Further

Table 2. Comparisons with other methods

Method	Features and extra resources used	F1
CNN	WordNet	82.7
SVM	POS, prefixes, morphological, WordNet, dependency parse, Levin classed, ProBank, FrameNet, NomLex-Plus, Google n-gram, paraphrases, TextRunner	82.2
RNN	-	74.8
	POS, NER, WordNet	77.6
MVRNN	-	79.1
	POS, NER, WordNet	82.4
Our model	parsing trees	84.8

Table 3. Classification results of different relationships

Relationship	P	R	F
Cause-Effect	95.43%	88.17%	91.65%
Component-Whole	80.45%	81.76%	81.10%
Content-Container	84.38%	77.14%	80.60%
Entity-Destination	95.21%	87.97%	91.45%
Entity-Origin	89.15%	85.82%	87.45%
Instrument-Agency	83.97%	76.61%	80.12%
Member-Collection	91.85%	76.70%	83.59%
Message-Topic	88.51%	74.76%	81.05%
Product-Producer	82.68%	78.60%	80.59%
Other	40.75%	71.43%	51.89%

investigation shows that for the relationships that model can classify better, there are usually some clear indicating words in the original sentences. For example, words "*cause*", "*caused*", and "*causes*" are often in the sentences where a "*cause-effect*" relationship holds. On the contrary, there are usually few of such kind of indicating words for the relationships that model classifies worse. As a result, max-pooling operation couldn't work well in such cases.

5 Conclusions and Future Work

In this paper, we propose a new relation classification method. The main contributions of our method are listed as follows.

First, our method uses syntactic tree structures in a deeper manner: the input sentence is parsed into a syntactic tree. And this tree is further decomposed into two sub-tree sequences and the convolution operations are performed on the sub-tree embeddings directly.

Second, we design two decomposition methods to guarantee the tree decomposition process performed in a reasonable way, which means that each of the generated subtrees has a relatively complete structure and can express a complete meaning.

However, there are still some other issues needed to be further investigated. For example, experimental results show that there are big performance gaps between different types of relationships, which should be further investigated in the future.

Acknowledgements. This work is supported by the National Natural Science Foundation of China (NSFC No. 61572120, 61672138 and 61432013).

References

Collobert, R., Weston, J., Bottou, L., Karlen, M., Kavukcuoglu, K., Kuksa, P.: Natural language processing (almost) from scratch. J. Mach. Learn. Res. **12**, 2493–2537 (2011)

Wang, L., Cao, Z., de Melo, G., Liu, Z.: Relation classification via multi-level attention CNNs. In: Proceedings of the 54th Annual Meeting of the Association for Computational Linguistics, pp. 1298–1307 (2016)

Cai, R., Zhang, X., Wang, H.: Bidirectional recurrent convolutional neural network for relation classification. In: Proceedings of the 54th Annual Meeting of the Association for Computational Linguistics, pp. 756–765 (2016)

Xu, K., Feng, Y., Huang, S., Zhao, D.: Semantic relation classification via convolutional neural networks with simple negative sampling. In: Proceedings of 2015 Conference on Empirical Methods in Natural Language Processing, pp. 536–540 (2015a)

Xu, Y., Mou, L., Li, G., Chen, Y., Peng, H., Jin, Z.: Classifying relations via long short term memory networks along shortest dependency paths. In: Proceedings of the 2015 Conference on Empirical Methods in Natural Language Processing, pp. 1785–1794 (2015b)

Zeng, D., Liu, K., Lai, S., Zhou, G., Zhao, J.: Relation classification via convolutional deep neural network. In: Proceedings of the 25th International Conference on Computational Linguistics, pp. 2335–2344 (2014)

Zhang, Z., Zhao, H., Qin, L.: Probabilistic graph-based dependency parsing with convolutional neural network. In: Proceedings of the 54th Annual Meeting of the Association for Computational Linguistics, pp. 1382–1392 (2016)

Socher, R., Bauer, J., Manning, C.D., Ng, A.Y.: Parsing with compositional vector grammars. In: Proceedings of the 51th Annual Meeting of the Association for Computational Linguistics, pp. 455–465 (2013a)

Socher, R., Perelygin, A., Wu, J.Y., Chuang, J., Manning, C.D., Ng, A.Y., Potts, C.: Recursive deep models for semantic compositionality over a sentiment treebank. In: Proceedings of the 2013 Conference on Empirical Methods in Natural Language Processing, pp. 1631–1642 (2013b)

dos Santos, C.N., Xiang, B., Zhou, B.: Classifying relations by ranking with convolutional neural networks. In: Proceedings of the 53rd Annual Meeting of the Association for Computational Linguistics, pp. 626–634 (2015)

Liu, Y., Wei, F., Li, S., Ji, H., Zhou, M., Wang, H.: A dependency-based neural network for relation classification. In: Proceedings of the 53rd Annual Meeting of the Association for Computational Linguistics, pp: 285–290 (2015)

Vu, N.T., Adel, H., Gupta, P., Schutze, H.: Combining recurrent and convolutional neural networks for relation classification. In: Proceedings of NAACL-HLT 2016, pp. 534–539 (2015)

Hashimoto, K., Miwa, M., Tsuruoka, Y., Chikayama, T.: Simple customization of recursive neural networks for semantic relation classification. In: Proceedings of the 2013 Conference on Empirical Methods in Natural Language Processing, pp. 1372–1376 (2013)

Socher, R., Huval, B., Manning, C.D., Ng, A.Y.: Semantic compositionality through recursive matrix-vector spaces. In: Proceedings of the 2012 Joint Conference on EMNLP and Computational Natural Language Learning, pp. 1201–1211 (2012)

Using Entity Relation to Improve Event Detection via Attention Mechanism

Jingli Zhang, Wenxuan Zhou, Yu Hong$^{(\boxtimes)}$, Jianmin Yao, and Min Zhang

Computer Science and Technology, Soochow University, Suzhou, Jiangsu, China
jlzhang05@gmail.com, chrisnotkris7@gmail.com, tianxianer@gmail.com,
{jyao,minzhang}@suda.edu.cn

Abstract. Identifying event instance in texts plays a critical role in the field of Information Extraction (IE). The currently proposed methods that employ neural networks have successfully solve the problem to some extent, by encoding a series of linguistic features, such as lexicon, part-of-speech and entity. However, so far, the entity relation hasn't yet been taken into consideration. In this paper, we propose a novel event extraction method to exploit relation information for event detection (ED), due to the potential relevance between entity relation and event type. Methodologically, we combine relation and those widely used features in an attention-based network with Bidirectional Long Short-term Memory (Bi-LSTM) units. In particular, we systematically investigate the effect of relation representation between entities. In addition, we also use different attention strategies in the model. Experimental results show that our approach outperforms other state-of-the-art methods.

Keywords: Event detection · Attention mechanisms · Entity relation

1 Introduction

Event extraction (EE) is an important task in IE. The purpose is to detect event triggers with specific types and their arguments. This paper only tackles event detection (ED) task, which is a crucial part of EE, focusing on identifying event triggers and their categories. Take the following sentence for example:

S1: *David Kaplan was killed by <u>sniper</u> fire in the <u>Balkans</u> in 1992.*

There is an *Attack* event is mentioned in **S1**. The *"fire"* is annotated as trigger in ACE-2005 corpus. Thus, an ED system should be able to identify the trigger as *"fire"* and assign it an event type **Attack**.

However, it might be easily misclassified as **End-Position** event in reality because *"fire"* is a polysemy. In this case, Liu et al. [15] utilized entity to reinforce the classification. Such as in **S1**, they proposed that considering the word *"sniper"*, which serves as an entity (*Attacker)* of the target event, to get more confidence in predicting the trigger as an **Attack** event.

© Springer Nature Switzerland AG 2018
M. Zhang et al. (Eds.): NLPCC 2018, LNAI 11108, pp. 171–183, 2018.
https://doi.org/10.1007/978-3-319-99495-6_15

Unfortunately, although the entity information is considered, there are still difficulties to identify event type correctly for some sentences. For example, consider the following two instances:

S2: *Swenson herself would also* **leave** *the NCA.*

S3: *It's the Iraqi ambassador's time to* **leave** *the United States.*

Table 1 lists the relevant information about S2 and S3. In S2, "*leave*" is a trigger of type *End-Position*. However, in S3, "*leave*" is a trigger of type *Transport*, which is more common than *End-Position*. In addition, the entities in S2 are *Non-Governmental* and *Individual*, which have no discrimination for identifying event type with entities (*Nation* and *Individual*) in S3. Because they all indicate institution and individual. However, if we consider the relation between the entities, which is *Membership* in S2 and *Located* in S3, we would have more confidence in predicting the *End-Position* event and the *Transport* event respectively. Such as in S2, *Membership* indicates the member relationship between two entities, then the probability that the sentence will contain an *End-Position* event will be higher than a *Transport* event.

Table 1. Event, Entity and Relation labels for the above two instances.

Instance	S2	S3
Event	*leave:* **End-Position**	*leave:* **Transport**
Entity1	*NCA:* **Non-Governmental**	*United States:* **Nation**
Entity2	*swenson:* **Individual**	*ambassador:* **Individual**
Relation	**Membership**	**Located**

In addition, we note that words in sentence have different contribution degrees for the correct recognition of trigger. Some words are important, yet others are meaningless. For example, in **S1**, "*killed*" and "*sniper*" provide more important clues than other words that an *Attack* event might happen. Thus, these words should get more attention than others. Guided by this, we employ attention mechanism to model the sentence. Attention values indicate the importance of different words in the sentence of predicting the event type.

To sum up, based on the entity relation information and different significant clues that different words can provide, we propose an attention-based Bi-LSTM model and utilize relation type embedding in the model to detect event. In summary, the contributions of this paper are as follows:

- We analyze the impact of entity relation information in ED task, and effectively merge it to the ED system.
- We propose an attention-based Bi-LSTM model, which aims to capture more important information within a sentence for ED task. In addition, we also investigate different attention strategies for the proposed model.

- We conduct extensive experiments on the ACE-2005 corpus. The experimental results show that our method outperforms other state-of-the-art methods.

The rest of the paper is organized as follows. Section 2 provides a brief overview of some event research works. Section 3 describes the ED task specifically. Section 4 gives a detailed introduction of our proposed method. Section 5 shows the experimental results and Sect. 6 concludes the paper.

2   Related Work

Event detection is an important subtask of event extraction [5]. Many approaches have been explored for ED. We divide these methods into feature-based methods and structure-based methods.

In feature-based methods, Ahn and David [1] leverage lexical, syntactic and external knowledge features to extract the event. Ji and Grishman [9] use rich evidence from related documents for the event extraction. Furthermore, Liao et al. [11] and Hong et al. [8] proposed the cross-event and cross-entity inference methods to capture more clues from the texts. Benefiting from the common modelling framework, these methods can achieve the fusion of multiple features, in addition, they can be used flexibly through feature selection. But feature engineering requires considerable expertise.

In structure-based methods, the use of neural network for ED has become a promising research. Some researchers like Nguyen et al. [18] and Chen et al. [3] learn continuous word representations and regard them as features to infer whether a word is a trigger or not by exploring neural network. Respectively, Nguyen et al. proposed a Convolutional Neural Network (CNN) with entity type information and word position information as extra features and Chen et al. presented a Dynamic Multi-pooling layer (DM-CNN) to capture information from sentence. Tang et al. [19] presented Bi-LSTM to model the preceding and following information of a word as people generally believe that LSTM is good at capturing long-term dependencies in a sequence. Some other models also effectively improve the performance of ED task, including Nguyen et al. [17] Bidirectional Recurrent Neural Network (Bi-RNN) and Feng et al. [6] Hybrid which combines the Bi-LSTM and CNN neural network.

3   Task Description

In this paper, we adopt the event extraction task defined in Automatic Content Extraction (ACE) evaluation. An event indicates a specific occurrence involving one or more participants. We introduce some ACE terminologies to facilitate the understanding of this task:

- **Event trigger**: the main word that most clearly expresses the occurrence of an event (typically, a verb or a noun).
- **Event argument**: the mentions that are involved in an event (participants).

- **Entity**: an object or a set of objects in one of the semantic categories.
- **Relation**: some relationship between entities in one sentence.

For easy to understand, we can see the instance of **S2**: an event extractor should detect an *End-Position* event, along with the event trigger *"leave"*. Moreover, the entities in this sentence are *NCA* and *swenson*. The relation between the entities is *Membership*. In this paper, we formalize the ED problem as a multi-class classification problem following previous work [5]. Given a sentence, we classify every token of the sentence into one of the predefined event classes or non-trigger class.

In addition, since entity recognition and relation detection are difficult task in ACE evaluation and not the focus in the event extraction task, we directly leverage the ground-truth entity and relation labels.

4   Our Approach

We model the ED task as a multi-class classification task. In detail, given a sentence, we treat every token in this sentence as a trigger candidate, and our goal is to classify each of these candidates into one of 34 classes.

In this section, we illustrate details of our approach, including the sentences representation which is the input of the model, the attention strategies, the use of Bi-LSTM which is to encode semantics of each word with its preceding and following information, as well as the other details in the model.

4.1   Input

We follow Chen et al. [3] to take all tokens of the sentence as the input. Before feeding tokens into the network, we transform each of them into a real-valued vector D. The vector is formed by concatenating a word embedding, an entity type embedding with a relation type embedding. As shown in Fig. 1.

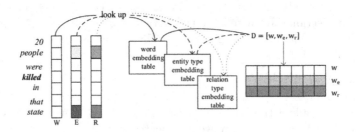

Fig. 1. Embedding.

Word Embedding. In this paper, we limit the context to a fixed length by trimming longer sentences and padding shorter sentences with special token. We let n be the fixed length. w_i is the current candidate trigger. So, we get the representation of the sentence: $W = \{w_1, w_2, ..., w_i, ..., w_n\}$. Then, looking up a pre-trained word embedding table to get the word embedding representation w. It is a fixed-dimensional real-valued vector which represents the hidden semantic properties of a token [4]. We use the Skip-gram model [16,20] to learn word embeddings on the NYT corpus[1].

Entity Type Embedding. Similarly, we limit the sentences to a fixed length n. However, we only label entities which have been annotated in corpus with specific symbols. Other non-entity words are labelled as 0. Thus, the entity representation as follows: $E = \{0, ...e_i, 0, ..., e_j, 0, ...\}$, where e_i is the i-th entity and e_j is the j-th entity. Then, we look up the entity type embedding table (initialized randomly). The result of the entity type embedding is w_e.

Relation Type Embedding. It is specially used to characterize the embedding of the ED model using relationship between two entities. For the sentence that has fixed length n, we set each word as 0 expect when there exists a word which is an entity e_i and is annotated relation type with another entity e_j in corpus, we set it as symbol r. In addition, we set the entity e_j as the same symbol with e_i. The relation type representation of this sentence is: $R = \{0, ..., r, 0, ..., r, 0...\}$. Similarly, we look up the relation type embedding table to get the relation type embedding w_r. We randomly initialize embedding vectors for each relation type (including the non-trigger type) and update them during training procedure.

We concatenate the above three embeddings as the input to the neural network. The process is shown in Fig. 1: $D = [w, w_e, w_r]$. We denote the final representation as $D = \{d_1, d_2, ...d_i, ...d_n\}$, where n is also the length of the sentence and d represents each word.

4.2 Attention Mechanism

In order to capture the information of important words as much as possible, and reduce the interference for modeling meaningless words, we leverage attention mechanism to the neural network. The specific attention structure is shown in Fig. 2. We follow Liu et al. [13] calculation method of attention value strictly. Given the sentence and its representation $D = \{d_1, d_2, ...d_i, ...d_n\}$, we treat each token as a candidate trigger, d_c represents the current candidate trigger. Firstly, we get the relatedness s_i between d_c and the other token representation d_i in the sentence by the following equation:

$$s_i = tanh(d_c{}^{\mathrm{T}} W d_i + b) \tag{1}$$

---

[1] https://catalog.ldc.upenn.edu/LDC2008T19.

where W is the weight matrix and b is the bias. Then, we calculate the importance p_i of each token in the sentence relative to the current candidate trigger.
Given all the importance weighs, we get the comprehensive information att_c
conveyed by D regarding the candidate trigger by computing the weighted sum:

$$p_i = \frac{exp(s_i)}{\sum_{k=1}^{n} exp(s_k)} \quad , \quad att_c = \sum_{i=1}^{n} p_i * d_i \tag{2}$$

Furthermore, we come up with two attention strategies according to the
different position of the attention mechanism in the model:

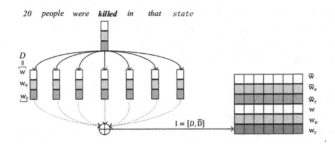

Fig. 2. Attention mechanism.

Att1: After obtaining the embedding D of the sentence, we apply the attention mechanism immediately. As shown in Fig. 2, through the above calculation
method, we get the new embedding representation $\overline{D}$ after assigning attention
weights. Then, we concatenate D and $\overline{D}$ as the input I to the Bi-LSTM.

Att2: We use embedding representation D as the input for Bi-LSTM firstly.
And we utilize hidden output H of Bi-LSTM as the input for attention mechanism. Similarly, we can obtain the new hidden representation $\overline{H}$ after attention
value calculation method. Then, we concatenate H and $\overline{H}$ for softmax.

4.3   Bi-LSTM

RNN with long short-term memory (LSTM) unit is adopted due to the superior
performance in a variety of NLP tasks [12,14]. Furthermore, Bi-LSTM is a type of
Bi-RNN, which can model word representation with its preceding and following
information simultaneously.

Bi-LSTM consists of input gate, forget gate and output gate. At step t, input
gate accepts a current input x_t, previous hidden state h_{t-1} and previous cell state
C_{t-1} as Eq. 3. i_t controls how much new information can be conveyed to the cell
state. $tildeC_t$ indicates new information at the current moment. Forget gate f_t
controls how much information of the previous cell moment can be conveyed

to the current moment. C_t represents updated cell state. Output gate gets the current hidden state h_t of the step t as Eq. 5.

$$i_t = \sigma(W_i) \cdot [h_{t-1}, x_t] + b_i, \qquad \tilde{C}_t = tanh(W_C \cdot [h_{t-1}, x_t] + b_c) \qquad (3)$$

$$f_t = \sigma(W_f) \cdot [h_{t-1}, x_t] + b_f, \qquad C_t = f_t * C_{t-1} + i_t * \tilde{C}_t \qquad (4)$$

$$o_t = \sigma(W_o) \cdot [h_{t-1}, x_t] + b_o, \qquad h_t = o_t * tanh(C_t) \qquad (5)$$

The details of our Bi-LSTM architecture for ED are shown in Fig. 3. The figure shows the case of **Att1**, so the input of Bi-LSTM consists of six embedding representation: $\{w, w_e, w_r, \overline{w}, \overline{w_e}, \overline{w_r}\}$. We can see that Bi-LSTM is composed of two LSTM neural networks, a forward F-LSTM to model the preceding contexts, and a backward Bi-LSTM to model the following contexts. We can get the hidden representation h_f and h_b via running F-LSTM and B-LSTM respectively. Then, we concatenate the hidden embedding h_f and h_b as the output $H = \{h_1, h_2, ..., h_i, ...h_n\}$ of Bi-LSTM.

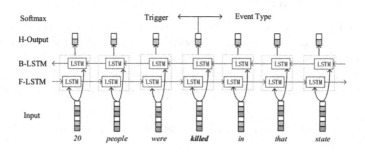

Fig. 3. Bi-LSTM architecture.

4.4   Output

AS mentioned earlier, we formulate ED as a multi-class classification problem. We predict each token of the sentence whether is an event trigger and assign event type to it. We use the hidden output H of the Bi-LSTM directly with **Att1** or combine H and $\overline{H}$ with **Att2** as the input to a softmax classifier. Thus, we can get the predicted probabilities of different types $P(y_j|x_i, \Theta)$, where Θ represents all the parameters of the model, and x_i is the i-th word and y_j is the j-th event type.

4.5   Training

We train the attention-based Bi-LSTM model by optimizing objective function which is defined as a multi-class cross-entropy loss:

$$L(\Theta) = -\sum_{k=1}^{N} \sum_{i=1}^{C} y_i log P(y_i|x_k, \Theta) + \lambda(\Theta) \qquad (6)$$

where N denotes the all candidate triggers in the training process, C denotes the all event classes, and y_i is the real event type of the word x_i. λ is the regularization parameter and Θ represents the all parameters in the model.

In addition, we train the network via stochastic gradient descent (SGD) [18] with shuffled mini-batches. The gradients are computed using back propagation.

5   Experiments

5.1   Dataset and Evaluation Metric

We use the ACE-2005 corpus in the experiments. For comparison purpose, we follow Li et al. [10] to select 529 articles in English as the training data set, 30 as development set and 40 for test.

Following the previous work [9], we use Precision (P), Recall (R) and F_1 score (F_1) as the evaluation metrics of our approach.

5.2   Hyperparameter Settings

The word embeddings are initialized with the 300-dimensional real-valued vectors. The entity type embeddings are specified as the 50-dimensional real-valued vectors. And the relation type embeddings are initialized with the 20-dimensional vectors. We follow Feng et al. [6] to set the dropout rate and the batch size. Table 2 shows the specific setting of parameters used in our experiments.

Table 2. Hyperparameters used in our experiments.

Parameters	Values	Parameters	Values
Word embedding	300-dimensional	Dropout rate	0.2
Entity embedding	50-dimensional	Learning rate	0.3
Relation embedding	20-dimensional	Hidden size	200
Batch size	10	Coefficient	10^{-3}

5.3   Compared Systems

The state-of-the-art models proposed in the past are compared with ours. We divide the models into three classes:

Feature Based Approaches: *Joint*: the method in Li et al. [10] which combines the local and global features and is based on the structured perceptron. *Cross-Entity*: Hong et al. [8] model, which employs the name entities as the additional discriminant features to aid event extraction.

External Resource Based Approaches: Include Chen et al. [2] *DMCNN-DS*, which utilizes world knowledge (Freebase) and linguistic knowledge (FrameNet) and Liu et al. [13] *GMLATT* which takes advantage of the multilingual information for the ED task.

Neural Network Based Approaches: *DMCNN*: the method in Chen et al. [3], which uses CNN to do automatical feature extraction. In addition, also including Nguyen et al. [17] *Bi-RNN*, Feng [6] *Bi-LSTM*, Feng [6] *Hybrid* model which combines Bi-LSTM and CNN and Liu [15] *ATT* which exploited argument information to improve ED via supervised attention mechanisms.

5.4 Experimental Results

Table 3 shows the performance of all methods for both trigger identification and type classification. It can be observed that our approach outperforms other models, with a performance gain of no less than 0.5% $F1$ on event type classification. The performance mainly benefits from the higher recalls which are 78.5% and 76.3% in two subtasks respectively. In addition, comparing the three experiments that we did (the last three rows in the table), we found that no matter whether we merge relation or use attention mechanism, the performance has been improved to a small extent compared to using Bi-LSTM alone. But when we integrate two methods with Bi-LSTM, the performance will be greatly improved, which is 73.9%. The better performance of our approach can be further explained by the following reasons:

Table 3. Performance of the all methods (n/a: the paper did't list results of this task).

Methods	Trigger identification			Type classification		
	P	R	F1	P	R	F1
Joint [10]	76.9	65.0	70.4	73.7	62.3	67.5
Cross-Entity [8]	n/a	n/a	n/a	72.9	64.3	68.3
DMCNN-DS [3]	79.7	69.6	74.3	75.7	66.0	70.7
GMLATT [13]	80.9	68.1	74.1	78.9	66.9	72.4
DMCNN [2]	80.4	67.7	73.5	75.6	63.6	69.1
Bi-RNN [17]	68.5	75.7	71.9	66.0	73.0	69.3
Bi-LSTM [6]	80.1	69.4	74.3	81.6	62.3	**70.6**
ATT [15]	n/a	n/a	n/a	78.0	66.3	71.7
Hybrid [6]	80.8	71.5	75.9	84.6	64.9	**73.4**
Bi-LSTM+Att1 (Ours)	74.5	75.1	74.7	72.1	72.6	72.3
Bi-LSTM+Re (Ours)	72.9	77.5	75.1	69.6	74.1	71.8
Re+Bi-LSTM+Att1 (Ours)	73.7	**78.5**	76.1	71.5	**76.3**	**73.9**

- Compared with feature based methods, such as *Joint, Cross-Event* and *Cross-Entity*, neural network based methods, such as *CNN* and *Bi-LSTM*, perform better because they can make better use of word semantic information and avoid the errors propagated from NLP tools. Moreover, *Bi-LSTM* performs better than *CNN* due to the former can capture more complete information of the whole sentence, which reduce the loss of information.

- Table 4 lists embedding types used in each method. We can see that relation type can provide richer information for ED than position (*PSN*) and dependency (*DEP*). Because we merge the relation type embedding to the model, the recall has improved significantly, which is higher 11.4% than *Hybrid* (add *PSN* embedding) and higher 3.3% than *Bi-RNN* (add *DEP* embedding).
- Attention mechanism can make certain words get higher attention, capture more accurate information and ignore the interference of meaningless words.
- We would like to believe that using entity relation and attention simultaneously can enhance the performance further. Due to we use not only entity embedding but also relation embedding which labels two entities in a sentence with the same relation type label, when we employ attention mechanism in model, entities is equivalent to get twice attention. Accordingly, model can better capture the information of key words.

Table 4. Embedding types (PSN: Position; ET: Entity Type; DEP: Dependency; RT: Relation Type).

Methods	Embedding types	Methods	Embedding types
DMCNN-DS	word, PSN	ATT	word, ET
GMLATT	word, ET, PSN	DMCNN	word, PSN
Bi-RNN	word, ET, DEP	Bi-LSTM	word, PSN
Hybrid	word, PSN	**Ours**	word, ET, RT

In order to further prove the rationality of the above explanations, we conduct two extra experiments to do detailed analysis. We use *TI* and *TC* to stand for F_1 score of Event Trigger Identification and Event Type Classification respectively.

Effect of Different Features. We conduct the experiments with *Bi-LSTM+Att1* to exploit the effects of different feature combinations. We set the word embedding as the baseline, and then add entity embedding and relation embedding step by step. Results are shown in Table 5.

According to the result, we can find that both entity embedding and relation embedding can yield effectively improvement. It seems that, entity embedding is more effective than relation embedding (by 0.5%) in type classification. However, relation type embedding is more effective for trigger identification than the former (by 0.6%). An intuitive explanation is that: we label the entities which process relationship with the same relation type symbol as the relation type representation. Although the same labels provide complement information for trigger identification, it also causes interference to classify trigger.

Moreover, using all as embedding, *TI* and *TC* are 76.1% and 73.9% respectively, which integrates the advantages of entity and relation embedding, and reaches the optimal performance. Such as **S2**, entity embedding can capture the important information of "*Swenson*" and "*NCA*" rather than other words in sentence. And relation embedding will provide necessary information (*Membership* between two entities) to classify *End-Position* event correctly.

Table 5. Performance on different features.

Methods	Features	TI	TC
Bi-LSTM+Att1	word embedding	74.4	71.4
	+entity embedding	74.7	72.3
	+relation embedding	75.3	71.8
	all	76.1	73.9

Effect of Different Attention Strategies. In order to verify whether the attention mechanism plays a critical role, and compare attention strategies, we designed three comparison experiments. Taking *Re+Bi-LSTM* as the baseline, we add *Att1* and *Att2* on the basis respectively. Notes that all three methods combine word, entity and relation embedding. The results are shown in Table 6.

From Table 6, it can be observed that $F1$ score reduces 1.3% on event type classification relative to the baseline when we place attention mechanism after Bi-LSTM (+Att2). However, when we add attention mechanism before entering the Bi-LSTM model (+Att1), the $F1$ score improves 2.1% compared to baseline. This may because: although Bi-LSTM can capture sentence information as much as possible, it still can't avoid the loss of some parts of the information, or change the importance of each word in the original sentence. Thus, employing attention after the Bi-LSTM will reduce the information of the incomplete sentence again. By contrast, applying the attention mechanism before Bi-LSTM is equivalent to process sentences with original complete information, which can improve the importance of keywords and reduce the interference of meaningless words. Hereafter, Bi-LSTM model further selects effective information of the sentence to capture the key information perfectly. Thus, +Att1 can effectively improve the performance of ED system.

Table 6. Performance on different attention strategies.

Methods	TI	TC
Re+Bi-LSTM	75.1	71.8
Re+Bi-LSTM+Att1	76.1	73.9
Re+Bi-LSTM+Att2	73.6	70.5

6 Conclusions

In this paper, we verified that integrating an attention mechanism before the Bi-LSTM neural network, which can assign different attention to words and better capture the key information of sentences. Furthermore, we first use the entity relation as the feature for ED, and we confirmed it can provide additional information for the ED task. In the future, we will further explore the relationship between entity relation and ED, to unite them into a supporting model.

Acknowledgements. This work was supported by the national Natural Science Foundation of China via Nos. 2017YFB1002104, 61672368 and 61672367.

References

1. Ahn, D.: The stages of event extraction. In: Proceedings of the Workshop on Annotating and Reasoning about Time and Events, pp. 1–8. Association for Computational Linguistics (2006)
2. Chen, Y., Liu, S., Zhang, X., Liu, K., Zhao, J.: Automatically labeled data generation for large scale event extraction. In: Proceedings of the 55th Annual Meeting of the Association for Computational Linguistics (Volume 1: Long Papers), vol. 1, pp. 409–419 (2017)
3. Chen, Y., Xu, L., Liu, K., Zeng, D., Zhao, J., et al.: Event extraction via dynamic multi-pooling convolutional neural networks. In: ACL (1), pp. 167–176 (2015)
4. Collobert, R., Weston, J.: A unified architecture for natural language processing: deep neural networks with multitask learning. In: Proceedings of the 25th International Conference on Machine Learning, pp. 160–167. ACM (2008)
5. Doddington, G.R., Mitchell, A., Przybocki, M.A., Ramshaw, L.A., Strassel, S., Weischedel, R.M.: The Automatic Content Extraction (ACE) program-tasks, data, and evaluation. In: LREC, vol. 2, p. 1 (2004)
6. Feng, X., Huang, L., Tang, D., Ji, H., Qin, B., Liu, T.: A language-independent neural network for event detection. In: Proceedings of the 54th Annual Meeting of the Association for Computational Linguistics (Volume 2: Short Papers), vol. 2, pp. 66–71 (2016)
7. Ghaeini, R., Fern, X., Huang, L., Tadepalli, P.: Event nugget detection with forward-backward recurrent neural networks. In: Proceedings of the 54th Annual Meeting of the Association for Computational Linguistics (Volume 2: Short Papers), vol. 2, pp. 369–373 (2016)
8. Hong, Y., Zhang, J., Ma, B., Yao, J., Zhou, G., Zhu, Q.: Using cross-entity inference to improve event extraction. In: Proceedings of the 49th Annual Meeting of the Association for Computational Linguistics: Human Language Technologies-Volume 1, pp. 1127–1136. Association for Computational Linguistics (2011)
9. Ji, H., Grishman, R.: Refining event extraction through cross-document inference. In: Proceedings of ACL-08: HLT, pp. 254–262 (2008)
10. Li, Q., Ji, H., Huang, L.: Joint event extraction via structured prediction with global features. In: ACL (1), pp. 73–82 (2013)
11. Liao, S., Grishman, R.: Using document level cross-event inference to improve event extraction. In: Proceedings of the 48th Annual Meeting of the Association for Computational Linguistics, pp. 789–797. Association for Computational Linguistics (2010)
12. Lin, Z., et al.: A structured self-attentive sentence embedding. arXiv preprint arXiv:1703.03130 (2017)
13. Liu, J., Chen, Y., Liu, K., Zhao, J.: Event detection via gated multilingual attention mechanism. Statistics **1000**, 1250 (2018)
14. Liu, P., Qiu, X., Chen, J., Huang, X.: Deep fusion LSTMs for text semantic matching. In: Proceedings of the 54th Annual Meeting of the Association for Computational Linguistics (Volume 1: Long Papers), vol. 1, pp. 1034–1043 (2016)
15. Liu, S., Chen, Y., Liu, K., Zhao, J.: Exploiting argument information to improve event detection via supervised attention mechanisms, vol. 1, pp. 1789–1797 (2017)

16. Mikolov, T., Chen, K., Corrado, G., Dean, J.: Efficient estimation of word representations in vector space. arXiv preprint arXiv:1301.3781 (2013)
17. Nguyen, T.H., Cho, K., Grishman, R.: Joint event extraction via recurrent neural networks. In: HLT-NAACL, pp. 300–309 (2016)
18. Nguyen, T.H., Grishman, R.: Event detection and domain adaptation with convolutional neural networks. In: ACL (2), pp. 365–371 (2015)
19. Tang, D., Qin, B., Liu, T.: Document modeling with gated recurrent neural network for sentiment classification. In: Proceedings of the 2015 Conference on Empirical Methods in Natural Language Processing, pp. 1422–1432 (2015)
20. Turian, J., Ratinov, L., Bengio, Y.: Word representations: a simple and general method for semi-supervised learning. In: Proceedings of the 48th Annual Meeting of the Association for Computational Linguistics, pp. 384–394. Association for Computational Linguistics (2010)

Five-Stroke Based CNN-BiRNN-CRF Network for Chinese Named Entity Recognition

Fan Yang[1], Jianhu Zhang[1], Gongshen Liu[1(✉)], Jie Zhou[1],
Cheng Zhou[2], and Huanrong Sun[2]

[1] School of Electric Information and Electronic Engineering,
Shanghai Jiaotong University, Shanghai, China
`{417765013yf,zhangjianhu3290,lgshen,sanny02}@sjtu.edu.cn`
[2] SJTU-Shanghai Songheng Content Analysis Joint Lab, Shanghai, China
`{zhoucheng,sunhuanrong}@021.com`

Abstract. Identifying entity boundaries and eliminating entity ambiguity are two major challenges faced by Chinese named entity recognition researches. This paper proposes a five-stroke based CNN-BiRNN-CRF network for Chinese named entity recognition. In terms of input embeddings, we apply five-stroke input method to obtain stroke-level representations, which are concatenated with pre-trained character embeddings, in order to explore the morphological and semantic information of characters. Moreover, the convolutional neural network is used to extract n-gram features, without involving hand-crafted features or domain-specific knowledge. The proposed model is evaluated and compared with the state-of-the-art results on the third SIGHAN bakeoff corpora. The experimental results show that our model achieves 91.67% and 90.68% F1-score on MSRA corpus and CityU corpus separately.

Keywords: CNN-BiRNN-CRF network
Stroke-level representations · N-gram features
Chinese named entity recognition

1 Introduction

Named entity recognition (NER) is one of the fundamental tasks in the field of natural language processing (NLP). It plays an important role in the development of information retrieval, relation extraction, machine translation, question answering systems and other applications. The task of NER is to recognize proper nouns or entities in the text and associate them with the appropriate types, such as the names of persons (PERs), organizations (ORGs), and locations (LOCs) [1]. Many researchers regard named entity recognition as a sequence labeling task. Traditional sequence labeling models are linear statistical models, including hidden markov model (HMM) [2], support vector machine (SVM) [3], maximum entropy (ME) [4] and conditional random field (CRF) [5,6].

© Springer Nature Switzerland AG 2018
M. Zhang et al. (Eds.): NLPCC 2018, LNAI 11108, pp. 184–195, 2018.
https://doi.org/10.1007/978-3-319-99495-6_16

With the development of word embedding technologies, neural networks have shown great achievements in NLP tasks. Character-based neural architecture has achieved comparable performance in English NER task [7–10]. Character-level features such as prefix and suffix could be exploited by the convolution neural network (CNN) or bidirectional long short-term memory (BiLSTM) structure, thus helping to capture deeper level of semantic meanings.

Compared with English NER, it is more difficult to identify Chinese entities, due to the attributes of Chinese words, such as the lack of word boundaries, the uncertainty of word length, and the complexity of word formation [11]. Inspired by the character embedding of English, some researches apply radical-level embedding of Chinese to improve the performance of word segmentation [12,13], part-of-speech (POS) tagging [14], or Chinese NER [15]. However, the acquired semantic information varies from the character splitting methods, which may lead to incomplete or biased results. Cao et al. [16] propose a stroke n-gram model by splitting each character into strokes, and each stroke is represented by an integer ranging 1 to 5. However, this method will bring about ambiguous information in the case of different characters with same type of strokes. As shown in Table 1, characters '天 (sky)' and '夫 (Husband)' are encoded into the same representation, so as to '由 (Reason)' and '申 (State)'. Therefore, it is essential to explore a general and effective way of semantic information extraction.

Recognizing entity boundaries is one of the most challenging problems faced by Chinese NER task, since entity boundary ambiguity can lead to incorrect entity identification. According to the word formation rules, many ORG entities contain the LOC and PER entities inside them. For example, '中国海军 (Chinese Navy)' is a ORG entity with nested LOC entity '中国 (China)'. Although the recognition of person and location names has achieved good results, the poor recognition of ORG entities still hinders the overall performance of named entity recognition [15]. Therefore, we need to pay more attention to the boundaries identification of ORG entities.

To address above difficulties and challenges, a five-stroke based CNN-BiRNN-CRF network (CBCNet hereafter) is proposed to optimize Chinese NER task. The main contributions of our model are summarized as follows:

1. A customized neural model is presented for Chinese NER, which conduces to the entity boundaries identification and entity ambiguity elimination, especially for the identification of ORG entities.

Table 1. Comparison of two character encoding methods.

Characters	Stroke n-gram [16]		Wubi method (Ours)	
	Decomposition	Code	Decomposition	Code
天 (Sky)	一一丿丶	1134	一大	GDI
夫 (Husband)	一一丿丶	1134	夫一一丶	GGGY
申 (State)	丨丁一一丨	25112	日丨	JHK
由 (Reason)	丨丁一一丨	25112	由一乙一	MHNG

2. We propose the five-stroke based representations, and integrate them into character embeddings to form the final inputs. Then the convolution neural network with different size of filters is employed to simulate traditional n-gram model, which helps to identify the prefix and suffix of Chinese named entities.
3. We conduct experiments on the third SIGHAN bakeoff NER task with two Chinese annotated corpora. Experimental results show that our model achieves comparable performance over other existing Chinese NER architectures.

2 Neural Model for Chinese NER

2.1 Overview of Proposed Architecture

Our CBCNet is built upon the character-based BiLSTM-CRF architecture, as shown in the Fig. 1. Instead of using the original pre-trained character embeddings as the final character representations, we construct a comprehensive character representation for each character in the input sentence. Firstly, we incorporate stroke embeddings into original character embeddings, to construct a comprehensive character representation for each character in the input sentence. Then, a convolutional layer and a pooling layer are applied to generate n-gram features contained character representations. After that, we feed character embeddings into a BiLSTM-CRF layer, to decode and predict the final tag sequence for the input sentence.

2.2 Stroke Embedding

In this paper, we use five-stroke character model input method (or Wubi method)[1] to encode input characters. Wubi method is an efficient encoding

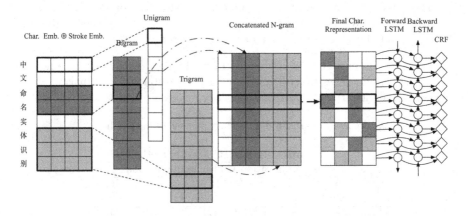

Fig. 1. The overall architecture of proposed CBCNet.

---

[1] https://en.wikipedia.org/wiki/Wubi_method.

Table 2. Five regions with corresponding strokes.

Regions	ASDFG	HJKLM	QWERT	YUIOP	XCVBN
Strokes	Horizontal (一)	Vertical (丨)	Left-falling (丿)	Right-falling (丶)	Hook (乙)

system which can represent each Chinese character with at most four Roman letters (keys), according to the structure of each character. In the rule of Wubi method, 25 keys are divided into five regions, and each region is assigned with a certain type of strokes, as shown in Table 2. Then, each key stands for a certain type of components that is similar to the basic stroke in its own region. Since "Z" is a wild card, it is not listed in the table. By this means, every Chinese character could be encoded with a Wubi representation. For example, ('中', '文', '命', '名', '实', '体', '识', '别') which means ('Chinese', 'named', 'entity', 'recognition') in English, can be encoded into ('KHK', 'YYGY', 'WGKB', 'QKF', 'PUDU', 'WSGG', 'YKWY', 'KEJH'). Moreover, unlike previous work [16], Wubi method is able to distinguish words with similar structure, as shown in Table 1.

Santos and Zadrozny [17] introduce a convolutional approach with character embeddings, to extract the most important morphological information from English words. A BiLSTM layer is employed to capture semantic information of Chinese characters by Dong et al. [15]. In this paper, we compare the two methods of stroke-level feature extraction, and select the most effective way of representation. Figure 2 depicts the convolutional approach [17] and recurrent approach [15] respectively of generating stroke embeddings of character ' '识' ' (Recognition).

For example, given a character x_i encoded with a sequence of Q Roman letters $\{r_1, r_2, ..., r_Q\}$ $(0 \leq Q \leq 4)$ under the look-up table of Wubi method[2], we first transform each letter r_q into a one-hot embedding. In the case of the

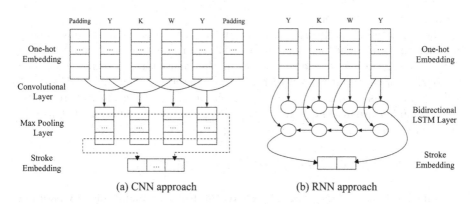

(a) CNN approach   (b) RNN approach

Fig. 2. Stroke embedding of character "识" (Recognition) with two approaches.

---

[2] https://github.com/yanhuacuo/98wubi-tables.

character with less than four Roman letters, we will randomly generate the initial embedding to ensure that each character has a four-dimensional stroke-level representation. During the training of the model, the stroke embeddings are continuously updated. Thus, the final stroke-level representation s_i of character x_i is defined as follows:

$$s_i = f(r_1, r_2, ..., r_Q) \tag{1}$$

where f is the function of a CNN or an RNN approach.

Pre-trained character embeddings are proved to be efficient over randomly initialized embeddings, since the former contain more contextual information. Therefore, we apply word2vec [18] to train our character-level embedding on Chinese Wikipedia backup dump. We concatenate the stroke embedding s_i to the character embedding v_i as the final representation c_i for each character:

$$c_i = v_i \oplus s_i \tag{2}$$

where $\oplus$ is a connection operator, and i indicates the ith character x_i in the sentence S.

2.3   Convolutional Layer

Inspired by Chen et al. [19], which use CNN to simulate a traditional discrete feature based model for POS tagging. In this paper, we use CNN to extract local information and n-gram features for the input characters, and we model different n-gram features by generating different feature map sets.

For example, $c_i \in R^d$ is the d-dimensional character representation corresponding to the ith character x_i in the sentence S. For a sentence with length l could be represented as:

$$c_{1:l} = c_1 \oplus c_2 \oplus ... \oplus c_l \tag{3}$$

where $c_{i:j}$ denotes the connection of $c_i, c_{i+1}, ..., c_j$.

The input of the network each time is an $l \times d$ matrix for one sentence. In order to ensure the integrity of the character, the width of convolutional filters is consistent with the dimension of the character representations. The convolution of the input matrix with filters are determined by the weights $W_k \in R^{kd \times N_k}$ and bias $b_k \in R^{N_k}$, where N_k is the number of k-gram feature maps. The k-gram feature map m_i^k can be generated from the combination of character representations $c_{i-\lfloor \frac{k-1}{2} \rfloor : i+\lceil \frac{k-1}{2} \rceil}$ according to the following formula:

$$m_i^k = tanh(W_k^T \cdot c_{i-\lfloor \frac{k-1}{2} \rfloor : i+\lceil \frac{k-1}{2} \rceil} + b_k) \tag{4}$$

The length and the order of character representations are maintained by padding zero to the input in the marginal case. The feature map sets matrix is $m \in R^{l \times \sum_{k=1}^{K} N_k} = \{m_1, m_2, ..., m_K\}$, where K is the maximum value of k in k-gram.

m_i is the concatenation of $m_i^k \in R^{l \times N_k}$:

$$m_i = m_i^1 \oplus m_i^2 \oplus ... \oplus m_i^K \tag{5}$$

Then, max-over-time pooling is applied to progressively reduce the spatial size of the representation and keep the most important features. Thus, the output sentence representations $c^* \in R^{l \times d} = \{c_1^*, c_2^*, ..., c_l^*\}$ can be generated after d-max pooling operation, where c_i^* is:

$$c_i^* = dmax\{m_i\} \tag{6}$$

2.4   Bidirectional LSTM Layer

The recurrent neural network (RNN) can effectively obtain the sequence information of the texts. As a special kind of RNN, the long short-term Memory (LSTM) network could not only solve the long-distance dependence problem of the sequence, but also effectively deal with the vanishing gradient or exploding gradient problem of RNN [15]. In order to make effective use of contextual information for a specific time frame, we use a bidirectional LSTM (BiLSTM) architecture. Thus, each hidden state h_t of BiLSTM can be formalized as a concatenation of the hidden states of forward and backward LSTMs:

$$h_t = \overrightarrow{h_t} \oplus \overleftarrow{h_t} \tag{7}$$

where $\overrightarrow{h_t}$ (or $\overleftarrow{h_t}$) can be generated from the multiplication of output gate result o_t and input character representation c_t^* at the specific time frame of t, and the calculation of o_t can refer to previous works [15].

2.5   CRF Layer

The linear CRF can obtain a globally optimal tag sequence by considering the relationship between adjacent tags. By combining BiLSTM layer and CRF layer, BiLSTM-CRF layer is able to efficiently make use of contextual features as well as sentence-level tag information. Given an input sentence $X = \{x_1, x_2, ..., x_l\}$, we consider the score matrix P, which is the output of BiLSTM layer, and the transition score matrix A. Thus, for a sequence of prediction results $y = \{y_1, y_2, ..., y_l\}$, the score of the sentence S along with a path of labels could be defined as:

$$Score(X, y) = \sum_{i=1}^{l} A_{y_i, y_{i+1}} + \sum_{i=1}^{l} P_{i, y_i} \tag{8}$$

where $A_{y_i, y_{i+1}}$ indicates the possibility of the transition from ith label to $i+1$th label for a pair of consecutive time steps, P_{i, y_i} is the score of the ith label of the ith input character.

For the decoding phase, Viterbi algorithm [20] is used to generate optimal tag sequence y^*, when maximizing the score $Score(X, y)$:

$$y^* = \underset{y \in Y_x}{\arg\max} \ \ Score(X, y). \tag{9}$$

where Y_x represents all the possible label sequences for sentence S.

3   Experimental Results and Analysis

3.1   Tagging Scheme

In this paper, we adopt the BIO (indicating Begin, Inside and Outside of the named entity) tagging set used in the third SIGHAN bakeoff [1], which followed by tags PER, ORG and LOC that denote persons, organizations and locations respectively. For instance, B-PER and I-PER denote the begin, inside part of a person's name respectively. O means that the character is not included in a named entity. We employ character-level precision (P), recall (R), and F1-score (F) as the evaluation metrics, same as the previous works [15].

3.2   Training

We use Tensorflow library to implement our neural network. Table 3 illustrates the hyper-parameters of all the experiments on different datasets. We train our network with the error back-propagation, and the network parameters are fine-tuned by back-propagating gradients. Adagrad algorithm [21] is used as the network optimizer. To accelerate network training on GPU, we adopt bucketing strategy [14], which usually implemented in seq2seq model, for the input sentences. That is, the sentences with similar lengths are grouped into the same buckets, and the sentences in the same buckets are padded into the same length. In order to reduce over-fitting during network training and improve the accuracy of neural network model, we apply dropout technique [22] before the BiLSTM layer with a probability of 0.5.

Table 3. Hyper-parameter settings.

Parameters	Details	Parameters	Details
Character embedding size	$d_c = 64$	Optimizer	Adagrad
Stroke embedding size	$d_s = 30$	Initial learning rate	$\alpha = 0.2$
Number of feature map sets	$K = 5$	Decay rate	0.05
Number of k-gram feature maps	$N_k = 100$	Dropout rate	0.5
LSTM dimensionality	$h = 200$	Batch size	10

Table 4. The statistics of NER training and testing corpora.

Copora	Training			Testing		
	Sentences	NEs	ORGs/LOCs/PERs	Sentences	NEs	ORGs/LOCs/PERs
MSRA	43907	75059	20584/36860/17615	3276	6190	1331/2886/1973
CityU	48169	112361	27804/48180/36377	6292	16407	4016/7450/4941

Sentences: Number of sentences; NEs: Number of named entities;
LOCs/PERs/ORGs: Number of location/person/organization names.

3.3 Dataset and Preprocessing

We conduct the experiments on the two corpora from the third SIGHAN bake-off [1]. The MSRA corpus is simplified Chinese character and the CityU corpus is traditional Chinese character, and both of them are in the CoNLL two column format. We convert CityU corpus simplified Chinese, so that the look-up table for pre-trained character embeddings, Wubi encoding method and other resources are compatible in the experiments. Table 4 shows the statistics of NER training and testing corpora of MSRA and CityU.

3.4 Evaluation of Different Components

We incrementally add each component on the BiLSTM-CRF with character embedding model, which is the baseline architecture in our comparison, to evaluate the impact of every component on the performance of our model. In general, experimental results given in Table 5 shows that "Ours" with stroke-level information and n-gram features significantly outperforms the baseline model, achieving 91.67% and 90.68% F1-score on MSRA and CityU separately.

To evaluate the effectiveness of two stroke-level feature extraction approaches, we test on the RNN and CNN separately, denoted as "+ Stroke Emb.(RNN)" and "+ Stroke Emb.(CNN)". Table 5 illustrates that CNN performs a little better than RNN. The reason is that CNN is an unbiased model, which treats fairly to each input in the same windows. While RNN is a biased model, in which later inputs are more dominant than earlier inputs. Thus, we

Table 5. Evaluation of different components on two corpora in F1-scores (%)

Variant	MSRA			CityU		
	P	R	F	P	R	F
BiLSTM-CRF (Char. Emb.)	90.98	89.97	90.47	91.32	88.49	89.88
+ Stroke Emb.(RNN)	92.36	90.18	91.25	91.61	88.77	90.16
+ Stroke Emb.(CNN)	92.34	90.31	91.31	91.57	88.96	90.24
+ CNN + Pooling	92.30	90.68	91.48	91.65	89.07	90.34
Ours	92.04	91.31	91.67	91.87	89.53	90.68

apply CNN approach to extract stroke-level features in our model. Moreover, compared with baseline, we could obtain relatively high performance in two corpora by incorporating stroke embeddings into character embeddings.

From Table 5 in the row "+ CNN + Pooling", we can see that notable improvement is achieved by combining 1-gram to K-gram (we set $K = 5$ for all the experiments.) features. This phenomenon is consistent with the length of the named entities, which is about two to five characters. It verifies that our CBCNet can efficiently identify entity boundaries and eliminate entity ambiguity by introducing n-gram features. Besides, the results imply that the n-gram features have a greater impact on model performance than stroke-level information.

Furthermore, it can be observed that the optimization of our model on the MSRA corpus is more significant than that on CityU corpus. The reason is that CityU corpus contains some English named entities, which are not sensitive to our two adopted measures.

3.5 Comparison with Previous Works

We compare proposed CBCNet with the reported results of several previous works on the third SIGHAN bakeoff corpora, as shown in the Tables 6 and 7. Each row represents the results of a NER model, including F1 scores for each entity category (PER-F, ORG-F, LOC-F) as well as the total precision (P), recall (R), and balanced F1-score (F).

Table 6. Evaluation of different models on MSRA corpus (%)

Model	MSRA					
	F-ORG	F-LOC	F-PER	P	R	F
CRF+Word.Emb. [5]	83.10	85.45	90.09	88.94	84.20	86.51
CRF+Char.Emb. [6]	81.96	90.53	82.57	91.22	81.71	86.20
Knowledge based [4]	85.90	90.34	96.04	92.20	90.18	91.18
Feature templates [23]	86.19	91.90	90.69	91.86	88.75	90.28
BiLSTM-CRF+Radi.Emb. [15]	87.30	92.10	91.77	91.28	90.62	90.95
Ours	**88.42**	**92.31**	**91.96**	**92.04**	**91.31**	**91.67**

Table 7. Evaluation of different models on CityU corpus (%)

Model	CityU					
	F-ORG	F-LOC	F-PER	P	R	F
CRF+Char.Emb. [6]	-	-	-	92.66	84.75	88.53
Knowledge based [4]	81.01	93.06	91.30	92.33	87.37	89.78
Ours	**83.99**	**93.76**	**91.63**	**91.87**	**89.53**	**90.68**

Conditional random fields (CRF) is one of the most popular and effective models for sequence labeling tasks [5,6]. Zhou et al. [5] used a word-level CRF-model with hand-crafted features incorporated, which won the first place in the third SIGHAN bakeoff Shared Tasks with 86.51% F1-score in MSRA corpus. Chen et al. [6] utilized character embeddings as the inputs to CRF model with 86.20% F1-score in MSRA corpus and 88.53% in CityU corpus, while the improvement of performance is still limited. By incorporating hand-crafted features [5,23] or knowledge bases [4] could relatively improve the performance. However, these methods may result in an inefficiency or failure when processing large-scale corpora or the datasets in other fields.

Dong et al. [15] implemented Chinese radical embedding into BiLSTM-CRF framework, achieving good performance. However, some characters cannot be split into radicals, leading to the failure of semantic feature extraction. Moreover, they ignore the characteristics of entities word formation. Compared with [15], we consider both the semantic information by using stroke embedding and contextual information by employing CBCNet model. In terms of MSRA corpus, experimental results indicate that proposed stroke-based CBCNet outperforms the best deep learning work [15] by +0.72 F1-score and best reported work [23] by +0.49 F1-score. For CityU corpus as shown in Table 7, compared with other works, our model obtains the best performances with 90.68% F1-score. Moreover, our model achieves significant improvement on ORG entities thanks to the extraction of n-gram features.

4   Related Works

Named entity recognition is a fundamental NLP task and studied by many researchers. Remarkable achievements have been made in the field of English NER through a variety of methods. An end-to-end architecture is implemented in [24], with a BiLSTM-CNNs-CRF model. Yang et al. [8] use transfer learning for jointly training the POS and NER tasks. Then, Liu et al. [9] enhance the neural framework by introducing a task-aware language model. Xu et al. [25] propose a FOFE-based strategy, which regards NER as a non-sequence labeling task. However, these approaches could not be simply transplanted into Chinese NER systems, due to the characteristics of Chinese named entities (NEs), which do not have word boundaries or case sensitivity.

BiLSTM-CRF architecture has been used in many Chinese sequence labeling problems. Peng and Dredze [26] adopt the model for jointly training Chinese word segmentation and named entity recognition. By extracting semantic information, the performance of labeling could be further improved. Dong et al. [15] combine radical embeddings with character embeddings in bidirectional LSTM-CRF model for Chinese NER, and show the efficiency. For the task of Chinese word segmentation and POS tagging, this fundamental structure also shows great performance as shown in [14]. He et al. [12] indicate that the performance of Chinese word segmentation could be boosted largely when tying subcharacters and character embeddings together.

5   Conclusions

In this paper we have presented a novel model for Chinese NER by considering the semantic information as well as n-gram features, without involving hand-crafted features or domain-specific knowledge. The empirical study shows the effectiveness of each components of our architecture. Experiments on two different corpora from the third SIGHAN bakeoff also indicate that our model achieves outstanding performance over other approaches. In the future, we would like to extend our model to other sequential labeling tasks, such as jointly learning the Chinese word segmentation, POS tagging and NER.

Acknowledgement. This research work has been funded by the National Natural Science Foundation of China (Grant No. 61772337, U1736207 and 61472248), the SJTU-Shanghai Songheng Content Analysis Joint Lab, and program of Shanghai Technology Research Leader (Grant No. 16XD1424400).

References

1. Levow, G.A.: The third international Chinese language processing bakeoff: word segmentation and named entity recognition. In: Proceedings of the Fifth SIGHAN Workshop on Chinese Language Processing, pp. 108–117 (2006)
2. Fu, G., Luke, K.K.: Chinese named entity recognition using lexicalized HMMs. ACM SIGKDD Explor. Newslett. **7**, 19–25 (2005)
3. Li, L., Mao, T., Huang, D., Yang, Y.: Hybrid models for Chinese named entity recognition. In: Proceedings of the Fifth SIGHAN Workshop on Chinese Language Processing, pp. 72–78 (2006)
4. Zhang, S., Qin, Y., Wen, J., Wang, X.: Word segmentation and named entity recognition for SIGHAN Bakeoff3. In: Proceedings of the Fifth SIGHAN Workshop on Chinese Language Processing, pp. 158–161 (2006)
5. Zhou, J., He, L., Dai, X., Chen, J.: Chinese named entity recognition with a multi-phase model. In: Proceedings of the Fifth SIGHAN Workshop on Chinese Language Processing, pp. 213–216 (2006)
6. Chen, A., Peng, F., Shan, R., Sun, G.: Chinese named entity recognition with conditional probabilistic models. In: Proceedings of the Fifth SIGHAN Workshop on Chinese Language Processing, pp. 173–176 (2006)
7. Chiu, J.P., Nichols, E.: Named entity recognition with bidirectional LSTM-CNNs. arXiv preprint arXiv:1511.08308 (2015)
8. Yang, Z., Salakhutdinov, R., Cohen, W.W.: Transfer learning for sequence tagging with hierarchical recurrent networks. arXiv preprint arXiv:1703.06345 (2017)
9. Liu, L., et al.: Empower sequence labeling with task-aware neural language model. arXiv preprint arXiv:1709.04109 (2017)
10. Dong, C., Wu, H., Zhang, J., Zong, C.: Multichannel LSTM-CRF for named entity recognition in Chinese social media. In: Sun, M., Wang, X., Chang, B., Xiong, D. (eds.) CCL/NLP-NABD -2017. LNCS (LNAI), vol. 10565, pp. 197–208. Springer, Cham (2017). https://doi.org/10.1007/978-3-319-69005-6_17
11. Duan, H., Zheng, Y.: A study on features of the CRFs-based Chinese named entity recognition. Int. J. Adv. Intell. **3**, 287–294 (2011)

12. He, H., et al.: Dual long short-term memory networks for sub-character representation learning. In: Latifi, S. (ed.) Information Technology - New Generations. AISC, vol. 738, pp. 421–426. Springer, Cham (2018). https://doi.org/10.1007/978-3-319-77028-4_55

13. Yu, J., Jian, X., Xin, H., Song, Y.: Joint embeddings of Chinese words, characters, and fine-grained subcharacter components. In: Proceedings of the 2017 Conference on Empirical Methods in Natural Language Processing, pp. 286–291 (2017)

14. Shao, Y., Hardmeier, C., Tiedemann, J., Nivre, J.: Character-based joint segmentation and POS tagging for Chinese using bidirectional LSTM-CRF. arXiv preprint arXiv:1704.01314 (2017)

15. Dong, C., Zhang, J., Zong, C., Hattori, M., Di, H.: Character-based LSTM-CRF with radical-level features for Chinese named entity recognition. In: Lin, C.-Y., Xue, N., Zhao, D., Huang, X., Feng, Y. (eds.) ICCPOL/NLPCC -2016. LNCS (LNAI), vol. 10102, pp. 239–250. Springer, Cham (2016). https://doi.org/10.1007/978-3-319-50496-4_20

16. Cao, S., Lu, W., Zhou, J., Li, X.: cw2vec: learning Chinese word embeddings with stroke n-gram information. (2018)

17. Santos, C.D., Zadrozny, B.: Learning character-level representations for part-of-speech tagging. In: Proceedings of the 31st International Conference on Machine Learning, pp. 1818–1826 (2014)

18. Mikolov, T., Sutskever, I., Chen, K., Corrado, G.S., Dean, J.: Distributed representations of words and phrases and their compositionality. In: Advances in Neural Information Processing Systems, pp. 3111–3119 (2013)

19. Chen, X., Qiu, X., Huang, X.: A feature-enriched neural model for joint Chinese word segmentation and part-of-speech tagging. arXiv preprint arXiv:1611.05384 (2016)

20. Forney, G.D.: The Viterbi algorithm. Proc. IEEE **61**(3), 268–278 (1973)

21. Duchi, J., Hazan, E., Singer, Y.: Adaptive subgradient methods for online learning and stochastic optimization. J. Mach. Learn. Res. **12**(Jul), 2121–2159 (2011)

22. Srivastava, N., Hinton, G., Krizhevsky, A., Sutskever, I., Salakhutdinov, R.: Dropout: a simple way to prevent neural networks from overfitting. J. Mach. Learn. Res. **15**(1), 1929–1958 (2014)

23. Zhou, J., Qu, W., Zhang, F.: Chinese named entity recognition via joint identification and categorization. Chinese J. Electron. **22**(2), 225–230 (2013)

24. Ma, X., Hovy, E.: End-to-end sequence labeling via bi-directional LSTM-CNNs-CRF. arXiv preprint arXiv:1603.01354 (2016)

25. Xu, M., Jiang, H., Watcharawittayakul, S.: A local detection approach for named entity recognition and mention detection. In: Proceedings of the 55th Annual Meeting of the Association for Computational Linguistics, vol. 1, pp. 1237–1247 (2017)

26. Peng, N., Dredze, M.: Improving named entity recognition for Chinese Social Media with word segmentation representation learning. arXiv preprint arXiv:1603.00786 (2016)

Learning BLSTM-CRF
with Multi-channel Attribute Embedding
for Medical Information Extraction

Jie Liu[(✉)], Shaowei Chen, Zhicheng He, and Huipeng Chen

College of Computer and Control Engineering, Nankai University, Tianjin, China
jliu@nankai.edu.cn, {chenshaowei,hezhicheng,chenhp}@mail.nankai.edu.cn

Abstract. In Recent years, medical text mining has been an active research field because of its significant application potential, and information extraction (IE) is an essential step in it. This paper focuses on the medical IE, whose aim is to extract the pivotal contents from the medical texts such as drugs, treatments and so on. In existing works, introducing side information into neural network based Conditional Random Fields (CRFs) models have been verified to be effective and widely used in IE. However, they always neglect the traditional attributes of data, which are important for the IE performance, such as lexical and morphological information. Therefore, starting from the raw data, a novel attribute embedding based MC-BLSTM-CRF model is proposed in this paper. We first exploit a bidirectional LSTM (BLSTM) layer to capture the context semantic information. Meanwhile, a multi-channel convolutional neural network (MC-CNN) layer is constructed to learn the relations between multiple attributes automatically and flexibly. And on top of these two layers, we introduce a CRF layer to predict the output labels. We evaluate our model on a Chinese medical dataset and obtain the state-of-the-art performance with 80.71% F1 score.

Keywords: Medical information extraction · Multi-channel Convolutional neural network

1 Introduction

Recently, online medical and health services have been rapidly developing, and a great deal of medical doctor-patient question and answer data has been accumulated on the Internet. Due to the great value and application potential of this information, text mining about online medical text data has been an active research field in recent years. A fundamental work of these studies is information extraction (IE), whose aim is to extract pivotal contents from the medical texts such as diseases, symptoms, medicines, treatments and checks. And these contents can be further used for other text mining works including information retrieve [20], Pharmacovigilance (PV) [2], and drug-drug interactions [19] tasks.

© Springer Nature Switzerland AG 2018
M. Zhang et al. (Eds.): NLPCC 2018, LNAI 11108, pp. 196–208, 2018.
https://doi.org/10.1007/978-3-319-99495-6_17

IE is an important research in natural language processing (NLP), which focuses on extracting knowledge from unstructured text [6]. Hand-crafted regular expressions, classifiers, sequence models and some other methods are always used in IE. For decades, Conditional Random Fields (CRFs) [13] have been widely considered as effective models. After that, to construct an end-to-end model which can automatically learn semantic relations between words without any hand-crafted features, neural networks have been introduced into CRF methods. Furthermore, the neural network based CRF models have achieved great success in IE tasks including name entity recognition (NER) [3,8], opinion extraction [10], and text chunking [12].

As a domain-specific task, medical IE is a challenging work because the online medical text data always has plenty of professional terminologies and noise. Thus, adopting simple neural network based models is not enough. In order to capture more information, many approaches have been proposed [1,11] to introduce side information and prior knowledge. However, depending on the LSTM-CRF structure, the existing methods always neglect the classical attributes of text such as syntax and morphology. Furthermore, these attribute are not difficult to obtain and can greatly improve the performance.

In this paper, we focus on the medical IE and aim to utilize the classical attributes of data to improve the performance. To achieve this, we propose a novel bidirectional LSTM-CRF model with multi-channel convolution neural network (MC-BLSTM-CRF), and introduce multiple attributes which covers aspects of lexical, morphological, and domain-specific information. These attributes play a strong guiding role in information extraction and are easy to obtain. A multi-channel convolution neural network (MC-CNN) is built to learn the hidden representations of multiple attributes automatically and flexibly. To evaluate our model, we construct a Chinese medical dataset which composed of doctor-patient question and answer data, and achieve the state-of-the-art result on it. Experimental results demonstrate that our approach can discover more meaningful contents than baseline methods. The main contributions of our work can be summarized as follows:

- We propose a MC-BLSTM-CRF model with multi-channel attribute embedding for medical IE task, which can model relations between tokens, multiple attributes and labels effectively.
- The proposed method has excellent extensibility and can flexibly capture meaningful information which is neglected by existing models.
- The experimental results on a Chinese medical IE dataset show that our model substantially outperforms other baseline methods.

The remainder of this paper is organized as follows. In Sect. 2, we discuss the related work. Section 3 introduces the details of the MC-BLSTM-CRF model. Section 4 discusses the experiments setting and results. Finally, we conclude our work in Sect. 5.

2   Related Work

Several works have been proposed to medical IE, and these works can be divided into two categories: traditional methods and neural network based methods.

Traditional IE methods such as Hidden Markov Models (HMM) and CRF [13,15,17,18] have achieved great performance and been widely used in various tasks including medical IE. For instance, Bodnari et al. (2013)[1] developed a supervised CRF model based on features and external knowledge for medical NER. Jochim and Deleris (2017) [11] proposed a constrained CRF approach to consider dependence relations and probability statements in the medical domain. However, these models rely heavily on feature engineering.

To capture features automatically, neural network based models have been proposed and been frequently adopted for IE tasks. Collobert et al. (2011) [4] used a CNN to capture relations between tokens over word embeddings with a CRF on top. Huang et al. (2015) [9] introduced a BLSTM-CRF model for word encoding and joint label decoding based on rich hand-crafted spelling features. Both Lample et al. (2016) [14] and Ma and Hovy (2016) [16] proposed a BLSTM-CRF model with character-level encoding. For medical IE tasks, Chalapathy et al. (2016) [2] and Zeng et al. (2017) [21] both adopted a BLSTM-CRF model to provide end-to-end recognition without hand-craft features. Dong et al.(2017) [5] presented a transfer learning based on BLSTM to employ domain knowledge for enhancing the performance of NER on Chinese medical records.

Although neural network based models are useful, side information and prior knowledge are also important to domain-specific IE. Many studies have proposed various approaches to introduce side information and prior knowledge [1,5], but they usually need to design complex features or train model on external corpora which make these methods more complicated and less scalable. Moreover, they always neglect the traditional attributes which is meaningful for IE.

Comparing with existing approaches, a MC-BLSTM-CRF model with multi-channel attribute embedding proposed by us can learn the relations between traditional attributes of raw data by devising a multi-channel CNN structure.

3   Methodology

In this section, we describe the components of our attribute embedding based MC-BLSTM-CRF model in details. Given a word sequence $X = \{x_1, x_2, \cdots, x_n\}$, its corresponding multi-attribute sequence $M = \{m_1, m_2, \cdots, m_n\}$ and label sequence $Y = \{y_1, y_2, \cdots, y_n\}$, the goal is extracting important information by modeling the conditional probability $P(Y \mid X)$. To achieve this, we need to consider the semantic relations between words, context relations between attributes and the transfer relations between labels. Thus, we propose an attribute embedding based MC-BLSTM-CRF model which introduces a multi-channel CNN based into a neural network based CRF method. Figure 1 illustrates the overall framework of the attribute embedding based MC-BLSTM-CRF.

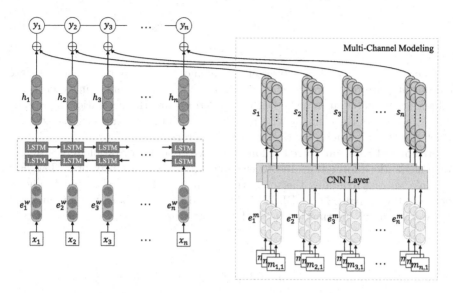

Fig. 1. The overall framework of MC-BLSTM-CRF.

To model semantic relations, each word x_i is mapped to a word embedding vector $e_i^w \in R^{d_w}$ firstly, where d_w is the dimension of word embedding. And then, we exploit a LSTM layer to obtain the hidden representation h_i which contains context semantic information by encoding embedding vector e_i^w:

$$h_i = F(e_i^w, h_{i-1}), \tag{1}$$

where F is the encoding function, and $h_i \in R^{d_h}$, where d_h is the dimension of hidden vectors.

3.1 Bidirectional LSTM

LSTM [7] is a type of RNN which can capture long-distance semantic relation by maintaining a memory cell to store context information. The memory cell is constantly updated in the encoding process, and the proportions of information are determined by three multiplicative gates including input gate, forget gate and output gate. Although various LSTM architectures have been explored, we adopt the basic LSTM model similar to [16]. Formally, the encoding process at the t-th time step is implemented as follow:

$$i_t = \sigma\left(W_{hi}h_{t-1} + W_{ei}e_t^w + b_i\right), \tag{2}$$

$$f_t = \sigma\left(W_{hf}h_{t-1} + W_{ef}e_t^w + b_f\right), \tag{3}$$

$$\widetilde{c}_t = tanh\left(W_{hc}h_{t-1} + W_{ec}e_t^w + b_c\right), \tag{4}$$

$$c_t = f_t \odot c_{t-1} + i_t \odot \widetilde{c}_t, \tag{5}$$

$$o_t = \sigma \left(W_{ho} h_{t-1} + W_{eo} e_t^w + b_o \right), \qquad (6)$$

$$h_t = o_t \odot tanh\left(c_t \right), \qquad (7)$$

where c_t, i_t, f_t, and o_t represent the memory cell, input gate, forget gate and output gate respectively. e_t^w and h_t donate the word embedding vector and hidden state vector at time t. Both σ and $tanh$ are the activation functions, and $\odot$ represents the element-wise product. W_* and b_* are network parameters which donate the weight matrices and bias vectors.

Although LSTM can solve the long-distance dependency problem, it still lose some semantic information due to the sequential encoding way of LSTM. For example, h_t only contains the semantic information before time step t. Therefore, a bidirectional LSTM (BLSTM) is needed to model both the forward and backward context information in the following form:

$$\begin{aligned}\overrightarrow{h_t} &= F\left(e_t^w, \overrightarrow{h_{t-1}} \right), \\ \overleftarrow{h_t} &= F\left(e_t^w, \overleftarrow{h_{t+1}} \right), \end{aligned} \qquad (8)$$

and the two hidden states are concatenated to obtain the final output as follow:

$$h_t = \left[\overrightarrow{h_t}, \overleftarrow{h_t} \right]. \qquad (9)$$

3.2 Multi-channel CNN

Due to professional terminologies existing in the medical text, adopting the simple neural network based CRF models to medical IE is not enough. Many studies tried to introduce side information and domain-specific knowledge into the neural network based models. However, they usually neglect the traditional attributes of text, which is important for medical IE, such as syntactic attribute. Therefore, we propose to integrate multiple attributes of words including syntactic, morphological, and semantic information into existing IE models.

To achieve this, we need to consider how to model context relations among multiple attributes. In previous IE studies, CNN [16] has been mainly used to encode character-level representation, and the benefit of CNN has been proved. Inspired by these studies, to model relations among attributes flexibly and effectively, a multi-channel CNN structure is construct in parallel with capturing semantic by LSTM. A CNN structure for an attribute is defined as a channel. Figure 2 shows the CNN structure of one channel.

Formally, given a multi-attribute sequence $m_i = \{m_{i,1}, m_{i,2}, \cdots, m_{i,k}\}$ of word x_i, where k denotes the number of channels and attribute categories. Firstly, we map each category of attributes to an embedding vector $e_{i,l}^m \in R^{d_m^l}$, where $e_{i,l}^m$ and d_m^l represent the l-th attribute embedding of word x_i and its dimension respectively.

Define $e_l^m = \left\{ e_{1,l}^m, e_{2,l}^m, \cdots, e_{n,l}^m \right\}$ to represent the l-th attribute embedding of a word sequence and extend its outer border to a padded sequence

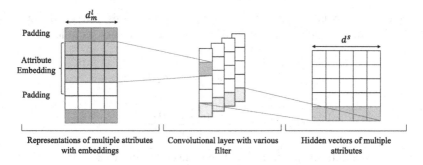

Fig. 2. The CNN structure of one channel. Different channels of CNN are used to capture context relations between different types of attributes.

$\{0_{\lfloor 1 \rfloor}, \cdots, 0_{\lfloor \frac{\beta}{2} \rfloor}, e^m_{1,l}, \cdots, e^m_{n,l}, 0_{\lfloor 1 \rfloor}, \cdots, 0_{\lfloor \frac{\beta}{2} \rfloor}\}$, where β is the size of CNN windows. The hidden representation of the l-th attribute can be obtained as follow:

$$s_{i,l} = W_l r_{i,l} + b_l, \tag{10}$$

$$r_{i,l} = \left\{ e^m_{i-\lfloor \frac{\beta}{2} \rfloor, l}, \cdots, e^m_{i,l}, \cdots, e^m_{i+\lfloor \frac{\beta}{2} \rfloor, l} \right\}, \tag{11}$$

where $r_{i,l}$ denotes the l-th attribute embedding of the current word, its left neighbors, and its right neighbors in convolution window. $s_{i,l}$ is the hidden representation of $e^m_{i,l}$. W_l and b_l are the network parameters which donate the weight matrices and bias vectors of l-th attribute. To control the proportions of various attributes in the final representations flexibly, we adopt a variety of weight matrices and bias vectors for different categories of attributes.

Finally, we obtain the complete representation of the i-th word by concatenating multi-attributes of each word and the hidden word vector as follow:

$$z_i = [h_i, s_{i,1}, s_{i,2}, \cdots, s_{i,k}]. \tag{12}$$

3.3   CRF

Sequence labeling for IE can be considered as a special classification problem. However, we cannot intuitively use a classifier because there are some dependencies across the output labels that can be overlooked by the classifier. For example, in NER tasks with BIO tagging scheme, I-LOC should not follow B-ORG.

Accordingly, we model the label sequence jointly with a CRF layer. Instead of predicting each label independently, CRF can model the relations between adjacent labels with a transition score and learn the interactions between a pair of token and label with a state score. Formally, given a hidden representation sequence $Z = \{z_1, z_2, \cdots, z_n\}$, and an output label sequence $Y = \{y_1, y_2, \cdots, y_n\}$, CRF is used to model the conditional probability $P(Y \mid X)$. The matrix of transition scores can be denoted by $A \in R^{k \times k}$ where k is the number of distinct labels, and the matrix of state scores can be denoted by $P \in R^{n \times k}$. Thus, we define the probability of a tag sequence as follows:

$$S(Z, y) = \sum_{i=1}^{n} A_{y_{i-1}, y_i} + \sum_{i=1}^{n} P_{i, y_i}, \tag{13}$$

$$p(y \mid Z) = \frac{\exp(S(Z, y))}{\sum_{\widetilde{y} \in Y_Z} \exp(S(Z, \widetilde{y}))}, \tag{14}$$

where Y_Z represents all possible label sequences for input Z. During training, we use the maximum conditional likelihood estimation for parameters learning:

$$\log(p(y \mid Z)) = S(Z, y) - \log \sum_{\widetilde{y} \in Y_Z} \exp(S(Z, \widetilde{y})). \tag{15}$$

While predicting, we search the output sequence which obtains the maximum conditional probability given by:

$$y^* = \mathrm{argmax}_{\widetilde{y} \in Y_Z} S(Z, \widetilde{y}), \tag{16}$$

with the Viterbi algorithm.

Finally, we construct our MC-BLSTM-CRF model with the above three layers. For each word, the hidden representation is obtained by a BLSTM layer, and the hidden vectors of multiple attributes are computed by a multi-channel CNN layer. On the top of the two layers, we integrate the hidden representations of word and multiple attributes, and feed the output vector into the CRF layer to jointly decode the best label sequence. We use the loss function of label predicting as the overall optimization objective.

4   Experiments

4.1   Datasets

To validate the effectiveness of our proposed MC-BLSTM-CRF model, we test the performance of our model on Chinese medical IE task. For this task, we construct a medical dataset with data crawled from an online medical platform, haodf.com. For our research, we design five types of labels for medical entities: disease (D), symptom (S), medicine (M), treatment (T), and check (C). Detailed statistics of the dataset are shown in Table 1. Considering that lengths of most entities are short, we adopt the BIO (Beginning, Inside, Outside) tagging scheme.

To ensure the reliability of our experimental results, we divide the dataset into training set, validation set and test set according to the proportion of 4 : 1 : 1, and use the 5-fold cross-validation.

Table 1. Statistics of the dataset. #Sent and #Token represent the number of sentences and words. #Entities, A#Entities and Avg L denote the total number of entities in each category, the average number of entities in each sentence and the average length of entities.

Diseases	#Sent	#Tokens
Gastritis	580	39312
Lung Cancer	513	34810
Asthma	690	45445
Hypertension	574	38564
Diabetes	560	31514

(a) Overall statistics of dataset.

Category	#Entities	A#Entities	Avg L
Disease	989	1.604	5.258
Symptom	1314	1.482	3.925
Medicine	1608	1.589	4.425
Treatment	500	0.590	4.694
Check	657	0.905	4.722

(b) Statistics of entities.

4.2 Attributes

In order to use the traditional attributes to help our model extract more meaningful contents better, we extract a series of attributes of raw data which are simple and easily accessible. The multiple attributes extracted by us can be divided into five categories as following, which cover three aspects of lexical, morphological, and domain-specific information.

- **POS attributes:** We use Ansj, which is an open source toolkit, to extract POS attributes. This kind of attributes represent the syntax information.
- **English acronym attributes:** In Chinese medical text, some professional terminologies are represented as English acronyms. And these acronyms can always provide some important information. Accordingly, we use "yes" or "no" to mark whether a word contains English acronyms.
- **Digital attributes:** Similar to English acronym, digits also play a unique role in sentences. We use "yes" or "no" to express if a word contains digits.
- **Suffix attributes:** In English IE, the suffix of words is often used to improve recognition performance, and the existing studies have proved the effectiveness of this operation. For Chinese medical IE, suffix information is also important. Thus, we choose the last characters of words as suffix attributes.
- **Body attributes:** Through observation, we found that medical entities are often related to body parts. Therefore, we build a dictionary of body parts and use "yes" or "no" to characterize the body attributes (whether

4.3 Experiment Setting

In our experiments, all embeddings are randomly initialized, and the dimensions of word embeddings, POS embeddings and suffix embeddings are set to 100, 40 and 50 respectively. Meanwhile, embeddings of English character, digital and body attributes are represented by One-Hot encoding with two dimensions. All the weight matrices are randomly initialized by sampling from (0.1, 0.1), and all the biases are initialized to zero. The size of the hidden units for word LSTM is set to 200, while the numbers of multi-channel CNN kernels are 90, 50, 2, 2 and

2 corresponding to POS, suffix, English character, number and body attributes respectively. The batch size and max iterations are set to 50 and 100.

We chose Adam as the optimization method for training with learning rates 0.015. And we adopt a dropout rate of 0.5 to mitigate overfitting.

For evaluation, we calculate the precision, recall, and F1 score with full word matching based method. This method means that an entity is correct only if all words of this entity are correctly predicted.

4.4 Experiment Results

To verify the performances of MC-BLSTM-CRF, we compare it against a variety of representative baselines. We can divide the baselines into two categories: traditional methods and neural network based methods. Details of baselines are shown as follows:

- **CRF:** CRF is a classical method for IE. For tagging processing, it can capture the transfer relations between labels, and relations between tokens and labels. With this method, we need to manually design features.
- **BLSTM:** LSTM is a variant of RNN network. It can capture long-term distance semantic relations among tokens automatically without any manual features as input. And then, the hidden vectors will be fed into a softmax layer for tag prediction. To learn forward and backward semantic information, we adopt bidirectional LSTM (BLSTM).
- **CNN:** CNN can capture the semantic relations between tokens by convolution. Compared with LSTM, it has better parallelism and flexibility. But it can only capture the semantic context feature in a certain window around a given token. We also construct a softmax layer to predict labels.
- **BLSTM-CRF:** BLSTM-CRF is a model that uses CRF to replace the softmax layer for labeling.
- **CNN-CRF:** This method is an extension of the CNN method which replaces the softmax layer with a CRF layer to retain the label relations.

Table 2 illustrates the performance of our models and baseline models on the Chinese medical dataset, and lists the best result on valid set (Dev), test set (Test) and the test result corresponding to the best valid set result (Dev-Test). We can see that BLSTM-CRF and CNN-CRF models outperform CRF model with 0.5% and 2.3% on Test F1 score, which proves that the neural network can capture semantic features effective without heavy hand-crafted features.

Meanwhile, both the BLSTM-CRF and CNN-CRF models are superior to BLSTM and CNN architectures with 0.45% on Dev-Test F1 score respectively, because the CRF layer can learn the transfer relations among output labels besides the relations between states and labels.

Despite BLSTM and CNN can both capture effective semantic information of words, the performance of CNN is still lower than BLSTM with about 2.0% on Dev-Test F1 score. The reason is that CNN can only consider the context within a certain kernel window, while BLSTM can retain all of the important

Table 2. Performance comparison among MC-BLSTM-CRF and baselines.

Model	Dev			Test			Dev-Test		
	Precision	Recall	F1	Precision	Recall	F1	Precision	Recall	F1
CRF	-	-	-	81.49	70.11	75.36	-	-	-
CNN	79.64	71.87	75.55	78.74	71.54	74.93	78.80	70.86	74.61
CNN-CRF	80.08	72.67	76.18	79.68	72.33	75.81	79.07	71.50	75.06
BLSTM	79.28	74.86	76.99	79.35	75.10	77.12	79.34	74.40	76.75
BLSTM-CRF	81.28	74.87	77.93	80.25	75.22	77.65	80.17	74.49	77.20
Ours Model	**82.82**	**79.60**	**81.18**	**82.56**	**79.55**	**81.01**	**82.45**	**79.05**	**80.71**

information in sentences with the memory cell. Therefore, this proves the validity of using LSTM for word-level encoding in our model.

Furthermore, compared with the baselines, we can find that the MC-BLSTM-CRF model proposed by us achieves higher scores with at least 3.2% on Dev, Test and Dev-Test F1 score. Therefore, it demonstrates that the attribute features captured by the multi-channel convolution layer can supplement extra information which is important to improve the performance.

4.5 Effectiveness Analysis

To demonstrate the effectiveness of multi-channel attribute embedding, we experiment with different attributes. Figure 3 shows the results. We can find that POS and suffix attributes play a important role in improving the performance of the model, while the English character, digital and body attributes have a slight promotion. We suppose that there are two reasons. Firstly, the POS and suffix attributes have more significant and diversified information. We extract 91 types of POS attributes and 11000 types of suffix attributes while other attributes are only "yes" or "no" value. And through a statistical analysis, we found that these two attributes have more indicative effect for information

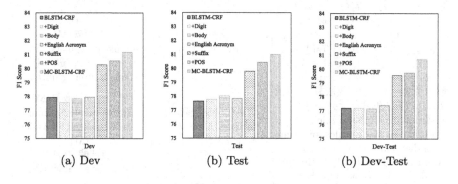

(a) Dev   (b) Test   (b) Dev-Test

Fig. 3. Attribute analysis of MC-BLSTM-CRF.

extraction. For instance, drugs usually appear after a verb or punctuation mark and ends with " 片 (tablet)", " 囊 (capsule)", and so on. Secondly, in the process of encoding, the semantic representation of POS and suffix attributes are richer. The embeddings of POS and suffix attributes are vectors with 40 and 50 dimensions respectively while embeddings of other attributes are one-hot vectors with 2 dimension.

In general, the results prove that our method can use the traditional attributes to enhance the semantic information which is helpful to the performance. Furthermore, attribute fusion can bring a great promotion to the model.

4.6   Case Study

For better understand what information learned from the multi-channel CNN layer can be supplemented to the BLSTM layer, we randomly pick out three sentences from our dataset and compare the recognition results of MC-BLSTM-CRF model with LSTM-CRF.

According to Table 3, we can find that the recognition result of our model is obviously better than that of BLSTM-CRF, and the improvement is three-fold: (1) the optimization of recognition boundary, (2) the recognition of medical terminologies, (3) the recognition of symptom descriptions. For example, in sentence 1, " 下肢血管b超检 (Ultrasonic artery examination)" can be recognized by MC-BLSTM-CRF, while BLSTM-CRF can introduce the noise word " 先行 (First)". For sentence 2 and 3, MC-BLSTM-CRF can find more medical terminologies like " 吉法酯片 (Gefarnate Tablets)" and symptom descriptions

Table 3. Case study.

Sentence	BLSTM-CRF	MFLSTM-CRF
建议先行下肢血管b超检查，看看有无动脉狭窄。 It is suggested that B Ultrasound should be checked first to see if there is any artery stenosis.	先行下肢血管b超检查(C) First ultrasonic artery examination(C) 动脉狭窄(D) Artery stenosis(D)	下肢血管b超检查(C) Ultrasonic artery examination(C) 动脉狭窄(D) Artery stenosis(D)
医生说是肠炎，给我开了腹可安，不管用，肚子经常咕噜响。 The doctor said it was enteritis and gave me Fukean Tablets. But it wasn't useful and my stomach often grunted.	肠炎(D) Enteritis(D)	肠炎(D) Enteritis(D) 腹可安(M) Fukean Tablets(M) 肚子经常咕噜响(S) Stomach often grunted(S)
服用说明：雷贝拉唑钠胶囊每日1次1次1片，吉法酯片一天3次一次2片。 Instructions: Rabeprazole Sodium Enteric-coated Capsules, one time a day and one piece a time. Gefarnate Tablets, three times a day and two pieces a time.	雷贝拉唑钠胶囊(M) Rabeprazole Sodium Enteric-coated Capsules (M)	雷贝拉唑钠胶囊(M) Rabeprazole Sodium Enteric-coated Capsules (M) 吉法酯片(M) Gefarnate Tablets(M)

like "肚子经常咕噜响 (Stomach often grunted)" consisting of multi-words than BLSTM-CRF.

5 Conclusion

In this paper, we proposed a attribute embedding based MC-BLSTM-CRF model for medical IE task. The main contribution of this model is to capture relations between attributes effectively and flexibly with a multi-channel CNN layer and use these attributes to improve recognition performance. Experimental results showed that our model outperforms the existing methods for IE. Meanwhile, the case study results showed our model's capability of learning domain-specific information which is helpful to improve recognition performance.

Acknowledgement. This research is supported by the National Natural Science Foundation of China under the grant No. U1633103 and 61502499, the Science and Technology Planning Project of Tianjin under the grant No. 17ZXRGGX00170, the Natural Science Foundation of Tianjin under the grant No. 18JCYBJC15800, and the Open Project Foundation of Information Technology Research Base of Civil Aviation Administration of China under the grant No. CAAC-ITRB-201601.

References

1. Bodnari, A., Deléger, L., Lavergne, T., Névéol, A., Zweigenbaum, P.: A supervised named-entity extraction system for medical text. In: Working Notes for CLEF 2013 Conference (2013)
2. Chalapathy, R., Borzeshi, E.Z., Piccardi, M.: An investigation of recurrent neural architectures for drug name recognition. In: Proceedings of the Seventh International Workshop on Health Text Mining and Information Analysis, pp. 1–5 (2016)
3. Chiu, J.P.C., Nichols, E.: Named entity recognition with bidirectional LSTM-CNNs. Computer Science (2015)
4. Collobert, R., Weston, J., Bottou, L., Karlen, M., Kavukcuoglu, K., Kuksa, P.P.: Natural language processing (almost) from scratch. J. Mach. Learn. Res. **12**, 2493–2537 (2011)
5. Dong, X., Chowdhury, S., Qian, L., Guan, Y., Yang, J., Yu, Q.: Transfer bidirectional LSTM RNN for named entity recognition in Chinese electronic medical records. In: 19th IEEE International Conference on e-Health Networking, Applications and Services, pp. 1–4 (2017)
6. Hassan, H., Awadallah, A.H., Emam, O.: Unsupervised information extraction approach using graph mutual reinforcement. In: EMNLP, pp. 501–508 (2006)
7. Hochreiter, S., Schmidhuber, J.: Long short-term memory. Neural Comput. **9**(8), 1735–1780 (1997)
8. Hu, Z., Ma, X., Liu, Z., Hovy, E.H., Xing, E.P.: Harnessing deep neural networks with logic rules. In: Proceedings of ACL (2016)
9. Huang, Z., Xu, W., Yu, K.: Bidirectional LSTM-CRF models for sequence tagging. Computer Science (2015)
10. Irsoy, O., Cardie, C.: Opinion mining with deep recurrent neural networks. In: Proceedings of EMNLP, pp. 720–728 (2014)

11. Jochim, C., Deleris, L.A.: Named entity recognition in the medical domain with constrained CRF models. In: Proceedings of ACL, pp. 839–849 (2017)
12. Kudoh, T., Matsumoto, Y.: Use of support vector learning for chunk identification. In: CoNLL, pp. 142–144 (2000)
13. Lafferty, J.D., McCallum, A., Pereira, F.C.N.: Conditional random fields: Probabilistic models for segmenting and labeling sequence data. In: Proceedings of the Eighteenth International Conference on Machine Learning, pp. 282–289 (2001)
14. Lample, G., Ballesteros, M., Subramanian, S., Kawakami, K., Dyer, C.: Neural architectures for named entity recognition. In: NAACL, pp. 260–270 (2016)
15. Luo, G., Huang, X., Lin, C., Nie, Z.: Joint entity recognition and disambiguation. In: Proceedings of the 2015 Conference on Empirical Methods in Natural Language Processing, pp. 879–888 (2015)
16. Ma, X., Hovy, E.H.: End-to-end sequence labeling via bi-directional LSTM-CNNs-CRF. In: Proceedings of ACL (2016)
17. Passos, A., Kumar, V., McCallum, A.: Lexicon infused phrase embeddings for named entity resolution. In: Proceedings of the Eighteenth Conference on Computational Natural Language Learning, pp. 78–86 (2014)
18. Ratinov, L., Roth, D.: Design challenges and misconceptions in named entity recognition. In: Proceedings of the Thirteenth Conference on Computational Natural Language Learning, pp. 147–155 (2009)
19. Segura-Bedmar, I., Martínez, P., de Pablo-Sánchez, C.: Using a shallow linguistic kernel for drug-drug interaction extraction. J. Biomed. Inform. **44**(5), 789–804 (2011)
20. Takaki, O., Murata, K., Izumi, N., Hasida, K.: A medical information retrieval based on retrievers' intentions. In: HEALTHINF 2011 - Proceedings of the International Conference on Health Informatics, pp. 596–603 (2011)
21. Zeng, D., Sun, C., Lin, L., Liu, B.: LSTM-CRF for drug-named entity recognition. Entropy **19**(6), 283 (2017)

Distant Supervision for Relation Extraction with Neural Instance Selector

Yubo Chen[1(✉)], Hongtao Liu[2], Chuhan Wu[1], Zhigang Yuan[1], Minyu Jiang[3], and Yongfeng Huang[1]

[1] Next Generation Network Lab, Department of Electronic Engineering,
Tsinghua University, Beijing, China
{yb-ch14,wuch15,yuanzg14}@mails.tsinghua.edu.cn, yfhuang@tsinghua.edu.cn
[2] Tianjin Key Laboratory of Advanced Networking,
School of Computer Science and Technology, Tianjin University, Tianjin, China
htliu@tju.edu.cn
[3] Fan Gongxiu Honor College, Beijing University of Technology, Beijing, China
ryancoper@emails.bjut.edu.cn

Abstract. Distant supervised relation extraction is an efficient method to find novel relational facts from very large corpora without expensive manual annotation. However, distant supervision will inevitably lead to wrong label problem, and these noisy labels will substantially hurt the performance of relation extraction. Existing methods usually use multi-instance learning and selective attention to reduce the influence of noise. However, they usually cannot fully utilize the supervision information and eliminate the effect of noise. In this paper, we propose a method called Neural Instance Selector (NIS) to solve these problems. Our approach contains three modules, a sentence encoder to encode input texts into hidden vector representations, an NIS module to filter the less informative sentences via multilayer perceptrons and logistic classification, and a selective attention module to select the important sentences. Experimental results show that our method can effectively filter noisy data and achieve better performance than several baseline methods.

Keywords: Relation extraction · Distant supervision
Neural Instance Selector

1 Introduction

Relation extraction is defined as finding relational sentences and specify relation categories from plain text. It is an important task in the natural language processing field, particularly for knowledge graph completion [7] and question answering [10]. Distant supervision for relation extraction aims to automatically label large scale data with knowledge bases (KBs) [14]. The labeling procedure is as follows: for a triplet (e_{head}, e_{tail}, r) in KB, all sentences (instances) that simultaneously mention head entity e_{head} and tail entity e_{tail} constitute a *bag* and are labeled as relation r.

© Springer Nature Switzerland AG 2018
M. Zhang et al. (Eds.): NLPCC 2018, LNAI 11108, pp. 209–220, 2018.
https://doi.org/10.1007/978-3-319-99495-6_18

However, distant supervised relation extraction is challenging because the labeling method usually suffers from the *noisy labeling problem* [17]. A sentence may not express the relation in the KB when mentioning two entities. Table 1 shows an example of the noisy labeling problem. Sentence 2 mentions *New Orleans* and *Dillard University* without expressing the relation */location/location/contains*. Usually, the existence of such noisy labels will hurt the performance of relation extraction methods. Thus, it's important to eliminate such noise when constructing relation extraction models.

Several approaches have been proposed to eliminate negative effects of noise instances. For example, Riedel et al. [17] proposed to use graphical model to predict which sentences express the relation based on the *at-least-once* assumption. Zeng et al. [24] propose to combine multi-instance learning [4] with Piecewise Convolutional Neural Networks (PCNNs) to choose the most likely valid sentence and predict relations. However, these methods ignore multiple informative sentences and only select one sentence from each bag for training. Therefore, they cannot fully exploit the supervision information.

In recent years, attention mechanism is introduced to this task to select information more effectively. Lin et al. [11] and Ji et al. [9] used bilinear and non-linear form attention respectively to assign higher weights to valid sentences and lower weights to invalid ones. Then the bag is represented as a weighted sum of all sentences' representations. However in these methods, the softmax formula of attention weights will assign positive weights to noisy data. These positive weights of noisy sentences violated the intuition that noisy sentences cannot provide relational information. Thus such attention based models can't fully eliminate the negative effect of noise.

In this paper we proposed a method called Neural Instance Selector (NIS) to further utilize rich supervision information and alleviate negative effect of noisy labeling problem. Our approach contains three modules. A sentence encoder transforms input texts into distributed vector representations with PCNN. A Neural Instance Selector filters less informative sentences with multilayer perceptrons and logistic classification. The NIS module can select multiple valid sentences, and exploit more information than MIL method. A selective attention module selects more important sentences with higher weights. In order to further

Table 1. An example of noisy labeling problem. The bold words are head/tail entities.

Triplet	Instances	Noisy?
(New Orleans, Dillard University, /location/location/contains)	1. Jinx Broussard, a communications professor at **Dillard University** in **New Orleans**, said ...	No
	2. When he came here in May 2003 to pick up an honorary degree from **Dillard University**, his dense schedule didn't stop him ... ever since he lived in **New Orleans** in the 1950's	Yes

eliminate noise effects than attention-based methods, we only assign attention weights to selected sentences. Experimental results on the benchmark dataset validate the effectiveness of our model.

2   Related Work

Early works focused on *feature-based* methods for relation extraction. GuoDong et al. [6] explored lexical and syntactic features with textual analysis and feed them into a SVM classifier. Bunescu et al. [3] connected weak supervision with multi-instance learning [4] and extend it to relation extraction. Riedel et al. [17] proposed *at-least-once* assumption to alleviate the wrong label problem. However, these methods lack the ability of fully utilizing supervision information and suppress noise. Besides, these methods cannot effectively use the contextual information.

Recent works attempt to use neural networks for *supervised* relation extraction. Socher et al. [19] represented words with vectors and matrices and use recursive neural networks to compose sentence representation. Zeng et al. [25], Nguyen et al. [16] and dos Santos et al. [18] extracted sentence level vector representation with CNNs. Other work adopted recurrent neural networks to this task [20, 26]. However, these methods need sentence-annotated data, which cannot be applied to large scale corpus without human annotation.

In order to apply neural networks to *distant supervision*, Zeng et al. [24] proposed PCNN to capture sentence structure information, and combined it with Multi-Instance Learning [4] (MIL) to select the sentence with the highest right probability as bag representation. Although proved effective, MIL suffers from *information loss problem* because it ignored the presence of more than one valid instances in most bags. Recently attention mechanism attracted a lot of interests of researchers [1,12,15,22]. Considering the flaw of MIL, Lin et al. [11] and Ji et al. [9] introduced bilinear and non-linear attention respectively into this task to make full use of supervision information by assigning higher weights to valid instances and lower weights to invalid ones. The two attention models significantly outperform MIL method. However, they suffer from *noise residue problem* because noisy sentences have harmful information but still have positive weights. The residue weights of noisy data mean that attention methods cannot fully eliminate the negative effects of noise.

Different from MIL and attention methods, we propose a method named NIS to further solve the information loss and noise residue problem. First, We use PCNNs [24] to learn sentence representations. Second, an NIS module takes all sentences' representations in a bag as input, and uses a MLP to capture the information of noise. Third, a logistic classifier takes MLP output to select valid sentences and filter noisy ones. The NIS module can alleviate information loss problem by retaining more than one valid sentences. Finally, we assign attention weights to selected sentences and use them to compute bag representation. In this way noise residue problem is reduced by avoiding assigning weights to unselected sentences. Experimental results show that the NIS module can alleviate these two problems and bring better performance to baseline models.

3   Methodology

In this section, we will introduce our method. Our framework contains three parts, which is shown in Fig. 1. We will introduce these parts each by each.

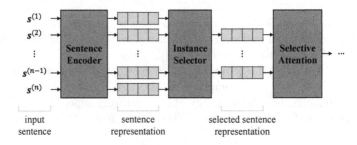

Fig. 1. Details of Instance Selector framework.

3.1   Sentence Encoder

Sentence encoder transforms the sentence into its distributed representation. First, words in a sentence are transformed into dense real-valued vectors. For word token w, we use pre-trained *word embeddings* as low dimension vector representation. Following Zeng et al. [24], we use *position embeddings* as extra position feature. We compute the relative distances between each word and two entity words, and transform them to real-valued vectors by looking up randomly-initialized embedding matrices. We denote the word embedding of word w by $\mathbf{w}_w \in \mathbb{R}^{d_w}$ and two position embeddings by $\mathbf{p}_w^{(1)}, \mathbf{p}_w^{(2)} \in \mathbb{R}^{d_p}$. The word representation $\mathbf{s}_w$ is then composed by horizontal concatenating word embeddings and position embeddings:

$$\mathbf{s}_w = [\mathbf{w}_w; \mathbf{p}_w^{(1)}; \mathbf{p}_w^{(2)}]. \quad \mathbf{s}_w \in \mathbb{R}^{(d_w + 2 \times d_p)} \tag{1}$$

Then, given a sentence and corresponding entity pair, we apply PCNN to construct a distributed representation of the sentence. Compared with common CNN, PCNN uses a piecewise max-pooling layer to capture sentence structure information. A sentence is divided into three segments by two entity words, then max-pooling is executed on each segment respectively. Following Zeng et al. [24], we apply tanh as activation function. We denote convolution kernel channels by c, and the output of PCNN by $\mathbf{f}^{(i)} \in \mathbb{R}^{3c}$.

3.2   Instance Selector

Although previous work yields high performance, there still exists some drawbacks. MIL suffers from *information loss problem* because it ignored multiple valid sentences and used only one sentence for representing a bag and training. Attention-based methods have *noise residue problem* because they assigned

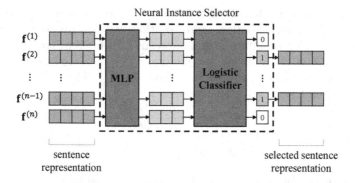

Fig. 2. The structure of NIS module. The 1s among the outputs of Logistic Noise Classifier indicate the corresponding sentence is valid, and 0s indicates invalid.

small but still positive weights to harmful noisy sentences, which means noise effects weren't completely removed.

In order to alleviate these two negative effects, we propose a method called **Neural Instance Selector (NIS)** to pick out more informative sentences. Figure 2 shows the structure of NIS. We use a small neural network to classify valid and invalid sentences. The core component of NIS is a Multilayer Perceptron (MLP) used for capturing information of noise. Then MLP output vectors are fed into a logistic noise classifier to produce sentence-level selection results. As shown in Fig. 2, NIS module has the ability of retaining multiple valid sentences, naturally reducing information loss problem. Then we alleviate noise residue problem by only assigning attention weights to selected sentences. The unselected noisy data will not be assigned weights, and will not participate in training process.

One alternative way for selecting instances is removing MLP and directly feeding PCNN output into logistic classifier [5]. However, we discover that this choice performs worse than NIS. This is because noise information is more complex than relation information, therefore requires deeper structure to be captured. The MLP can improve the non-linear fitting ability of instance selector. Also, the sentence level classifier has many alternatives. We conduct experiments on logistic classifier and two-class softmax classifier, and choose logistic classifier because of its better performance.

3.3 Selective Attention

The object of attention mechanism is to learn higher weights for more explicit instances and lower weights for less relevant ones. Attention-based models represent the ith bag M_i (with label y_i) as a real-valued vector $\mathbf{r}_i$. We denote the jth sentence's representation in the ith bag as $\mathbf{f}_i^{(j)}$. Previous work used bilinear [11] and non-linear form [9] attention. Considering computational efficiency and effectiveness, we choose the non-linear form in our method, denoted as **APCNN**.

Intuitively, relation information is useful when recognizing informative sentences. So we introduce relation representation and concatenate it with sentence representation to compute attention weights. Inspired by translation-based knowledge graph methods [2, 21], the relation r is represented as the difference vector of entity word embeddings: $\mathbf{v}_{rel} = \mathbf{w}_{e_1} - \mathbf{w}_{e_2}$. Then $\alpha^{(j)}$ is computed through a hidden layer:

$$\alpha^{(j)} = \frac{\exp(e^{(j)})}{\sum_k \exp(e^{(k)})}, \tag{2}$$

$$e^{(j)} = \mathbf{W_a}^T(\tanh[\mathbf{f}_i^{(j)}; \mathbf{v}_{rel}]) + b_a. \tag{3}$$

With attention weights $\alpha^{(j)}$ computed, M_i is represented by $\mathbf{r}_i = \sum_{j=1}^{|M_i|} \alpha^{(j)} \mathbf{f}_i^{(j)}$. The bag representation is fed into a softmax classifier to predict relations and compute cross-entropy objective function $J(\theta) = -\sum_{i=1}^{T} \log p(y_i|\mathbf{o}_i; \theta)$:

4   Experiments

4.1   Dataset and Evaluation Metrics

We evaluate our NIS mechanism on the dataset developed by Riedel et al. [17] by aligning Freebase triplets with the New York Times (NYT) corpus. The training data is aligned to the years of 2005–2006 of the NYT corpus, and the testing to year of 2007. The dataset contains 53 relations (including 'NA' for no relation) and 39,528 entity pairs. Training data includes 522,611 sentences, the test set includes 172,448 sentences.

Following Lin et al. [11], we evaluate our method in the held-out evaluation. It provides an approximate measure of precision without time-consuming manual evaluation. We report the aggregate precision/recall curve and Precision@N in our experiments.

4.2   Parameter Settings

In our experiments, we use *word2vec*, proposed by Mikolov et al. [13], to pre-train word embeddings on NYT corpora. We select the dimension of word embedding d_w among {50, 100, 200, 300}, the dimension of position embedding d_p among {5, 10, 20}, the number of feature maps c among {100, 200, 230}, batch size among {50, 100, 150, 160, 200}. The best configurations are: $d_w = 50$, $d_p = 5$, $c = 230$, the batch size is 100. We choose MLP hidden layers' dimensions as [512, 256, 128, 64]. We use dropout strategy [8] and Adadelta [23] to train our models. For training, we set the iteration number over the training set as 20, and decay the learning rate every 10 epochs.

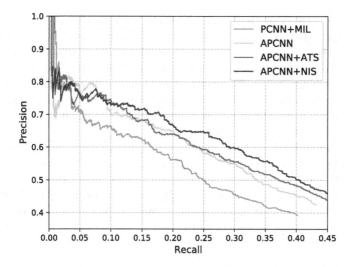

Fig. 3. Aggregate precision/recall curves for PCNN+MIL, APCNN, APCNN+ATS, APCNN+NIS. APCNN denotes the non-linear attention method proposed by Ji et al. [9]. We choose $0.8 \times \max(attention\ weight)$ as APCNN+ATS threshold with the highest performance in our experiments.

4.3 Performance Evaluation

We compare our method with two previous works: **PCNN+MIL** [24] selects the sentence with the highest right probability as bag representation; **APCNN** [9] use non-linear attention to assign weights to all sentences in a bag. In order to prove the superiority of our NIS module, we propose a more intuitive and simpler way for instance selection: we set a threshold on attention weights and filter sentences with lower weights than threshold. We denote this method as **Attention Threshold Selector (ATS)**. We adopt both ATS and NIS to APCNN to demonstrate the effectiveness of instance selectors, denoted as **APCNN+ATS** and **APCNN+NIS** respectively. Figure 3 shows the aggregated precision/recall curves, and Table 2 shows the Precision@N with $N = \{100, 200, 500\}$ of our approaches and all the baselines. From Fig. 3 and Table 2 we have the following observations:

Table 2. Precision@N of PCNN+MIL, APCNN, APCNN+ATS, APCNN+NIS.

Precision@N (%)	Top 100	Top 200	Top 500	Average
PCNN+MIL	71.72	67.84	61.62	66.89
APCNN	78.79	76.38	66.33	73.83
APCNN+ATS	73.74	76.38	65.53	71.88
APCNN+NIS	**78.79**	**76.38**	**69.94**	**75.04**

1. For both ATS and NIS, the instance selector methods outperform PCNN+MIL. It indicates that instance selectors can alleviate information loss problem because it can pick out more than one valid sentences in a bag.
2. Figure 3 shows that the instance selectors bring better performance compared with APCNN for both ATS and NIS method on high recall range. This is because the attention weights are assigned only to selected sentences. It indicates that our method can reduce noise residue problem because the weights of unselected sentences are masked as zero.
3. The NIS method achieves the highest precision over most of the entire recall range compared to other methods including the ATS. It indicates that NIS method can effectively eliminate negative effects of insufficient information utilization and residue noisy weights. It also proves that NIS is better than ATS at filtering noise because the MLP provides deeper structure to handle the complexity of noise information.

4.4   Effectiveness of NIS Module

The NIS module we propose has independent parameters, thus can be adapted to various kinds of neural relation extraction methods. To further demonstrate its effectiveness on different methods, we (1) replace MIL module of PCNN+MIL with our NIS module (denoted as **PCNN+NIS**). Note that this method is a **sentence-level** extraction, different from all other settings; (2) replace APCNN and APCNN+NIS's non-linear attention with bilinear form attention [11] (denoted as **PCNN+ATT** and **PCNN+ATT+NIS** respectively). We report the aggregated precision/recall curves of all the NIS methods in Fig. 4. From Fig. 4 we can see that:

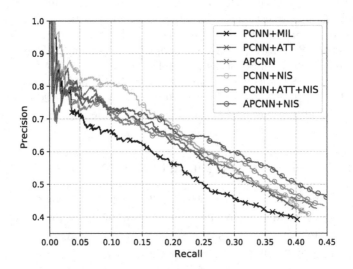

Fig. 4. Aggregate precision/recall curves of PCNN+MIL, PCNN+ATT, APCNN, PCNN+NIS, PCNN+ATT+NIS, APCNN+NIS.

1. Models with NIS module outperform all the corresponding baseline methods, proving its effectiveness and robustness on different structures.
2. PCNN+NIS model has better performance than PCNN+MIL. It indicates the role of NIS module in filtering noise. Lin et al. [11] have proved that sentence-level PCNN has worse performance than PCNN+MIL because it ignores the effect of noise. But our sentence-level PCNN+NIS method defeats PCNN+MIL, which means NIS module can filter out most noise sentences and improve sentence-level performance.
3. PCNN+NIS model also outperforms attention-based models. This is actually a comparison between hard selection and soft selection strategy. The result demonstrates that NIS's hard selection strategy can effectively reduce the negative effects of residue weights brought by soft attention strategy.

4.5   Analysis of ATS Threshold

Although not so powerful as NIS method, ATS method still brings improvement to APCNN model. However, ATS method needs a fine-tuned threshold to achieve its best performance. Higher thresholds bring back information loss problem because more informative sentences are neglected. Lower thresholds bring back noise residue problem because more noisy sentences are selected and assigned weights. We conduct experiments on ATS with different thresholds. For clarify, we use a histogram to approximate precision/recall curves of different thresholds, shown in Fig. 5.

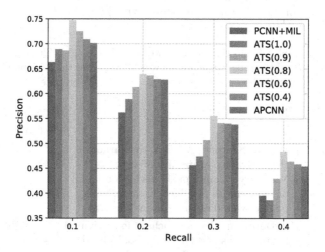

Fig. 5. Aggregate precision/recall histogram of ATS with different thresholds. ATS(α) means the threshold is $\alpha \times \max(attention\ weights)$. APCNN is equivalent to ATS(0). ATS(1.0) means only select the sentence with maximum attention weight as bag representation, similar to PCNN+MIL. We use histogram for clarify because some of the curves are too close.

In our experiments, the best model is ATS(0.8). With higher thresholds (ATS(1.0) and ATS(0.9)), the precisions decline significantly because less informative sentences are utilized. ATS(1.0) selects only the sentence with maximum attention weight to train. Similar to MIL select strategy, ATS(1.0) also has similar performance to PCNN+MIL. With lower thresholds (ATS(0.6) and ATS(0.4)), the performance decreases slightly, close to APCNN model (equivalent to ATS(0)). The reason is that when threshold is lower, more invalid sentences are involved in training, which means the noise effects cannot be fully eliminated. The change of performance with the threshold perfectly shows the impact of information loss and noise residue on relation extraction. It also proves the superiority of NIS because it provides deeper structure to capture the complex information of noise and doesn't require fine-tuned threshold.

4.6 Case Study

Table 3 shows an example of selection result and attention weights of a bag. The bag contains three instances in which the 1st instance is invalid. The remaining instances are informative because they all contain significantly keywords that express the corresponding relation */people/person/place_lived*. APCNN assigns bigger weight to the 1st sentence (invalid) than the 2nd sentence (valid). Therefore, the big noise residue will substantially hurt performance. Our NIS module correctly selects last 2 sentences as valid ones. The selection results shows NIS's ability of filtering noise sentences.

With the help of NIS, attention mechanism only assigns weights to selected sentences. APCNN+NIS assigns a very high weight to the 3rd sentence because the appearance of *lived* strongly indicates the *place_lived* relation. The 2nd

Table 3. An example of selection result and attention weight. The bold strings are head/tail entities, and the red strings are keywords to predict the relation. The relation */place_lived* corresponds the */people/person/place_lived* in Freebase.

Triplet	Instances	APCNN+NIS		APCNN
		Select.	Att.	
(Jane Jacobs, Toronto, /place_lived)	1. Alice Sparberg Alexiou, the author of the biography "**Jane Jacobs**: Urban Visionary" . . . in a panel discussion based on the work of Ms. Jacobs, the urban planner who died in April in **Toronto**	0	-	0.3503
	2. Dovercourt has a penchant for arriving at rock clubs and bars with books by the famed urban critic **Jane Jacobs**, who has made **Toronto** her home for nearly 40 years	1	0.2890	0.0875
	3. **Jane Jacobs**, the activist who took him on, now lives in **Toronto**	1	0.7110	0.5623

sentence has *home*, but the semantic is not strong enough. The attention weights demonstrates that our attention module is able to selectively focus on more relevant sentences.

5 Conclusion

Distant supervision for relation extraction is an efficient method to find relational sentences in very large corpus without manual annotation. Existing methods suffers from information loss and noise residue problem. We proposed a method named NIS to alleviate these two negative effects simultaneously. We use a sentence encoder to transform input texts to vector representation, an NIS module to select multiple valid sentences and an attention module to assign weights to selected sentences. We conduct experiments on a widely used dataset and experimental results validate the effectiveness of our method.

References

1. Bahdanau, D., Cho, K., Bengio, Y.: Neural machine translation by jointly learning to align and translate. arXiv preprint arXiv:1409.0473 (2014)
2. Bordes, A., Usunier, N., Garcia-Duran, A., Weston, J., Yakhnenko, O.: Translating embeddings for modeling multi-relational data. In: Advances in Neural Information Processing Systems, pp. 2787–2795 (2013)
3. Bunescu, R.C., Mooney, R.J.: A shortest path dependency kernel for relation extraction. In: Proceedings of the Conference on Human Language Technology and Empirical Methods in natural Language Processing, pp. 724–731. Association for Computational Linguistics (2005)
4. Dietterich, T.G., Lathrop, R.H., Lozano-Pérez, T.: Solving the multiple instance problem with axis-parallel rectangles. Artif. intell. **89**(1–2), 31–71 (1997)
5. Feng, J., Huang, M., Zhao, L., Yang, Y., Zhu, X.: Reinforcement learning for relation classification from noisy data (2018)
6. GuoDong, Z., Jian, S., Jie, Z., Min, Z.: Exploring various knowledge in relation extraction. In: Proceedings of the 43rd Annual Meeting on Association for Computational Linguistics, pp. 427–434. Association for Computational Linguistics (2005)
7. Han, X., Liu, Z., Sun, M.: Neural knowledge acquisition via mutual attention between knowledge graph and text (2018)
8. Hinton, G.E., Srivastava, N., Krizhevsky, A., Sutskever, I., Salakhutdinov, R.R.: Improving neural networks by preventing co-adaptation of feature detectors. arXiv preprint arXiv:1207.0580 (2012)
9. Ji, G., Liu, K., He, S., Zhao, J., et al.: Distant supervision for relation extraction with sentence-level attention and entity descriptions. In: AAAI, pp. 3060–3066 (2017)
10. Lee, C., Hwang, Y.G., Jang, M.G.: Fine-grained named entity recognition and relation extraction for question answering. In: Proceedings of the 30th Annual International ACM SIGIR Conference on Research and Development in Information Retrieval, pp. 799–800. ACM (2007)
11. Lin, Y., Shen, S., Liu, Z., Luan, H., Sun, M.: Neural relation extraction with selective attention over instances. In: Proceedings of the 54th Annual Meeting of the Association for Computational Linguistics (Volume 1: Long Papers), vol. 1, pp. 2124–2133 (2016)

12. Luong, M.T., Pham, H., Manning, C.D.: Effective approaches to attention-based neural machine translation. arXiv preprint arXiv:1508.04025 (2015)

13. Mikolov, T., Chen, K., Corrado, G., Dean, J.: Efficient estimation of word representations in vector space. arXiv preprint arXiv:1301.3781 (2013)

14. Mintz, M., Bills, S., Snow, R., Jurafsky, D.: Distant supervision for relation extraction without labeled data. In: Proceedings of the Joint Conference of the 47th Annual Meeting of the ACL and the 4th International Joint Conference on Natural Language Processing of the AFNLP: Volume 2, vol. 2, pp. 1003–1011. Association for Computational Linguistics (2009)

15. Mnih, V., Heess, N., Graves, A., et al.: Recurrent models of visual attention. In: Advances in Neural Information Processing Systems, pp. 2204–2212 (2014)

16. Nguyen, T.H., Grishman, R.: Relation extraction: perspective from convolutional neural networks. In: Proceedings of the 1st Workshop on Vector Space Modeling for Natural Language Processing, pp. 39–48 (2015)

17. Riedel, S., Yao, L., McCallum, A.: Modeling relations and their mentions without labeled text. In: Balcázar, J.L., Bonchi, F., Gionis, A., Sebag, M. (eds.) ECML PKDD 2010. LNCS (LNAI), vol. 6323, pp. 148–163. Springer, Heidelberg (2010). https://doi.org/10.1007/978-3-642-15939-8_10

18. Santos, C.N.d., Xiang, B., Zhou, B.: Classifying relations by ranking with convolutional neural networks. arXiv preprint arXiv:1504.06580 (2015)

19. Socher, R., Huval, B., Manning, C.D., Ng, A.Y.: Semantic compositionality through recursive matrix-vector spaces. In: Proceedings of the 2012 Joint Conference on Empirical Methods in Natural Language Processing and Computational Natural Language Learning, pp. 1201–1211. Association for Computational Linguistics (2012)

20. Sorokin, D., Gurevych, I.: Context-aware representations for knowledge base relation extraction. In: Proceedings of the 2017 Conference on Empirical Methods in Natural Language Processing, pp. 1784–1789 (2017)

21. Wang, Z., Zhang, J., Feng, J., Chen, Z.: Knowledge graph and text jointly embedding. In: Proceedings of the 2014 Conference on Empirical Methods in Natural Language Processing (EMNLP), pp. 1591–1601 (2014)

22. Xu, K., Ba, J., Kiros, R., Cho, K., Courville, A., Salakhudinov, R., Zemel, R., Bengio, Y.: Show, attend and tell: Neural image caption generation with visual attention. In: International Conference on Machine Learning, pp. 2048–2057 (2015)

23. Zeiler, M.D.: Adadelta: an adaptive learning rate method. arXiv preprint arXiv:1212.5701 (2012)

24. Zeng, D., Liu, K., Chen, Y., Zhao, J.: Distant supervision for relation extraction via piecewise convolutional neural networks. In: Proceedings of the 2015 Conference on Empirical Methods in Natural Language Processing, pp. 1753–1762 (2015)

25. Zeng, D., Liu, K., Lai, S., Zhou, G., Zhao, J.: Relation classification via convolutional deep neural network. In: Proceedings of COLING 2014, the 25th International Conference on Computational Linguistics: Technical Papers, pp. 2335–2344 (2014)

26. Zhou, P., Shi, W., Tian, J., Qi, Z., Li, B., Hao, H., Xu, B.: Attention-based bidirectional long short-term memory networks for relation classification. In: Proceedings of the 54th Annual Meeting of the Association for Computational Linguistics (Volume 2: Short Papers), vol. 2, pp. 207–212 (2016)

Complex Named Entity Recognition via Deep Multi-task Learning from Scratch

Guangyu Chen[1], Tao Liu[1], Deyuan Zhang[2(✉)], Bo Yu[3], and Baoxun Wang[3]

[1] School of Information, Renmin University of China, Beijing, China
{hcs,tliu}@ruc.edu.cn
[2] School of Computer, Shenyang Aerospace University, Shenyang, China
dyzhang@sau.edu.cn
[3] Tricorn (Beijing) Technology Co., Ltd, Beijing, China
{yubo,wangbaoxun}@trio.ai

Abstract. Named Entity Recognition (NER) is the preliminary task in many basic NLP technologies and deep neural networks has shown their promising opportunities in NER task. However, the NER tasks covered in previous work are relatively simple, focusing on classic entity categories (Persons, Locations, Organizations) and failing to meet the requirements of newly-emerging application scenarios, where there exist more informal entity categories or even hierarchical category structures. In this paper, we propose a multi-task learning based subtask learning strategy to combat the complexity of modern NER tasks. We conduct experiments on a complex Chinese NER task, and the experimental results demonstrate the effectiveness of our approach.

Keywords: Complex named entity recognition · Multi-task learning Deep learning

1 Introduction

Nowadays, many basic NLP technologies have been utilized in the newly-emerging application scenarios, among which the Named Entity Recognition (NER) models are believed to be of paramount importance for locating the essential information slots and predicting user intentions in the task-oriented AI products, especially the ones with speech interface such as conversational agents[1], smart speakers[2].

This work is supported by visiting scholar program of China Scholarship Council and National Natural Science Foundation of China (Grant No. 61472428 and No. U1711262). The work was done when the first author was an intern in Tricorn (Beijing) Technology Co., Ltd.
Bo Yu is currently working in Baidu, Inc.

[1] https://dueros.baidu.com/.
[2] https://developer.amazon.com/alexa.

© Springer Nature Switzerland AG 2018
M. Zhang et al. (Eds.): NLPCC 2018, LNAI 11108, pp. 221–233, 2018.
https://doi.org/10.1007/978-3-319-99495-6_19

Compared to traditional NER tasks focusing on the classic entity categories involving names of persons, locations, organizations, etc. [4,11,21], the NER modules for the task-oriented scenarios of new AI products are facing the more complex situation with more informal entity categories or even hierarchical category structures (Fig. 1 gives an example). The difference is basically brought by the requirements of practical task-oriented systems which take spoken language as the interactive interface, since it is much more difficult to parse spoken language sentences to detect slots or predict intentions. More importantly, in the task-oriented NER scenarios, the complex entity categories significantly increase the difficulty of human annotation. Consequently, the amount of high-quality annotated datasets generally can not be guaranteed, which blocks the NER models from achieving satisfying performance with no doubt.

For the task-oriented NER models which take spoken languages as input, the limitation of the amount of human-annotated data is a severe problem that should be handled first. From the perspective of the principle of NER, this task objectively keeps the correlation with the other NLP tasks such as word segmentation, part-of-speech (POS) tagging, etc. More importantly, the NER model is very possible to benefit from the performance improvements on such tasks, since they are logistically the basis of the NER task. Consequently, it is fairly reasonable to conduct the multi-task learning procedure upon a basic shared learnable component, which can be updated in accordance with each training iteration of each task. This architecture becomes more practicable due to the natural characteristics of deep learning models, since the parameter sharing and fine-tuning mechanisms are suitable for building trainable shared layers providing implicit representations of linguistic knowledge.

In this paper, we propose a learning strategy to combat the complex NER task by firstly dividing the NER task into fine-grained subtasks (according to domain affiliation) then integrating these subtasks into the multi-task learning process and finally training them from scratch. A key aspect of our idea is that these fine-grained subtasks will get better performance without the disturbance coming from other domains.

Fig. 1. An example of the complex NER task facing by new AI products. All the six entity names in this figure belong to the same domain of 'Location' in traditional NER tasks. It demonstrates the complex NER is a challenging learning problem. On the one hand, the model not only needs to classify which domain (Person, Location, Organization) the named entity belongs to, but also needs to dive down and get the correct fine-grained category. On the other hand, the boundaries between these categories may be blurry. For example, "Lijiang" is an administrative division, but in a specific context it may refer to the "Old Town of Lijiang" which is a scenery spot.

The remaining part of this paper is organized as follows. Section 2 surveys the related studies. Our proposed methodology is presented in Sect. 3. The experimental results are given and analyzed in Sect. 4. Finally, we conclude our work in Sect. 5.

2   Related Work

NER is a challenging learning problem considering the amount of supervised training data available and the few constraints on the kinds of words that can be names. As a result, orthographic features and language-specific knowledge resources, such as gazetteers, are widely used to improve the NER tasks' performance [6,16,22]. However, this practice ruins the possibility of training the NER task from scratch, since the language-specific resources and features are costly to obtain when facing new language or new domains.

Multi-task learning (MTL) has led to success in many applications of language processing [5] by sharing representations between related tasks [1,3]. Compared with single-task learning, the architecture commonly used in MTL has shared bottom layers and several individual top layers for each specific task. By jointly (or iteratively) training on related tasks, the representation capacity of the shared layers are enhanced. On MTL for sequence labeling tasks, Ando [1] proposed a multi-task joint training framework that shares structural parameters among multiple tasks, and improved the performance on various tasks including NER. Collobert [6] presented a task independent convolutional network and employed multi-task joint training to improve the performance of chunking. The BiLSTM-CRFs neural architecture proposed by Lample [15] achieved state-of-the-art results [23]. These previous works exclusively focused on the traditional named entity recognition, however, the named entity categories are much more complicated in practical applications nowadays. We argue that it would be beneficial to take apart the original complex NER task into fine-grained subtasks. In the training process, each subtask is trained independently, while in the testing process, these subtasks will be executed simultaneously and the results of these subtasks will be integrated to produce the final result. This perspective also makes it easy to add these subtasks into MTL's iterative training procedure with an end-to-end manner. To the best of our knowledge, the work that is closet to ours is [19], which focuses on domain adaptation with a simple NER classification category, but their work does not explore the possibility of applying it to the more complicated NER task.

3   Multi-task Learning Architecture for NER

In this section, we first provide a brief description of LSTMs used in our architecture, and then present a baseline model for single sequence tagging tasks based on Bidirectional LSTM (Bi-LSTM) recurrent units. Finally we elaborate on the iteration training procedure of MTL and the details about NER subtasks training strategy.

3.1 Basic Bi-LSTM Structure

Recurrent neural networks are a sort of neural networks taking the input data in the form of vector sequence $(x_1, x_2, \ldots, x_n)$ and generating the sequence output $(h_1, h_2, \ldots, h_t)$ which represents the context information encoded in every step of the input vectors. It has been shown the traditional RNNs tend to be biased toward the recent tokens, which caused the failure of learning long-term dependencies [2]. Long Short-term Memory networks (LSTMs) [12] are put forward to combat the above problem by applying additional gates to control the proportion of input given to the memory cell, and the proportion of the previous state to forget. There are several architectures of LSTM units. We used the following implementations [9]:

$$f_t = \sigma(W_f x_t + U_f h_{t-1} + b_f) \tag{1}$$

$$i_t = \sigma(W_i x_t + U_i h_{t-1} + b_i) \tag{2}$$

$$o_t = \sigma(W_o x_t + U_o h_{t-1} + b_o) \tag{3}$$

$$c_t = f_t \odot c_{t-1} + i_t \odot \tanh(W_c x_t + U_c h_{t-1} + b_c) \tag{4}$$

$$h_t = o_t \odot \tanh(c_t) \tag{5}$$

where $\odot$ denotes element-wise product and σ is the sigmoid logistic function (defined as $\sigma = 1/(1 + e^{-x})$).

For an input sequence $(x_1, x_2, \ldots, x_n)$ containing n characters, the output $\overrightarrow{h_t}$ encodes the left context information of character t in the input sequence, which means the network models the input sentence only in forward direction. It is usually helpful by adding another LSTM reading reversely and concatenating both outputs as the final output $h_t = [\overrightarrow{h_t}, \overleftarrow{h_t}]$. This is referred as the bidirectional LSTM [10]. It has been demonstrated that this bidirectional architecture will improve the performance in many sequence labeling tasks, thus it has become the common building component in such tasks and is extensively used in many sequence tagging tasks such as Chinese word segmentation (SEG), POS tagging and NER.

3.2 Single Task Training

We start with a basic neural network (NN) model for training single sequence tagging task in this subsection, then move on to the structure for iterative training in Sect. 3.3.

The NN model for single sequence tagging tasks can be seen in Fig. 2. For an input character sequence $X = \{w_1, w_2, \ldots, w_n\}$, it can be transformed to a sequence of character vectors with length n by performing embedding lookup operation on each character w_i. Then, this vector sequence will be fed to the Bi-LSTM layer. Finally, a fully connected layer and a softmax layer will be used to produce the final tagging result. This structure can be easily applied to many sequence tagging tasks in an end-to-end manner and have the capacity of being integrated into the MTL environment.

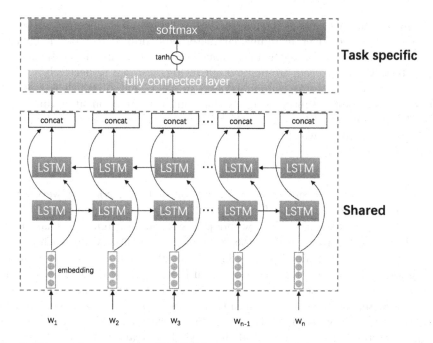

Fig. 2. The architecture of our Bi-LSTM recurrent network for single sequence tagging task. Character embeddings successively pass though a bidirectional LSTM layer, a fully connected layer and a softmax layer to generate the predicted outputs. This architecture can be employed directly in the MTL environment by simply dividing it into shared parts and task specific parts which marked by the red box and blue box above. Figure 3 gives the further detail in MTL aspects.

We employ mini-batch Adam optimizer [13] to train our neural network in an end-to-end manner with back propagation. In order to train the model with batched inputs, input sentences will be tailed to a fixed length before the embedding lookup operation, thus the lengths of input sentences for the Bi-LSTM layer are equal. When performing evaluation for the NER task, the Viterbi algorithm [8] is used to decode the most probable tag sequence.

3.3   Multi-Task Training Scheme

Multi-task training is desirably leveraged to boost model performance, since different sequence tagging tasks (such as SEG, POS and NER) in the same language share language-specific characteristics and should learn similar underlying representation. To apply the basic NN model mentioned above to the multi-task mechanism, we divide the parameters (W_t) of single task t into two sets (task specific set and shared set):

$$W_t = W_{tshare} \cup W_{tspec} \tag{6}$$

according to different roles of the parameters. We share all the parameters below the fully connected layer including character embeddings to learn language-specific characteristics for all the tasks, and the rest are regarded as task specific parameters. These two sets are marked out in Figs. 2 and 3 respectively.

There are two patterns commonly used in MTL, one of which is training different tasks jointly (optimize more than one loss function at a time), for example, training the SEG task and the NER task simultaneously by maximizing a weighted joint objective [18]:

$$Loss_{joint} = \lambda Loss_{SEG} + Loss_{NER} \tag{7}$$

where λ trades off between better segmentation or better NER. However, this approach needs additional modifications to the model architecture and it is hard to combine too many tasks. Thus, we adopt the iterative training mechanism of MTL. For each iteration, all the tasks are trained sequentially. For each task t, we first load previous trained parameters for initialization. The shared parameter set is initialized with the latest W_{tshare} if there exists one. The task specific parameter set is initialized with the latest W_{tspec} of the same task if there exists one. Otherwise, we make a random initialization of parameters. Then, we perform gradient descent steps to update model parameters W_t until the model performance tends to be stable, and then switch to training the next task. In this way, the models for all the tasks are gradually optimized through the iterative training process and the performance for each task is improved. We show the effect of this iterative training mechanism in the experiment.

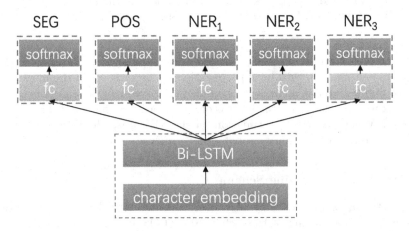

Fig. 3. The architecture used for NER subtask training. In the MTL training process, the parts marked by blue boxes are task-specific and the parts located in the red box is shared by all the tasks. To combat the complexity of the NER task, we divide the NER task into three fine-grained subtasks (NER_{Travel}, $NER_{Shopping}$, $NER_{Entertainment}$) according to the domain affiliation.

3.4 NER Subtasks Training Scheme

To reduce the complexity of the complex NER task, we divided it into three fine-grained subtasks (NER_{Travel}, $NER_{Shopping}$, $NER_{Entertainment}$) by splitting the dataset according to the domain affiliation (Sect. 4 presents the detail of data processing). Since the training data for each NER subtask is isolate, it is possible to treat them as independent sequence tagging tasks and combine them into the iterative training process of MTL. This design captures two intuitions. First, each fined-grained subtask only focuses on a single domain and will get better performance since the labels needed to predict are downsized, without the disturbance of other domains. Second, since SEG and POS tasks may bring fundamental and latent contributions to NER task, it could be helpful to enhance the representation capacity of shared layers by training with SEG and POS tasks. After adding these tasks, the final architecture is shown in Fig. 3.

During evaluation, we employ these NER models on the same test set, collect the results and perform merging operations:

(1) Ideally, these models' predictions will be non-intersected, for example the named entity "Beijing Hotel" belongs to the Travel domain, thus it will only get the named label from NER_{Travel}. The other two subtasks will neglect this entity, thus this merging operation can be performed by simply merging theses models' predictions.

(2) However, due to the complexity of our NER task, there is a possibility that these models may given different tagging results for the same entity which can not be merged, e.g. for named entity "Shaolin Temple", NER_{Travel} marks it as SCENE (scenery spot) while $NER_{Entertainment}$ marks it as FILM. And in more complex situations, these results can be overlapping. In these cases, the tagging result with higher sentence probability (generated by the Viterbi decoding operation) will be chosen as the final result. We find there are rare situations (about 0.5% of total entities) that need this further merging operation from experiments, which demonstrates theses subtasks have the capacity of concentrating on entities of their own domains.

4 Experiments and Analysis

In this section, we first introduce the datasets used in SEG, POS and NER tasks. Then we present the details of tagging schemes and embedding settings. Finally we give the results and the analysis of our experiment.

4.1 Datasets

We consider three tasks: sentence segmentation, part-of-speech tagging, and complex named entity recognition. All these tasks are in Chinese, and each task corresponds to a dataset. Chinese Treebank 9.0[3] datasets are used for SEG and

---

[3] https://catalog.ldc.upenn.edu/ldc2016t13.

Table 1. The NER tags for each domain and their proportions in total named entities.

Domain	Tag	Entity number	Percentage (%)
Travel	CATER	88944	37.1
	HOTEL	9186	3.8
	SCENE	36641	15.3
Shopping	PROD_BRAND	17934	7.5
	PROD_TAG	44138	18.4
Entertainment	TV	19052	7.9
	FILM	16125	6.7
	MUSIC	5032	2.1
	ENT_OTHER	2822	1.2

POS tagging, and each contains 13.2k training sentences. An internal dataset crawled from Chinese forums and Chinese News websites is used for complex NER. It is preprocessed with data cleaning, and labeled with nine entity types[4] covering domains of travel, shopping and entertainment. The NER dataset contains 189.7k sentences, and each sentence only belongs to one domain. Entity distribution of this dataset is shown in Table 1. Note that the datasets for SEG (POS) and NER are drawn from different distributions, yet share many commonalities and still make contributions to each task. To combat the complexity of the complex NER task, we propose the subtask strategy. It consists of the following steps:

(1) Hold out 10% of the NER dataset for testing. The remaining part NER_{remain} is used for further process.
(2) Split NER_{remain} into three subsets according to the domain affiliation, since each sentence belongs to one domain in our NER dataset.
(3) Balance the dataset of each subtask by equally adding sentences that drawn from the other two datasets with the out-of-domain labels transformed to "O". The motivation is to prevent biased model for each subtask if the training data is only drawn from one domain.
(4) For each subset, split the data using an 8:1 for training and validation.

Table 2 shows the tags and volumes of datasets for all subtasks after processing with the subtask strategy.

4.2   Tagging Schemes

The goal of NER is to assign a named entity label to each character of a sentence. In our NER task, we used a "BMESO" tagging format, where B-label and E-label represent the beginning and ending of a named entity respectively, M-label

---

[4] CATER,HOTEL,SCENE,PROD_TAG,PROD_BRAND,FILM,MUSIC,TV, ENT_OTHER.

Table 2. Tags contained in each subtask and the sentence volume of each dataset. Note that these three tasks share the same test set and the domain-unrelated tags will be transformed to tag "O" (for Others) during testing.

Subtask	Tag	Train set	Validation set	Test set
NER_{Travel}	CATER HOTEL SCENE	151751	18968	18967
$NER_{Shopping}$	PROD_BRAND PROD_TAG	77071	9633	18967
$NER_{Entertainment}$	ENT_OTHER FILM MUSIC TV	61456	7681	18967

represents the remaining section of an entity except both ends, and the S-label represents a singleton entity. For the character that do not belongs to any entity type, we use "O" as its label. Sentences can also be tagged in a more concise format as "BMO" if we throw away the information about singleton entity and entity ending. However, works in [7,20] showed that using a more expressive tagging scheme like "BMESO" improves the model performance marginally. We employ this expressive format in our experiments.

4.3 Preprocessing and Pretrained Embeddings

In the iterative training process, to make it possible for sharing the character embedding layer between different tasks, we use the same vocabulary of size 10571 covering most of the Chinese characters and punctuations. During training, the numbers and English words in the corpus will be replaced with "_DIGIT" and "_ALPHABET", and for the characters out of this vocabulary, we use "_UNK" to represent them. To initialize the lookup table, we use pretrained character embeddings which are trained by the word2vec [17] tool. We observe improvements using pretrained embeddings compared with randomly initialized ones and finally set the embedding dimension to 256 for improving the model's learning capability.

4.4 Results and Analysis

We consider three baselines, the first baseline is a linear chain Conditional Random Field (CRF) [14] commonly used in sequence labeling task. The second baseline is the single NN model mentioned in Sect. 3.2. The last baseline integrates the single NN model into the MTL pattern mentioned in Sect. 3.3. Compared with the proposed fine-grained model, the difference is that it takes the complex NER task as a whole. Table 3 presents the test results for complex NER in terms of precision, recall and F1 score. Table 4 gives the details of F1 results in each domain. Table 5 shows the F1 scores achieved by the proposed fine-grained model in each training iteration.

As shown in Table 3, the Single Model (Method 2) gets a lower precision but achieves a higher recall, which results in a marginal increase (+0.72) in F1 compared to the CRF model. For the MTL based models (Method 3 and 4), both

Table 3. Statistics of NER results on test dataset. The methods 3 and 4 use the iterative training scheme of MTL. All these models haven't used external labeled data such as gazetteers and knowledge bases.

Method	Precision (%)	Recall (%)	F1 (%)
1 CRF	82.78	74.75	78.56
2 Single Model	80.59	78.00	79.28
3 Single Model with MTL	82.04	78.78	80.37
4 Fine-grained Model with MTL	**83.56**	**81.57**	**82.55**

Table 4. F1 results in domains of *Travel*, *Shopping* and *Entertainment*. The fine-grained model (Method 4) beats all the others in all domains, and the improvements are more than 2% compared with the integrated model (Method 3). It demonstrates that this fine-grained training scheme eliminates the interference from the other domains which helps the model learn better.

Method	F1 (%)		
	NER_{Travel}	$NER_{Shopping}$	$NER_{Entertainment}$
1 CRF	81.31	79.29	68.36
2 Single Model	82.53	78.71	69.52
3 Single Model with MTL	83.53	79.77	71.17
4 Fine-grained Model with MTL	**85.81**	**81.82**	**73.40**

Table 5. The evaluation F1 achieved by the Fine-grained Model with MTL. The last row displays the max improvement gained in iteration processes, which shows the MTL scheme helps improve models' performance.

Iteration	F1 (%)				
	SEG	POS	NER_{Travel}	$NER_{Shopping}$	$NER_{Entertainment}$
1	95.44	89.86	85.87	81.89	72.12
2	95.51	90.07	85.79	82.10	72.84
3	95.66	**90.10**	86.17	82.17	73.42
4	**95.74**	90.01	**86.18**	**82.94**	**73.52**
Improvement	0.3	0.24	0.31	1.05	1.4

of them improve performances over the Single Model. Knowing that the SEG and POS tasks help the model better learn effective representations, the Single Model with MTL (Method 3) gives us an increase of +1.81 and the Fine-grained Model gives us the biggest improvement of +3.99, in terms of F1 score. In the MTL settings, the Fine-grained Model outperforms the Single Model with MTL, which demonstrates that the subtask training strategy does benefit the performance of complex NER. This can be further confirmed by the experimental results in

Table 4, where the proposed fine-grained model beats all the other baselines in all domains.

Considering the results in Table 5, we find that SEG, POS and NER_{Travel} get small improvements compared with $NER_{Shopping}$ and $NER_{Entertainment}$. The reason is as follows. The tasks of SEG and POS are much easier than NER, thus their top limits are easy to achieve and the room for further improvement is also limited. For NER_{Travel}, the data volume of this task is the largest in the three subtasks (nearly twice the data volume of the others), which means it can be trained more sufficiently thus leaving less room for improvement. Among these subtasks, $NER_{Entertainment}$ gets the worst F1 score, which is mainly because the film entities and TV entities have similar contexts, and it is hard to discriminate them without the help of knowledge bases. However, $NER_{Entertainment}$ achieves the highest F1 improvement in the MTL process, which verifies the efficiency of MTL framework.

5 Conclusions

In this paper, we have proposed a learning strategy to combat the complex NER task by first dividing the NER task into fine-grained subtasks (according to domain affiliation), then integrate these subtasks into a multi-task learning process and finally train from scratch. The experimental results show that these fine-grained subtasks will get better results without the disturbance from other domains.

In this work, we mainly focus on the complex NER tasks which only cover three domains. In the future, we will test this strategy on NER tasks with more domains. Furthermore, it will be interesting to apply our approach to other languages.

References

1. Ando, R.K., Zhang, T.: A framework for learning predictive structures from multiple tasks and unlabeled data. J. Mach. Learn. Res. **6**, 1817–1853 (2005). http://dl.acm.org/citation.cfm?id=1046920.1194905
2. Bengio, Y., Simard, P., Frasconi, P.: Learning long-term dependencies with gradient descent is difficult. IEEE Trans. Neural Netw. **5**(2), 157–166 (1994). https://doi.org/10.1109/72.279181
3. Caruana, R.: Multitask learning. Mach. Learn. **28**(1), 41–75 (1997)
4. Chieu, H.L., Ng, H.T.: Named entity recognition: a maximum entropy approach using global information. In: Proceedings of the 19th International Conference on Computational Linguistics, COLING 2002, vol. 1, pp. 1–7. Association for Computational Linguistics, Stroudsburg (2002)
5. Collobert, R., Weston, J.: A unified architecture for natural language processing: deep neural networks with multitask learning. In: Proceedings of the 25th International Conference on Machine Learning, ICML 2008, pp. 160–167. ACM, New York (2008). https://doi.org/10.1145/1390156.1390177, https://doi.acm.org/10.1145/1390156.1390177

6. Collobert, R., Weston, J., Bottou, L., Karlen, M., Kavukcuoglu, K., Kuksa, P.: Natural language processing (almost) from scratch. J. Mach. Learn. Res. **12**, 2493–2537 (2011). http://dl.acm.org/citation.cfm?id=1953048.2078186
7. Dai, H.J., Lai, P.T., Chang, Y.C., Tsai, R.T.H.: Enhancing of chemical compound and drug name recognition using representative tag scheme and fine-grained tokenization. J. Cheminform. **7**(Suppl 1), S14–S14 (2015). https://doi.org/10.1186/1758-2946-7-S1-S14, http://www.ncbi.nlm.nih.gov/pmc/articles/PMC4331690/. 1758-2946-7-S1-S14[PII]
8. Forney, G.D.: The viterbi algorithm. Proc. IEEE **61**(3), 268–278 (1973). https://doi.org/10.1109/PROC.1973.9030
9. Gers, F.A., Schmidhuber, J., Cummins, F.: Learning to forget: continual prediction with LSTM. Neural Comput. **12**, 2451–2471 (1999)
10. Graves, A., Schmidhuber, J.: Framewise phoneme classification with bidirectional LSTM and other neural network architectures. Neural Networks **18**(5), 602–610 (2005). https://doi.org/10.1016/j.neunet.2005.06.042, http://www.sciencedirect.com/science/article/pii/S0893608005001206. iJCNN 2005
11. Grishman, R., Sundheim, B.: Design of the MUC-6 evaluation. In: Proceedings of the 6th Conference on Message Understanding, MUC6 1995, pp. 1–11. Association for Computational Linguistics, Stroudsburg (1995)
12. Hochreiter, S., Schmidhuber, J.: Long short-term memory. Neural Comput. **9**(8), 1735–1780 (1997). https://doi.org/10.1162/neco.1997.9.8.1735
13. Kingma, D., Ba, J.: Adam: A Method for Stochastic Optimization (2014)
14. Lafferty, J.D., McCallum, A., Pereira, F.C.N.: Conditional random fields: probabilistic models for segmenting and labeling sequence data. In: Proceedings of the Eighteenth International Conference on Machine Learning, ICML 2001, pp. 282–289. Morgan Kaufmann Publishers Inc., San Francisco (2001). http://dl.acm.org/citation.cfm?id=645530.655813
15. Lample, G., Ballesteros, M., Subramanian, S., Kawakami, K., Dyer, C.: Neural architectures for named entity recognition. In: Proceedings of the 2016 Conference of the North American Chapter of the Association for Computational Linguistics: Human Language Technologies, pp. 260–270. Association for Computational Linguistics (2016)
16. Lin, D., Wu, X.: Phrase clustering for discriminative learning. In: Proceedings of the Joint Conference of the 47th Annual Meeting of the ACL and the 4th International Joint Conference on Natural Language Processing of the AFNLP: Volume 2 - Volume 2, ACL 2009, pp. 1030–1038. Association for Computational Linguistics, Stroudsburg (2009). http://dl.acm.org/citation.cfm?id=1690219.1690290
17. Mikolov, T., Chen, K., Corrado, G., Dean, J.: Efficient estimation of word representations in vector space. CoRR abs/1301.3781 (2013). http://arxiv.org/abs/1301.3781
18. Peng, N., Dredze, M.: Learning word segmentation representations to improve named entity recognition for chinese social media. CoRR abs/1603.00786 (2016). http://arxiv.org/abs/1603.00786
19. Peng, N., Dredze, M.: Multi-task domain adaptation for sequence tagging. In: Proceedings of the 2nd Workshop on Representation Learning for NLP, pp. 91–100. Association for Computational Linguistics (2017). http://aclweb.org/anthology/W17-2612
20. Ratinov, L., Roth, D.: Design challenges and misconceptions in named entity recognition. In: Proceedings of the Thirteenth Conference on Computational Natural Language Learning, CoNLL 2009, pp. 147–155. Association for Computational Linguistics, Stroudsburg (2009). http://dl.acm.org/citation.cfm?id=1596374.1596399

21. Tjong Kim Sang, E.F., De Meulder, F.: Introduction to the CoNLL-2003 shared task: Language-independent named entity recognition. In: Proceedings of the Seventh Conference on Natural Language Learning at HLT-NAACL 2003, CONLL 2003, vol. 4, pp. 142–147. Association for Computational Linguistics, Stroudsburg (2003)
22. Turian, J., Ratinov, L., Bengio, Y.: Word representations: a simple and general method for semi-supervised learning. In: Proceedings of the 48th Annual Meeting of the Association for Computational Linguistics, ACL 2010, pp. 384–394. Association for Computational Linguistics, Stroudsburg (2010). http://dl.acm.org/citation.cfm?id=1858681.1858721
23. Yang, Z., Salakhutdinov, R., Cohen, W.W.: Multi-task cross-lingual sequence tagging from scratch. CoRR abs/1603.06270 (2016). http://arxiv.org/abs/1603.06270

Machine Learning for NLP

Hierarchical Attention Based Semi-supervised Network Representation Learning

Jie Liu$^{(\boxtimes)}$, Junyi Deng, Guanghui Xu, and Zhicheng He

College of Computer and Control Engineering, Nankai University, Tianjin, China
jliu@nankai.edu.cn, {dengjunyi,xugh,hezhicheng}@mail.nankai.edu.cn

Abstract. Network Embedding is a process of learning low-dimensional representation vectors of nodes by comprehensively utilizing network characteristics. Besides structure properties, information networks also contain rich external information, such as texts and labels. However, most of the traditional learning methods do not consider this kind of information comprehensively, which leads to the lack of semantics of embeddings. In this paper, we propose a Semi-supervised Hierarchical Attention Network Embedding method, named as SHANE, which can incorporate external information in a semi-supervised manner. First, a hierarchical attention network is used to learn the text-based embeddings according to the content of nodes. Then, the text-based embeddings and the structure-based embeddings are integrated in a closed interaction way. After that, we further introduce the label information of nodes into the embedding learning, which can promote the nodes with the same label closed in the embedding space. Extensive experiments of link prediction and node classification are conducted on two real-world datasets, and the results demonstrate that our method outperforms other comparison methods in all cases.

Keywords: Network representation learning
Hierarchical attention network · Semi-supervised learning

1 Introduction

Information network is a data form with rich structure and semantic information. With the prevalence of various social media, massive social networks have attracted a lot of researchers' attention. The applications of information network include various aspects, such as node classification, community detection, and content recommendation. Network representation learning is the foundation of these network applications. Different from the one-hot vectors, network representation can map each node into a low-dimensional, dense and real-valued vector, thus avoiding the effect of sparsity.

Most of the studies on network representation are based on network structure information, such as the sequences generated by network nodes [5,14], the

© Springer Nature Switzerland AG 2018
M. Zhang et al. (Eds.): NLPCC 2018, LNAI 11108, pp. 237–249, 2018.
https://doi.org/10.1007/978-3-319-99495-6_20

first-order and second-order proximities [18], and the adjacency matrix [4]. With further research, the external information of nodes are considered to improve the quality of embeddings, such as text information [9,17,19,22] and label information [10,20]. The introduction of text feature can enrich the semantics of nodes and improve the performance of representation learning. It is noteworthy that people usually write sentences first, and then compose the whole document with multiple sentences. However, when considering the text information of a network, existing works usually obtain the text feature matrix of the nodes based on words, which ignores the hierarchical structure. In order to incorporate the document structure (document consists of sentences, sentences consist of words), it is necessary to obtain document representations in a hierarchical way. Besides, different words and sentences contain varying amounts of information, even the same words in different sentences can play different roles. So how to make a difference between different components of nodes' content is a practical problem which needs to be solved. In addition to the text information, label is another important attribute of network nodes, and it is a kind of significant supervised information on directing practical tasks such as classification. Making full use of this supervised information will further enrich the network embeddings [10,20]. However, since the network is usually large-scale, there are still a lot of unlabeled nodes, thus the rational use of labeled data and unlabeled data is important for network representation learning.

In view of the above problems, we propose a hierarchical structure based semi-supervised network representation learning method, Semi-supervised Hierarchical Attention Network Embedding (SHANE), which can learn the hierarchical relational network embeddings by integrating text and label features of nodes. In SHANE, we adopt a hierarchical attention structure to extract text features at different levels [23], which can model the hierarchical semantic information of network. Meanwhile, label information is utilized in a semi-supervised manner to make full use of both labeled data and unlabeled data. We apply the proposed model to link prediction and node classification. The experiment results show that the proposed model outperforms all the comparison methods. Our contributions are summarized as follows:

- We propose a SHANE model, which can integrate structures, texts, and labels of nodes together, and learn network embeddings in a semi-supervised manner.
- We use hierarchical attention network to model the nodes' text features, which can capture the semantic features more granularly.
- We extensively evaluate our representations with multiple tasks on two real-world citation networks. Experimental results prove the effectiveness of the proposed model.

2   Model

The overall architecture of the proposed model is shown in Fig. 1. It consists of Word Encoder, Sentence Encoder, and Node Encoder. Word Encoder and Sentence Encoder constitute the text-based representation learning process, while

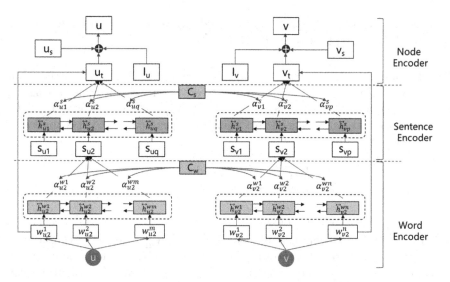

Fig. 1. The illustration of SHANE model.

the Node Encoder combines structure-based embedding, text-based embedding, and label information together. We describe the details of different components in the following sections.

2.1 Problem Formulation

First of all, we introduce the related notions and define the problem formally. Let $G = (V, E, T, L)$ denotes a given information network, where V is the set of nodes, E is the edge set that indicates the relation between nodes, T denotes the text information of nodes and L is the label information of nodes. Each edge $e_{u,v} \in E$ represents the relationship between two nodes (u, v). The text information of node u is $D_u = (S_{u1}, S_{u2}, \cdots, S_{uq})$, where S_{ui} is the ith sentence of u and q is the sentences number of u. $S_{ui} = (w_{ui}^1, w_{ui}^2, \cdots, w_{ui}^m)$, where w_{ui}^j is the jth word of sentence S_{ui} and m is the words number of sentence S_{ui}. The label information of u is l_u.

Given an information network, the goal of our model is to learn a low-dimensional vector **u** for each node u, that can integrate its structure, text and label information.

2.2 Text-Based Representation Learning

As mentioned above, the text information of nodes usually has a natural hierarchical structure. That is, each document contains multiple sentences, and each sentence contains multiple words. Empirically, each word and sentence are of different importance in a document, and learning all sentences and words indiscriminately will lose the focus of text content. So we use a hierarchical attention network [23] to learn the text-based embedding $\mathbf{u}_t$ for each node u, and we describe the learning process in details as follow.

Word Encoder. Assume that u contains q sentences, and each sentence contains m words. We can get the word sequence of sentence S_{ui} by table looking-up, so the sentence can be expressed as $S_{ui} = (\mathbf{w}^1_{ui}, \mathbf{w}^2_{ui}, \cdots, \mathbf{w}^m_{ui})$, where $\mathbf{w}^j_{ui} \in R^d$ is a d-dimensional embedding vector. Then a bidirectional GRU [1] is applied to encode the word sequences as:

$$
\begin{aligned}
\overrightarrow{\mathbf{h}}^{wj}_{ui} &= \overrightarrow{GRU}(\mathbf{w}^j_{ui}), j \in [1, m], \\
\overleftarrow{\mathbf{h}}^{wj}_{ui} &= \overleftarrow{GRU}(\mathbf{w}^j_{ui}), j \in [m, 1].
\end{aligned}
\tag{1}
$$

The word annotation $\mathbf{h}^{wj}_{ui}$ of $\mathbf{w}^j_{ui}$ should contain two directions of information, which can be simply obtained by concatenating $\overrightarrow{\mathbf{h}}^{wj}_{ui}$ and $\overleftarrow{\mathbf{h}}^{wj}_{ui}$. Considering that words contribute differently to the sentence representation, the attention mechanism is used to identify the importance of words, and the operations can be expressed as follows:

$$
\begin{aligned}
\mathbf{g}_{ij} &= \tanh\left(W_w \mathbf{h}^{wj}_{ui} + \mathbf{b}_w\right), \\
\alpha^{wj}_{ui} &= \frac{\exp(\mathbf{g}^T_{ij} C_w)}{\sum(\exp(\mathbf{g}^T_{ij} C_w))}, \\
\mathbf{s}_{ui} &= \sum_j \alpha^{wj}_{ui} \mathbf{h}^{wj}_{ui},
\end{aligned}
\tag{2}
$$

where $\mathbf{s}_{ui}$ is the embedding of the ith sentence of node u, C_w is the global word-level context vector, and α_{ij} is a normalized importance weight used to fuse word annotations to get the representation of sentence.

Sentence Encoder. Sentence encoder is similar to the word encoder except that the objects are sentences, so we omit the equations due to lack of space. Similar bidirectional GRU and attention layers are applied to the sentence encoding process, and then we can get the text embedding $\mathbf{u}^h_t$ encoded by hierarchical attention network.

To avoid the deviation of the learned representation from the original text, after getting the embedding from hierarchical attention network, we add it with another vector $\mathbf{u}^a_t$, which is the mean of word embeddings of this node. Then, we can get the text-based representation $\mathbf{u}_t$ of node u.

$$
\mathbf{u}_t = \mathbf{u}^h_t + \mathbf{u}^a_t.
\tag{3}
$$

Overall, two layers of bidirectional GRUs extract the latent features of words and sentences, in which the word-level attention is used to capture the lexical features, and the sentence-level attention is used to capture the textual features. Therefore, the hierarchical learning method can obtain text information with different granularities.

2.3 Structure-Based Representation Learning

In addition to the text-based embeddings discussed above, the structures of nodes are also crucial information of the network. Structure features reflect the connection characteristics of nodes. In general, two nodes with an edge between them are similar in structure. Therefore, while getting the text-based embeddings of nodes, we also learn a network structure based embedding u_s for each node. In order to comprehensively learn the node representations, it is necessary to consider the correlation between structure features, the relationship between text features, and their interactions.

Following CANE [19], we set the log-likelihood functions of each part as follows:

$$L_{ss}(u) = \sum_{e_{u,v} \in E} w_{u,v} \log p(\mathbf{v}_s^u \mid \mathbf{u}_s^v),$$

$$L_{tt}(u) = \sum_{e_{u,v} \in E} w_{u,v} \log p(\mathbf{v}_t \mid \mathbf{u}_t),$$

$$L_{st}(u) = \sum_{e_{u,v} \in E} w_{u,v} \log p(\mathbf{v}_s^u \mid \mathbf{u}_t),$$

$$L_{ts}(u) = \sum_{e_{u,v} \in E} w_{u,v} \log p(\mathbf{v}_t \mid \mathbf{u}_s^v),$$

$$(4)$$

where v is a node connected with u. $w_{u,v}$ is the weight of the edge between node u and v. $\mathbf{u}_s^v$ is the structure-based embedding of u when it connects with v. The uses of symbols of v are analogous to u. Thus, we can comprehensively model the interaction between u and v through Eq. 4. For each $\mathbf{u} \in \{\mathbf{u}_s^v, \mathbf{u}_t\}$ and $\mathbf{v} \in \{\mathbf{v}_s^u, \mathbf{v}_t\}$, the conditional probability of $\mathbf{v}$ generated by $\mathbf{u}$ is defined through a softmax function:

$$p(\mathbf{v} \mid \mathbf{u}) = \frac{\exp(\mathbf{u}^T \cdot \mathbf{v})}{\sum_{z \in V} \exp(\mathbf{u}^T \cdot \mathbf{z})}.$$

$$(5)$$

The structure-based embeddings are free parameters to learn, and the text-based embeddings are obtained through the method described in the previous section. Note that the structure-based embeddings of u are different according to the node it connects, and the motivation of this setting is that a node has different connection characteristics when connected with different nodes. The final structure-based embedding is the mean of them:

$$\mathbf{u}_s = \frac{1}{|E_u|} \sum_{e_{u,v} \in E} \mathbf{u}_s^v,$$

$$(6)$$

where $|E_u|$ is the edges number of u.

2.4 Semi-supervised Hierarchical Attention Network Embedding

Label is another valuable external information of nodes. Nodes with the same label may also be similar in representations. Thus in this section, we incorporate label information into the learning process. However, the label information of a network in the real world is mostly incomplete, and only a subset of nodes have the corresponding class labels. Therefore, we design our model under a semi-supervised manner so that it can make full use of labeled and unlabeled nodes simultaneously.

Firstly, for the unlabeled nodes, we only consider its structure and text features. So we add the log-likelihood functions in Eq. 4 together to get the objective function of unlabeled nodes:

$$
\begin{aligned}
L_{unlabel}(u^u) = \alpha \cdot L_{ts}(u^u) + \beta \cdot L_{tt}(u^u) + \theta \cdot L_{st}(u^u) \\
+ \gamma \cdot L_{ts}(u^u),
\end{aligned}
\tag{7}
$$

where $u^u \in L_u$ and L_u represents the unlabeled node subset, and $\alpha, \beta, \theta, \gamma$ control the weights of each part.

For the label matching loss of the nodes, we map the node embeddings into the label space by using a fully-connected layer. Then we can get the nodes' predicted label distributions. The purpose of label matching loss is to minimize the distance between predicted label distribution and ground truth distribution.

$$
L_{match}(u^l) = -l_u \log p(\widehat{l_u} \mid u^l) + \Omega,
\tag{8}
$$

where $u^l \in L_l$, and L_l represents the node subset with label information. l_u is the ground truth and $\widehat{l_u}$ is the predicted label distribution. Ω is the regularizing term for the parameters in Eq. 8. Then the objective function of labeled node u^l can be denoted as follow:

$$
\begin{aligned}
L_{label}(u^l) = \alpha \cdot L_{ts}(u^l) + \beta \cdot L_{tt}(u^l) + \theta \cdot L_{st}(u^l) \\
+ \gamma \cdot L_{ts}(u^l) - \lambda L_{match}(u^l),
\end{aligned}
\tag{9}
$$

where λ is the weight of label matching loss. Therefore, the overall objective function of SHANE can be defined as:

$$
L = \sum_{u^l \in L_l} L_{label}(u^l) + \sum_{u^u \in L_u} L_{unlabel}(u^u).
\tag{10}
$$

2.5 Model Optimization

In order to maximize the objective function, we need to calculate the conditional probabilities, which have an expensive computational cost. So we employ the negative sampling technology [12] to reduce the calculation cost. For each $\mathbf{u} \in \{\mathbf{u}_s^v, \mathbf{u}_t\}$ and $\mathbf{v} \in \{\mathbf{v}_s^u, \mathbf{v}_t\}$, the objective functions in Eq. 4 can be transformed as follow:

$$
\log \sigma(u^T \cdot v) + \sum_{i=1}^{k} E_{z \sim P(v)}[\log \sigma(u^T \cdot z)],
\tag{11}
$$

where k is the number of negative samples, and σ is the sigmoid function. We set $P(v) \propto d_v^{3/4}$ as proposed in [12], where d_v is the degree of node v. So in the process of optimization, we replace the corresponding parts of Eq. 10 with the form of Eq. 11. Then, we use Adam to optimize the whole objective function.

3 Experiments

In this section, experiments are performed to verify the effectiveness of the proposed method, including link prediction and node classification.

3.1 Dataset

We conduct our experiments on two citation networks that are commonly used in network representation learning:

- **Cora** is a citation network. We adopt the version processed by CANE [19]. The network contains 2277 machine learning papers in 7 categories and there are 5214 edges between them. The text features of nodes are the abstracts of these papers.
- **DBLP** is a computer science bibliography. It contains 30422 nodes in 4 research areas with the same setting as that of [13], and the edge number is 41206. Abstracts of these papers are treated as text information as well.

3.2 Baseline

To investigate the performance of the proposed model, we compare our model with 7 state-of-the-art methods, including structure-only models, text-only models, structure-text models and structure-text-label models. The details of the comparison methods are described as follow:

- **DeepWalk** [14] is a structure-only model that employs random walk and Skip-Gram [12] to learn the embeddings of nodes.
- **LINE** [18] can learn nodes embeddings of large-scale networks by considering the first-order and second-order proximities and it is a structure-only model.
- **node2vec** [5] is an improved method of DeepWalk with a biased random walk procedure, which only considers the structure information of network.
- **Doc2vec** [8] learns the embeddings of documents by predicting the co-occurrence probability of words, and it is a pure text representation learning method.
- **TADW** [22] is a structure-text model, which learns the structure features and text features of the network by matrix decomposition.
- **CANE** [19] can learn the content aware embeddings of nodes, and it introduce text information into the learning process.
- **TriDNR** [13] is a network representation learning method that considers the structure, text and label information of nodes simultaneously.

3.3 Link Prediction

Link prediction is an important applications of network representation learning. The primary purpose of this task is to predict whether there is an edge between two nodes in the network. In practical applications, it can be used for recommendation tasks, such as book recommendation. We adopt AUC [6] to evaluate the performance of link prediction. When the AUC is higher than 0.5, it indicates that the similarity of the connected nodes is higher than the unconnected nodes. So higher AUC means better performance. It can be calculated as follows:

$$AUC = \frac{\sum_{i \in positive} rank_i - \frac{M(1+M)}{2}}{M \times N}, \tag{12}$$

where M and N are the numbers of the connected node pairs and the unconnected node pairs, respectively. In the evaluation process, we calculate the similarities of the node pairs and rank them. The $rank_i$ is the number of correct orders and $positive$ is the collection of connected node pairs in the test set.

In order to verify the effectiveness of each model on link prediction task, the models are trained with different proportions of edges in the network. The training proportion of edges ranges from 15% to 95%, and the experimental results on the two datasets are shown in Tables 1 and 2. It is worth noting that HANE is a simplified form of the model proposed in this paper, which is designed to verify the effect of the introduction of label information. HANE doesn't consider the label information of nodes. Besides, since Doc2vec is a

Table 1. Link prediction performance on Cora.

Training edge	15%	25%	35%	45%	55%	65%	75%	85%	95%
DeepWalk	56.0	63.0	70.2	75.5	80.1	85.2	85.3	87.8	90.3
LINE	55.0	58.6	66.4	73.0	77.6	82.8	85.6	88.4	89.3
node2vec	55.9	62.4	66.1	75.0	78.7	81.6	85.9	87.3	88.2
TADW	86.6	88.2	90.2	90.8	90.0	93.0	91.0	93.4	92.7
CANE	86.8	91.5	92.2	93.9	94.6	94.9	95.6	96.6	97.7
TriDNR	83.7	84.7	85.2	85.5	85.8	85.9	86.3	87.2	87.7
HANE	**93.0**	**94.1**	94.8	95.0	95.5	95.8	96.2	97.5	98.3
SHANE	92.7	93.3	**95.1**	**95.7**	**96.0**	**96.5**	**96.9**	**97.8**	**98.5**

Table 2. Link prediction performance on DBLP.

Training edge	15%	25%	35%	45%	55%	65%	75%	85%	95%
DeepWalk	71.2	72.8	74.3	74.5	74.6	75.5	81.4	81.7	82.4
LINE	57.4	60.5	63.2	66.0	66.4	68.8	67.2	68.2	69.9
node2vec	67.0	79.2	84.4	88.1	90.0	91.8	93.2	94.1	95.4
TADW	72.0	78.5	88.5	86.0	87.9	89.2	90.5	91.3	93.4
CANE	91.0	92.2	94.5	94.6	94.8	94.9	95.2	95.7	96.2
TriDNR	86.2	86.0	86.6	86.7	87.2	87.8	88.4	88.6	90.7
HANE	92.4	**93.6**	**95.3**	95.6	**96.3**	**96.5**	96.7	97.0	97.4
SHANE	**92.6**	93.4	**95.3**	**96.1**	**96.3**	96.4	**96.8**	**97.2**	**97.9**

text-only method, so we do not analyze it in this part of the experiment. From Tables 1, 2, we have the following observations:

- Structure-text methods outperform structure-only methods, especially when the reserving proportion of the edge is small. This phenomenon shows that the introduction of text features can significantly improve the quality of embeddings and capture the internal relationship between nodes better.
- The performances of all methods increase with the training ratio of edges, but the methods considering text information are relatively stable. It proves that adequate structure information is conducive to the representation learning, and it also shows that the introduction of text information can supplement the lack of network structure information.
- Either HANE or SHANE performs better than other comparison methods on Core and DBLP, which proves the effectiveness of the hierarchical structure based method proposed in this paper, and our methods can be well adapted to different scales of networks.
- According to the comparison between HANE and SHANE, the introduction of label information achieves a slight improvement in link prediction. It shows that the label information is not the primary factor in the nodes relation learning on these datasets.

3.4   Node Classification

Node classification is also a typical application of network representation learning. While link prediction can evaluate the ability of models to learn the connection characteristics, node classification can verify the ability to capture the group characteristics of nodes. To reduce the influence of the differences between classifiers, we adopt a standard linear SVM on the embeddings learned by all the methods. We use the Macro-F1 score [11] as the evaluation metric, and the higher Macro-F1 means the better performance of classification. In order to study the performance of models under different label completeness, the classification experiment is conducted under the condition of retaining different ratios of labeled nodes. For the unsupervised models, the changes in the ratio of labeled data are reflected in varying the amount of labeled data used for training classifiers. Experimental results on the two datasets are shown in Tables 3 and 4. From these tables, we have following observations:

- Generally speaking, the performances of structure-text models are superior to the text-only method, and both of them perform better than the structure-only methods, which proves that the text information is critical when learning the group characteristics of the nodes.
- Both TriDNR and SHANE introduce the label and text information into the learning process, but the classification performances of SHANE are better than TriDNR. It shows that the way of introducing external information can also affect the performance of network representation learning.

– The experimental results on two datasets show that the proposed model exhibits consistent superior performance to other comparison methods, which proves the effectiveness of the SHANE model. It can efficiently capture the properties of nodes and improve the quality of the network embeddings.

Table 3. Node classification performance on Cora.

Ratio of labeled nodes	10%	30%	50%	70%
DeepWalk	0.446	0.635	0.697	0.733
LINE	0.259	0.299	0.331	0.353
node2vec	0.715	0.761	0.784	0.793
Doc2vec	0.530	0.617	0.654	0.670
TADW	0.413	0.781	0.838	0.852
CANE	0.825	0.861	0.863	0.871
TriDNR	0.655	0.677	0.714	0.744
SHANE	**0.852**	**0.871**	**0.873**	**0.886**

Table 4. Node classification performance on DBLP.

Ratio of labeled nodes	10%	30%	50%	70%
DeepWalk	0.379	0.454	0.459	0.461
LINE	0.328	0.362	0.371	0.372
node2vec	0.448	0.473	0.475	0.476
Doc2vec	0.574	0.598	0.604	0.605
TADW	0.660	0.687	0.697	0.699
CANE	0.801	0.810	0.817	0.822
TriDNR	0.724	0.742	0.747	0.748
SHANE	**0.806**	**0.811**	**0.821**	**0.850**

Figure 2 shows how the performances of the proposed methods change over the proportion of labeled nodes on Cora. The training proportion of the labeled data ranges from 10% to 70%. It can be seen from the figure that the performances of the models are improved when the proportion of the labeled data increases, but SHANE always performs better than HANE. This phenomenon illustrates that the semi-supervised learning method proposed in this paper is effective. The introduction of label information is beneficial to capture the characteristics of nodes, thus improving the quality of network embeddings.

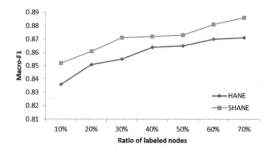

Fig. 2. Performance variation on different training ratio of labeled data.

4   Related Work

Recently, more and more studies focus on how to learn effective network embeddings. The related methods can be divided into two main categories, including the methods only considering network structure, and the methods introducing external information. As a classic structure-only method, DeepWalk [14] learns embeddings of network by performing truncated random walks over networks. On the basis of DeepWalk, Grover et al. proposed node2vec [5] which extends DeepWalk by modifying the random walk strategy. LINE [18] preserves both the first-order proximity and the second-order proximity of the network. These methods can capture the network structure features well, but there is a lack of understanding the semantics of nodes. In order to enrich the semantics of embeddings, many methods introduce external information into the process of learning. PTE [17], CANE [19], and TADW [22] introduce the content of nodes to enrich the network representations. In addition to text, there is also some other external information that can be considered, such as MMDW [20] and DDRW [10]. Although these methods use external information, they do not take into account the hierarchical structure of node content which is an important feature of nodes.

Text representations learning methods are needed when considering the text information of the network. Traditional text representation learning methods, such as LDA [3] and NMF [2], learn the representation of text from the perspective of topic distribution. In recent years, due to the rapid development of neural network and deep learning, text representation learning methods based on the neural network have made significant progress, such as CNNs [7] and LSTM [16] to learn text representations. Recently, attention mechanism is widely used in Natural Language Processing tasks, and Bahdanau et al. first introduced it to NLP for machine translation task [1]. After that, attention mechanism has been widely applied to various applications, such as syntactic parsing [21] and natural language Q&A [15]. The hierarchical attention network [23] proposed by Yang et al. takes into account the hierarchical structure of documents, and applies two levels of attention mechanisms at the word and sentence-level, respectively, which improves the performance of text classification.

5 Conclusion

In this paper, we proposed a semi-supervised hierarchical attention network embedding method, i.e. SHANE. It integrates rich external information into the learning process. The proposed SHANE leverages hierarchical attention network to learn the text-based embedding, which can effectively model the hierarchical structure of the text. Through a semi-supervised learning framework, the embeddings of nodes can be learned by incorporating structure, text and label information together. Extensive experiments conducted on two citation datasets demonstrate the effectiveness and superiority of the proposed model.

Acknowledgement. This research is supported by the National Natural Science Foundation of China under the grant No. U1633103 and 61502499, the Science and Technology Planning Project of Tianjin under the grant No. 17ZXRGGX00170, the Natural Science Foundation of Tianjin under the grant No. 18JCYBJC15800, and the Open Project Foundation of Information Technology Research Base of Civil Aviation Administration of China under the grant No. CAAC-ITRB-201601.

References

1. Bahdanau, D., Cho, K., Bengio, Y.: Neural machine translation by jointly learning to align and translate. CoRR abs/1409.0473 (2014)
2. Berry, M.W., Browne, M., Langville, A.N., Pauca, V.P., Plemmons, R.J.: Algorithms and applications for approximate nonnegative matrix factorization. Comput. Stat. Data Anal. **52**(1), 155–173 (2007)
3. Blei, D.M., Ng, A.Y., Jordan, M.I.: Latent Dirichlet allocation. J. Mach. Learn. Res. **3**, 993–1022 (2003)
4. Cao, S., Lu, W., Xu, Q.: GraRep: learning graph representations with global structural information. In: Proceedings of the 24th ACM International on Conference on Information and Knowledge Management, pp. 891–900 (2015)
5. Grover, A., Leskovec, J.: node2vec: scalable feature learning for networks. In: Proceedings of KDD, pp. 855–864 (2016)
6. Hanley, J.A., McNeil, B.J.: The meaning and use of the area under a receiver operating characteristic (ROC) curve. Radiology **143**(1), 29–36 (1982)
7. Kalchbrenner, N., Grefenstette, E., Blunsom, P.: A convolutional neural network for modelling sentences. In: Proceedings of ACL, pp. 655–665 (2014)
8. Le, Q.V., Mikolov, T.: Distributed representations of sentences and documents. In: Proceedings of the 31st International Conference on Machine Learning, pp. 1188–1196 (2014)
9. Li, H., Wang, H., Yang, Z., Liu, H.: Effective representing of information network by variational autoencoder. In: Proceedings of IJCAI, pp. 2103–2109 (2017)
10. Li, J., Zhu, J., Zhang, B.: Discriminative deep random walk for network classification. In: Proceedings of ACL (2016)
11. Manning, C.D., Raghavan, P., Schütze, H.: Introduction to Information Retrieval. Cambridge University Press, Cambridge (2008)
12. Mikolov, T., Sutskever, I., Chen, K., Corrado, G.S., Dean, J.: Distributed representations of words and phrases and their compositionality. In: Proceedings of NIPS, pp. 3111–3119 (2013)

13. Pan, S., Wu, J., Zhu, X., Zhang, C., Wang, Y.: Tri-party deep network representation. In: Proceedings of IJCAI, pp. 1895–1901 (2016)
14. Perozzi, B., Al-Rfou, R., Skiena, S.: DeepWalk: online learning of social representations. In: Proceedings of KDD, pp. 701–710 (2014)
15. Sukhbaatar, S., Szlam, A., Weston, J., Fergus, R.: End-to-end memory networks. In: Proceedings of NIPS, pp. 2440–2448 (2015)
16. Tai, K.S., Socher, R., Manning, C.D.: Improved semantic representations from tree-structured long short-term memory networks. In: Proceedings of ACL, pp. 1556–1566 (2015)
17. Tang, J., Qu, M., Mei, Q.: PTE: predictive text embedding through large-scale heterogeneous text networks. In: Proceedings of the 21st ACM SIGKDD International Conference on Knowledge Discovery and Data Mining, pp. 1165–1174 (2015)
18. Tang, J., Qu, M., Wang, M., Zhang, M., Yan, J., Mei, Q.: LINE: large-scale information network embedding. In: Proceedings of WWW, pp. 1067–1077 (2015)
19. Tu, C., Liu, H., Liu, Z., Sun, M.: CANE: context-aware network embedding for relation modeling. In: Proceedings of ACL, pp. 1722–1731 (2017)
20. Tu, C., Zhang, W., Liu, Z., Sun, M.: Max-margin DeepWalk: discriminative learning of network representation. In: Proceedings of IJCAI, pp. 3889–3895 (2016)
21. Vinyals, O., Kaiser, L., Koo, T., Petrov, S., Sutskever, I., Hinton, G.E.: Grammar as a foreign language. In: Proceedings of NIPS, pp. 2773–2781 (2015)
22. Yang, C., Liu, Z., Zhao, D., Sun, M., Chang, E.Y.: Network representation learning with rich text information. In: Proceedings of IJCAI, pp. 2111–2117 (2015)
23. Yang, Z., Yang, D., Dyer, C., He, X., Smola, A.J., Hovy, E.H.: Hierarchical attention networks for document classification. In: Proceedings of the NAACL, pp. 1480–1489 (2016)

Joint Binary Neural Network for Multi-label Learning with Applications to Emotion Classification

Huihui He and Rui Xia$^{(\boxtimes)}$

School of Computer Science and Engineering,
Nanjing University of Science and Technology, Nanjing, China
hehuihui1994@gmail.com, rxia@njust.edu.cn

Abstract. Recently the deep learning techniques have achieved success in multi-label classification due to its automatic representation learning ability and the end-to-end learning framework. Existing deep neural networks in multi-label classification can be divided into two kinds: binary relevance neural network (BRNN) and threshold dependent neural network (TDNN). However, the former needs to train a set of isolate binary networks which ignore dependencies between labels and have heavy computational load, while the latter needs an additional threshold function mechanism to transform the multi-class probabilities to multi-label outputs. In this paper, we propose a joint binary neural network (JBNN), to address these shortcomings. In JBNN, the representation of the text is fed to a set of logistic functions instead of a softmax function, and the multiple binary classifications are carried out synchronously in one neural network framework. Moreover, the relations between labels are captured via training on a joint binary cross entropy (JBCE) loss. To better meet multi-label emotion classification, we further proposed to incorporate the prior label relations into the JBCE loss. The experimental results on the benchmark dataset show that our model performs significantly better than the state-of-the-art multi-label emotion classification methods, in both classification performance and computational efficiency.

Keywords: Sentiment analysis · Emotion classification
Multi-label classification

1 Introduction

Multi-label emotion classification, is a sub-task of the text emotion classification, which aims at identifying the coexisting emotions (such as joy, anger and anxiety, etc.) expressed in the text, has gained much attention due to its wide potential applications. Taking the following sentence
Example 1: *"Feeling the warm of her hand and the attachment she hold to me, I couldn't afford to move even a little, fearing I may lost her hand"*

© Springer Nature Switzerland AG 2018
M. Zhang et al. (Eds.): NLPCC 2018, LNAI 11108, pp. 250–259, 2018.
https://doi.org/10.1007/978-3-319-99495-6_21

for instance, the co-existing emotions expressed in it contain *joy*, *love*, and *anxiety*.

Traditional multi-label emotion classification methods normally utilize a two-step strategy, which first requires to develop a set of hand-crafted expert features (such as bag-of-words, linguistic features, emotion lexicons, etc.), and then makes use of multi-label learning algorithms [5,14,16,17,22] for multi-label classification. However, the work of feature engineering is labor-intensive and time-consuming, and the system performance highly depends on the quality of the manually designed feature set. In recent years, deep neural networks are of growing attention due to their capacity of automatically learn the internal representations of the raw data and integrating feature representation learning and classification into one end-to-end framework.

Existing deep learning methods in multi-label classification can be roughly divided into two categories:

- Binary relevance neural network (BRNN), which constructs an independent binary neural network for each label, where multi-label classification is considered as a set of isolate binary classification tasks and the prediction of the label set is composed of independent predictions for individual labels.
- Threshold dependent neural network (TDNN), which normally constructs one neural network to yield the probabilities for all labels via a softmax function, where the probabilities sum up to one. Then, an additional threshold mechanism (e.g., the calibrated label ranking algorithm) is further needed to transform the multi-class probabilities to multi-label outputs.

The structures of BRNN and TDNN are shown in Fig. 1(a) and (b), respectively.

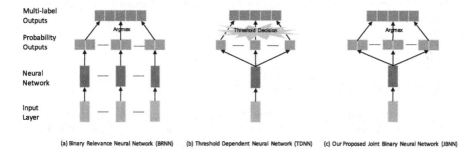

Multi-label Outputs

Probability Outputs

Neural Network

Input Layer

(a) Binary Relevance Neural Network (BRNN)   (b) Threshold Dependent Neural Network (TDNN)   (c) Our Proposed Joint Binary Neural Network (JBNN)

Fig. 1. Different ways of constructing neural networks for multi-label classification.

However, both kinds of methods have their shortcomings. The former one, BRNN, usually known in the literature as binary relevance (BR) transformation [12], not only ignores dependencies between labels, but also consumes much more resources due to the need of training a unique classifier and make prediction for each label. The latter one, TDNN, although has only one neural network, can only yield the category probabilities of all class labels. Instead, it needs an additional threshold function mechanism to transform the category probabilities to multi-label outputs. However, building an effective threshold function is also full of challenges for multi-label learning [4,7,10,15,20].

In this paper, we propose a simple joint binary neural network (JBNN), to address these two problems. We display the structure of JBNN in Fig. 1(c). As can be seen, in JBNN, the bottom layers of the network are similar to that in TNDD. Specifically, we employ a Bidirectional Long Short-Term Memory (Bi-LSTM) structure to model the sentence. The attention mechanism is also constructed to get the sentence representation. After that, instead of a softmax function used in TDNN, we feed the representation of a sentence to multiple logistic functions to yield a set of binary probabilities. That is, for each input sentence, we conduct multiple binary classifications synchronously in one neural network framework. Different from BRNN, the word embedding, LSTMs, and the sentence representation are shared among the multiple classification components in the network. Moreover, the relations between labels are captured based on a joint binary learning loss. Finally, we convert the multi-variate Bernoulli distributions into multi-label outputs, the same as BRNN. The JBNN model is trained based on a joint binary cross entropy (JBCE) loss. To better meet the multi-label emotion classification task, we further proposed to incorporate the prior label relations into the JBCE loss. We evaluate our JBNN model on the widely-used multi-label emotion classification dataset Ren-CECps [9]. We compare our model with both traditional methods and neural networks. The experimental results show that:

- Our JBNN model performs much better than the state-of-the-art traditional multi-label emotion classification methods proposed in recent years;
- In comparison with the BRNN and TDNN systems, our JBNN model also shows the priority, in both classification performance and computational efficiency.

2   Model

2.1   Joint Binary Neural Network

A Bi-LSTM structure is first employed to model the sentence. On the basis of Bi-LSTM, we propose our Joint Binary Neural Network (JBNN) for multi-label emotion classification. The structure of JBNN is shown in Fig. 2.

Before going into the details of JBNN, we first introduce some notations. Suppose $E = \{e_1, e_2, \ldots, e_m\}$ is a finite domain of possible emotion labels. Formally, multi-label emotion classification may be defined as follows: giving the dataset $D = \{(x^{(k)}, y^{(k)}) \mid k = 1, \ldots, N\}$ where N is the number of examples in the D. Each example is associated with a subset of E and this subset is described as an m-dimensional vector $y^{(k)} = \{y_1, y_2, \ldots, y_m\}$ where $y_j^{(k)} = 1$ only if sentence $x^{(k)}$ has emotion label e_j, and $y_j^{(k)} = 0$ otherwise. Given D, the goal is to learn a multi-label classifier that predicts the label vector for a given example. An example is a sentence in emotion classification.

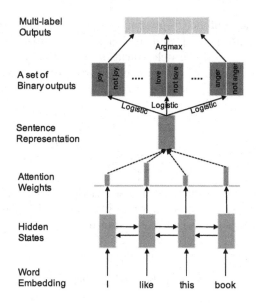

Fig. 2. Overview of the joint binary neural network.

As shown in Fig. 2, in JBNN, each word is represented as a low dimensional, continuous and real-valued vector, also known as word embedding [2,6]. All the word vectors are stacked in a word embedding matrix $L_w \in R^{d \times |V|}$, where d is the dimension of word vector and $|V|$ is vocabulary size. After we feed word embedding to Bi-LSTM, we can get hidden states $[h_1, h_2, \ldots, h_n]$ for a sentence as the initial representation.

Since not all words contribute equally to the representation of the sentence, we adopt the attention mechanism [1,18] to extract such words that are important to the meaning of the sentence. Assume h_t is the hidden states outputted in Bi-LSTM. We use an attention function to aggregate the initial representation of the words to form the attention vector v, also called sentence representation. Firstly, we use

$$u_t = \tanh (wh_t + b), \qquad (1)$$

as a score function to calculate the importance of h_t in the sentence, where w and b are weight matrix and bias respectively. Then we get a normalized importance weight α_t for the sentence through a softmax function:

$$\alpha_t = \frac{\exp(u_t^T u_1)}{\sum_t \exp(u_t^T u_1)}. \qquad (2)$$

After computing the word attention weights, we can get the final representation v for the sentence using equation:

$$v = \sum_t \alpha_t h_t. \qquad (3)$$

After getting the sentence representation v, traditional Bi-LSTM based classification model normally feed v into a softmax function to yield multi-class probabilities for multi-class classification. Our JBNN model differs from the standard model in that, we feed the feature vector v to C logistic functions, instead of a softmax function, to predict a set of binary probabilities $\{p(y_j = 1 \mid x), j = 1, \ldots, C\}$.

$$p(y_j = 1 \mid x) = p_j = \frac{1}{1 + e^{w_j v + b_j}}, \tag{4}$$

$$p(y_j = 0 \mid x) = 1 - p_j, \tag{5}$$

where w_j and b_j are the parameters in j-th logistic component.

Each component will receive a binary probabilities which determines whether this label is True or False in the current instance (i.e., whether the label belongs to the instance):

$$\hat{y}_j = \arg\max_{y_j} p(y_j \mid x). \tag{6}$$

At last, we concatenate $\hat{y}_j$ to form the final predictions $\hat{y} = [\hat{y}_1, \ldots, \hat{y}_C]$.

2.2 Joint Binary Cross Entropy Loss with Label Relation Prior

The JBNN model can be trained in a supervised manner by minimizing the following Joint Binary Cross Entropy (JBCE) loss function:

$$L = -\sum_{j}^{C} \Big(y_j \log p_j + (1 - y_j) \log(1 - p_j) \Big) + \lambda \|\theta\|^2, \tag{7}$$

where λ is the weight for L_2-regularization, and θ denotes the set of all parameters. Note that different from the standard cross entropy loss defined in a multi-class classification task, our JBCE loss is defined in a set of binary classification tasks.

To better meet the multi-label emotion classification task, inspired by [22], we further proposed to incorporate the prior label relations defined in the Plutchik's wheel of emotions [8] into the JBCE loss.

Plutchik's psychoevolutionary theory of emotion is one of the most influential classification approaches for general emotional responses. He considered there to be eight primary emotions: anger, fear, sadness, disgust, surprise, anticipation, trust, and joy. The wheel Plutchik's is used to illustrate different emotions in a compelling and nuanced way. It includes several typical emotions and its eight sectors indicate eight primary emotion dimensions arranged as four pairs of opposites.

In the emotion wheel, emotions sat at opposite end have an opposite relationship, while emotions next to each other are more closely related. As shown in Fig. 3, we followed [22] by measuring the relations $w_{s,t}$ between the s-th and t-th emotions based on the angle between them.

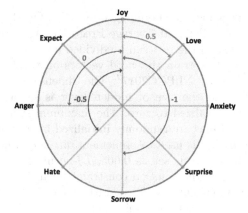

Fig. 3. Plutchik's wheel of emotions.

- In case of emotion pairs with $180°$ (*i.e.*, opposite to each other), define $w_{s,t} = -1$;
- In case of emotion pairs with $90°$, define $w_{s,t} = 0$;
- In case of emotion pairs with $45°$, define $w_{s,t} = 0.5$;
- In case of emotion pairs with $135°$, define $w_{s,t} = -0.5$.

On this basis, the union loss function is defined as:

$$L = -\sum_{j=1}^{C} \left(y_j \log p_j + (1 - y_j) \log(1 - p_j) \right)$$
$$+ \lambda_1 \sum_{s,t} w_{s,t}(p_s - p_t)^2 + \lambda_2 ||\theta||^2. \tag{8}$$

The behind motivation is that if two emotions (such as joy and love) have a high positive correlation, we hope the prediction on the two emotions remain similar. On the contrary, if two emotions (such as joy and sorrow) have a high negative correlation, we hope the predictions on the two emotions remain different.

3   Experiments

3.1   Experimental Settings

We conduct the experiments on the Ren-CECps corpus [9] which was widely used in multi-label emotion classification. It contains 35,096 sentences selected from Chinese blogs. Each sentence is annotated with 8 basic emotions, such as *anger, anxiety, expect, hate, joy, love, sorrow* and *surprise*.

Due to the inherent differences in classification problems, common metrics for multi-label classification are different from those used in single-label classification. In this study, five popular evaluation metrics are adopted in the multi-label classification experiment include Hamming Loss (HL), One-Error (OE), Coverage (Co), Ranking Loss (RL), and Average Precision (AP) [21]. Hamming loss is a label-based metric, and the rest can be divided into ranking-based metrics.

We utilize word2vec[1] to train the word vectors on the 1.1 million Chinese Weibo corpora provided by NLPCC2017[2]. The dimension of word embedding vectors is set as 200 and the size of hidden layer is set as 100. All out-of-vocabulary words are initialized to zero. The maximum sentence length is 90. All weight matrices and bias are randomly initialized by a uniform distribution $U(-0.01, 0.01)$. TensorFlow is used to implement our neural network model. In model training, learning rate is set as 0.005, L_2-norm regularization is set as 1e−4, the parameter λ_1 in the emotion constraint term is set as 1e−3. We use the stochastic gradient descent (SGD) algorithm and Adam update rule with shuffled mini-batch for parameter optimization.

3.2   Comparison with Traditional Multi-label Learning Models

In this section, we compare JBNN with six strong multi-label learning models for multi-label emotion classification, namely EDL [22], ML-KNN [21], Rank-SVM [21], MLLOC [3], ECC [11], LIFT [19]. For each algorithm, ten-fold cross validation is conducted.

Table 1 shows the experimental results of the proposed method in comparison with the six strong multi-label learning methods. The two-tailed t-tests with 5% significance level are performed to see whether the differences between JBNN and the compared models are statistically significant. We can find that the MLLOC method is the worst, and the ECC method performs better than MLLOC. The experimental performance of MLKNN and LIFT is similar, while the performance of RankSVM is slightly worse than them. Among these traditional multi-label learning models, EDL performs the best. However, our model improves the EDL method with an impressive improvement in all kinds of evaluation metrics, *i.e.*, 10.02% reduction in RL, 4.60% reduction in HL, 12.04%

Table 1. Experimental results in comparison with traditional multi-label learning methods (mean ± std). '↓' means 'the smaller the better'. '↑' means 'the larger the better'. Boldface highlights the best performance. '•' indicates significance difference.

Algorithm	RL(↓)	HL(↓)	OE(↓)	Co(↓)	AP(↑)
ECC [11]	0.3281 ± 0.0659•	0.1812 ± 0.0940•	0.6969 ± 0.0598•	2.7767 ± 0.0876•	0.5121 ± 0.0892•
MLLOC [3]	0.4742 ± 0.0734•	0.1850 ± 0.0659•	0.6971 ± 0.0924•	3.6994 ± 0.0764•	0.4135 ± 0.0568•
ML-KNN [21]	0.2908 ± 0.0431•	0.2459 ± 0.0781•	0.5339 ± 0.0954•	2.4480 ± 0.0981•	0.5917 ± 0.0742•
Rank-SVM [21]	0.3055 ± 0.0579•	0.2485 ± 0.0458•	0.5603 ± 0.0921•	2.5861 ± 0.0777•	0.5738 ± 0.0892•
LIFT [19]	0.2854 ± 0.0427•	0.1779 ± 0.0597•	0.5131 ± 0.0666•	2.4267 ± 0.0492•	0.5979 ± 0.0891•
EDL [22]	0.2513 ± 0.0560•	0.1772 ± 0.0568•	0.5239 ± 0.0945•	2.1412 ± 0.0235•	0.6419 ± 0.0235•
JBNN (Our approach)	**0.1511 ± 0.0030**	**0.1312 ± 0.0009**	**0.4035 ± 0.0073**	**1.7864 ± 0.0193**	**0.7171 ± 0.0041**

---

[1] https://code.google.com/archive/p/word2vec/.
[2] http://www.aihuang.org/p/challenge.html.

reduction in OE, 35.48% reduction in Co and 7.52% increase in AP. In short, it can be observed that our JBNN approach performs consistently the best on all evaluation measures. The improvements are all significant in all situations.

3.3 Comparison with Two Types of Neural Networks (BRNN and TDNN)

These models usually utilize neural networks to automatically extract features of sentence and obtain final results. In this section, we compare our proposed JBNN with two major neural networks for multi-label classification, namely BRNN and TDNN, with multi-label classification performance and computational efficiency. We implement all these approaches based on the same neural network infrastructure, use the same 200-dimensional word embeddings, and run them on the same machine. The details of implement are as follows:

- **BRNN** is implemented by constructing multiple binary neural networks, as shown in Fig. 1(a), based on Bi-LSTM and attention mechanism.
- **TDNN** is implemented using the method in [13], which used a neural network based method to train one multi-class classifier and c binary classifiers to get the probability values of the c emotion labels, and then leveraged Calibrated Label Ranking (CLR) method to obtain the final emotion labels.

Classification Performance. In Table 2, we report the performance of JBNN, BRNN and TDNN models. From this table, we can see that our JBNN model performs significantly better than BRNN among all five kinds of evaluation metrics. Compared with the TDNN, our JBNN model is much better in Ranking Loss, Hamming Loss, One-Error, Average Precision. In general, our JBNN model performs better than both BRNN and TDNN models. The improvements according to two-tailed t-test are significant.

Computational Efficiency. We also report the size of parameters and runtime cost of BRNN, TDNN and JBNN in Table 3. From Table 3, we can find that our JBNN model is much simpler than BRNN and TDNN. For example, our JBNN model only has 0.28 M parameters, while BRNN has 2.53M parameters and TDNN has 2.81M parameters. As for runtime cost, we can see that BRNN and TDNN are indeed computationally expensive. Our JBNN model is almost 8 times faster than BRNN and 9 times faster than TDNN in model training. In summary, our JBNN model has significantly priority against BRNN and TDNN in computation efficiency.

Table 2. Experimental results in comparison with two types of neural networks methods (mean $\pm$ std). '$\downarrow$' means 'the smaller the better'. '$\uparrow$' means 'the larger the better'. Boldface highlights the best performance. '•' indicates significance difference.

Algorithm	RL($\downarrow$)	HL($\downarrow$)	OE($\downarrow$)	Co($\downarrow$)	AP($\uparrow$)
BRNN	0.1612 $\pm$ 0.0051•	0.1346 $\pm$ 0.0015•	0.4243 $\pm$ 0.0073•	1.8779 $\pm$ 0.0371•	0.7017 $\pm$ 0.0054•
TDNN	0.1532 $\pm$ 0.0040•	0.1334 $\pm$ 0.0013•	0.4148 $\pm$ 0.0098•	1.7922 $\pm$ 0.0299	0.7115 $\pm$ 0.0060•
JBNN	**0.1511 $\pm$ 0.0030**	**0.1312 $\pm$ 0.0009**	**0.4035 $\pm$ 0.0073**	**1.7864 $\pm$ 0.0193**	**0.7171 $\pm$ 0.0041**

Table 3. Computational Efficiency of different neural networks. Params means the number of parameters, while Time cost means runtime (seconds) of each training epoch.

Algorithm	Params($\downarrow$)	Time cost($\downarrow$)
BRNN	2.53M	265 s
TDNN	2.81M	305 s
JBNN	**0.28M**	**35 s**

4   Conclusion

In this paper, we have proposed a joint binary neural network (JBNN) model for multi-label emotion classification. Unlike existing multi-label learning neural networks, which either needs to train a set of binary networks separately (BRNN), or although model the problem within a multi-class network, an extra threshold function is needed to transform the multi-class probabilities to multi-label outputs (JDNN), our model is an end-to-end learning framework that integrates representation learning and multi-label classification into one neural network. Our JBNN model is trained on a joint binary cross entropy (JBCE) loss. Furthermore, the label relation prior is also incorporated to capture the correlation between emotions. The experimental results show that our model is much better than both traditional multi-label emotion classification methods and the representative neural network systems (BRNN and TDNN), in both multi-class classification performance and computational efficiency.

Acknowledgments. The work was supported by the Natural Science Foundation of China (No. 61672288), and the Natural Science Foundation of Jiangsu Province for Excellent Young Scholars (No. BK20160085).

References

1. Bahdanau, D., Cho, K., Bengio, Y.: Neural machine translation by jointly learning to align and translate. arXiv preprint arXiv:1409.0473 (2014)
2. Bengio, Y., Ducharme, R., Vincent, P., Jauvin, C.: A neural probabilistic language model. J. Mach. Learn. Res. **3**, 1137–1155 (2003)
3. Huang, S.J., Zhou, Z.H., Zhou, Z.: Multi-label learning by exploiting label correlations locally. In: AAAI, pp. 949–955 (2012)
4. Lenc, L., Král, P.: Deep neural networks for Czech multi-label document classification. arXiv preprint arXiv:1701.03849 (2017)
5. Li, S., Huang, L., Wang, R., Zhou, G.: Sentence-level emotion classification with label and context dependence. In: ACL, pp. 1045–1053 (2015)
6. Mikolov, T., Sutskever, I., Chen, K., Corrado, G.S., Dean, J.: Distributed representations of words and phrases and their compositionality. In: NIPS, pp. 3111–3119 (2013)

7. Nam, J., Kim, J., Loza Mencía, E., Gurevych, I., Fürnkranz, J.: Large-scale multi-label text classification — revisiting neural networks. In: Calders, T., Esposito, F., Hüllermeier, E., Meo, R. (eds.) ECML PKDD 2014. LNCS (LNAI), vol. 8725, pp. 437–452. Springer, Heidelberg (2014). https://doi.org/10.1007/978-3-662-44851-9_28

8. Plutchik, R.: Chapter 1 - a general psychoevolutionary theory of emotion. Elsevier Inc. (1980)

9. Quan, C., Ren, F.: Sentence emotion analysis and recognition based on emotion words using ren-cecps. Int. J. Adv. Intell. **2**(1), 105–117 (2010)

10. Read, J., Perez-Cruz, F.: Deep learning for multi-label classification. arXiv preprint arXiv:1502.05988 (2014)

11. Read, J., Pfahringer, B., Holmes, G., Frank, E.: Classifier chains for multi-label classification. In: Buntine, W., Grobelnik, M., Mladenić, D., Shawe-Taylor, J. (eds.) ECML PKDD 2009. LNCS (LNAI), vol. 5782, pp. 254–269. Springer, Heidelberg (2009). https://doi.org/10.1007/978-3-642-04174-7_17

12. Spyromitros, E., Tsoumakas, G., Vlahavas, I.: An empirical study of lazy multilabel classification algorithms. In: Darzentas, J., Vouros, G.A., Vosinakis, S., Arnellos, A. (eds.) SETN 2008. LNCS (LNAI), vol. 5138, pp. 401–406. Springer, Heidelberg (2008). https://doi.org/10.1007/978-3-540-87881-0_40

13. Wang, Y., Feng, S., Wang, D., Yu, G., Zhang, Y.: Multi-label Chinese microblog emotion classification via convolutional neural network. In: Li, F., Shim, K., Zheng, K., Liu, G. (eds.) APWeb 2016. LNCS, vol. 9931, pp. 567–580. Springer, Cham (2016). https://doi.org/10.1007/978-3-319-45814-4_46

14. Wang, Y., Pal, A.: Detecting emotions in social media: a constrained optimization approach. In: IJCAI, pp. 996–1002 (2015)

15. Xu, G., Lee, H., Koo, M.W., Seo, J.: Convolutional neural network using a threshold predictor for multi-label speech act classification. In: BigComp, pp. 126–130 (2017)

16. Xu, J., Xu, R., Lu, Q., Wang, X.: Coarse-to-fine sentence-level emotion classification based on the intra-sentence features and sentential context. In: CIKM, pp. 2455–2458 (2012)

17. Yan, J.L.S., Turtle, H.R.: Exposing a set of fine-grained emotion categories from tweets. In: IJCAI, p. 8 (2016)

18. Yang, Z., Yang, D., Dyer, C., He, X., Smola, A.J., Hovy, E.H.: Hierarchical attention networks for document classification. In: HLT-NAACL, pp. 1480–1489 (2016)

19. Zhang, M.L., Wu, L.: Lift: multi-label learning with label-specific features. IEEE Trans. Pattern Anal. Mach. Intell. **37**(1), 107–120 (2015)

20. Zhang, M.L., Zhou, Z.H.: Multilabel neural networks with applications to functional genomics and text categorization. IEEE Trans. Knowl. Data Eng. **18**(10), 1338–1351 (2006)

21. Zhang, M.L., Zhou, Z.H.: A review on multi-label learning algorithms. IEEE Trans. Knowl. Data Eng. **26**(8), 1819–1837 (2014)

22. Zhou, D., Zhang, X., Zhou, Y., Zhao, Q., Geng, X.: Emotion distribution learning from texts. In: EMNLP, pp. 638–647 (2016)

Accelerating Graph-Based Dependency Parsing with Lock-Free Parallel Perceptron

Shuming Ma$^{(\boxtimes)}$, Xu Sun, Yi Zhang, and Bingzhen Wei

MOE Key Lab of Computational Linguistics, School of EECS,
Peking University, Beijing, China
{shumingma,xusun,zhangyi16,weibz}@pku.edu.cn

Abstract. Dependency parsing is an important NLP task. A popular approach for dependency parsing is structured perceptron. Still, graph-based dependency parsing has the time complexity of $O(n^3)$, and it suffers from slow training. To deal with this problem, we propose a parallel algorithm called parallel perceptron. The parallel algorithm can make full use of a multi-core computer which saves a lot of training time. Based on experiments we observe that dependency parsing with parallel perceptron can achieve 8-fold faster training speed than traditional structured perceptron methods when using 10 threads, and with no loss at all in accuracy.

Keywords: Dependency parsing · Lock-free · Structured perceptron

1 Introduction

Dependency parsing is an important task in natural language processing. It tries to match head-child pairs for the words in a sentence and forms a directed graph (a dependency tree). Former researchers have proposed various models to deal with this problem [1,11].

Structured perceptron is one of the most popular approaches for graph-based dependency parsing. It is first proposed by Collins [3] and McDonald et al. [9] first applied it to dependency parsing. The model of McDonald is decoded with an efficient algorithm proposed by Eisner [5] and they trained the model with structured perceptron as well as its variant Margin Infused Relaxed Algorithm (MIRA) [4,16]. It proves that MIRA and structured perceptron are effective algorithms for graph-based dependency parsing. McDonald and Pereira [11] extended it to a second-order model while Koo and Collins [6] developed a third-order model. They all used perceptron style methods to learn the parameters.

Recently, many models applied deep learning to dependency parsing. Titov and Henderson [17] first proposed a neural network model for transition-based dependency parsing. Chen and Manning [2] improved the performance of neural network dependency parsing algorithm while Le and Zuidema [7] improved

© Springer Nature Switzerland AG 2018
M. Zhang et al. (Eds.): NLPCC 2018, LNAI 11108, pp. 260–268, 2018.
https://doi.org/10.1007/978-3-319-99495-6_22

the parser with Inside-Outside Recursive Neural Network. However, those deep learning methods are very slow during training [13].

To address those issues, we hope to implement a simple and very fast dependency parser, which can at the same time achieve state-of-the-art accuracies. To reach this target, we propose a lock-free parallel algorithm called lock-free parallel perceptron. We use lock-free parallel perceptron to train the parameters for dependency parsing. Although lots of studies implemented perceptron for dependency parsing, rare studies try to implement lock-free parallel algorithms. McDonald et al. [10] proposed a distributed perceptron algorithm. Nevertheless, this parallel method is not a lock-free version on shared memory systems. To the best of our knowledge, our proposal is the first lock-free parallel version of perceptron learning.

Our contribution can be listed as follows:

- The proposed method can achieve 8-fold faster speed of training than the baseline system when using 10 threads, and without additional memory cost.
- We provide theoretical analysis of the parallel perceptron, and show that it is convergent even with the worst case of full delay. The theoretical analysis is for general lock-free parallel perceptron, not limited by this specific task of dependency parsing.

2   Lock-Free Parallel Perceptron for Dependency Parsing

The dataset can be denoted as $\{(x_i, y_i)\}_{i=1}^n$ while x_i is input and y_i is correct output. GEN is a function which enumerates a set of candidates $GEN(x)$ for input x. $\Phi(x, y)$ is the feature vector corresponding to the input output pair (x, y). Finally, the parameter vector is denoted as α.

In structured perceptron, the score of an input output pair is calculated as follows:

$$s(x, y) = \Phi(x, y) \cdot \alpha \tag{1}$$

The output of structured perceptron is to generate the structure y' with the highest score in the candidate set $GEN(x)$.

In dependency parsing, the input x is a sentence while the output y is a dependency tree. An edge is denoted as (i, j) with a head i and its child j. Each edge has a feature representation denoted as $f(i, j)$ and the score of edge can be written as follows:

$$s(i, j) = \alpha \cdot f(i, j) \tag{2}$$

Since the dependency tree is composed of edges, the score are as follows:

$$s(x, y) = \sum_{(i,j) \in y} s(i, j) = \sum_{(i,j) \in y} \alpha \cdot f(i, j) \tag{3}$$

$$\Phi(x, y) = \sum_{(i,j) \in y} f(i, j) \tag{4}$$

Algorithm 1. Lock-free parallel perceptron

1: **input**: Training examples $\{(x_i, y_i)\}_{i=1}^n$
2: **initialize**: $\alpha = 0$
3: **repeat**
4:     **for all** Parallelized threads **do**
5:         Get a random example (x_i, y_i)
6:         $y' = argmax_{z \in GEN(x)} \Phi(x, y) \cdot \alpha$
7:         if $(y' \neq y)$ then $\alpha = \alpha + \Phi(x, y) - \Phi(x, y')$
8:     **end for**
9: **until** Convergence
10:
11: **return** The averaged parameters α^*

The proposed lock-free parallel perceptron is a variant of structured perceptron [12,14]. We parallelize the decoding process of several examples and update the parameter vector on a shared memory system. In each step, parallel perceptron finds out the dependency tree y' with the highest score, and then updates the parameter vector immediately, without any lock of the shared memory. In typical parallel learning setting, the shared memory should be locked, so that no other threads can modify the model parameter when this thread is computing the update term. Hence, with the proposed method the learning can be fully parallelized. This is substantially different compared with the setting of McDonald et al. [10], in which it is not lock-free parallel learning.

3   Convergence Analysis of Lock-Free Parallel Perceptron

For lock-free parallel learning, it is very important to analyze the convergence properties, because in most cases lock-free learning leads to divergence of the training (i.e., the training fails). Here, we prove that lock-free parallel perceptron is convergent even with the worst case assumption. The challenge is that several threads may update and overwrite the parameter vector at the same time, so we have to prove the convergence.

We follow the definition in Collins's work [3]. We write $\overline{GEN(x)}$ as all incorrect candidates generated by input x. We define that a training example is separable with margin $\delta > 0$ if $\exists U$ with $\|U\| = 1$ such that

$$\forall z \in \overline{GEN(x)}, U \cdot \Phi(x, y) - U \cdot \Phi(x, z) \geq \delta \tag{5}$$

Since multiple threads are running at the same time in lock-free parallel perceptron training, the convergence speed is highly related to the delay of update. Lock-free learning has update delay, so that the update term may be applied on an "old" parameter vector, because this vector may have already been modified by other threads (because it is lock-free) and the current thread does not know that. Our analysis show that the perceptron learning is still convergent, even with the worst case that all of the k threads are delayed. To our knowledge, this is the first convergence analysis for lock-free parallel learning of perceptrons.

We first analyze the convergence of the worse case (full delay of update). Then, we analyze the convergence of optimal case (minimal delay). In experiments we will show that the real-world application is close to the optimal case of minimal delay.

3.1 Worst Case Convergence

Suppose we have k threads and we use j to denote the j'th thread, each thread updates the parameter vector as follows:

$$y'_j = \operatorname*{argmax}_{z \in GEN(x)} \Phi_j(x,y) \cdot \alpha \tag{6}$$

Recall that the update is as follows:

$$\alpha^{i+1} = \alpha^i + \Phi_j(x,y) - \Phi_j(x,y'_j) \tag{7}$$

Here, y'_j and $\Phi_j(x,y)$ are both corresponding to j^{th} thread while α^i is the parameter vector after i^{th} time stamp.

Since we adopt lock-free parallel setting, we suppose there are k perceptron updates in parallel in each time stamp. Then, after a time step, the overall parameters are updated as follows:

$$\alpha^{t+1} = \alpha^t + \sum_{j=1}^{k} (\Phi_j(x,y) - \Phi_j(x,y'_j)) \tag{8}$$

Hence, it goes to:

$$U \cdot \alpha^{t+1} = U \cdot \alpha^t + \sum_{j=1}^{k} U \cdot (\Phi_j(x,y) - \Phi_j(x,y'_j))$$

$$\geq U \cdot \alpha^t + k\delta$$

where δ is the separable margin of data, following the same definition of Collins [3]. Since the initial parameter $\alpha = 0$, we will have that $U \cdot \alpha^{t+1} \geq tk\delta$ after t time steps. Because $U \cdot \alpha^{t+1} \leq \|U\|\|\alpha^{t+1}\|$, we can see that

$$\|\alpha^{t+1}\| \geq tk\delta \tag{9}$$

On the other hand, $\|\alpha^{t+1}\|$ can be written as:

$$\|\alpha^{t+1}\|^2 = \|\alpha^t\|^2 + \|\sum_{j=1}^{k} (\Phi_j(x,y) - \Phi_j(x,y'_j))\|^2$$

$$+ 2\alpha^t \cdot (\sum_{j=1}^{k} (\Phi_j(x,y) - \Phi_j(x,y'_j)))$$

$$\leq \|\alpha^t\|^2 + k^2 R^2$$

where R is the same definition following Collins [3] such that $\Phi(x, y) - \Phi(x, y'_j) \leq R$. The last inequality is based on the property of perceptron update such that the incorrect score is always higher than the correct score (the searched incorrect structure has the highest score) when an update happens. Thus, it goes to:

$$\|\alpha^{t+1}\|^2 \leq tk^2R^2 \tag{10}$$

Combining Eqs. 10 and 9, we have:

$$t^2k^2\delta^2 \leq \|\alpha^{t+1}\|^2 \leq tk^2R^2 \tag{11}$$

Hence, we have:

$$t \leq R^2/\delta^2 \tag{12}$$

This proves that the lock-free parallel perceptron has bounded number of time steps before convergence even with the worst case of full delay, and the number of time steps is bounded by $t \leq R^2/\delta^2$ in the worst case. The worst case means that the parallel perceptron is convergent even if the update is extremely delayed, such that k threads are updating based on the same old parameter vector.

3.2 Optimal Case Convergence

In practice the worst case of extremely delayed update is not probable to happen, or at least not always happening. Thus, we expect that the real convergence speed should be much faster than this worst case bound. The optimal bound is as follows:

$$t \leq R^2/(k\delta^2) \tag{13}$$

This bound is derived when the parallel update is not delayed (i.e., the update of each thread is based on a most recent parameter vector). As we can see, in the optimal case we can get k times speed up by using k threads for lock-free parallel perceptron training. This can achieve full acceleration of training by using parallel learning.

4 Experiments

4.1 Dataset

Following prior work, we use English Penn TreeBank (PTB) [8] to evaluate our proposed approach. We follow the standard split of the corpus, using section 2-21 as training set, Sect. 22 as development set, and Sect. 23 as final test set. We implement two popular model of graph-based dependency parsing: first-order model and second-order model. We tune all of the hyper parameters in development set. The features in our model can be found in McDonald et al. [9, 11]. Our baselines are traditional perceptron, MST-Parser [9][1], and the locked version of parallel perceptron. All of the experiment is conducted on a computer with the Intel(R) Xeon(R) 3.0 GHz CPU.

---

[1] www.seas.upenn.edu/~strctlrn/MSTParser/MSTParser.html.

4.2   Results

Table 2 shows that our lock-free method can achieve 8-fold faster speed than the baseline system, which is better speed up when compared with locked parallel perceptron. For both 1st-order parsing and 2nd-order parsing, the results are consistent that the proposed lock-free method achieves the best rate of speed up. The results show that the lock-free parallel peceptron in real-world applications is near the optimal case theoretical analysis of low delay, rather than the worst case theoretical analysis of high delay.

The experimental results of accuracy are shown in Table 1. The baseline MST-Parser [9] is a popular system for dependency parsing. Table 1 shows that our method with 10 threads outperforms the system with single-thread. Our lock system is slightly better than MST-Parser mainly because we use more feature including distance based feature – our distance features are based on larger size of contextual window.

Figure 1 shows that the lock-free parallel perceptron has no loss at all on parsing accuracy on both 1st-order and 2nd-order parsing setting, in spite of the substantial speed up of training.

Figure 2 shows that the method can achieve near linear speed up, and with almost no extra memory cost.

Table 1. Accuracy of baselines and our method.

Models	1st-order	2nd-order
MST Parser	91.60	92.30
Locked Para-Perc	91.68	**92.55**
Lock-free Para-Perc 5-thread	91.70	**92.55**
Lock-free Para-Perc 10-thread	**91.72**	92.53

Table 2. Speed up and time cost per pass of our algorithm.

Models	1st-order	2nd-order
Structured Perc	$1.0 \times (449\,\text{s})$	$1.0 \times (3044\,\text{s})$
Locked Para-Perc	$5.1 \times (88\,\text{s})$	$5.0 \times (609\,\text{s})$
Lock-free Para-Perc 5-thread	$4.3 \times (105\text{s})$	$4.5 \times (672\,\text{s})$
Lock-free Para-Perc 10-thread	$\mathbf{8.1 \times (55\ s)}$	$\mathbf{8.3 \times (367\ s)}$

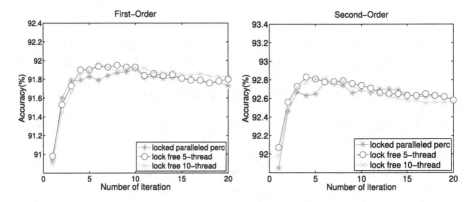

Fig. 1. Accuracy of different methods for dependency parsing.

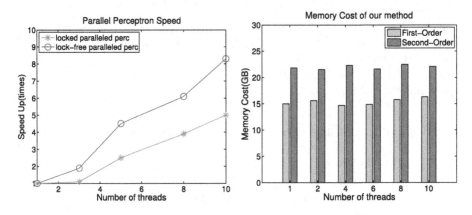

Fig. 2. Speed up and memory cost of different methods for dependency parsing.

5   Conclusions

We propose lock-free parallel perceptron for graph-based dependency parsing. Our experiment shows that it can achieve more than 8-fold faster speed than the baseline when using 10 running threads, and with no loss in accuracy. We also provide convergence analysis for lock-free parallel perceptron, and show that it is convergent in the lock-free learning setting. The lock-free parallel perceptron can be directly used for other structured prediction NLP tasks.

Acknowledgements. The authors would like to thank the anonymous reviewers for insightful comments and suggestions on this paper. This work was supported in part by National Natural Science Foundation of China (No. 61673028).

References

1. Bohnet, B.: Very high accuracy and fast dependency parsing is not a contradiction. In: Proceedings of the 23rd International Conference on Computational Linguistics, pp. 89–97 (2010)
2. Chen, D., Manning, C.D.: A fast and accurate dependency parser using neural networks. In: Proceedings of the 2014 Conference on Empirical Methods in Natural Language Processing (EMNLP), pp. 740–750 (2014)
3. Collins, M.: Discriminative training methods for hidden Markov models: Theory and experiments with perceptron algorithms. In: Proceedings of the ACL-2002 Conference on Empirical Methods in Natural Language Processing-Volume 10, pp. 1–8. Association for Computational Linguistics (2002)
4. Crammer, K., Singer, Y.: On the algorithmic implementation of multiclass kernel-based vector machines. J. Mach. Learn. Res. **2**, 265–292 (2002)
5. Eisner, J.: Three new probabilistic models for dependency parsing: an exploration. In: Proceedings of the 16th Conference on Computational Linguistics, pp. 340–345 (1996)
6. Koo, T., Collins, M.: Efficient third-order dependency parsers. In: Proceedings of the 48th Annual Meeting of the Association for Computational Linguistics, pp. 1–11 (2010)
7. Le, P., Zuidema, W.: The inside-outside recursive neural network model for dependency parsing. In: Proceedings of the 2014 Conference on Empirical Methods in Natural Language Processing (EMNLP), pp. 729–739 (2014)
8. Marcus, M.P., Marcinkiewicz, M.A., Santorini, B.: Building a large annotated corpus of English: the penn treebank. Comput. Linguist. **19**(2), 313–330 (1993)
9. McDonald, R., Crammer, K., Pereira, F.: Online large-margin training of dependency parsers. In: Proceedings of the 43rd Annual Meeting on Association for Computational Linguistics, pp. 91–98 (2005)
10. McDonald, R., Hall, K., Mann, G.: Distributed training strategies for the structured perceptron. In: Human Language Technologies: The 2010 Annual Conference of the North American Chapter of the Association for Computational Linguistics, pp. 456–464. Association for Computational Linguistics (2010)
11. McDonald, R.T., Pereira, F.C.N.: Online learning of approximate dependency parsing algorithms. In: 11st Conference of the European Chapter of the Association for Computational Linguistics (2006)
12. Sun, X.: Towards shockingly easy structured classification: a search-based probabilistic online learning framework. Technical report, arXiv:1503.08381 (2015)
13. Sun, X.: Asynchronous parallel learning for neural networks and structured models with dense features. In: COLING (2016)
14. Sun, X., Matsuzaki, T., Okanohara, D., Tsujii, J.: Latent variable perceptron algorithm for structured classification. In: Proceedings of the 21st International Joint Conference on Artificial Intelligence (IJCAI 2009), pp. 1236–1242 (2009)
15. Sun, X., Ren, X., Ma, S., Wang, H.: meProp: sparsified back propagation for accelerated deep learning with reduced overfitting. In: Proceedings of the 34th International Conference on Machine Learning, ICML 2017, Sydney, NSW, Australia, 6–11 August 2017, pp. 3299–3308 (2017)
16. Taskar, B., Klein, D., Collins, M., Koller, D., Manning, C.D.: Max-margin parsing. In: Proceedings of the 2004 Conference on Empirical Methods in Natural Language Processing (EMNLP), vol. 1, p. 3 (2004)

17. Titov, I., Henderson, J.: A latent variable model for generative dependency parsing. In: Proceedings of the 10th International Conference on Parsing Technologies, pp. 144–155 (2007)
18. Wei, B., Sun, X., Ren, X., Xu, J.: Minimal effort back propagation for convolutional neural networks. CoRR **abs/1709.05804** (2017)

Memory-Based Matching Models for Multi-turn Response Selection in Retrieval-Based Chatbots

Xingwu Lu[1], Man Lan[1,2(✉)], and Yuanbin Wu[1,2(✉)]

[1] School of Computer Science and Software Engineering,
East China Normal University, Shanghai 200062, People's Republic of China
51174506023@stu.ecnu.edu.cn, {mlan,ybwu}@cs.ecnu.edu.cn
[2] Shanghai Key Laboratory of Multidimensional Information Processing,
Shanghai, China

Abstract. This paper describes the system we submitted to Task 5 in NLPCC 2018, i.e., Multi-Turn Dialogue System in Open-Domain. This work focuses on the second subtask: Retrieval Dialogue System. Given conversation sessions and 10 candidates for each dialogue session, this task is to select the most appropriate response from candidates. We design a memory-based matching network integrating sequential matching network and several NLP features together to address this task. Our system finally achieves the precision of 62.61% on test set of NLPCC 2018 subtask 2 and officially released results show that our system ranks 1st among all the participants.

Keywords: Multi-turn conversation · Response selection
Neural networks

1 Introduction

Recently, more and more attention is paying to building open domain chatbots that can naturally converse with humans on vary topics. Existing work on building chatbots includes generation-based methods [1–3] and retrieval-based methods [4–7]. Compared to generation-based chatbots, retrieval-based chatbots enjoy the advantages of informative and fluent responses, because they select a proper response for the current conversation from a repository.

Different from the single-turn conversation, multi-turn conversation needs to consider not only the matching between the response and the input query but also matching between the response and context in previous turns. The challenges of the task are how to identify important information in previous utterances and properly model the utterances relationships to ensure the consistency of conversation.

There have been many attempts to address these challenges where the state-of-the-art methods include dual LSTM [4], Multi-View LSTM [6], Sequential

© Springer Nature Switzerland AG 2018
M. Zhang et al. (Eds.): NLPCC 2018, LNAI 11108, pp. 269–278, 2018.
https://doi.org/10.1007/978-3-319-99495-6_23

Matching Network (SMN) [7] and so on. Among them, SMN improves the leveraging of contextual information by matching a response with each utterance in the context on multiple levels of granularity with a convolutional neural network, and then accumulates the matching vectors into a chronological order through a recurrent neural network to model sequential relationships among utterances. Although SMN model has achieved remarkable results, there are still problems of inconsistency between response and context. On the one hand, the context of the dialogue may sometimes be complex. For example, some utterances are interrelated and some are even reversed. On the other hand, important context cues require global information to be captured.

In this work, based on SMN, we develop a novel way of applying multiple-attention mechanism, which is proven to be effective in multiple tasks such as sentiment analysis [8,9], dependency parsing [10] and coherence modeling [11]. Different from the SMN that only considers the sequential relationships of the context, our method also synthesizes important features in complex context. Besides, considering the effectiveness of NLP features in some retrieval tasks [12], we also design several effective NLP features. Specifically, our framework first adopts matching vectors to produce the memory. After that, we pay multiple attentions on the memory and nonlinearly combine the attention results with a recurrent network, i.e. Long Short-term Memory (LSTM) [13] networks. Finally, we combine the output of the LSTM network with the output of SMN and NLP features to calculate the final matching score. Our system finally achieves the precision of 62.61% on the test set of NLPCC 2018 Task 5[1] and ranks 1st among all the participants.

The rest of this paper is structured as follows: we describe the system architecture and detailed modules in Sect. 2, and present the experimental results in Sect. 3. Finally, Sect. 4 presents our conclusion and future work.

2   The Approach

2.1   Model Overview

The architecture of our model is shown in Fig. 1, which consists of three main modules, i.e., SMN, MBMN and NLP. The left red wire frame is sequential matching network (SMN) module, which is based on [7]. It is designed to identify important information in previous utterances and model the sequential relationships in context. Considering that SMN may not capture implied and complex contextual features, we design the memory-based matching network (MBMN) module, i.e., the middle green wire frame. As shown in the right blue wire frame, we also design several effective NLP features in NLP features module since it is proved to be effective in some retrieval tasks. Finally, we concatenate the outputs of these three modules in the matching prediction layer to calculate the final matching score. Next we give detailed description.

---

[1] http://tcci.ccf.org.cn/conference/2018/dldoc/taskgline05.pdf

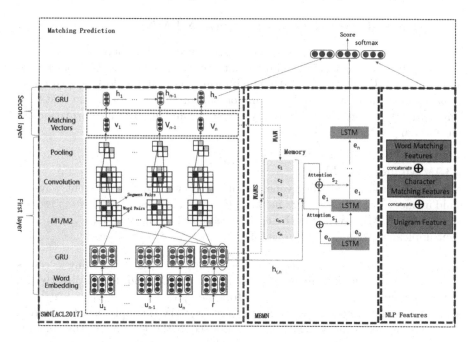

Fig. 1. System architecture of our approach. The dotted lines on MBMN module indicate the memory building is alternative.

2.2   Sequential Matching Network (SMN)

We follow [7] and design the SMN module. The source code[2] of SMN is released by [7]. This module has two advantages: (1) identify and extract semantic structures that are useful for response selection in each utterance and (2) model chronological relationships of the context. As shown on the left of Fig. 1, this module is divided into two layers: utterance-response matching layer (first layer) and matching accumulation layer (second layer). The two layers reflect the above two advantages respectively and their implementations are shown as follows:

- *Utterance-Response Matching Layer:* This layer matches a response candidate with each utterance in the context on a word level and a segment level, and the important matching information from the two levels is distilled by convolution, pooling and then encoded into a matching vector.
- *Matching Accumulation Layer:* We feed the matching vectors into the matching accumulation layer where they are accumulated in the hidden states of a recurrent neural network with gated recurrent units (GRU) [14] following the chronological order of the utterances in the context.

---

[2] https://github.com/MarkWuNLP/MultiTurnResponseSelection.

2.3 Memory-Based Matching Network (MBMN)

The SMN model only considers semantic structures and chronological relationships in utterances, ignoring the important features in complex context. Herein, we design MBMN module to distill cue information that should be captured by global context information and some important contextual information that have long-distance dependence on the query. These cue and important information are captured and retained by memory.

2.3.1 Memory Building

In order to explore the effectiveness of memory, we use two different ways to build memory: *matching vectors memory* (MVM) and *sequence matching vectors memory* (SMVM). We define the representation of memory as $[c_1, \ldots, c_n]$ and their implementation are shown as follows:

- *Matching Vectors Memory (MVM):* Suppose that matching vectors $[v_1, \ldots, v_n]$ is the output of the first layer in SMN module, we directly use the matching vectors as memory vectors, i.e., $[c_1, \ldots, c_n] = [v_1, \ldots, v_n]$.
- *Sequence Matching Vectors Memory (SMVM):* MVM simply uses the matching vectors as memory, which ignores the sequential features in the context. Considering the sequential features dominate in dialogue utterances, we use the hidden states of final GRU in second layer and utterance GRU in first layer of SMN module to build the memory. Then, $[c_1, \ldots, c_n]$ is defined as

$$c_i = tanh(W_{1,1}h_{u_i,n_u} + W_{1,2}h_i + b_i) \tag{1}$$

where $W_{1,1} \in \mathbb{R}^{q \times p}$, $W_{1,2} \in \mathbb{R}^{q \times q}$ and $b_1 \in \mathbb{R}^q$ are parameters, p is the hidden size of utterance GRU, q is the hidden size of final GRU, h_i and h_{u_i,n_u} are the i-th hidden states of final GRU and the final hidden state of the i-th utterance GRU respectively.

2.3.2 Multiple Attentions on Memory

To accurately select the candidate response, it is essential to: (1) correctly distill the related context information from its utterance-response matching memory; and (2) appropriately manufacture such information as the input of the matching prediction. We employ multiple attentions to fulfil the first aspect, and a recurrent network for the second aspect which nonlinearly combines the attention results with LSTMs.

Particularly, we employ a LSTM to update the episode e (i.e., hidden state of LSTM) after each attention. Let e_{t-1} denote the episode at the previous time and s_t is the current information attended from the memory C, and the process of updating e_t is as follows:

$$i_t = \sigma(W_i s_t + U_i e_{t-1}) \tag{2}$$

$$f_t = \sigma(W_f s_t + U_f e_{t-1}) \tag{3}$$

$$o_t = \sigma(W_o s_t + U_o e_{t-1}) \tag{4}$$

$$g = tanh(W_c s_t + U_c e_{t-1}) \tag{5}$$

$$C_t = f \odot C_{t-1} + i \odot g \tag{6}$$

$$h_t = o_t \odot tanh(C_t) \tag{7}$$

where $W_i, W_f, W_o, W_c \in \mathbb{R}^{h \times c}$, $U_i, U_f, U_o, U_c \in \mathbb{R}^{h \times h}$ are parameters, h is the hidden size of LSTM and c is the size of memory vector c_i, a zero vector is denoted as e_0.

For calculating the attended information s_t at time t, the input of an attention layer includes the memory slices $c_i (1 \leq i \leq N)$, N is the number of utterances, the previous episode e_{t-1} and $h_{r,n}$, which is the final hidden state of the response GRU in the first layer of SMN module. We first calculate the attention score of each memory slice c_i as follows:

$$g_i^t = v^T tanh(W_c c_i + W_e e_{t-1} + W_r h_{r,n} + b_{attn}) \tag{8}$$

where W_c, W_e, W_r and b_{attn} are parameters.

Then we calculate the normalized attention score of each memory slice as:

$$\alpha_{t_i} = \frac{exp(g_i^t)}{\sum_{j=1}^{T} exp(g_j^t)} \tag{9}$$

Finally, the inputs to a LSTM at time t are the episode e_{t-1} at time $t-1$ and the content s_t, which is read from the memory as:

$$s_t = \sum_{i=1}^{N} \alpha_{t_i} c_i \tag{10}$$

2.4 NLP Features

This task provides 10 candidate responses corresponding to the context in test dataset and participants are required to rerank the candidates and return the top one as a proper response to the context. We connect all utterances as a post and measure the matching level of the post and its candidate response. We design several traditional NLP features to capture the relevance between the post context and their candidate response. The details of these features are shown as follows:

- *Word Matching Feature:* Word is the basic unit of sentence and the matching of word level benefits the matching of sentence level. Given the post and response as A and B, we record the matching information using the following ten measure functions: $|A|$, $|B|$, $|A \cap B|$, $|A \cap B|/|A|$, $|A \cap B|/|B|$, $|A - B|/|A|$, $|A - B|/|B|$, $|A \cap B|/|A \cup B|$, $|A \cup B| - |A \cap B|/|A \cup B|$, $||A| - |B||$, where $|A|$ stands for the number of non-repeated words in A, $|A - B|$ means the number of non-repeated words found in A but not in B, $|A \cap B|$ stands for the set size of non-repeated words found in both A and B, and $|A \cup B|$ means the set size of shared words found in either A or B.

- *Character Matching Feature:* Similar to word matching, all sentences are treated as the set of single-character representations, then we use above ten measure functions to represent character matching.
- *Unigram Feature:* We extract unigram to represent each sentence, and each vector stores the corresponding TF-IDF of the words in the sentence. We adopt kernel functions to calculate sentence pair matching score. Here we use two types of kernel functions: linear and non-linear. The liner functions contain *Cosine, Chebyshev, Manhattan,* and *Euclidean.* And the non-liner functions contain *Polynomial, Sigmoid* and *Laplacian.*

2.5   Matching Prediction

The representations of above three modules described in Sects. 2.2, 2.3 and 2.4 are concatenated (denoted as $[p_1, p_2, p_3]$) to calculate the final matching score $g(u, r)$. We define u_i represents a conversation context, r_i represents a response candidate and $y_i \in \{0, 1\}$ denotes label. Then we use *softmax* to obtain the final matching score $g(u, r)$ as follows:

$$g(u, r) = softmax(W_2[p_1, p_2, p_3] + b_2), \qquad (11)$$

where W_2 and b_2 are parameters.

We learn $g(u, r)$ by minimizing *cross entropy* with dataset D. Let Θ denotes the parameters, then the objective function $L(D, \Theta)$ of learning is formulated as:

$$-\sum_{i=1}^{|D|}[y_i log(g(u_i, r_i)) + (1 - y_i)(1 - log(g(u_i, r_i)))] \qquad (12)$$

2.6   Parameter Learning

All models are implemented using Tensorflow. We train word embeddings on the training data using word2vec [15] and the dimensionality of word vectors is set as 200. As previous works did [7], we set the hidden size of utterance GRU and response GRU as 200, window size of convolution and pooling as (3, 3), the number of feature maps as 8 and the dimensionality of matching vectors as 50. Different from [7], we tune the hidden size of final GRU in second layer of SMN module in [50, 100, 200, 300] and choose 200 finally. The LSTM in the MBMN module uses a hidden size of 200. We try the number of attention cycles in [1, 3, 5, 7, 9] and set 5 finally. The parameters are updated by stochastic gradient descent with Adam algorithm [16] and the initial learning rate is 0.001. We employ early stop as a regularization strategy. Models are trained in mini-batches with a batch size of 256. Hyperparameters are chosen using the validation set.

3   Experiments

3.1   Datasets

Specifically, this task provides $5,000,000$ conversation sessions containing context, query and reply as the training set and extra $10,000$ conversation sessions only contain context and query as the test set. Participants are required to select a appropriate reply from 10 candidates corresponding to the sessions in test set. Examples of the datasets are shown in Table 1.

Table 1. Data format of multi-turn response selection examples.

Training Set	Test Context	Test candidates
Context: 谢谢你所做的一切 Context: 你开心就好 Context: 开心 Context: 嗯因为你的心里只有学习 Query: 某某某，还有你 Reply: 这个某某某用的好	Context: 你能看下去么 Context: 昨晚已经看完了 Context: 我看睡着了 Context: 爱情啊菇凉 Query: 太静谧了这电影	(1) 对对对，超级喜欢 (2) 到底是为什么 (3) 我也是后学的 (4) 静静的去感受 (5) 关注啦 (6) 呵呵，有意思 (7) 我建议以后单眼皮一对你发嗲，你就跟她对发，哈哈 (8) 不要得瑟你 (9) 感觉好冷 (10) 有钱淫

We randomly split the data into 4,960,000/40,000 for training/validation. For each dialogue in training and validation set, we take the reply as a positive response for the previous turns as a context and randomly sample another response from the 5 million data as a negative response. The ratio of the positive and the negative is 1:1 in training set, and 1:9 in validation set. The word-segmentation is obtained with jieba[3]. We set the maximum context length (i.e., number of utterances) as 10. We pad zeros if the number of utterances in a context is less than 10, otherwise we keep the last 10 utterances. Table 2 gives the statistics of the training set, validation set and test set.

Table 2. Statistics of the training set, validation set and test set.

	Train	Val	Test
# context-response pairs	9.92M	400K	100K
# candidates per context	2	10	10
# positive candidates per context	1	1	1
Max. # turns per context	86	50	34
Avg. # turns per context	3.10	3.07	3.10
Max. # words per utterance	135	93	94
Avg. # words per utterance	5.97	6.22	6.28

To evaluate the performance, given 10 candidates, we calculate precision at top 1.

---

[3] https://github.com/fxsjy/jieba.

3.2 Experiments on Training Data

In order to explore the effectiveness of each module, we perform a series of experiments. Table 3 lists the comparison of different modules on training set. We observe the following findings:

(1) The MBMN(SMVM) performs the best among all single models. The performance of MBMN(MVM) is lower than SMN. The possible reason may be that SMN captures the sequential relationship of utterances in the context and sequential relationship plays a dominant role in this dialogues context.
(2) The memory-based matching modules are quite effective. The combined model MBMN(MVM)+SMN performs better than any single model. It indicates that the memory-based matching module is able to distill the cue information captured by global information in complex context rather than sequential context alone.
(3) The performance of model MBMN(MVM)+SMN is comparable to that of MBMN(SMVM)+SMN. It shows that the MBMN(SMVM) model itself has taken advantage of sequential features and its combination with SMN may not significantly improve the performance.
(4) The combination of three modules, i.e., MBMN(MVM)+SMN+NLP, achieves the best performance, which proves the effectiveness of our designed NLP features.

Therefore, the system configuration for our final submission is the combined model of MBMN(MVM)+SMN+NLP.

Table 3. Performance of different modules on validation set. (MVM) means the model based on matching vectors memory, (SMVM) means the model based on sequence matching vectors memory and "+" means module combination.

	Model	Precision (%)
Single model	NLP features	39.67
	SMN [ACL2017]	61.76
	MBMN(MVM)	60.03
	MBMN(SMVM)	**61.97**
Combined model	MBMN(MVM)+SMN	62.11
	MBMN(SMVM)+SMN	62.08
	MBMN(MVM)+SMN+NLP	**62.26**
	MBMN(SMVM)+SMN+NLP	62.16

3.3 Results on Test Data

Table 4 shows the results of our system and the top-ranked systems provided by organizers for this Retrieval Dialogue System task. Our system finally achieves the precision of 62.61% on the test set and ranks 1st among all the participants. This result validates the effectiveness of our model.

Table 4. Performance of our system and the top-ranked systems in terms of precision (%). The numbers in the brackets are the official rankings.

Team ID	Precision (%)
ECNU	62.61 (1)
wyl_buaa	59.03 (2)
YiwiseDS	26.68 (3)

4   Conclusion

In this paper, we design three modules of sequential matching network, memory-based matching network and NLP features to perform multi-turn response selection in retrieval dialogue system. The system performance ranks 1st among all the participants. In future work, we consider to design more effective memory to incorporate the location and inner semantic information of context in dialogues.

Acknowledgements. The authors would like to thank the task organizers for their efforts, which makes this event interesting. And the authors would like to thank all reviewers for their helpful suggestions and comments, which improve the final version of this work. This work is supported by the Science and Technology Commission of Shanghai Municipality Grant (No. 15ZR1410700) and the open project of Shanghai Key Laboratory of Trustworthy Computing (No. 07dz22304201604).

References

1. Shang, L., Lu, Z., Li, H.: Neural responding machine for short-text conversation. arXiv preprint arXiv:1503.02364 (2015)
2. Sordoni, A., Galley, M., Auli, M., Brockett, C., Ji, Y., Mitchell, M., Nie, J.-Y., Gao, J., Dolan, B.: A neural network approach to context-sensitive generation of conversational responses. arXiv preprint arXiv:1506.06714 (2015)
3. Serban, I.V., Sordoni, A., Bengio, Y., Courville, A.C., Pineau, J.: Building end-to-end dialogue systems using generative hierarchical neural network models. In: AAAI, vol. 16, pp. 3776–3784 (2016)
4. Lowe, R., Pow, N., Serban, I., Pineau, J.: The Ubuntu dialogue corpus: a large dataset for research in unstructured multi-turn dialogue systems. Computer Science (2015)
5. Yan, R., Song, Y., Wu, H.: Learning to respond with deep neural networks for retrieval-based human-computer conversation system. In: Proceedings of the 39th International ACM SIGIR Conference on Research and Development in Information Retrieval, pp. 55–64. ACM (2016)
6. Zhou, X., Dong, D., Wu, H., Zhao, S., Yu, D., Tian, H., Liu, X., Yan, R.: Multi-view response selection for human-computer conversation. In: Conference on Empirical Methods in Natural Language Processing, pp. 372–381 (2016)
7. Wu, Y., Wu, W., Xing, C., Zhou, M., Li, Z.: Sequential matching network: a new architecture for multi-turn response selection in retrieval-based chatbots (2017)
8. Tang, D., Qin, B., Liu, T.: Aspect level sentiment classification with deep memory network, pp. 214–224 (2016)

9. Chen, P., Sun, Z., Bing, L., Yang, W.: Recurrent attention network on memory for aspect sentiment analysis. In: Conference on Empirical Methods in Natural Language Processing, pp. 452–461 (2017)

10. Zhang, Z., Liu, S., Li, M., Zhou, M., Chen, E.: Stack-based multi-layer attention for transition-based dependency parsing. In: Conference on Empirical Methods in Natural Language Processing, pp. 1677–1682 (2017)

11. Logeswaran, L., Lee, H., Radev, D.: Sentence ordering and coherence modeling using recurrent neural networks (2017)

12. Tay, Y., Phan, M.C., Tuan, L.A., Hui, S.C.: Learning to rank question answer pairs with holographic dual LSTM architecture (2017). https://doi.org/10.1145/3077136.3080790. arXiv arXiv:1707 (2017)

13. Hochreiter, S., Schmidhuber, J.: Long short-term memory. Neural Comput. **9**(8), 1735–1780 (1997)

14. Chung, J., Gulcehre, C., Cho, K., Bengio, Y.: Empirical evaluation of gated recurrent neural networks on sequence modeling. arXiv preprint arXiv:1412.3555 (2014)

15. Mikolov, T., Sutskever, I., Chen, K., Corrado, G.S., Dean, J.: Distributed representations of words and phrases and their compositionality. In: Advances in Neural Information Processing Systems, pp. 3111–3119 (2013)

16. Kingma, D.P., Ba, J.: Adam: a method for stochastic optimization. arXiv preprint arXiv:1412.6980 (2014)

NEUTag's Classification System for Zhihu Questions Tagging Task

Yuejia Xiang[1(✉)], HuiZheng Wang[1], Duo Ji[2], Zheyang Zhang[1], and Jingbo Zhu[1]

[1] Natural Language Processing Laboratory,
Northeastern University, Shenyang, China
xiangyuejia@qq.com,
{wanghuizhen, zhujingbo}@mail.neu.edu.cn,
hldnpqzzy@sina.com
[2] Criminal Investigation Police University of China, Shenyang, China
18640037173@168.com

Abstract. In the multi-label classification task (Automatic Tagging of Zhihu Questions), we present a classification system which includes five processes. Firstly, we use a preprocessing step to solve the problem that there is too much noise in the training dataset. Secondly, we choose several neural network models which proved effective in text classification task. Then we introduce k-max pooling structure to these models to fit this task. Thirdly, in order to obtain a better performance in ensemble process, we use an experiment-designing process to obtain classification results that are not similar to each other and all achieve relatively high scores. Fourthly, we use an ensemble process. Finally, we propose a method to estimate how many labels should be chosen. With these processes, our F1 score achieves 0.5194, which ranked No. 3.

Keywords: Multi-label classification · Question tagging · Ensemble learning

1 Introduction

In the automatic tagging task of Zhihu questions, we need to pick out at most 5 labels out of a set which contains more than 25000 labels. Tagging a label of one question means the question belongs to a class which is corresponding with this label. So, we use term 'label' as a synonym for term 'class' in this paper. And main difficulties of this task are shown as follows.

- Zhihu question texts which contain a large number of non-subject-related terms and other noise are too informal to be analysis.
- There are too many classes and the semantic gaps between some classes are so narrow, which observably increase the difficulty of classification.
- As numbers of labels of different questions are frequently different, it is difficult to estimate how many labels should be chosen for a certain question in order to reach a better performance.

© Springer Nature Switzerland AG 2018
M. Zhang et al. (Eds.): NLPCC 2018, LNAI 11108, pp. 279–288, 2018.
https://doi.org/10.1007/978-3-319-99495-6_24

We use a preprocessing process to reduce the adverse effect of noise. This process includes three parts: word segmentation, data cleaning and long-text truncation. Because the neural networks based classification model have been corroborated superior to some traditional classification models [1], notably superior when facing tasks with lots of categories [1–4], we select several neural network based models: RNNText [5], CNNText [6], RCNNText [7], and fastText [8]. What's more, in order to make these neural network-based models more suitable for our task, we introduce k-max pooling structure to them to build a baseline.

In order to obtain better classification results, we use an ensemble process. As we all know, an outstanding ensemble effect requires many classification results that not similar to each other and all achieve relatively high scores [9]. So we use some methods to design experiments, in order to find some results that meet our needs. Moreover, we propose a method to measure the differences between classification results, which can be used to guide the design of experiments.

Finally, we propose a method to estimate how many labels should be chosen. Our method is a post-processing process, as the method forecast the number via the analysis of relations between statistical characteristics of classification results and the number of labels.

The paper is organized as follows. Section 2 introduces details of our system. Section 3 contains our experimental results and analysis. Section 4 includes summarization and future works. Finally, we express our thanks and appreciations.

2 Our System

2.1 Preprocessing

Our preprocessing process includes word segmentation, data cleansing and long-text truncation. Moreover, we divide data cleansing into two parts: a stopword process that is used to filter useless words and a rule-based matching process which is designed to clean up rubbish information such as web sites, page formats, etc. What's more, because neural network models do not perform well in the classification of text which is very long [10], we introduce long-text truncation to suffer this disadvantage [7], while some texts in our training dataset reach tens of times the length of average (Fig. 1).

2.2 Baseline

In order to build our baseline, we choose four models: RNNText [5], CNNText [6], RCNNText [7] and fastText [8], which have been proved effective in text classification tasks. And then make a small change to their structures. The change is using k-max pooling structures to replace max pooling structures [11], in order to get better adaptation to milt-label classification task.

All models we used could be unified with one process: after using embedding layers, they extract context information by CNN-based, RNN-based or NN-based structures, and then they use fully connected layers, named as classifiers to produce classification results. The general structures of these models are shown in Fig. 2.

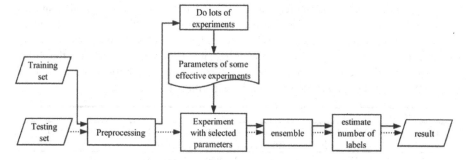

Fig. 1. This is a flow chart of our system, where the solid arrows represent the training process, while the dotted arrows represent the testing process.

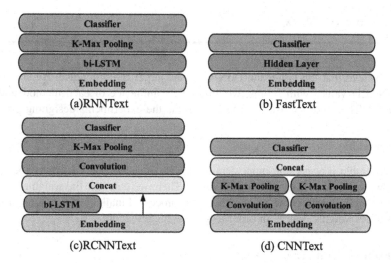

Fig. 2. General structures of models in our baseline

2.3 Design Experiments

In order to find out some classification results that perform well in the following ensemble process, we propose the experiment designing process. In the process, we design lots of experiments by methods listed as follows.

- Use different models in our baseline: RNNText, CNNText, RCNNText or fastText.
- Use different input forms: char-level or word-level. (Different from a char-level method that directly splits original text into character [2], our char-level method splits the text which underwent the word segmentation process and the preprocessing process).
- Use different parameters of model structures: the number of hidden layers, the size of hidden layer units and the batch size.
- Use different training dataset: using additional data or not, a random proportion of data will not be used.

And then we choose classification results that not similar to each other and all achieve relatively high scores, for obtain an outstanding ensemble effect [9]. Moreover, we propose a method to evaluate the differences between classification results, which can be used to guide the designing of experiments. This method evaluates the differences by the difference of normalized accuracy rate distribution on each label (DoNARD), and the algorithm is shown in Algorithm 1.

Algorithm 1

Input: A: experiment result A's accuracy rate on all labels, B: experiment result B's accuracy rate on all labels, T: the number of labels, w: 100000

Output: DoNARD: a number denote the difference between experiment results

begin
for: i = 1, 2, ..., T do:
$$x_i = \frac{A_i}{\sum_{t=1}^{T} A_t}, y_i = \frac{B_i}{\sum_{t=1}^{T} B_t}$$
DoNARD = $w \times \sum_{t=1}^{T} (x_i - y_i)^2$
end.

We believe that the larger the DoNARD, the larger the difference. By comparing the DoNARD values with scores, we analyze the influence of each method that has been listed. Therefore we use DoNARD to guide the experiments designing process.

2.4   Ensemble

In the ensemble process, the input data is the classification result of our experiment. Firstly, we will assign a weight to each result. Then, we consider the weighted sum of these results as the outcome of our ensemble process. Finally, we use a program to choose these weights' value, in order to get the highest score.

2.5   Estimate Number of Labels

To estimate how many labels should be chosen for each problem is quite difficult in multi-label classification task. Firstly, because probabilities of each label in different questions are tremendous difference, we failed to find out a threshold to judge whether a label should be chosen or not. Secondly, we do not simply estimate the number of labels appeared in the standard answer, but also need to consider the performance of the classification result. For example, top 5 labels of a question are shown in Table 1, and standard labels are 'Healthy', 'Life' and 'Bodybuilding'. The scores for selecting k-top labels are shown in Table 2.

Table 1. Top five labels of the question

Label	Healthy	Bodybuilding	Motion	Travel	Life
Probability	1.841	1.648	1.520	0.9165	0.8426

Table 2. Scores for selecting k-top labels

K-top	1	2	3	4	5
Score	0.6206	0.8000	0.6942	0.6206	0.7843

In this case, when selecting top five or top two labels, we can get higher scores than select top three labels.

In our method, we need to calculate that how many labels are predicted for each question can help us get the highest score in training dataset. After analysis the relationship between statistics of classification results and the optimal number of labels, we find that the sum of probability of top five labels is positive correlation to the optimal number. So we propose a method based on the top five label-probabilities' sum (T5LPS) to estimate the number of labels.

3   Results

3.1   Dataset Sources

Our training dataset includes 721,608 questions from official training dataset and 350,000 additional questions from Zhihu website. Each question includes a title, a description and some labels. There are 25,551 different labels in our dataset.

3.2   Performance Evaluation Indicators

This task uses the positional weighted precision to evaluate performance. Let $correct\_num_{p_i}$ denotes the count of predicted tags which are correct at position i, $predict\_num_{p_i}$ denotes the count of predicted tags at position i and $ground\_truth\_num$ denotes the count of correct tags.

$$P = \frac{\sum_{i=1}^{5} correct\_num_{p_i}/\log(i+2)}{\sum_{i=1}^{5} predict\_num_{p_i}/\log(i+2)}$$

$$R = \frac{\sum_{i=1}^{5} correct\_num_{p_i}}{ground\_truth\_num}$$

$$F_1 = 2 \times \frac{P \times R}{P+R}$$

3.3   Preprocessing

In this part, we compare the effect of each preprocessing method. Firstly, we show the effect of word segmentation which is based on jieba segmentation tool in Table 3. Secondly, we show more experiment results in Table 4.

Table 3. Effect of word segmentation

Model	Score before seg	Score after seg	Score improving
RCNNText	0.3413	0.3455	0.0042
RNNText	0.3598	0.3661	0.0063
fastText	0.2775	0.3583	0.0808
CNNText	0.3213	0.3337	0.0124

Table 4. More experiment results in preprocessing

Experiment	Score change	Score
baseline		0.4070
baseline + word-level	0.0063	0.4133
baseline + word-level + unified expression of all number	−0.0012	0.4121
baseline + word-level + clean all punctuation	0.0054	0.4187
baseline + word-level + stopword	0.0035	0.4168
baseline + word-level + rule-based matching	0.0021	0.4154
baseline + word-level + simplify traditional forms of characters	0.0029	0.4162
baseline + word-level + long-text truncation	0.0011	0.4144
baseline + preprocessing	0.0243	0.4313

We can see that, after segmentation the system score (the highest score of models in our baseline) increases 0.0063. And we find that fastText is more sensitive to segmentation than either RNNText or CNNText, while RCNNText's sensitivity is the least.

The reason is that, as we analysis, single character carries little information which benefits classification, while the fastText's structure do not have Excellent abstract ability provided by deep networks, fastText is more sensitive. [12] Another reason is that the char-level text's length is much bigger than word-level text's. Thus, char-level is more difficult to be learnt [13].

In Table 4, we only show the best result of each method in preprocessing, for example, in stopword method we use a manually selected stopword dictionary with the help of TF/IDF and in the long-text truncation method we use a value which is three times the value of question texts' average length as a truncation threshold. From the experiment results, we get some conclusions that are shown as follows.

- Replacing numbers with uniform expression would lower score, via some labels are sensitive to the value of numbers.
- The method which cleans all punctuations promotes the score, because punctuations do not contribute to classification in this task and there are a lot of useless punctuations which are used as emoticons in the text.
- As long-text truncation brings an improvement, we consider that words in the tail of a long text have little contribution to classification and bring adverse effect to our system because they are so long [13].
- The combination of various preprocessing methods brings an additional score of 0.0030.

3.4   Baseline

We try to apply k-max pooling structure to RNNText, CNNText, RCNNText and fastText models. And experiments are shown in Table 5 that used same preprocessing, based on whole training dataset and additional dataset.

Table 5. Scores of experiments

	RNNText	CNNText	RCNNText	fastText
Max pooling	0.4623	0.4174	0.4557	0.3251
2-max pooling	0.4632	0.4349	0.4425	0.2682
3-max pooling	0.4584	0.4197	0.4441	0.2682

We find that Max Pooling structure performance best in RCNNText and fastText, while 2-max Pooling structure performance best in RNNText and CNNText. But 3-max Pooling structure performance worst. So we introduce 2-max Pooling structure to RNNText and CNNText models. The reason of this phenomenon needs further study whether k-max pooling structure benefit from an better expression ability in multi-label classification task [5].

We can see that 2-max pooling structure is effective in some models,

3.5   Design Experiments

With methods listed in Sect. 2.3, we design hundreds of experiments. Firstly, we check the effectiveness of our char-level method in Table 6. Then with the help of DoNARD method, we analyze the influence on classification results of each method in Table 7. Finally, we show several models in Table 8, with which we obtaining an optimal ensemble effect.

Table 6. Effect of our char-level method

Experiments	Score
RNNText + char-level [2]	33.02
RNNText + our char-level method	35.95

Table 7. The analysis of various changes on RNNText

Change of original model	DoNARD	Ensemble's effect
Using word-level	1.585	0.0107
Double hidden layer size	6.395	0.0082
Adding a hidden layer	0.716	0.0055
Using additional data	2.402	0.0236
30% random data will not be use	1.535	0.0076

Table 8. Several models which are used in our ensemble process

Experiment	Shortened form	Score
fastText + word-level + batch size*0.25 + additional data	FW1	0.4763
RNNText + word-level + hidden size*2 + additional data	LW1	0.4704
RNNText + word-level + additional data	LW2	0.4701
RNNText + char-level + additional data	LW3	0.4561
RNNText + word-level	LW4	0.4352
RCNNText + word-level + randomly not use 30% data	RW1	0.4350
RNNText + char-level + dim*2	LC1	0.4302

Our char-level method is effective, because it removes the noise in original texts, that achieves a better result.

We find that, when the DoNARD is in the range of about (1, 3), the effect of ensemble is better. We consider that if the DoNARD is too large, it means one result is much worse than another, so the effect of ensemble is poor. And if the DoNARD is too small, this indicates that these two results are too similar, so the ensemble not works well. When we use changes such as word-level, size of hidden layer and external dataset, we achieve better performance of ensemble process, so we designed more experiments with these changes.

After our analysis, there are two conclusions which are shown as follows. Firstly, compared 'RNNText + word-level' with 'RNNText + word-level + additional data', we find that using additional data is effective. It can improve about 0.0396 score. Secondly, as fastText achieves the best performance in our experiments, we guess that the hierarchical softmax structure in word2vet benefits most in fastText model and this still needs further works [12].

3.6   Ensemble

Different from translation task, where an ensemble method based on checkpoints of one experiment can yield a boost of performance [14], in this task the ensemble method is useless, as shown in Table 9.

Table 9. Ensemble method based on checkpoints of one experiment

Checkpoint	Checkpoint-1	Checkpoint-2	Checkpoint-3	Checkpoint-4	Ensemble
Score	0.4208	0.4356	0.4267	4229	0.4354

So, we use an ensemble method based on results (best checkpoints) of several experiments. After searching the best weights of classification results in the range of [0.2, 5], we get the highest score which reaches 0.4954. And we show the weights of all classification results in Table 10 with the progressive ensemble performance after ensemble each classification result.

Table 10. Weight of each classification result

Experiment	FW1	LW3	LW1	LW4	LW2	RW1	LC1
Weight	1.12	1.08	1.06	1.06	1.00	1.00	0.96
Ensemble's effect of each step		0.009961	0.005085	0.001124	0.001067	0.001038	0.0008841

We find that the optimal weights that we searched in the range of [0.2, 5] are all close to 1, and the highest score only has 0.0003 higher than using weights that all equal 1. Therefore, it suggests that we should focus on the process of designing experiments instead of focus on searching optimal weights.

3.7 Estimate Number of Labels

With the help of our method, we estimate the number of labels based on T5LPS values, the score improves about 0.0240. Details are shown in Table 11, in which we use T to denote T5LPS value.

Table 11. Details of T5LPS method

Conditions	T > 13	13 > T > 10	10 > T > 6	6 > T > 3	3 > T
Number of Labels	5	4	3	2	1

We expect that the performance can still be improved if we find out better statistics than T5LPS. However, this statistical requires manual screening which is expensive. And we found that the parameters in Table 11 need to be re-tuned manually on training datasets to achieve the best performance for different ensemble results.

4 Conclusions

In our experiments, the effect of all processes evaluated by F1 score is show as follows. 0.0139 from using 2-maxPool structure, 0.0243 from using preprocessing, 0.0450 from designing experiment (including 0.0396 from using additional data), 0.0191 from using ensemble process and 0.0240 from estimating the number of labels. And our conclusions are listed as follows.

- The preprocessing has a significant effect, because reducing noises in texts and shortening the length of texts are beneficial for classification.
- The 2-max pooling structure is effective in multi-label classification tasks.
- The method we proposed to measure the differences between models is useful to guide the designing of experiments.
- The method we proposed to estimate the number of labels is important, as it can promote the performance of the system effectively.
- Using additional training data can improve the performance of classification remarkably.

Acknowledgements. This work was supported in part by the National Project (2016YFB0801306) and the open source project (PyTorchText in GitHub). The authors would like to thank anonymous reviewers, Le Bo, Jiqiang Liu, Qiang Wang, YinQiao Li, YuXuan Rong and Chunliang Zhang for their comments.

References

1. Saha, A.K., Saha, R.K., Schneider., K.A.: A discriminative model approach for suggesting tags automatically for stack overflow questions. In: 10th IEEE Working Conference on Mining Software Repositories, San Francisco, pp. 73–76 (2013)
2. Yang, Z., Yang, D., Dyer, C., He, X., Smola A., Hovy, E.: Hierarchical attention networks for document classification. In: Conference of the North American Chapter of the Association for Computational Linguistics: Human Language Technologies, New Orleans, pp. 1480–1489 (2017)
3. Conneau, A., Schwenk, H., Cun, Y.L.: Very Deep Convolutional Networks for Text Classification. arXiv preprint arXiv:1606.01781 (2017)
4. Johnson, R., Zhang, T.: Semi-supervised convolutional neural networks for text categorization via region embedding. Adv. Neural. Inf. Process. Syst. **28**, 919–927 (2015)
5. Zhou, Y., Xu, B., Xu, J., Yang, L., Li, C., Xu, B.: Compositional recurrent neural networks for Chinese short text classification. In: 2016 IEEE, Omaha, pp. 137–144 (2016)
6. Kim, Y.: Convolutional Neural Networks for Sentence Classification. Eprint Arxiv (2014)
7. Lai, S., Xu, L., Liu, K., Zhao, J.: Recurrent convolutional neural networks for text classification. In: AAAI Conference on Artificial Intelligence, Austin, pp. 2267–2273 (2015)
8. Jouling, A., Grave, E., Bojanowshi, P., Mikolov, T.: Bag of Tricks for Efficient Text Classification. arXiv preprint arXiv:1607.01759 (2016)
9. Zhou, Z.H.: Machine Learning. Tsinghua University Press, Beijing (2016)
10. Sundermeyer, M., SchlÜter, R., Ney, H.: LSTM neural networks for language modeling. Interspeech **31**(43), 601–608 (2012)
11. Li, W., Wu, Y.: Multi-level gated recurrent neural network for dialog act classification. In: COLING 2016, Osaka, pp. 1970–1979 (2016)
12. Peng, H., Li, J.X., Song, Y.Q., Liu, Y.P.: Incrementally learning the hierarchical softmax function for neural language models. In: 2016, AAAI, Feinikesi (2016)
13. Kalchbrenner, N., Grefenstette, E., Blunsom, P.: A Convolutional Neural Network for Modelling Sentences. arXiv preprint arXiv:1404.2188 (2014)
14. Sennrich, R., Haddow, B., Birch, A.: Edinburgh neural machine translation systems for WMT 16. WMT16 Shared Task System Description (2016)

Machine Translation

Otem&Utem: Over- and Under-Translation Evaluation Metric for NMT

Jing Yang[1,2], Biao Zhang[1], Yue Qin[1], Xiangwen Zhang[1], Qian Lin[1], and Jinsong Su[1,2(✉)]

[1] Xiamen University, Xiamen, China
{zb,qinyue,xwzhang,linqian17}@stu.xmu.edu.cn
[2] Provincial Key Laboratory for Computer Information Processing Technology, Soochow University, Suzhou, China
jingy@stu.xmu.edu.cn, jssu@xmu.edu.cn

Abstract. Although neural machine translation (NMT) yields promising translation performance, it unfortunately suffers from over- and under-translation issues [31], of which studies have become research hotspots in NMT. At present, these studies mainly apply the dominant automatic evaluation metrics, such as BLEU, to evaluate the overall translation quality with respect to both adequacy and fluency. However, they are unable to accurately measure the ability of NMT systems in dealing with the above-mentioned issues. In this paper, we propose two quantitative metrics, the *Otem* and *Utem*, to automatically evaluate the system performance in terms of over- and under-translation respectively. Both metrics are based on the proportion of mismatched n-grams between gold reference and system translation. We evaluate both metrics by comparing their scores with human evaluations, where the values of Pearson Correlation Coefficient reveal their strong correlation. Moreover, in-depth analyses on various translation systems indicate some inconsistency between BLEU and our proposed metrics, highlighting the necessity and significance of our metrics.

Keywords: Evaluation metric · Neural machine translation Over-translation · Under-translation

1 Introduction

With the rapid development of deep learning, the studies of machine translation have evolved from statistical machine translation (SMT) to neural machine translation (NMT) [28,29]. Particularly, the introduction of attention mechanism [1] enables NMT to significantly outperform SMT. By now, attentional NMT has dominated the field of machine translation and continues to develop, pushing the boundary of translation performance.

Despite of its significant improvement in the translation quality, NMT tends to suffer from two specific problems [31]: (1) over-translation where some words

© Springer Nature Switzerland AG 2018
M. Zhang et al. (Eds.): NLPCC 2018, LNAI 11108, pp. 291–302, 2018.
https://doi.org/10.1007/978-3-319-99495-6_25

are unnecessarily translated for multiple times, and (2) under-translation where some words are mistakenly untranslated. To address these issues, researchers often learn from the successful experience of SMT to improve NMT [7,9,31,33]. In these studies, the common practice is to use the typically used translation metrics, such as BLEU [24], METEOR [2] and so on, to judge whether the proposed models are effective. However, these metrics are mainly used to measure how faithful the candidate translation is to its gold reference in general, but not for any specific aspects. As a result, they are incapable of accurately reflecting the performance of NMT models in addressing the drawbacks mentioned previously.

Let us consider the following example:

- **Source Sentence:** *tā hūyù měiguó duì zhōngdōng hépíng yào yǒu míngquè de kànfǎ, bìng wèi cǐ fāhuī zuòyòng, yǐ shǐ liánhéguó yǒuguān juéyì néng dédào qièshí zhíxíng.*
- **Reference 1:** *he urged that the united states maintain a clear notion of the peace in the middle east and play its due role in this so that the un resolutions can be actually implemented.*
- **Reference 2:** *he urged u.s. to adopt a clear position in the middle east peace process and play its role accordingly. This is necessary for a realistic execution of united nations' resolutions.*
- **Reference 3:** *he called for us to make clear its views on mideast peace and play its role to ensure related un resolutions be enforced.*
- **Reference 4:** *he called on the us to have a clear cut opinion on the middle east peace, and play an important role on it and bring concrete implementation of relative un resolutions.*
- **Candidate 1:** *he called on the united states to have a clear view on peace in the middle east peace and play a role in this regard so that the relevant un resolutions can be effectively implemented.* (**BLEU = 45.83**)
- **Candidate 2:** *he called on the united states to have a clear view on in the middle east and play a role in this regard so that the relevant un resolutions can be effectively implemented.* (**BLEU = 46.33**)

Obviously, two candidate translations have different translation errors. Specifically, in Candidate 1, the Chinese word "*hépíng*" is over-translated, and thus "*peace*" appears twice in Candidate 1. In contrast, in Candidate 2, "*hépíng*" is under-translated, for its translation is completely omitted. However, the BLEU metric is unable to distinguish between these two kinds of translation errors and assigns similar scores to these two candidates. This result strongly indicates the incapability of BLEU in detecting the over- and under-translation phenomena. Therefore, it is significant for NMT to explore better translation quality metric specific to over-translation and under-translation.

In this paper, we propose two novel automatic evaluation metrics: "*Otem*" short for <u>o</u>ver-<u>t</u>ranslation <u>e</u>valuation <u>m</u>etric and "*Utem*" short for <u>u</u>nder-<u>t</u>ranslation <u>e</u>valuation <u>m</u>etric, to assess the abilities of NMT models in dealing with over-translation and under-translation, respectively. Both metrics count the lexical differences between gold reference and system translation, and provide

Reference Candidate

BLEU	A vs. A+C
OTEM	C vs. A+C
UTEM	B vs. A+B

Fig. 1. Intuitive comparison of BLEU, OTEM and UTEM. We use gray circle to illustrate the gold reference (left) and candidate translation (right). Capital "A" denotes the matched n-grams, while capital "B" and "C" denotes the mismatched parts.

quantitative measurement according to the proportion of mismatched n-grams. Figure 1 shows the intuitive comparison among BLEU, OTEM and UTEM. The BLEU calculates the precision of matched n-grams (A) over the whole candidate translation ($A + C$). By contrast, the OTEM focuses on the proportion of repeated n-grams in the candidate translation (C) over the whole candidate ($A + C$), and the UTEM estimates the proportion of untranslated n-grams in the reference (B) over the whole reference ($A + B$). Clearly, BLEU is correlated with both UTEM and OTEM but incapable of inferring them.

To evaluate the effectiveness of our proposed metrics, we conducted translation experiments on Chinese-English translation using various SMT and NMT systems. We draw the following two conclusions: (1) There exists strong correlations between the proposed metrics and human evaluation measured by the Pearson Correlation Coefficient, and (2) The significant improvement in terms of BLEU score doesn't imply the same improvement in OTEM and UTEM, by contrast, our proposed metrics can be used as supplements to the BLEU score. Moreover, further analysis shows the diverse characteristics of the NMT systems based on different architectures.

2   Related Work

Usually, most of the widely-used automatic evaluation metrics are used to perform the overall evaluation of translation quality. On the whole, these metrics can be divided into the following three categories: (1) **The Lexicon-based Metrics** are good at capturing the lexicon or phrase level information but can not adequately reflect the syntax and semantic similarity [2,4,5,8,15,16,23,24,27]; (2) **The Syntax/Semantic-based Metrics** exploit the syntax and semantic similarity to evaluate translation quality, but still suffer from the syntax/semantic parsing of the potentially noisy machine translations [11,17–20,22,30,34]; (3) **The Neural Network-based Metrics** mainly leverage the embeddings of the candidate and its references to evaluate the candidate quality [3,12,13].

Since our metrics involve n-gram matching, we further discuss the two subclasses in the first aspect: (1) **Evaluation Metrics based on N-gram Matching**. By utilizing the n-gram precisions between candidate and references, the F-measure, the recall and so on, these metrics attain the goal to evaluate the overall quality of candidate [2,4,5,8,16,24]. (2) **Evaluation Metrics based on Edit Distance**. The core idea of these metrics [15,23,25,27] is to calculate the edit distance required to modify a candidate into its reference, which can reflect the discrepancy between a candidate and its references.

Our work is significantly different from most of the above-mentioned studies, for we mainly focus on the over- and under-translation issues, rather than measuring the translation quality in terms of adequacy and fluency. The work most closely related to ours is the N-gram Repetition Rate (N-GRR) proposed by Zhang et al. [35], which merely computes the portion of repeated n-grams for over-translation evaluation. Compared with our metrics, the OTEM in particular, N-GRR is much simpler for it completely ignores the n-gram distribution in gold references and doesn't solve length bias problem. To some extent, OTEM can be regarded as a substantial extension of N-GRR.

Meanwhile, the metrics proposed by Popovic and Ney [25] also evaluate the MT translation on different types of errors such as missing words, extra words and morphological errors based on edit distance. However, its core idea extends from WER and PER, and it only takes the word-level information into consideration, while the length bias problem can't be solved similarly. The evaluation of addition and omission can be seen as the simplified 1-gram measurement of OTEM and UTEM theoretically. In addition, Malaviya et al. [21] also presented two metrics to account for over- and under-translation in MT translation. Unlike our model, however, the problem of length bias was also not solved in this work.

3   Our Metrics

In this section, we give detailed descriptions of the proposed metrics. The ideal way to assess over-translation or under-translation problems is to semantically compare a source sentence with its candidate translation and record how many times each source word is translated to the target word, which unfortunately is shown to be trivial. Therefore, here we mainly focus on the study of simple but effective automatic evaluation metrics for NMT specific to over- and under-translation.

Usually, a source sentence can be correctly translated into diverse target references which differ in word choices or in word orders even using the same words. Besides that, there are often no other significant differences among the n-gram distributions of these target references. If the occurrence of a certain n-gram in the generated translation is significantly greater than that in all references, we can presume that the generated translation has the defect of over-translation. Similarly, if the opposite happens, we can assume that under-translation occurs in the generated translation. Based on these analyses, we follow Papineni et al. [24] to design OTEM and UTEM on the basis of the lexical matchings between candidate translations and gold references:

$$Otem/Utem := LP * \exp \left(\sum_{n=1}^{N} w_n \log mp_n \right), \tag{1}$$

where LP indicates a factor of length penalty, N is the maximum length of the considered n-grams, and mp_n denotes the proportion of the mismatched n-grams contributing to the metric by the weight w_n. It should be noted that here we directly adapt the weight definition of BLEU [24] to ours, leaving more sophisticated definitions to future work. Specifically, we assume that different n-grams share the same contribution to the metric so that w_n is fixed as $\frac{1}{N}$. Although this formulation looks very similar to the BLEU, the definitions of BP and p_n, which lie at the core of our metrics, differs significantly from those of BLEU and mainly depend on the specific proposed metrics. We elaborate more on these details in the following subsections.

3.1   Otem

As described previously, when over-translation occurs, the candidate translation generally contains many repeated n-grams. To capture this characteristic, we define mp_n to be the proportion of these over-matched n-grams over the whole candidate translation as follows:

$$mp_n = \frac{\sum\limits_{C \in \{Candidates\}} \sum\limits_{n\text{-}gram \in C} Count_{over}(n\text{-}gram)}{\sum\limits_{C' \in \{Candidates\}} \sum\limits_{n\text{-}gram' \in C'} Count_{cand}(n\text{-}gram)}, \tag{2}$$

where $\{Candidates\}$ denotes the candidate translations of a dataset, $Count_{over}(\cdot)$ calculates the over-matched times of the n-gram from the candidate translation, and $Count_{cand}(\cdot)$ records the occurrence of the n-gram in the candidate translation. When referring to $Count_{over}(\cdot)$, we mainly focus on two kinds of over-matched n-grams: (1) the n-gram which occurs in both reference and candidate, and its occurrence in the latter exceeds that in the former; and (2) the n-gram that occurs only in candidate, and its occurrence exceeds 1.

Moreover, we define the over-matched times of $n\text{-}gram$ as follows:

$$\begin{cases} Count_{cand}(n\text{-}gram) - Count_{ref}(n\text{-}gram), & \text{if } Count_{cand}(n\text{-}gram) > Count_{ref}(n\text{-}gram) > 0; \\ Count_{cand}(n\text{-}gram) - 1, & \text{if } Count_{cand}(n\text{-}gram) > 1 \text{ and } Count_{ref}(n\text{-}gram) = 0; \\ 0, & \text{otherwise,} \end{cases}$$

$$\tag{3}$$

where $Count_{ref}(n\text{-}gram)$ denotes the count of $n\text{-}gram$ in its reference. When multiple references are available, we choose the minimum $Count_{over}(n\text{-}gram)$ for this function, as we argue that a n-gram is not over-matched as long as it is not over-matched in any reference. Back to the Candidate 1 mentioned in Sect. 1, $Count_{cand}(\text{``peace''})$ is 2, while $Count_{ref}(\text{``peace''})$ in all references is 1, and thus $Count_{over}(\text{``peace''})$ is calculated as 1.

Another problem with over-translation is that candidates tend to be longer because many unnecessary n-grams are generated repeatedly, which further causes the calculation bias in OTEM. To remedy this, we introduce the length penalty LP to penalize long translations. Formally,

$$LP = \begin{cases} 1, & \text{if } c < r; \\ e^{1-\frac{r}{c}}, & \text{otherwise,} \end{cases} \tag{4}$$

where c and r denote the length of candidate translation and its reference respectively. For multiple references, we select the one whose length is closest to the candidate translation, following Papineni et al. [24].

3.2 Utem

Different from OTEM, UTEM assesses the degree of omission in the candidate translation for a source sentence. Whenever under-translation occurs, some n-grams are often missed compared with its reference. Therefore, we define mp_n to be the proportion of these under-matched n-grams over the reference as follows:

$$mp_n = \frac{\sum\limits_{\mathcal{R} \in \{References\}} \sum\limits_{n\text{-}gram \in \mathcal{R}} Count_{under}\,(n\text{-}gram)}{\sum\limits_{\mathcal{R}' \in \{References\}} \sum\limits_{n\text{-}gram' \in \mathcal{R}'} Count_{ref}\,(n\text{-}gram)}, \tag{5}$$

where $\{References\}$ indicates the gold references from a dataset, and $Count_{ref}(\cdot)$ counts the occurrence of n-gram in the reference.

Note that the above formula only deals with one reference for each source sentence, however, both numerator and denominator in Eq. 5 suffer from the selection bias problem, when there are multiple references. In this case, we employ a default strategy to preserve the minimum $Count_{under}(\cdot)$ value as well as the maximum $Count_{ref}(\cdot)$ value for each n-gram based on an optimistic scenario.

As for $Count_{under}(\cdot)$, we mainly consider two types of under-matched n-grams: (1) the n-gram that occurs in both reference and candidate, and its occurrence in the former exceeds that in the latter; and (2) the n-gram that appears only in reference. Furthermore, we calculate their $Count_{under}(\cdot)$ as follows:

$$\begin{cases} Count_{ref}(n\text{-}gram) - Count_{cand}(n\text{-}gram), & \text{if } Count_{ref}(n\text{-}gram) > Count_{cand}(n\text{-}gram); \\ 0, & \text{otherwise.} \end{cases} \tag{6}$$

In this way, the more parts are omitted in translation, the larger $Count_{under}(\cdot)$ will be, which as expected can reflect the under-translation issue. Still take the Candidate 2 described in Sect. 1 as an example, we find that $Count_{ref}(\text{``peace''})$ is 1, so, $Count_{under}(\text{``peace''})$ is computed as 1.

Furthermore, when some source words or phrases are untranslated, the resulting candidate translation generally tends to be shorter. Accordingly, we also leverage the length penalty LP to penalize short translations, i.e.

$$LP = \begin{cases} 1, & \text{if } c > r; \\ e^{1-\frac{c}{r}}, & \text{otherwise.} \end{cases} \tag{7}$$

where the definitions of c and r are the same as those in Eq. 4.

4 Experiments

We evaluated our proposed metrics on Chinese-English translation task.

4.1   Datasets and Machine Translation Systems

We collected 1.25M LDC sentence pairs with 27.9M Chinese words and 34.5M English words as the training corpus. Besides, we chose the NIST 2005 dataset as the validation set and the NIST 2002, 2003 and 2004 datasets as the test sets. Each source sentence in these datasets is annotated with four different references.

For the sake of efficiency, we only kept the sentences of length within 50 words to train NMT models. In this way, there are 90.12% of parallel sentences were involved in our experiments. As for the data preprocessing, we segmented Chinese words using *Stanford Word Segmenter*[1], and English tokens via *Byte Pair Encoding* (BPE) [26]. We set the vocabulary size to 30K for NMT model training. For all the out-of-vocabulary words in the corpus, we replaced each of them with a special token UNK. Finally, our vocabularies covered 97.4% Chinese words and 99.3% English words of the corpus.

We carried out experiments using the following state-of-the-art MT systems:

(1) **PbSMT and HieSMT**: We trained a phrase-based (*PbSMT*) [14] and a hierarchical phrase-based (*HieSMT*) [6] SMT system using *MOSES* with default settings. *GIZA++* and *SRILM* are used to generate word alignments and 5-gram language model respectively.

(2) **RNNSearch**: a re-implementation of the attention-based NMT model [1] based on dl4mt tutorial. We set word embedding size as 620, hidden layer size as 1000, learning rate as 5×10^{-4}, batch size as 80, gradient norm as 1.0, and beam size as 10. All the other settings are the same as in [1].

(3) **Coverage**: an enhanced RNNSearch equipped with a coverage mechanism [31]. We used the same model settings as in the above RNNSearch.

(4) **FairSeq**: a convolutional sequence-to-sequence learning system [10]. We used 15 convolutional encoder and decoder layers with a kernel width of 3, and set all embedding dimensions to 256. Others were kept as default.

(5) **Transformer**: model [32] reimplemented by Tsinghua NLP group[2]. We trained the base Transformer using 6 encoder and decoder layers with 8 heads, and set batch size as 128.

4.2   Comparison with Human Translation

In theory, our metrics are capable of distinguishing human translation with no over- and under-translation issues from the machine translated ones that may suffer from these issues. To verify this, we collected the translations produced by RNNSearch and one of four references of each source sentence in NIST 2002 dataset. We compare them by calculating the mismatch proportion mp_n against three other gold references. Figure 2 shows the results.

Not surprisingly, with the increase of n-gram length, the proportion of over-matched n-grams drops gradually. This is reasonable because long n-grams are

---

[1] https://nlp.stanford.edu/software/segmenter.html.
[2] https://github.com/thumt/THUMT.

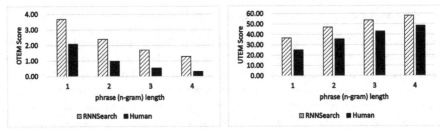

(a) OTEM distinguish human translation (b) UTEM distinguish human translation
from RNNSearch translation   from RNNSearch translation

Fig. 2. Comparison between RNNSearch and human translation on NIST 2002 dataset
in terms of mismatch proportion mp_n, where n ranges from 1 to 4. (a) is for OTEM,
and (b) is for UTEM.

more difficult to be generated repeatedly. By contrast, the proportion of under-
matched n-grams grows steadily. The underlying reason is that long n-grams
tend to be more difficult to be matched against the reference. No matter how
long the n-gram is, our OTEM metric assigns significantly greater scores to the
human translations than the machine translated ones. Meanwhile, the scores of
our UTEM metric on the human translations are also significantly less than those
of machine translation. Besides, it is clear that both OTEM and UTEM metrics
show great and consistent difference between the evaluation score of RNNSearch
and human translation, strongly indicating their ability in differentiating human
translations from the machine translated ones.

4.3 Human Evaluation

In this section, we investigate the effectiveness of OTEM and UTEM by measuring their correlation and consistency with human evaluation. Existing manual labeled dataset is usually annotated with respect to faithfulness and fluency, rather than over- and under-translation. To fill this gap, we first annotate
a problem-specific evaluation dataset. Then we examine our proposed metrics
using the Pearson Correlation Coefficient (Pearson's r).

Data Annotation. Following the similar experimental setup in [24], we used
the NIST 2002 dataset for this task. In order to avoid selection bias problem, we
randomly sampled five groups of source sentences from this dataset. Each group
contains 50 sentences paired with candidate translations generated by different
NMT systems (including RNNSearch, Coverage, FairSeq and Transformer). In
total, this dataset consists of 1000 Chinese-English sentence pairs.

We arranged two annotators to rate translations in each group from 1 (almost
no over- or under-translation issue) to 5 (serious over- or under-translation issue),
and average their assigned scores to the candidate translation as the final manually annotated score. The principle of scoring is the ratio of over-translated

or under-translated word occurrence in candidate translations. It is to be noted that this proportion has a certain subjective influence on scoring according to the length of the candidate and source sentence. For example, with the same over-translated number of words in the candidate (e.g. 5 words), the score can change from 2 (few words have been over-translated) to 4 (a large number of words have been over-translated) for a long sentence with the length of 60 words and a short sentence with the length of 10 words.

Correlation with Human Evaluation. We collected the annotated sentence pairs for each NMT system, and summarized the average manually annotated score with the corresponding OTEM and UTEM in Fig. 3. We find that both OTEM and UTEM are positively correlated with the manually annotated score. To further verify this observation, we computed the Pearson's r for both metrics, where the value is 0.9461 and 0.8208 for OTEM ($p < 0.05$) and UTEM ($p < 0.05$), respectively. These Pearson's r values strongly suggest that our proposed metrics are indeed highly consistent with human judgment (notice that lower OTEM and UTEM score indicates a better translation).

We also provide comparison between manually annotated score and BLEU score in Fig. 4. Obviously, BLEU score demonstrates rather weak association

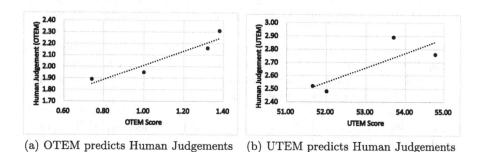

(a) OTEM predicts Human Judgements   (b) UTEM predicts Human Judgements

Fig. 3. Correlation between human judgment and OTEM, UTEM. Clear positive correlation is observed for both metrics.

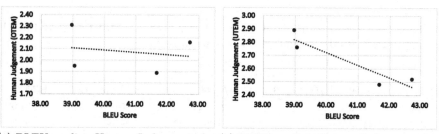

(a) BLEU predicts Human Judgements for (b) BLEU predicts Human Judgements for
Over-translation   Under-translation

Fig. 4. Correlation between human judgment and BLEU.

with the over-translation. By contrast, its correlation with the under-translation is much stronger. We conjecture that this is because some important clauses are left untranslated, leading to the occurrence of under-translation, and in consequence, the generated translation usually suffers from unfaithfulness issue, a critical aspect for BLEU evaluation. In addition, we also calculated the corresponding Pearson's r between the manually annotated scores and BLEU. The value of Pearson's r for the over- and under-translation is -0.1889 and -0.9192, of which p values are larger than 0.05, indicating that the negative correlation is not significant. In other words, BLEU score is incapable of fully reflecting the over- and under-translation issues.

4.4 Analysis on MT Systems

We summarize the BLEU, OTEM, UTEM scores for different MT systems in Table 1. Particularly, we show OTEM-2(2-gram) rather than OTEM-4(4-gram) because of data sparsity issue.

From Table 1, although all NMT systems outperform all SMT systems with respect to the BLEU score, we observe that for OTEM score, almost all SMT systems outperform all NMT systems. We contribute this to the hard-constraint coverage mechanism in SMT which disables the decoder to repeatedly translate the same source phrases. Sharing similar strength with the coverage mechanism in SMT, Coverage yields substantial improvements over the RNNSearch. It is very interesting that although FairSeq and Transformer produce very similar BLEU scores, Transformer achieves significantly better OTEM scores than FairSeq. We argue that this is because attention in Transformer builds up strong dependencies with both source and previous target words, while convolution in FairSeq can only capture local dependencies.

We can also discover that in terms of UTEM score, all MT systems show similar results, although NMT systems remarkably outperform SMT systems regarding the BLEU score. Through the coverage mechanism, SMT can successfully enforce the translation of each source word. On the contrary, Coverage fails to share this strength. The underlying reason is complicated, which, we argue, requires much more efforts.

Table 1. Case-insensitive BLEU-4/OTEM-2/UTEM-4 score on NIST Chinese-English translation task. **Bold** highlights the best result among all systems.

Model	Dev	MT02	MT03	MT04
PbSMT	33.09/1.00/56.41	34.50/0.84/52.96	33.46/0.80/55.46	35.23/0.98/55.36
HierSMT	34.18/0.78/**55.77**	36.21/0.66/**52.14**	34.44/0.58/54.88	36.91/0.74/**54.62**
RNNSearch	34.72/2.05/60.31	37.95/1.67/54.68	35.23/2.08/56.88	37.32/1.78/58.49
Coverage	35.02/1.30/61.06	38.40/1.01/55.09	36.18/1.48/55.96	37.92/1.27/57.90
FairSeq	38.84/1.84/58.65	**41.90**/1.24/53.79	**40.67**/1.81/54.71	42.32/1.89/55.57
Transformer	**38.90**/1.02/57.76	41.33/0.77/52.79	40.62/0.94/**54.26**	**42.74**/0.83/55.56

Overall, different MT systems show different characteristics with respect to over- and under-translation. BLEU score itself can hardly reflect all these above observations, which highlights the necessity of our work.

5 Conclusion

In this paper, we have proposed two novel evaluation metrics, OTEM and UTEM, to evaluate the performance of NMT systems in dealing with over- and under-translation issues, respectively. Although our proposed metrics are based on lexical matching, they are highly correlated to human evaluation, and very effective in detecting the over- and under-translation occurring in the translations produced by NMT systems. Moreover, experimental results show that the coverage mechanism, CNN-based FairSeq and attention-based Transformer possess specific architectural advantages on overcoming these undesired defects.

Acknowledgement. The authors were supported by Natural Science Foundation of China (No. 61672440), the Fundamental Research Funds for the Central Universities (Grant No. ZK1024), Scientific Research Project of National Language Committee of China (Grant No. YB135-49), and Research Fund of the Provincial Key Laboratory for Computer Information Processing Technology in Soochow University (Grant No. KJS1520). We also thank the reviewers for their insightful comments.

References

1. Bahdanau, D., Cho, K., Bengio, Y.: Neural machine translation by jointly learning to align and translate. In: ICLR (2015)
2. Banerjee, S., Lavie, A.: METEOR: an automatic metric for MT evaluation with improved correlation with human judgments. In: ACL Workshop (2005)
3. Chen, B., Guo, H.: Representation based translation evaluation metrics. In: ACL (2015)
4. Chen, B., Kuhn, R.: AMBER: a modified BLEU, enhanced ranking metric. In: WMT (2011)
5. Chen, B., Kuhn, R., Foster, G.: Improving AMBER, an MT evaluation metric. In: WMT (2012)
6. Chiang, D.: Hierarchical phrase-based translation. Comput. Linguist. **33**(2), 201–228 (2007)
7. Cohn, T., Hoang, C.D.V., Vymolova, E., Yao, K., Dyer, C., Haffari, G.: Incorporating structural alignment biases into an attentional neural translation model. In: NAACL (2016)
8. Doddington, G.: Automatic evaluation of machine translation quality using n-gram co-occurrence statistics. In: HLT (2002)
9. Feng, S., Liu, S., Yang, N., Li, M., Zhou, M., Zhu, K.Q.: Improving attention modeling with implicit distortion and fertility for machine translation. In: COLING (2016)
10. Gehring, J., Auli, M., Grangier, D., Yarats, D., Dauphin, Y.N.: Convolutional sequence to sequence learning. In: ICML (2017)
11. Giménez, J., Màrquez, L.: Linguistic features for automatic evaluation of heterogenous MT systems. In: WMT (2007)

12. Gupta, R., Orasan, C., van Genabith, J.: ReVal: a simple and effective machine translation evaluation metric based on recurrent neural networks. In: EMNLP (2015)

13. Guzmán, F., Joty, S., Màrquez, L., Nakov, P.: Pairwise neural machine translation evaluation. In: ACL (2015)

14. Koehn, P., Och, F.J., Marcu, D.: Statistical phrase-based translation. In: NAACL (2003)

15. Leusch, G., Ueffing, N., Ney, H.: A novel string to string distance measure with applications to machine translation evaluation. In: Mt Summit IX (2003)

16. Lin, C.Y.: ROUGE: a package for automatic evaluation of summaries. In: ACL Workshop (2004)

17. Liu, D., Gildea, D.: Syntactic features for evaluation of machine translation. In: ACL Workshop (2005)

18. Lo, C.k., Tumuluru, A.K., Wu, D.: Fully automatic semantic MT evaluation. In: WMT (2012)

19. Lo, C.k., Wu, D.: MEANT: an inexpensive, high-accuracy, semi-automatic metric for evaluating translation utility based on semantic roles. In: ACL (2011)

20. Lo, C.k., Wu, D.: MEANT at WMT 2013: a tunable, accurate yet inexpensive semantic frame based MT evaluation metric. In: WMT (2013)

21. Malaviya, C., Ferreira, P., Martins, A.F.: Sparse and constrained attention for neural machine translation. arXiv preprint arXiv:1805.08241 (2018)

22. Mehay, D.N., Brew, C.: BLEUÂTRE: flattening syntactic dependencies for MT evaluation. In: MT Summit (2007)

23. Nießen, S., Och, F.J., Leusch, G., Ney, H.: An evaluation tool for machine translation: fast evaluation for MT research. In: LREC (2000)

24. Papineni, K., Roukos, S., Ward, T., Zhu, W.: BLEU: a method for automatic evaluation of machine translation. In: ACL (2002)

25. Popović, M., Ney, H.: Towards automatic error analysis of machine translation output. Comput. Linguist. **37**(4), 657–688 (2011)

26. Sennrich, R., Haddow, B., Birch, A.: Neural machine translation of rare words with subword units. In: ACL (2016)

27. Snover, M., Dorr, B., Schwartz, R., Micciulla, L., Makhoul, J.: A study of translation edit rate with targeted human annotation. In: AMTA (2006)

28. Sundermeyer, M., Alkhouli, T., Wuebker, J., Ney, H.: Translation modeling with bidirectional recurrent neural networks. In: EMNLP (2014)

29. Sutskever, I., Vinyals, O., Le, Q.V.: Sequence to sequence learning with neural networks. In: NIPS (2014)

30. Owczarzak, K., van Genabith, J., Way, A.: Dependency-based automatic evaluation for machine translation. In: SSST (2007)

31. Tu, Z., Lu, Z., Liu, Y., Liu, X., Li, H.: Modeling coverage for neural machine translation. In: ACL (2016)

32. Vaswani, A., et al.: Attention is all you need. In: NIPS (2017)

33. Yang, Z., Hu, Z., Deng, Y., Dyer, C., Smola, A.: Neural machine translation with recurrent attention modeling. In: EACL (2017)

34. Yu, H., Wu, X., Xie, J., Jiang, W., Liu, Q., Lin, S.: RED: a reference dependency based MT evaluation metric. In: COLING (2014)

35. Zhang, B., Xiong, D., Su, J.: A GRU-gated attention model for neural machine translation. arXiv preprint arXiv:1704.08430 (2017)

Improved Neural Machine Translation with Chinese Phonologic Features

Jian Yang[1], Shuangzhi Wu[2], Dongdong Zhang[3],
Zhoujun Li[1(⊠)], and Ming Zhou[3]

[1] Beihang University, Beijing, China
{yangjian123,lizj}@buaa.edu.cn
[2] Harbin Institute of Technology, Harbin, China
v-shuawu@microsoft.com
[3] Microsoft Researcher Asian, Beijing, China
{dozhang,mingzhou}@microsoft.com

Abstract. Chinese phonologic features play an important role not only in the sentence pronunciation but also in the construction of a native Chinese sentence. To improve the machine translation performance, in this paper we propose a novel phonology-aware neural machine translation (PA-NMT) model where Chinese phonologic features are leveraged for translation tasks with Chinese as the target. A separate recurrent neural network (RNN) is constructed in NMT framework to exploit Chinese phonologic features to help facilitate the generation of more native Chinese expressions. We conduct experiments on two translation tasks: English-to-Chinese and Japanese-to-Chinese tasks. Experimental results show that the proposed method significantly outperforms state-of-the-art baselines on these two tasks.

Keywords: Neural Machine Translation · Chinese phonology

1 Introduction

Neural Machine Translation (NMT) with the attention-based encoder-decoder framework [2] has been proved to be the most effective approach to machine translation tasks on many language pairs [2,13,21,24]. In a conventional NMT model, an encoder reads in source sentences of variable lengths, and transforms them into sequences of intermediate hidden vector representations. With weighted attention operations, the hidden vectors are combined and fed to the decoder to generate target translations.

There have been much work to improve the performance of NMT models, such as exploring novel network architectures [7,22] and introducing prior knowledge of syntax information [4,6,12,23]. As the translation accuracy of NMT increases along with new algorithms and models proposed, it still suffers from the challenge of generating idiomatic and native translation expressions on target languages. Intuitively, native expressions may relate to phonologic knowledge of a

© Springer Nature Switzerland AG 2018
M. Zhang et al. (Eds.): NLPCC 2018, LNAI 11108, pp. 303–315, 2018.
https://doi.org/10.1007/978-3-319-99495-6_26

language beyond the surface form of words. In terms of Chinese, linguists pointed that Chinese phonologic features play an important role in both the sentence pronunciation and the construction of native Chinese sentences [25]. For instance, in the translation example of Fig. 1, the meaning of the verb phrase 'raise money' can be literally represented in Chinese as "筹集 (raise) 钱 (money)", "筹集 (raise) 资金 (money)" or "筹 (raise) 钱 (money)". But the last two Chinese expressions are more native than the first one. The reason is that, from Chinese phonologic perspective, the verb-object pair is more common to have the same number of syllables. Therefore, the disyllable-disyllable collocation "筹集 (raise) 资金 (money)" and the monosyllable-monosyllable collocation "筹 (raise) 钱 (money)" acted as verb-object pairs in the references appear more native than the disyllable-monosyllable collocation "筹集 (raise) 钱 (money)" in the NMT baseline. There was previous work applying phonologic features to significantly improve the performance of tasks such as name entity recognition [3]. But Chinese phonologic features have not been explored in translation tasks when Chinese is the target language. In this paper, we propose a novel phonology-aware neural machine translation (PA-NMT) model where Chinese phonologic features are taken into account for translation tasks with Chinese as target. A PA-NMT model encodes source inputs with bi-directional RNNs and associates them with target word prediction via attention mechanism as in most NMT models, but it comes with a new decoder which is able to leverage Chinese phonologic features to help facilitate the generation of target Chinese sentences. Chinese phonology is equivalently represented by Chinese Pinyin, which includes syllable structure (the sequence of Chinese words) and intonation. Intonation mainly consists of high-level tone(first tone), rising tone(second tone), low tone(third tone), falling tone(fourth tone) and neutral tone. Our new decoder in PA-NMT consists of two RNNs. One is to generate the sequence of translation words, and the other is to produce the corresponding Chinese phonologic features, which are further used to help the selection of translation candidates from a phonological perspective.

Source:	They planned to raise money for the project.						
Baseline:	他们 (They)	计划 (plan)	为 (for)	这一 (this)	项目 (item)	筹集 (raise)	钱 (money)
Reference1:	他们 (They)	计划 (plan)	为 (for)	这一 (this)	项目 (item)	筹集 (raise)	资金 (money)
Reference2:	他们 (They)	计划 (plan)	为 (for)	这一 (this)	项目 (item)	筹 (raise)	钱 (money)

Fig. 1. The meaning of the translation in NMT baseline is correct, but its expression is not native comparing to the references.

We evaluate our method on publicly available data sets with English-Chinese and Japanese-Chinese translation tasks. Experimental results show that our model significantly improves translation accuracy over the conventional NMT baseline systems.

The major contribution of our work is two folds:

(1) We propose a new PA-NMT model to leverage Chinese phonologic features. Our PA-NMT can encode target phonologic features and use them to help rank the translation candidates, so that the translation results of PA-NMT can be more native and accurate.

(2) Our PA-NMT model can achieve high quality results on both English-Chinese and Japanese-Chinese translation tasks, on publicly available data sets.

2 Background: Neural Machine Translation

Neural Machine Translation (NMT) is an end-to-end paradigm [2,7,22] which directly models the conditional translation probability $P(Y|X)$ of the source sentence X and the target translation Y. It usually consists of three parts: the encoder, attention mechanism and the decoder.

For the RNN based NMT model, the RNN encoder bidirectionally encodes the source sentence into a sequence of context vectors $H = h_1, h_2, h_3, ..., h_m$, where m is the source sentence length and $h_i = [\boldsymbol{h}_i, h_i]$, $\boldsymbol{h}_i$ and h_i are calculated by two RNNs from left-to-right and right-to-left respectively. The RNN can be a Gated Recurrent Unit (GRU) [5] or a Long Short-Term Memory (LSTM) [9] in practice. In this paper, we use GRU for all RNNs.

Based on the encoder states, the decoder generates target translations with length n word by word with probability

$$P(Y|X) = \prod_{j=1}^{n} P(y_j|y_{<j}, H) \tag{1}$$

The probability $P(y_j|y_{<j}, H)$ for the jth target word is computed by

$$P(y_j|y_{<j}, H) = g(s_j, y_{j-1}, c_j) \tag{2}$$

where g is a nonlinear, potentially multi-layered, function that outputs the probability of y_j, s_j is the j-th hidden state of decoder RNN, computed by

$$s_j = f_{RNN}(y_{j-1}, s_{j-1}, c_j)$$

c_j is the source context which is calculated by the attention mechanism. The attention mechanism is proposed to softly align each decoder state with the encoder states, where the attention score a_{jk} is computed to explicitly quantify how much each source word contributes to the target word by the following equations,

$$a_{jk} = \frac{\exp(e_{jk})}{\sum_{d=1}^{m} \exp(e_{jd})} \tag{3}$$

The calculation for e_{jk} can be in several ways [14], in this paper we compute e_{jk} by

$$e_{jk} = v_a^T \tanh(W_a s_{j-1} + U_a h_k) \tag{4}$$

where v_a, W_a, U_a are the weight matrix. The final source context c_j is the weighted sum of all encoder states

$$c_j = \sum_{k=1}^{n} a_{jk} h_k. \tag{5}$$

3   Phonology-Aware Neural Machine Translation Model

A phonology-aware neural machine translation (PA-NMT) model is an extension to the conventional NMT model augmented with Chinese phonologic features. The Chinese phonology is equivalently represented by Chinese Pinyin, so we model the phonologic features by Chinese Pinyin associated with tones. For a Chinese sentence, it has a corresponding Pinyin sequence with the same length. $\mathcal{P}_{T-2}, \mathcal{P}_{T-1}$Given a source sentence $X = x_1, x_2, .., x_m$, its target translation $Y = y_1, y_2, .., y_n$ and Y's Pinyin sequence $\mathcal{P} = \mathcal{P}_1, \mathcal{P}_2, .., \mathcal{P}_n$, the goal of our model is to use $\mathcal{P}$ to help the generation of Y. Figure 2 sketches the global overview of our PA-NMT model. Our model first encode source words in the conventional way as described in Sect. 2. In decoder, we use two recurrent neural networks (RNN) to model phonologic features and generate target words. At each timestep, the word RNN generates a list of translation candidates and the phonetic RNN helps to re-score the translation candidates based on phonologic features. Specially, we leverage two attention models for the two decoder RNNs. One is for modeling the phonetic features, the other is used for target word generation.

4   Model Encoder

Our encoder follows the standard RNN encoder [2] (left part in Fig. 2), which bidirectionally reads the input sequence and generate a sequence of context vectors $H = h_1, h_2, h_3, ..., h_m$, where m is the source sentence length and $h_i = [\boldsymbol{h}_i, h_i]$, $\boldsymbol{h}_i$ and h_i are calculated by two RNNs as described in Sect. 2.

4.1   Phonetic-Aware Decoder

Unlike the standard decoder [2], we use phonetic RNN to read Pinyins of history words and rescore current translation candidates as shown in Fig. 2 right top part. The right bottom part is a standard RNN decoder. We map the word dictionary to a Pinyin dictionary one by one, thus the words and Pinyins can

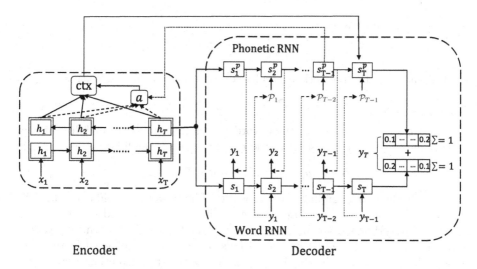

Fig. 2. Overview of PA-NMT model. The phonetic RNN only takes the Pinyin of history words as input and helps re-score the translation candidates. The word RNN is a standard RNN decoder with attention model which is omitted in the figure to simplify readability.

be aligned in their dictionaries. During decoding we force the phonetic RNN to read Pinyins of previous words generated by the word RNN,

$$y_i^{\mathcal{P}} = \sigma_{\mathcal{P}}(y_{i-1}) \tag{6}$$

where $\sigma_{\mathcal{P}}$ is the function which map predicted word to Pinyin which has the most probability. Thus the decoding procedure of the two RNNs can be aligned.

Although both RNNs have separate parameters, word RNN is in a coherent feature with phonetic RNN. By denoting the hidden state of phonetic decoder as $s_i^{\mathcal{P}}$, the calculation in phonetic RNN is as follows,

$$s_i^{\mathcal{P}} = GRU^{\mathcal{P}}(s_{i-1}^{\mathcal{P}}, y_{i-1}^{\mathcal{P}}, c_i^{\mathcal{P}}) \tag{7}$$

where $c_i^{\mathcal{P}}$ is the source context vector and $y_{i-1}^{\mathcal{P}}$ is the Pinyin of previous word y_{i-1}. The context vector $c_i^{\mathcal{P}}$ depends on the source states $[h_1, ..., h_m]$ and is calculated by the attention model,

$$c_i^{\mathcal{P}} = \sum_{j=1}^{last} a_{ij}^{\mathcal{P}} \cdot h_j \tag{8}$$

The weight $a_{ij}^{\mathcal{P}}$ of each annotation h_j is computed by

$$a_{ij}^{\mathcal{P}} = \frac{\exp(e_{ij})}{\sum_{k=1}^{last} \exp(e_{ik})} \tag{9}$$

where e_{ij} is computed as

$$e_{ij}^P = v_P^T \tanh(W_P[s_{i-1}^P; h_j]) \tag{10}$$

With s_i^P and c_i^P, the phonetic can calculate a probability list by softmax as

$$p(\mathcal{P}_i|\mathcal{P}_1, ..., \mathcal{P}_{i-1}, x) = g(\mathcal{P}_{i-1}, s_i, c_i^P) \tag{11}$$

We rescore the prediction of a new word by adding the log probability of the two softmax list. Thus the score of the new translation candidate y_i is

$$\text{score} = p(y_j|y_{<j}, X) + \alpha \log(\mathcal{P}_j|\mathcal{P}_{<j}, X) \tag{12}$$

where $\mathcal{P}_j$ is the corresponding Pinyin for the Chinese word y_j, and α is a hyper-parameter which control the importance of Chinese phonologic decoder. We set α to 0.5 in experiments. We find that when α is too big, it will make model worse. When α approximates zero, our model fails to extract phonological feature. In order to better preserve the Pinyin information, we use two attention parameters to store the Chinese character and pinyin alignment information respectively.

In our model, one RNN extracting phonological features interacts with another RNN in two aspects. For every timestep, word RNN generates next input of phonetic RNN which is converted from word to Pinyin to keep consistent with phonetic RNN. While predicting next word, phonetic RNN evaluate results of word RNN to generate final predicted candidates.

4.2 Chinese Polyphone Disambiguation

When mapping Chinese words to Pinyins, the major problem is that there could be multiple polyphone candidates for one Chinese word. Given a Chinese sentence, its Pinyin sequence expression is usually deterministic based on the context of the whole sentence. In our work, to align the Pinyin sequence and word sequence, the Pinyins are generated from the words with context-free information. To make disambiguation of Pinyin generation, we heuristically map each Chinese word to the Pinyin with the highest probability in terms of statistics over a Chinese monolingual corpus of 20 million sentences.

4.3 Model Training

Different form the objective function to train the conventional NMT model, for our joint PA-NMT model, we use the sum of log-likelihoods of word sequence and Pinyin sequence as our objective function:

$$J(\theta) = \sum_{(X,Y,\mathcal{P}) \in D} \log P(\mathcal{P}|X) + \log P(Y|X) \tag{13}$$

Thus, our training data format is (source sentence, (target Chinese sentence, target Chinese Pinyin)). In this way, we incorporate the phonological features into Chinese sentence generation.

5 Experiment

5.1 Setup

In the English-Chinese task, we use a subset from LDC corpus[1] which has around 2.6M sentence pairs from News domain. We use NIST 2008 as testset which has 4 references for each source sentence. We also make several other testsets by reversing the direction of Chinese-English sets NIST 2003, NIST 2005 and NIST 2012, as these sets all have four English reference, we just use the first reference as English source sentence. We use the WMT2009 English-Chinese set for development.

In the Japanese-Chinese task, we use 2.87M sentence pairs from ASPEC Japanese-Chinese corpus [15][2]. The development data contains 1, 784 sentences, and the test data contains 1, 812 sentences with single reference per source sentence. Both source and target language are tokenized with our in-house tools.

For the training data of target Chinese sentences in both the English-Chinese task and the Japanese-Chinese task, we covert them into Chinese Pinyin using our in-house implemented tool based on a statistical translation model with the accuracy of above 90%.

In the neural network training, the vocabulary size is limited to 30 K high frequent words for both source and target languages. All low frequent words are normalized into a special token unk and post-processed by following the work in [14]. The size of word embedding and transition action embedding is set to 512. The dimensions of the hidden states for all RNNs are set to 1024. All model parameters are initialized randomly with Gaussian distribution [8] and trained on a NVIDIA Tesla 1080 GPU. The stochastic gradient descent (SGD) algorithm is used to tune parameters with a learning rate of 1.0. The batch size is set to 128. In the update procedure, Adadelta [26] algorithm is used to automatically adapt the learning rate. The beam sizes for both word prediction and transition action prediction are set to 8 in decoding.

The baselines in our experiments is a neural translation system, denoted by RNNsearch which is an in-house implementation of the attention-based neural machine translation model [2] using the same parameter settings as our PA-NMT model. The evaluation results are reported with the word level and character level case-insensitive IBM BLEU-4 [17] denoted as **word-BLEU** and **char-BLEU** respectively. A statistical significance test is performed using the bootstrap resampling method proposed by [11] with a 95% confidence level.

5.2 Evaluation Results

We first evaluate our method on the English-Chinese translation task. The evaluation results over all test sets against baselines are listed in bottom part

---

[1] LDC2002E17, LDC2002E18, LDC2003E07, LDC2003E14, LDC2005E83, LDC2-005T06, LDC2005T10, LDC2006E17, LDC2006E26, LDC2006E34, LDC2006E85, LDC2006E92, LDC2006T06, LDC2004T08, LDC2005T10.

[2] http://orchid.kuee.kyoto-u.ac.jp/ASPEC/.

Table 1. Evaluation results on English-Chinese and Japanese-Chinese translation tasks with word-BLEU% and char-BLEU% metrics. The "Average" row in the English-Chinese part refers to the averaged result of all test sets. The numbers in bold indicate statistically significant difference ($p < 0.05$) from baselines.

Japanese-Chinese				
	RNNsearch		PA-NMT	
	word-BLEU	char-BLEU	word-BLEU	char-BLEU
dev	33.41	44.49	**34.03**	**45.23**
devtest	33.38	44.35	**34.26**	**45.26**
test	33.53	44.57	**34.19**	**45.26**
English-Chinese				
	RNNsearch		PA-NMT	
	word-BLEU	char-BLEU	word-BLEU	char-BLEU
NIST2003	18.38	30.79	**19.34**	**32.07**
NIST2005	17.02	28.87	**17.90**	**30.17**
NIST2008	22.71	34.30	**23.94**	**35.10**
NIST2012	13.51	23.33	**14.37**	**24.52**
Avg.	17.91	29.32	18.89	30.47

of Table 1. From the table, our PA-NMT outperforms RNNsearch on all the test sets on both word- and char- BLEU scores, where our model surpasses the baseline most on NIST 2008 set with 1.23 and 0.80 more scores on the two metrics. In terms of the average word-BLEU scores, our model outperforms the baseline by 0.98 BLEU points. And on the average char-BLEU scores, our model also outperforms the baseline by 1.14 BLEU points which shows our proposed phonology-aware NMT model performs much better than traditional sequence-to-sequence NMT model.

We also report results on the Japanese-Chinese translation task. The top part of Table 1 shows the comparison results with the evaluation metrics of word- and char- BLEU. From the table, our method outperforms the NMT baseline on the three datasets in terms of both word- and char- BLEUs.

5.3   Case Study

We sampled some translation examples from the test sets to make case study of how our method can improve the English-Chinese translation task. In the examples in Table 2, there are merely two phonology aspects we investigated: auxiliary word and syllable repetition.

Auxiliary word Structural auxiliary words have almost no actual grammatical meaning but play a role in the language structure. They just express the sound of the utterance or make the syllable of the language symmetry. In modern Chinese,

Table 2. Translation examples of RNNsearch and our PA-NMT on English-Chinese translation task. RNNsearch fails to generate pure Chinese sentences. Whereas with the help of the phonologic knowledge, our PA-NMT can get much better translations.

source	we cannot leave a heavy burden for later generations : this includes the issue of the rapid expansion of the population
RNNsearch	我们 不能 留给 后人 的 沉重 负担 : 这 包括 人口 的 快速 发展 的 问题 wǒmen búnéng liúgěi hòurén de chénzhòng fùdān : zhè bāokuò rénkǒu de kuàisù fāzhǎn de wèntí
PA-NMT	我们 不能 留给 后人 沉重 负担 : 这 包括 人口 急速 膨胀 的 问题 wǒmen búnéng liúgěi hòurén chénzhòng fùdān : bāokuò rénkǒu jísù péngzhàng de wèntí
reference	我们 不可 为 后代 人 留下 沉重 的 负担 , 包括 人口 急速 膨胀 的 问题 wǒmen búkě wéi houdài rén liúxià chénzhòng de fùdān , bāokuò rénkǒu jísù péngzhàng de wèntí
source	London underground train derailed , injuring 37 people
RNNsearch	伦敦 地下 火车 出轨 , 伤 人 37 人 lúndūn dìxià huǒchē chūguǐ , shāng rén sānshí qī rén
PA-NMT	伦敦 地下 火车 出轨 事故 造成 37 人 受伤 lúndūn dìxià huǒchē chūguǐ shìgù zàochéng sānshíqī rén shòushāng
reference	伦敦 地铁 列车 出轨 三十七 人 受伤 lúndūn dìtiě lièchē chūguǐ sānshíqī rén shòushāng
source	zhoushan is one of chinese cities suffering from a serious water shortage
RNNsearch	舟山市 是 一个 严重 缺 水 的 城市 之一 zhōushānshì shì yīgè yánzhòng quē shuǐ de chéngshì zhīyī
PA-NMT	舟山 是 中国 城市 缺水 较为 严重的 城市 之一 zhōushān shì zhōngguó chéngshì quēshuǐ jiàowéi yánzhòng de chéngshì zhīyī
reference	舟山市 是 中国 缺 水 较为 严重 的 城市 之一 zhōushān shì zhōngguó quē shuǐ jiàowéi yánzhòng de chéngshì zhīyī

"的 (de)" is one of the important structural auxiliary words. But if the word "的 (de)" in a incorrect position, it will disturb syllables of the whole sentence.

As shown in the first example of Table 2, "我们不能留给后人的沉重负担" can be represented by Pinyin as "wǒmen búnéng liúgěi hòurén de chénzhòng fùdān". "的 (de)" splits the Pinyin sequence into two parts "wǒmen búnéng liúgěi hòurén de" and "chénzhòng fùdān". The former part modifies the latter part. It will lead a unbalanced pronunciation for native speakers, because both "wǒmen búnéng liúgěi hòurén de" and "chénzhòng fùdān" have too many syllables which will make the whole sentence unstable and lead difficulty of speaking. However, in the sentence "我们不能留给后人沉重负担 (búnéng liúgěi hòurén chénzhòng fùdān)", "沉重 (chénzhòng)" modify "负担 (fùdān)". From semantic perspective, both "wǒmen búnéng liúgěi hòurén de chénzhòng fùdān" and "wǒmen búnéng liúgěi hòurén chénzhòng de fùdān" have the same meaning in

Chinese. "沉重 (chénzhòng)" and "负担 (fùdān)" are both two-syllable word, "沉重 (chénzhòng) 负担 (fùdān)" will be more easy to read. Our model's translation is "我们 (we) 不能 (cannot) 给 (give) 后人 (later generations) 留下 (leave) 沉重 (heavy) 的负担 (burden)" and it's Pinyin is "wǒmen búnéng gěi hòuràn liúxià chénzhòng fùdān". The sentence don't have "(的)de" in a incorrect position of the whole sentence which ensure keeping the original meaning unchanged and make pronunciation of sentence more fluent and authentic in Chinese.

Syllable repetition Syllable repetition also affect the rhythm of sentences. In second example, "injuring 37 people" is translated into " 伤人 (injuring) 人 (people)" in RNNsearch and "造成 (make) 人 (people) 受伤 (injured)" in our model. We are not used to speaking continuous "ren" in Chinese because it can break the rhythm of whole sentence. Pinyin of "伤人 (injuring) 人 (people)" is "shāngrén sānshíqī rén". The Pinyin sequence is spoken as two parts "shāngrén" which is word and "sānshíqī rén" which has four syllables. Two adjacent sequence both have "人 (ren)" as their ending. On the one hand, Except some special usage, we are not used to it. Generally speaking, we can speak "造成 (make) 人 (people) 受伤 (injured)" or "人 (people) 受伤 (injured)" which avoid continuously use the same monosyllable word in our daily. On the other hand, "造成 (make) 人 (people) 受伤 (injured)" has three parts "造成 (make)" , "人 (people)" and "受伤 (injured)". They all have similar number of syllables. We are accustomed to using this style which make the whole sentence are more rhythmic.

In the last example, we can see the RNNsearch's translation express the same meaning "一 (one)" twice. But this kind of expression of Chinese is illegal in grammar. The first "一个 (one)" and "之一 (one of)" behind it together is regarded as an adjective and play a role in attribute. The first "一个 (one)" will result in a semantic and phonological repetition in Chinese. The RNNsearch consider more on semantic information. PA-NMT consider both semantic feature and phonological feature. In most real cases, Chinese native speaker rarely express the similar words twice despite that they play different roles in the sentence. Hence, the Chinese sentence only keeps "之一 (one of)" translated by PA-NMT. We can see that PA-NMT avoid the repetition of pronunciation.

These examples mainly reflects in the usage of syllable structure in Chinese. Our model generates more native Chinese sentence with considering the combination of syllables and repetition of several syllables. Because Chinese characters are monosyllables, the syllables in sentences are very important in English-Chinese translation which considers balancing syllables in sentence. Each word only has each syllable. Hence, Pinyin is suitable for model to collect phonological features. By introducing Chinese phonologic feature, our model can learn the both semantic and phonetic information.

6   Related Work

Recently, neural machine translation (NMT) has achieved better performance than SMT in many language pairs [13,16,19,24,27]. A lot of work has been done to incorporate linguistic knowledge into NMT models [4,6,12,23]. A tree-to-sequence attentional NMT model is proposed in [6] where source-side HPSG tree was used. [4] leveraged the phrase-structure trees as in the Penn Chinese Treebank as prior knowledge for NMT inputs. They proposed a tree-coverage model to let the attention depend on the source-side syntax. In these models, the source dependency structure is used. For the target side, [1] proposed to replace the target sentence with the linearized, lexicalized constituency tree. [23] proposed to jointly learn target translation and dependency parsing.

Many languages like Russian use morphological information to improve translation quality in recent work [20]. Chinese linguistic features have been leveraged to help NLP tasks. For example, Chinese radicals were used as additional features to improve machine translation [10,18]. Chinese phonologic features were explored to address named entity recognition problem [3]. Different from those work, in this paper we propose to involve Chinese target phonetic features into NMT model to help translate pure and native Chinese sentences.

7 Conclusion

In this paper, we propose a novel phonology-aware neural machine translation (PA-NMT) model where Chinese phonologic features are used for translation tasks with Chinese as target. Our model encodes these features in the NMT decoder by another recurrent neural network (RNN), aiming to help facilitate the generation of target Chinese sentences. Our method tries to collect and use phonological features to optimize language model which is different from RNNsearch. Experimental results show that our method can boost the translation generation and achieve significant improvements on the translation quality of NMT systems. Along this research direction, in future work we will try to integrate other prior knowledge, such as semantic information, into NMT systems.

Acknowledgments. This work was supported in part by the Natural Science Foundation of China (Grand Nos. U1636211,61672081,61370126), and Beijing Advanced Innovation Center for Imaging Technology (No. BAICIT-2016001) and National Key R&D Program of China (No. 2016QY04W0802).

References

1. Aharoni, R., Goldberg, Y.: Towards string-to-tree neural machine translation. In: Proceedings of the 55th Annual Meeting of the Association for Computational Linguistics (Volume 2: Short Papers), pp. 132–140. Association for Computational Linguistics, Vancouver, Canada, July 2017. http://aclweb.org/anthology/P17-2021
2. Bahdanau, D., Cho, K., Bengio, Y.: Neural machine translation by jointly learning to align and translate. In: ICLR 2015 (2015)
3. Bharadwaj, A., Mortensen, D., Dyer, C., Carbonell, J.: Phonologically aware neural model for named entity recognition in low resource transfer settings. In: Proceedings of the 2016 Conference on Empirical Methods in Natural Language Processing, pp. 1462–1472 (2016)

4. Chen, H., Huang, S., Chiang, D., Chen, J.: Improved neural machine translation with a syntax-aware encoder and decoder. In: Proceedings of the 55th Annual Meeting of the Association for Computational Linguistics (Volume 1: Long Papers), pp. 1936–1945. Association for Computational Linguistics, Vancouver, Canada, July 2017. http://aclweb.org/anthology/P17-1177

5. Cho, K., et al.: Learning phrase representations using RNN encoder-decoder for statistical machine translation. In: Proceedings of ENMLP 2014, October 2014

6. Eriguchi, A., Hashimoto, K., Tsuruoka, Y.: Tree-to-sequence attentional neural machine translation. In: Proceedings of ACL 2016, August 2016

7. Gehring, J., Auli, M., Grangier, D., Yarats, D., Dauphin, Y.N.: Convolutional sequence to sequence learning. arXiv preprint arXiv:1705.03122 (2017)

8. Glorot, X., Bengio, Y.: Understanding the difficulty of training deep feedforward neural networks. Aistats **9**, 249–256 (2010)

9. Hochreiter, S., Schmidhuber, J.: Long short-term memory. Neural Comput. **9**(8), 1735–1780 (1997)

10. Zhang, J., Matsumoto, T.: Improving character-level Japanese-Chinese neural machine translation with radicals as an additional input feature (2003)

11. Koehn, P.: Statistical significance tests for machine translation evaluation. In: EMNLP, pp. 388–395. Citeseer (2004)

12. Li, J., Xiong, D., Tu, Z., Zhu, M., Zhang, M., Zhou, G.: Modeling source syntax for neural machine translation. In: Proceedings of the 55th Annual Meeting of the Association for Computational Linguistics (Volume 1: Long Papers), pp. 688–697. Association for Computational Linguistics, Vancouver, Canada, July 2017. http://aclweb.org/anthology/P17-1064

13. Luong, T., Pham, H., Manning, C.D.: Effective approaches to attention-based neural machine translation. In: Proceedings of EMNLP 2015, September 2015

14. Luong, T., Sutskever, I., Le, Q., Vinyals, O., Zaremba, W.: Addressing the rare word problem in neural machine translation. In: Proceedings of ACL 2015, July 2015

15. Nakazawa, T., et al.: ASPEC: Asian scientific paper excerpt corpus. In: Chair, N.C.C., et al. (eds.) Proceedings of the Ninth International Conference on Language Resources and Evaluation (LREC 2016), pp. 2204–2208. European Language Resources Association (ELRA), Portoroz, Slovenia, May 2016

16. Neubig, G.: Lexicons and minimum risk training for neural machine translation: NAIST-CMU at WAT2016. In: Proceedings of the 3rd Workshop on Asian Translation (WAT2016), Osaka, Japan, December 2016

17. Papineni, K., Roukos, S., Ward, T., Zhu, W.J.: BLEU: a method for automatic evaluation of machine translation. In: Proceedings of ACL 2002 (2002)

18. Kuang, S., Han, L.: Apply Chinese radicals into neural machine translation: deeper than character level

19. Shen, S., et al.: Minimum risk training for neural machine translation. In: Proceedings of ACL 2016, August 2016

20. Song, K., Zhang, Y., Zhang, M., Luo, W.: Improved English to Russian translation by neural suffix prediction. In: Proceedings of the Thirty-Second AAAI Conference on Artificial Intelligence (2018)

21. Tu, Z., Lu, Z., Liu, Y., Liu, X., Li, H.: Modeling coverage for neural machine translation. In: Proceedings of ACL 2016, August 2016

22. Vaswani, A., et al.: Attention is all you need. In: Advances in Neural Information Processing Systems, pp. 6000–6010 (2017)

23. Wu, S., Zhang, D., Yang, N., Li, M., Zhou, M.: Sequence-to-dependency neural machine translation. In: Proceedings of the 55th Annual Meeting of the Association for Computational Linguistics (Volume 1: Long Papers), pp. 698–707. Association for Computational Linguistics, Vancouver, Canada, July 2017. http://aclweb.org/anthology/P17-1065

24. Wu, Y., et al.: Google's neural machine translation system: bridging the gap between human and machine translation. arXiv preprint arXiv:1609.08144 (2016)

25. Xia, L.: A brief discussion on phonology and rhythm beauty in English-Chinese translation (2003). https://wenku.baidu.com/view/c3666404a200a6c30c225901020 20740be1ecd76.html

26. Zeiler, M.D.: ADADELTA: an adaptive learning rate method. arXiv preprint arXiv:1212.5701 (2012)

27. Zhang, B., Xiong, D., Su, J., Duan, H., Zhang, M.: Variational neural machine translation. In: Proceedings of EMNLP 2016, November 2016

Coarse-To-Fine Learning for Neural Machine Translation

Zhirui Zhang[1(✉)], Shujie Liu[2], Mu Li[2], Ming Zhou[2], and Enhong Chen[1]

[1] University of Science and Technology of China, Hefei, China
zrustc11@gmail.com, cheneh@ustc.edu.cn
[2] Microsoft Research Asia, Beijing, China
{shujliu,mingzhou}@microsoft.com, limugx@outlook.com

Abstract. In this paper, we address the problem of learning better word representations for neural machine translation (NMT). We propose a novel approach to NMT model training based on coarse-to-fine learning paradigm, which is able to infer better NMT model parameters for a wide range of less-frequent words in the vocabulary. To this end, our proposed method first groups source and target words into a set of hierarchical clusters, then a sequence of NMT models are learned based on it with growing cluster granularity. Each subsequent model inherits model parameters from its previous one and refines them with finer-grained word-cluster mapping. Experimental results on public data sets demonstrate that our proposed method significantly outperforms baseline attention-based NMT model on Chinese-English and English-French translation tasks.

Keywords: Neural machine translation · Coarse-to-fine learning
Hierarchical cluster

1 Introduction

As a recently proposed novel approach to machine translation, and despite its short history [2,7,14,29], neural machine translation (NMT) has been making rapid progress from catching up with statistical machine translation (SMT) [3,6, 15] to outperforming it by significant margins on many language pairs [10,18,30, 31,34]. Aside from better translation performance, NMT also demonstrates other appealing properties such as little requirements for human feature engineering or prior domain knowledge, so it is also drawing attention from researchers working on other NLP tasks [24,27,32].

Much recent work in the literature focuses on addressing the issue of restricted vocabulary size in NMT systems. Popular NMT system implementations employ moderate-sized vocabularies typically containing most frequent 30K–80K words, and map all the other words to a single <unk> label. Luong et al. [19] proposed a method which uses lexicon look-up to replace generated <unk> labels in target translations. This method solves part of the problem,

© Springer Nature Switzerland AG 2018
M. Zhang et al. (Eds.): NLPCC 2018, LNAI 11108, pp. 316–328, 2018.
https://doi.org/10.1007/978-3-319-99495-6_27

but the translation still cannot be well recovered when the unknown word rate is high, due to the fact that too many words with distinct usages sharing a single <unk> label leads to a substantial amount of ambiguities. Jean et al. [13] tackled the small vocabulary size limit with an efficient softmax approximation algorithm, which enables to use very large vocabulary in NMT systems. Although this method effectively reduces the unknown word rate and brings further improvement to translation accuracy, we note that the inclusion of more words in a larger vocabulary intensifies the challenge of learning accurate usage for the less-frequent words, even if they are not viewed as unknown words. For example, the Chinese word 窜改 (alter), which appears near the tail of a 50K-word vocabulary in terms of frequency, is such a long-tail less-frequent word. Due to its small number of occurrences in the training data, the learnt representation in a conventional NMT model is very likely to overfit to its specific usage in the training corpus, and as a result usually left ignored in unseen contexts during decoding. Figure 1 shows an incorrect translation example caused by this word.

Input: 他 **窜改** 老师 与 学生 对话 的 录音 .
Output: He teacher the recording of teacher and student conversation .
Reference: He <u>tempered with</u> the recording of conversation between the teacher and the student .

Fig. 1. Example of incorrect translation of less-frequent word.

In this paper, we present a novel NMT training method based on coarse-to-fine paradigm, which is able to learn better NMT model parameters for less-frequent words that do not have sufficient usage coverage in the training data. The presented method is inspired by a common linguistic observation that a group of words belonging to the same syntactic/semantic class, for instance, *large, enormous, gigantic, mammoth*, tend to share certain properties such as collocations and translations, and are expected to be close to each other in embedding space. This gives the opportunity that if we can assign a less-frequent word to an appropriate class whose representation can be more accurately learned, it could benefit from inheriting part of the class' representation which generalizes better to unseen contexts. Our proposed method works as follows: at first, source and target words are grouped into a set of hierarchical tree-structured clusters based on bilingual data, then a sequence of NMT models are learned based on sets of clusters at different levels of the clustering tree with finer and finer granularity. When training each model, the training data is first transformed such that all words are replaced with their corresponding clusters at the specified hierarchical level. Every cluster's representation is initialized with its parent cluster's representation learned by the previous model, then the standard NMT training process is performed to refine the model parameters.

We conduct experiments on public Chinese-English and English-French translation data sets. Experimental results demonstrate that our proposed method significantly outperforms baseline attention-based NMT model on these two translation tasks.

2   Neural Machine Translation

In this work, we concentrate on applying our coarse-to-fine learning method to sequence-to-sequence NMT models. In particular, we follow the neural machine translation architecture proposed by Bahdanau et al. [2].

Neural machine translation system is implemented as an encoder-decoder framework with recurrent neural networks (RNN), which can be Gated Recurrent Unit (GRU) [7] or Long Short-Term Memory (LSTM) [12] networks in practice. The encoder reads in the source sentence $X = (x_1, x_2, \dots, x_T)$ and transforms it into a sequence of hidden states $h = (h_1, h_2, \dots, h_T)$, using a bi-directional recurrent neural network. The decoder uses another recurrent neural network to generate a corresponding translation $Y = (y_1, y_2, \dots, y_{T'})$ based on the encoded sequence of hidden state h. At each time i, the conditional probability of each word y_i from a target vocabulary V_y is computed by

$$p(y_i | y_{<i}, h) = g(y_{i-1}, z_i, c_i) \tag{1}$$

where z_i is the i_{th} hidden state of the decoder and is calculated conditional on the previous hidden state z_{i-1}, previous word y_{i-1} and the source context vector c_i:

$$z_i = \text{RNN}(z_{i-1}, y_{i-1}, c_i) \tag{2}$$

In attention-based NMT, the context vector c_i is a weighted sum of the hidden states $(h_1, h_2, \dots, h_T)$ with the coefficients $\alpha_1, \alpha_2, \dots, \alpha_T$ computed by

$$\alpha_t = \frac{\exp(a(h_t, z_{i-1}))}{\sum_k \exp(a(h_k, z_{i-1}))} \tag{3}$$

where a is a feed-forward neural network with a single hidden layer.

The whole model is jointly trained to maximize the conditional log-probability of the correct translation given a source sentence with respect to the parameters θ of the model:

$$\theta^* = \arg\max_\theta \sum_{n=1}^{N} \sum_{i=1}^{|y^n|} \log p(y_i^n | y_{<i}^n, x^n) \tag{4}$$

where (x^n, y^n) is the n-th training pair of sentences, and $|y^n|$ is the length of the n-th target sentence y^n.

Note that in this model, the dominant parts of the parameters θ are word embedding matrices and weight matrix for the output layer. All of them are closely related to representations of source and target words, therefore learning accurate parameters for them plays a critical role in searching for good NMT models.

3 Coarse-To-Fine Learning for NMT

Conceptually there are two major steps in our coarse-to-fine learning method: constructing a hierarchical cluster tree and learning a sequence of gradually refined NMT models. Figure 2 shows the overview framework of our approach. Based on bilingual data, a set of cluster hierarchies $\{H_0, \ldots, H_l\}$ is formed with increasing granularity and finally expands to the full vocabulary V. $M_0, \ldots, M_l$ are NMT models which use $H_0, \ldots, H_l$ as vocabularies at different level respectively and trained by bilingual data. The following of this section details how these two tasks are performed.

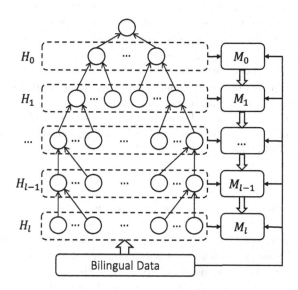

Fig. 2. The coarse-to-fine learning framework for neural machine translation.

3.1 Hierarchical Clustering

In this paper, we adopt the agglomerative hierarchical clustering algorithm to build cluster hierarchies for a given set of words.

Agglomerative hierarchical clustering algorithm works in a bottom-up manner. It starts with every word as a singleton cluster:

$$C_0 = \{a_0 = \{w_0\}, a_1 = \{w_1\}, ..., a_n = \{w_n\}\} \tag{5}$$

where C_0 is the set of initial clusters, a_i stands for cluster i, w_i denotes word in V and $n = |V|$ is the vocabulary size. Then the algorithm merges pairs of clusters step by step, until all clusters have been merged into a single cluster that contains all words. Specifically, at each step k, we have the set of clusters $C_k = \{..., a_u, ..., a_v, ...\}$. We calculate the similarity for each pair of clusters in

C_k and combine two closest clusters a_u, a_v to form a new cluster $a' = (a_u, a_v)$. The new set of clusters C_{k+1} can be represented as:

$$C_{k+1} = (C_k \setminus \{a_u, a_v\}) \cup \{a'\} \qquad (6)$$

It can be easily seen that each combination reduces the number of clusters by one. So this clustering algorithm needs n steps to finish the entire procedure in total, and we have $|C_k| = n - k$.

The similarity between two clusters is measured by the cosine metric of cluster embeddings. At first, cluster embeddings in C_0 are initialized with word embeddings, which are trained from bilingual data with an improved skip-gram model proposed by Luong et al. [17]. In the following steps, the embedding of a new cluster is computed as the average of its two sub-clusters, so embedding of every cluster can be computed in a bottom-up order.

Apparently, the clustering process described above generates too many cluster sets, and it is not necessary to use all of them. Instead, before starting NMT model training, a subset of the agglomerative hierarchical clustering results needs to be selected for actual model refinement purpose.

Concretely, $H_0, \ldots, H_l$ are selected in a way that the number of clusters will grow at a geometric rate γ. Let $n_0 = |H_0|$ be the size of initial cluster H_0, H_i can be determined by the following condition

$$H_i = C_k, \quad n_0 \gamma^i = |C_k| \qquad (7)$$

For the last cluster set H_l, as a special case, we have $H_l = C_0 = V$ while $n_0 \gamma^l \geq |C_0|$.

Note that the selected cluster sets $H_0, \ldots, H_l$ remain to be a tree structure with each cluster set H_i representing one hierarchy of the tree. For any cluster $c_p \in H_j$, there must be a parent cluster $c_q \in H_i$ satisfying $c_p \subseteq c_q$ if $j > i$.

In NMT task, the above-mentioned process is extended to support to use hierarchical clusters to refine vocabularies on both source and target side. First, we build two cluster trees, S and T for source and target words respectively, then each hierarchy of the final cluster tree is constructed by combining the corresponding hierarchy of these two cluster trees: $H_i = (S_i, T_i)$.

3.2   NMT Model Refinement

When NMT model M_{i-1} finishes training, model M_i will be learned based on the selected cluster set H_i. The learning process mostly follows the standard training procedure, but it differs from conventional NMT training in two aspects.

The first difference is the requirement for vocabulary mapping, because model M_i is expected to be trained on the vocabulary defined by H_i instead of the original vocabulary V. So a pre-processing step is needed to convert every word token in the training data into its corresponding cluster. Let (x^n, y^n) and (cx^n, cy^n) denote a word sentence pair and its cluster sentence pair respectively, and θ_i

denote the model parameters of model M_i, the objective function of NMT model training should be updated to be

$$\theta_i^* = \arg\max_{\theta_i} \sum_{n=1}^{N} \sum_{j=1}^{|y^n|} \log p(cy_j^n | cy_{<j}^n, cx^n)$$

The second difference is related to the model parameter initialization. In a conventional NMT model, all parameters are randomly initialized with some heuristics [11]. But in the coarse-to-fine learning process, only the first model M_0 is initialized in this way. All the subsequent models inherit their parameters from its previous model, that is, M_{i+1}'s parameters will be initialized with ones of M_i.

Not all parameters in model M_{i+1} can be inherited from M_i directly because their parameter structures are not fully compatible. M_{i+1} uses a larger vocabulary and thus has more parameters. Extra parameters in M_{i+1} belong to 3 categories: source word embedding, target word embedding, and weight matrix of output layer.

Our solution to this problem is to leverage the inclusion relations between clusters in H_i and H_{i+1}. The basic principle is that all sub-clusters in H_{i+1} inherit parameters of the same category from their parent cluster in H_i. Suppose $E(H_i)$ and $E(H_{i+1})$ are embedding matrices of M_i and M_{i+1}, $W_o(H_i)$ and $W_o(H_{i+1})$ denote weight matrices of output layers of M_i and M_{i+1}, and c_q is parent cluster of $\{c_{p_1}, c_{p_2}, c_{p_3}\}$. Formally, for any cluster $c_p \in H_{i+1}$, and its parent cluster $c_q \in H_i$, we have

$$E(H_{i+1})[c_p] = E(H_i)[c_q] \tag{8}$$
$$W_o^T(H_{i+1})[c_p] = W_o^T(H_i)[c_q] \tag{9}$$

Note that Eq. 8 works for both source and target clusters, while Eq. 9 is only applied to target clusters.

We notice that changing vocabulary and migrating related parameters during model transition could lead to temporary deviations in model prediction, but the deviations will be automatically fixed by later training process.

We use a validation set D to determine when to transit model learning from M_i to M_{i+1}. For each epoch during the training process, we check the perplexity change ratio ΔPPL from the last epoch: if ΔPPL is smaller than a pre-specified threshold α, the training for M_i finishes and M_{i+1} is started in the next epoch.

Algorithm 1 shows the overall training procedure. Lines 2–6 perform model initialization—except for the first model M_0, every other model is initialized with its previous model and parameter transformation function Γ defined in Eqs. 8 and 9. Word-cluster mapping is done in line 7, and lines 8–15 handle the learning of model M_i over training data T, in which α is the threshold for minimum perplexity reduction.

Algorithm 1. Coarse-To-Fine Training Algorithm for NMT

Input : Bilingual data $T = \{(x^n, y^n)\}$;
Validation set D;
Cluster hierarchies $H_0, \ldots, H_l$;
Output: A sequence of NMT models $M_0, \ldots, M_l$;

1 **for** $i \leftarrow 0$ **to** l **do**
2   **if** $i == 0$ **then**
3 | Initialize θ_0 in M_0 ;
4   **else**
5 | $\theta_i = \Gamma(\theta_{i-1}, H_{i-1}, H_i)$;
6   **end**
7   $\{(cx^n, cy^n)\} = \text{Map}(\{(x^n, y^n)\}, H_i)$;
8   **for** $e \leftarrow 0$ **to** max_epoch **do**
9     $\theta_j^e = \arg\max_{\theta_j} \sum_T \log p(cy^n | cx^n)$;
10    $ppl^e = \text{CalcPerpelxity}(D, \theta_j^e)$;
11    $\Delta PPL = \frac{ppl^{e-1} - ppl^e}{ppl^{e-1}}$;
12   **if** $\Delta PPL < \alpha$ **then**
13 | break ;
14   **end**
15 **end**
16 **end**

---

4 Experiments

4.1 Setup

We evaluate our approach on two translation tasks: Chinese-English and English-French. In all experiments, we use BLEU [20] as the automatic metric for translation quality evaluation.

Dataset. For Chinese-English translation, we select our training data from LDC collection which consists of 5.2M sentence pairs with 102.1M Chinese words and 107.7M English words respectively. NIST OpenMT 2006 evaluation set is used as validation set, and NIST 2003, NIST 2005, NIST 2008 datasets as test sets.

For English-French translation, we choose a subset of the WMT 2014 training corpus used in Jean et al. [13]. This training corpus contains 12M sentence pairs with 304M English words and 348M French words. The concatenation of news-test 2012 and news-test 2013 is used as the validation set and news-test 2014 as the test set.

For each language pair, both source and target words are grouped into a cluster hierarchy respectively with agglomerative hierarchical clustering algorithm based on word embeddings. We utilize improved skip-gram model proposed by Luong et al. [17] to train word embedding on bilingual data.

Training Setting. We limit the vocabulary to contain up to 80 K most frequent words on both the source and target side, and convert remaining words into the <unk> token. In practice, we note that some of the most frequent words such as functional words, cannot gain benefit from the coarse-to-fine learning process, so we keep the 5,000 most frequent words to be singleton clusters throughout model refinement process, and all the hierarchical clustering and cluster set selection tasks are only performed on the remaining part of the vocabulary.

We adopt the RNNSearch model proposed by Bahdanau et al. [2] as our baseline, which uses a single layer GRU for encoder and decoder. The dimension of word embedding (for both source and target words) is set to 512 and the size of hidden layer is set to 1024. The matrix and vector parameters are initialized using a normal distribution with a mean of 0 and a variance of $\sqrt{6/(d_{row} + d_{col})}$, where d_{row} and d_{col} are the number of rows and columns in the structure [11]. Each NMT model is trained on a Tesla K40m GPU and optimized with the Adadelta [35] algorithm with mini-batch size set to 80. At test time, beam search is employed to find the best translation with beam size 12 and translation probabilities normalized by the length of the candidate translations. In post-processing step, we follow the work of Luong et al. [19] to handle <unk> replacement. Other hyper-parameters used in clustering and model refinement set as $\alpha = 0.05$, $n_0 = 100$ and $\gamma = 10$. In addition, we define every 1M sentences as an epoch in coarse-to-fine training process.

4.2 Results on Chinese-English Translation

Table 1 shows the evaluation results from different models on NIST datasets, in which CTF-NMT represents our coarse-to-fine methods for NMT training. In addition, we also compare our method with sub-word models - Byte Pair Encoding (BPE) [26][1]. All the results are reported based on case-insensitive BLEU.

We can observe that CTF-NMT can bring significant improvement across different test sets. These results demonstrate that coarse-to-fine training process can learn better NMT model parameters for less-frequent words so that NMT

Table 1. Case-insensitive BLEU scores (%) on Chinese-English translation. The "Average" denotes the average results of all datasets.

System	NIST2006	NIST2003	NIST2005	NIST2008	Average
RNNSearch	36.97	39.17	38.97	29.35	36.11
RNNSearch + BPE	37.58	39.73	39.87	30.48	36.92
CTF-NMT	39.14	41.69	41.02	32.66	38.63
CTF-NMT + BPE	**39.72**	**42.20**	**42.24**	**32.90**	**39.26**

---

[1] We learn BPE models on pre-processed source and target sentences respectively with 78K merge operations.

can yield higher quality translations. Besides, our approach achieves 1.71 points BLEU improvement than RNNSearch+BPE on average. Since BPE method splits up all words to sub-word units and expects to learn better representation for similar words that share some sub-word units, there still exist plenty syntactic or semantic similar words that do not share any sub-word units, like apple and orange. Our approach uses pre-trained word embedding to better characterize relations between these words and leverage it in NMT training, thus NMT can learn better representation for similar words. Actually, our approach also can be complementary to BPE method. We apply this method in the data preprocessed by BPE method, called CTF-NMT + BPE. In this way, another 0.63 BLEU points improvement can be achieved, which adds up to 3.15 points BLEU improvement over baseline NMT model on average. This confirms the effectiveness of combining our method with sub-word models.

4.3 Results on English-French Translation

For English-French translation task, in addition to the baseline RNNSearch system, we also include results from other existing NMT systems. Experiment results are shown in Table 2. In order to be comparable with other work, all the results are reported based on case-sensitive BLEU.

First, we can see that the baseline NMT model with 80K vocabulary achieves comparable results with Jean et al. [13], which use a larger vocabulary. Also, our CTF-NMT significantly outperforms baseline NMT model with 1.34 points on test set, while achieves 0.52 points improvement than RNNSearch+BPE. When we combine our approach with BPE method, we obtain the best BLEU score 36.12 in Table 2. We believe our approach can get more improvements with deep model in future experiments.

Table 2. Case-sensitive BLEU scores (%) on English-French translation. The "PosUnk" denotes Luong et al. [19]'s technique of handling rare words. The "MRT" denotes minimum risk training proposed in Shen et al. [28]. The "LAU" represents Linear Associative Unit proposed in Wang et al. [33].

System	Architecture	Vocab Size	Test
Sutskever et al. [29]	LSTM with 4 layers	80K	30.59
Luong et al. [19]	LSTM with 6 layers + PosUnk	40K	32.70
Shen et al. [28]	Gated RNN with search + PosUnk + MRT	30K	34.23
Jean et al. [13]	Gated RNN with search + PosUnk + LV	500K	34.60
Wang et al. [33]	LAU with 4 layers	30k	35.10
Zhou et al. [37]	LSTM with 16 layers + F-F connections	30k	35.90
RNNSearch	Gated RNN with search + PosUnk	80K	34.33
RNNSearch + BPE	Gated RNN with search + BPE	80K	35.15
CTF-NMT	Gated RNN with search + PosUnk	80K	35.67
CTF-NMT + BPE	Gated RNN with search + BPE	80K	**36.12**

Figure 3 shows both perplexity and translation BLEU changes at different stages of model training for two translation tasks. To make model training with different cluster hierarchies comparable, we use word-level perplexity, which can be computed by the assumption that the probability of all words in one cluster is uniform. The BLEU is also computed at word level. We replace the generated target cluster with a word which has highest unigram probability in the cluster. From Fig. 3, it can be seen that the coarse-to-fine learning method performs consistently better (for both perplexity and BLEU) than the baseline NMT model throughout the model training process. Another observation is that, compared with the baseline system, the coarse-to-fine method needs to learn from similar amount of data (and similar training time) to achieve peak translation accuracy on validation set.

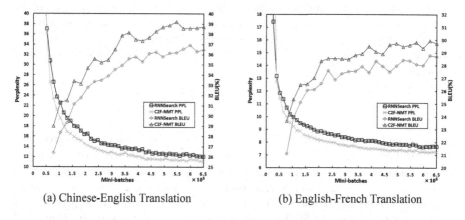

(a) Chinese-English Translation   (b) English-French Translation

Fig. 3. The perplexity (PPL) and BLEU scores on Chinese-English and English-French validation sets for RNNSearch and CTF-NMT as training progresses.

5 Related Work

This has been a long history that coarse-to-fine method is used in computer vision research, such as face detection [9] and object recognition [21]. This method has also been successfully applied to NLP tasks such as syntactic parsing [22]. Charniak et al. [4] propose a multilevel coarse-to-fine PCFG parsing algorithm, aiming at improving the efficiency of search for the best parse. Petrov et al. [23] propose a coarse-to-fine approach to statistical machine translation. They utilize an encoding-based language projection in conjunction with order-based projections to achieve speed-ups in decoding.

As a new paradigm for MT, neural machine translation has drawn more and more attention from a wide range of researchers. Resolving the OOV issue in NMT system is one of the focuses. One line of efforts [13,19] concentrated on rare words that do not exist in the system vocabulary. Jean et al. [13] explore the way based on importance sampling to directly use large vocabulary.

Luong et al. [19] propose replacement methods to handle rare words. In another direction, Costajussa et al. [8] and Sennrich et al. [26] propose character-based or subword-based neural machine translation to tackle the rare words problem. The character-based or subword-based encoding, from certain perspective, performs implicit clustering on words and affixes, and it is especially useful for morphologically rich languages such as German and Russian.

Recently, Arthur et al. [1] propose to incorporate external resources into NMT systems. Their approach employs external translation lexicons to rectify the probability distribution of rare words in the output layer. Zhang et al. [36] propose a method that leverages synthesized data to incorporate bilingual dictionaries in NMT systems, following previous work of exploiting large-scale monolingual data [5,25]. Li et al. [16] propose another method for OOV translation in NMT system: OOV words are replaced with similar in-vocabulary words during training and decoding, and the replaced words are recovered based on alignment information in decoding. Theoretically, their method can be used in invocabulary less-frequent words, but it is usually difficult to determine the set of words to be replaced, and requirement for accurate similar words brings more complexity to the training.

6   Conclusion

In this paper, we have presented a novel coarse-to-fine learning framework for neural machine translation. With the help of hierarchical clusters of words, our proposed method constructs a sequence of NMT models where each model refines its previous one. The key step is that each subsequent model inherits its model parameters according to cluster hierarchical relations, so that more precise representations can be learnt for less-frequent words in the vocabulary. Empirical evaluations are conducted in Chinese-English and English-French translation tasks on public available data sets. Experimental results demonstrate that our proposed method significantly outperforms baseline attention-based NMT model on these tasks.

In the future work, we plan to extend our approach to other NLP tasks and sequence-to-sequence models. Another direction we are interested in is to explore the possibility to leverage the coarse-to-fine method in incremental NMT model learning to speed-up the training process.

References

1. Arthur, P., Neubig, G., Nakamura, S.: Incorporating discrete translation lexicons into neural machine translation. In: EMNLP (2016)
2. Bahdanau, D., Cho, K., Bengio, Y.: Neural machine translation by jointly learning to align and translate. CoRR abs/1409.0473 (2014)
3. Brown, P.F., Pietra, S.D., Pietra, V.J.D., Mercer, R.L.: The mathematics of statistical machine translation: Parameter estimation. Computational Linguistics (1993)
4. Charniak, E., et al.: Multilevel coarse-to-fine PCFG parsing. In: HLT-NAACL (2006)

5. Cheng, Y., et al.: Semi-supervised learning for neural machine translation. In: ACL (2016)
6. Chiang, D.: Hierarchical phrase-based translation. Computational Linguistics (2007)
7. Cho, K., et al.: Learning phrase representations using RNN encoder-decoder for statistical machine translation. In: EMNLP (2014)
8. Costa-jussà, M.R., Fonollosa, J.A.R.: Character-based neural machine translation. In: ACL (2016)
9. Fleuret, F., Geman, D.: Coarse-to-fine face detection. Int. J. Comput. Vis. **41**, 85–107 (2001)
10. Gehring, J., Auli, M., Grangier, D., Yarats, D., Dauphin, Y.: Convolutional sequence to sequence learning. In: ICML (2017)
11. Glorot, X., Bengio, Y.: Understanding the difficulty of training deep feedforward neural networks. In: AISTATS (2010)
12. Hochreiter, S., Schmidhuber, J.: Long short-term memory. Neural computation (1997)
13. Jean, S., Cho, K., Memisevic, R., Bengio, Y.: On using very large target vocabulary for neural machine translation. In: ACL (2015)
14. Kalchbrenner, N., Blunsom, P.: Recurrent continuous translation models. In: EMNLP (2013)
15. Koehn, P., Och, F.J., Marcu, D.: Statistical phrase-based translation. In: HLT-NAACL (2003)
16. Li, X., Zhang, J., Zong, C.: Towards zero unknown word in neural machine translation. In: IJCAI (2016)
17. Luong, T., Pham, H., Manning, C.D.: Bilingual word representations with monolingual quality in mind. In: HLT-NAACL (2015)
18. Luong, T., Pham, H., Manning, C.D.: Effective approaches to attention-based neural machine translation. In: EMNLP (2015)
19. Luong, T., Sutskever, I., Le, Q.V., Vinyals, O., Zaremba, W.: Addressing the rare word problem in neural machine translation. In: ACL (2015)
20. Papineni, K., Roucos, S.E., Ward, T., Zhu, W.J.: BLEU: a method for automatic evaluation of machine translation. In: ACL (2002)
21. Pedersoli, M., Vedaldi, A., Gonzàlez, J., Roca, F.X.: A coarse-to-fine approach for fast deformable object detection. In: CVPR (2011)
22. Petrov, S.: Coarse-to-fine natural language processing. In: Theory and Applications of Natural Language Processing (2009)
23. Petrov, S., Haghighi, A., Klein, D.: Coarse-to-fine syntactic machine translation using language projections. In: EMNLP (2008)
24. Rush, A.M., Chopra, S., Weston, J.: A neural attention model for abstractive sentence summarization. In: EMNLP (2015)
25. Sennrich, R., Haddow, B., Birch, A.: Improving neural machine translation models with monolingual data. In: ACL (2016)
26. Sennrich, R., Haddow, B., Birch, A.: Neural machine translation of rare words with subword units. In: ACL (2016)
27. Shang, L., Lu, Z., Li, H.: Neural responding machine for short-text conversation. In: ACL (2015)
28. Shen, S., et al.: Minimum risk training for neural machine translation. In: ACL (2016)
29. Sutskever, I., Vinyals, O., Le, Q.V.: Sequence to sequence learning with neural networks. In: NIPS (2014)

30. Tu, Z., Lu, Z., Liu, Y., Liu, X., Li, H.: Modeling coverage for neural machine translation. In: ACL (2016)
31. Vaswani, A., et al.: Attention is all you need. In: NIPS (2017)
32. Vinyals, O., Kaiser, L., Koo, T., Petrov, S., Sutskever, I., Hinton, G.E.: Grammar as a foreign language. In: NIPS (2015)
33. Wang, M., Lu, Z., Zhou, J., Liu, Q.: Deep neural machine translation with linear associative unit. In: ACL (2017)
34. Wu, Y., et al.: Google's neural machine translation system: bridging the gap between human and machine translation. CoRR abs/1609.08144 (2016)
35. Zeiler, M.D.: ADADELTA: an adaptive learning rate method. CoRR abs/1212.5701 (2012)
36. Zhang, J., Zong, C.: Bridging neural machine translation and bilingual dictionaries. CoRR abs/1610.07272 (2016)
37. Zhou, J., Cao, Y., Wang, X., Li, P., Xu, W.: Deep recurrent models with fast-forward connections for neural machine translation. In: TACL (2016)

Source Segment Encoding for Neural Machine Translation

Qiang Wang[1,2(✉)], Tong Xiao[1,2], and Jingbo Zhu[1,2]

[1] Natural Language Processing Lab, Northeastern University, Shenyang, China
{xiaotong,zhujingbo}@mail.neu.edu.cn
[2] NiuTrans Inc., Shenyang, China
wangqiangneu@gmail.com

Abstract. Sequential word encoding lacks explicit representations of structural dependencies (e.g. tree, segment) over the source words in neural machine translation. Instead of using source syntax, in this paper we propose a source segment encoding (SSE) approach to modeling source segments in encoding process by two methods. One is to encode off-the-shelf n-grams of the source sentence into original source memory. The other is to jointly learn an optimal segmentation model with the translation model in an end-to-end manner without any supervision of segmentation. Experimental results show that the SSE method yields an improvement of 2.1+ BLEU points over the baselines on the Chinese-English translation task.

Keywords: Source segment encoding · Structure learning
Neural machine translation

1 Introduction

Neural machine translation (NMT) exploits an encoder-decoder framework to model the whole translation process in an end-to-end fashion, and has achieved state-of-the-art performance in many language pairs [17,19,22]. For the encoder, a popular way is to treat the source sentence as a sequence of words. In a view point of memory network [18], the encoder reads the source sentence, and then builds a source memory where each memory cell is corresponding to a source word, referred to as a *word-level cell*.

Recent studies suggested that the sequential word encoding lacks explicit representations of the structural dependencies (e.g. tree, segment) among the source words [4,7,8,13]. Many studies resort to source syntax to improve word-level representation by enhancing word embedding [13,16], guiding encoding order [4], or learning latent graph structure of the source-side [7]. Most of these works are required to prepare a good source parser in advance, which is scarce for some languages and may cause the error propagation for the downstream applications. Alternatively, [8] propose to model latent source-side segment in the attention layer of NMT. But their method slows down the system significantly.

© Springer Nature Switzerland AG 2018
M. Zhang et al. (Eds.): NLPCC 2018, LNAI 11108, pp. 329–340, 2018.
https://doi.org/10.1007/978-3-319-99495-6_28

In this paper, we develop a *source segment encoding* (SSE) approach to enhancing original word-level representation by using source segment. In form, the source segment consists of a subsequence of consecutive source words[1]. The segment has the advantage of putting more emphasis on local dependencies over the words, which is proved helpful in NMT [8]. In SSE, we propose two methods to incorporate the source segment in encoding. One is to directly encode off-the-shelf n-grams of the source sentence into the source memory, where a n-gram is equivalent to a segment. However, the size of segments explodes as n gets larger. To alleviate this problem, we present the other method which jointly learns a segmentation model with the translation model to capture an optimal segmentation of the source sentence. The segmentation model is trained end-to-end without any supervisions of segmented sentences. Afterwards, the source memory is enhanced by combining the original word-level cells with the representations of all segmentations (referred to as *segment-level cells*). In addition, our model is light and requires almost no modification of the standard decoder network. We evaluate our model on Chinese→English translation task. Experimental results on various test sets show that our model yields an average improvement of 2.1+ BLEU points over the baseline.

2   Attention-Based NMT

Given a source sentence $X = (x_1, \ldots, x_{L_s})$, and a target sentence $Y = (y_1, \ldots, y_{L_t})$, the translation probability $P(Y|X)$ can be decomposed by the chain rule:

$$P(Y|X) = \prod_{t=1}^{L_t} P(y_t|y_{<t}, X) \tag{1}$$

where $y_{<t} = (y_1, \ldots, y_{t-1})$ denotes the previous translated sequence. The NMT directly models the conditional probability as:

$$P(y_t|y_{<t}, X) = \phi(H, z_{t-1}, e'_{y_{t-1}}) \tag{2}$$

where H is the source memory of X, z_{t-1} is the target hidden state at the decoding time step $t - 1$, $e'_{y_{t-1}}$ is the target word embedding of the previous generated word, $\phi(\cdot)$ is the function of predicting the next target word.

Following the attention-based model presented in [1], we model H using a bidirectional recurrent neural network (bi-RNN) consisting of a forward RNN and a backward RNN [15] to represent a source sentence as a sequence of memory cells. More formally, $H = (h_1, \ldots, h_{L_s})$, where $h_i = [\overrightarrow{h}_i; \overleftarrow{h}_i]$ is a memory cell constructed by the concatenation of the forward annotation vector $\overrightarrow{h}_i$ and the backward annotation vector $\overleftarrow{h}_i$:

$$\begin{aligned} \overrightarrow{h}_i &= \overrightarrow{f}(e_{x_i}, \overrightarrow{h}_{i-1}) \\ \overleftarrow{h}_i &= \overleftarrow{f}(e_{x_i}, \overleftarrow{h}_{i+1}) \end{aligned} \tag{3}$$

---

[1] Actually, the basic unit can be smaller than word, e.g. subword or character. We use word as the basic unit of source language in this paper.

where $\overrightarrow{f}(\cdot)$ and $\overleftarrow{f}(\cdot)$ are two gated recurrent units (GRUs) [2], e_{x_i} is the source word embedding of word x_i.

At the decoding time step t, the function ϕ generates the distribution of next target word using a conditional GRU[2]:

$$\phi(H, z_{t-1}, e'_{y_{t-1}}) \propto g(e'_{y_{t-1}}, z_t, c_t) \tag{4}$$

where $g(\cdot)$ is a two-layer feedforward neural network, and c_t is the context vector as the source condition linking up the encoder and the decoder. A two-layer GRU is used to calculate z_t. The first GRU layer produces the intermediate state $\tilde{z}_t$ based on the input $(z_{t-1}, e'_{y_{t-1}})$, and the second GRU layer produces the current state z_t with the input $(\tilde{z}_t, c_t)$. c_t is defined as a weighted sum of each cell in H:

$$c_t = \sum_{j=1}^{L_s} a_{t,j} * h_j \tag{5}$$

where $a_{t,j}$ is the alignment weight of the j-th source word and t-th target word. $a_{t,j}$ is normalized over $(x_1, \ldots, x_{L_s})$ with a single-layer feedforward neural network r:

$$a_{t,j} = \frac{exp\{r(\tilde{z}_t, h_j)\}}{\sum_{k=1}^{L_s} exp\{r(\tilde{z}_t, h_k)\}} \tag{6}$$

3   Source Segment Encoding

For the conventional attention model in NMT (as described in Sect. 2), every memory cell h_i makes a contribution to the context vector c_t by matching its own state h_i with the decoding state $\tilde{z}_t$, which is a case of content-based addressing [6]. Intuitively, more cells can produce more substantial context due to various views provided by different cells. However, for the standard model, the source memory is built by word-level cells and the number of memory cells is limited to be equal to the source words count. On the other hand, a segment corresponds to a block of memory cells in a memory network. Introducing segment also makes sense as a segment contains more hierarchical and structural information than an independent cell. As a result, in our approach, we extend the source memory by incorporating the segment-level cells.

Given a source sentence X, the set of all segments in X is $S(X) = \{X_i^j\}$, where $X_i^j = (x_i, \ldots, x_j)$, $1 \leq i \leq j \leq L_s$. Taking computing cost into consideration, we choose a subset $\tilde{S}(X) \subset S(X)$ to delegate the set of whole segments, where $|\tilde{S}(X)| = m$. Then we define a function $\psi(\cdot)$ to encode every segment $S_k \in \tilde{S}(X)$ into a vector s_k, where $|s_k| = |h_i| = 2d$, and d is the dimension of $\overrightarrow{h}_i$. Let $H = (h_1, \ldots, h_{L_s})$ be the baseline word-level cells as described in

---

[2] We follow the implementation in *dlmt*, referred to https://github.com/nyu-dl/dl4mt-tutorial/blob/master/docs/cgru.pdf.

Sect. 2, and $S = (s_1, \ldots, s_m)$ be the segment-level cells generated by $\psi(\cdot)$, we concatenate all encodings together as the final source memory H^*:

$$
\begin{aligned}
H^* &= [H; S] \\
&= [(h_1, \ldots, h_{L_s}); (s_1, \ldots, s_m)]
\end{aligned}
\tag{7}
$$

The advantage is that the number of the cells can be arbitrary, and is not restricted to the length of the source sequence. Note that as we do not change the model parameters of the original encoding, we can reuse the baseline model when training the SSE model. Then the attention model computes the alignment weights as usual but with the new source memory $H^* = (h_1^*, \ldots, h_{L_s+m}^*)^3$:

$$
a_{t,j}^* = \frac{exp\{r(\tilde{z}_t, h_j^*)\}}{\sum_{k=1}^{L_s+m} exp\{r(\tilde{z}_t * h_k^*)\}}
\tag{8}
$$

where

$$
c_t^* = \sum_{j=1}^{L_s+m} a_{t,j}^* * h_j^*
\tag{9}
$$

In the following we describe two methods for producing $S = (s_1, \ldots, s_m)$.

3.1 N-Gram-Based SSE

In phrase-based Statistical Machine Translation [9], all source phrases limited by a max length are memorized explicitly in a big phrase table. When translating a sentence, the decoder accesses the table and then generates corresponding translations. Inspired by this, a direct way to construct $\tilde{S}(X)$ is to use a variety of off-the-shelf n-grams. More specifically, given the order of n-gram (i.e., n), we can generate all possible segments $\tilde{S}_{ng}(X) = \{x_i^j\}$ subject to $j - i + 1 \leq n$.

Next, the core problem is how to represent a segment. Instead of encoding a segment using the last hidden state of recurrent neural network [2],

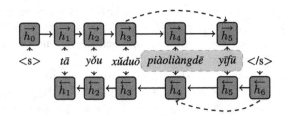

Fig. 1. An example of RNN-MINUS for encoding a segment. The dotted rounded rectangle denotes the encoded segment. <s> and </s> represent the beginning and ending of a sentence respectively, with dummy vectors.

---

[3] We share the same model parameters for aligning the word-level cells and the segment-level cells. Independent parameters may bring further improvement, which needs our further investigation.

we follow the span encoding in syntactic parsing [3,21]. Here we refer to this method as RNN-MINUS. Given the bidirectional RNN encoding $H = \{[\overrightarrow{h}_1; \overleftarrow{h}_1], \ldots, [\overrightarrow{h}_{L_s}; \overleftarrow{h}_{L_s}]\}$, RNN-MINUS encodes a segment x_i^j as:

$$\psi_{ng}(x_i^j) = [\overrightarrow{h}_j - \overrightarrow{h}_{i-1}; \overleftarrow{h}_i - \overleftarrow{h}_{j+1}] \tag{10}$$

For the beginning and ending of the sequence, we add dummy vectors $\overrightarrow{h}_0 = 0$ and $\overleftarrow{h}_{L_s+1} = 0$ to make Eq. (10) feasible. See Fig. 1 for the RNN-MINUS encoding of an example sequence.

The idea behind RNN-MINUS is simple: assuming that the information before entering a segment is I_s (i.e., information of $\{x_1, \ldots, x_{s-1}\}$), and the information after passing through this segment is I_e (i.e., information of $\{x_1, \ldots, x_e\}$), then we can regard the information offered by this segment as $I_e - I_s$. In NMT, we generally regard the hidden state of bidirectional RNN in each time step as the corresponding encoded information. Therefore, for segment x_i^j, the left-to-right information offered by the forward RNN can be represented as $\overrightarrow{h}_j - \overrightarrow{h}_{i-1}$. Likewise, the right-to-left information for x_i^j is represented as $\overleftarrow{h}_i - \overleftarrow{h}_{j+1}$.

It is worth noting that although RNN-MINUS generates encodings for each individual segment, it is still context dependent. As $[\overrightarrow{h}_i; \overleftarrow{h}_i]$ encodes the left and right contexts of the position i, the value of the subtraction of these vectors may vary in different contexts. In other words, the same segment can have different representations when the surrounding context changes. In addition, RNN-MINUS is based on the existing encoding representations with no increase in model size.

3.2   Joint-Learning-Based SSE

The n-gram-based SSE is straightforward but the size of used segments scales linearly with the order of n. In consequence, the encoding of segments consumes more memory space (especially for GPU) and slows down the system. As a result, we propose to an end-to-end joint learning of both source segmentation model and translation model, which can learn a *latent and optimal* segmentation of a source sentence rather than accessing all possible segments.

To determine a segment, we define two tags B and M for each position in the source sequence like [14], where B denotes the beginning of a segment, and M denotes the case of the middle. Then we build an identifier layer on top of the bi-RNN encoder to estimate the probability of the identity tag of each position, which can be regarded as a sequence labeling problem with two tags in each position. In this work we model this problem using a uni-directional GRU layer $Layer_{gru}$ followed by a two-layer feedforward neural network $Layer_{fnn}$. Figure 2 illustrates the network architecture of the identifier layer. The final result is a scalar denoting the probability of B for the position j, which is computed as:

$$P(B|j) = sigmoid(W_1 * o_j + b_1)$$
$$o_j = tanh(W_2 * v_j + b_2) \tag{11}$$
$$v_j = gru(h_j, v_{j-1})$$

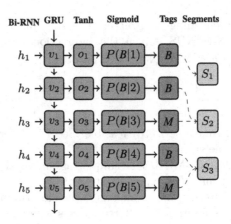

Fig. 2. An example of learning latent segments in the identifier layer. The tag of each position is induced from existing encoding representations by bi-RNN. S denotes the learned segments. In this case, our model compresses the source sequence with 5 words into 3 segments.

where $o_j \in \mathbb{R}^{d_i \times 1}$, $v_j \in \mathbb{R}^{d_i \times 1}$ are the outputs of the first layer of $Layer_{fnn}$ and $Layer_{gru}$ respectively. d_i is the dimension of the hidden states in the identifier layer. $W_1 \in \mathbb{R}^{1 \times d_i}$, $b_1 \in \mathbb{R}^1$, $W_2 \in \mathbb{R}^{d_i \times d_i}$, $b_2 \in \mathbb{R}^{d_i}$ are model parameters in $Layer_{fnn}$. $P(M|j)$ can be obtained by $1 - P(B|j)$. Then, we infer the tag for each position as follows:

$$T(j) = \begin{cases} B \ j = 1 \\ B \ j \neq 1, P(B|j) \geq P(M|j) \\ M \ j \neq 1, P(B|j) < P(M|j) \end{cases} \tag{12}$$

To prevent the sequence of illegal tags, we always assign B to the beginning position. After having the tag for each position, we can take every word sequence between two Bs as an identified segment. For example, the tag sequence $(B_1, B_2, M_3, M_4, B_5)$ contains 3 segments $(1, 1), (2, 4), (5, 5)$. Obviously, the obtained segments essentially define a segmentation of the sentence.

However, unfortunately, we cannot use RNN-MINUS directly to represent the learned segments as the increased model parameters in identifier layer are not reachable during back-propagation[4]. To learn these parameters, we follow the idea used in the local attention model [12]. We explicitly make these parameters

---

[4] These parameters control the decisions of segmentation, and are not differentiable with respect to the loss of translation, which is a similar problem to hard attention model [24]. The hard attention model picks up a determined source word to align, whereas our model chooses a determined segmentation of source sentence.

part of the translation model, and define the encoding of x_i^j with its boundary confidence, as follows:

$$\psi_{jl}(x_i^j) = [\beta \overrightarrow{h}_j - \alpha \overrightarrow{h}_{i-1}; \alpha \overleftarrow{h}_i - \beta \overleftarrow{h}_{j+1}]$$
$$\alpha = P(B|i) \tag{13}$$
$$\beta = P(M|j)$$

In this model, α and β are the left-boundary confidence and the right-boundary confidence of a segment respectively. By using the encoding method defined in Eq. 13, the increased parameters as part of the model can be learned in the standard back-propagation procedure. The boundary confidence also can be seen as a special case of dropout [10], which can alleviate the segmentation errors to some extent and improve the robustness of our segmentation model.

4   Experiments

4.1   Setup

We evaluated our proposed approach on word-based Chinese→English translation task. We used part bitext provided within NIST12 OpenMT[5] and we chose NIST 2006 (MT06) as the validation set, and 2004 (MT04), 2005 (MT05), 2008 (MT08) as the test sets. All the sentences of more than 50 words were filtered out. Data on both sides was tokenized by an in-house implement, where the Chinese-side data was segmented based on n-gram language model. The resulting training data consisted of 1.85M sentence pairs. We limited the vocabularies to the most frequent 30K words in Chinese and English. All the out-of-vocabulary words were replaced with $<UNK>$.

Table 1. Translation results (BLEU score) on Chinese→English tasks. *WC* denotes the standard encoding of word-level cells, while *SC-ngram* and *SC-joint* denotes segment-level cells using our n-gram-based SSE and joint-learning-based SSE respectively.

♯	System	Valid. MT06	Test MT04	MT05	MT08	Ave.
1	PBSMT	32.09	36.65	31.30	25.99	31.31
2	NMT baseline (WC)	36.88	43.14	36.02	29.57	36.24
3	+ SC-ngram (n=1)	38.31	44.55	37.20	30.15	37.30
4	+ SC-ngram (n=4)	38.48	44.67	37.58	30.28	37.51
5	+ SC-joint	**39.26**	**45.69**	**38.17**	**31.19**	**38.35**
6	SC-ngram (n=1)	37.24	43.65	35.73	29.56	36.31
7	SC-ngram (n=4)	38.67	44.95	37.81	30.63	37.80
8	SC-joint	20.39	23.32	18.26	14.59	18.72

---

[5] LDC2000T46,   LDC2000T47,   LDC2000T50,   LDC2003E14,   LDC2005T10, LDC2002E18, LDC2007T09, LDC2004T08.

The sizes of both source and target word embedding were set to 512. We set $d = 1024$ and $d_i = 100$ respectively. We followed [17] and used the same dropout mask at each time step, with the dropout probability of 0.1 for full words and 0.2 for other layers. We trained all the NMT models using stochastic gradient algorithm Adadelta [25] with mini-batch size of 80. The baseline NMT models were tuned for 10 epochs and then finetuned by fixing the both source and target embeddings for 10 epochs. Our models were further tuned and finetuned based on the well-tuned baseline NMT model. At test time, we employed beam search with the beam size of 12. All the translation results were evaluated by case-insensitive BLEU-4 metric using *mteval-v13a.pl*. In our experiments, we compared our method with two baselines learned on the same bilingual training data. One is the phrase-based system provided within the NiuTrans open-source SMT toolkit [23]. The other is a standard attention-based NMT system using bidirectional RNN as encoder.

5   Results and Analysis

5.1   Evaluation of Translations

Table 1 shows the BLEU scores in different settings. We can see that all the NMT systems benefit from the combination of our segment-level cells (♯3–5). It confirms that explicitly incorporating bigger linguistic units in encoding helps. This result also agrees with the findings in [4].

In particular, the best result is obtained by combining the word-level cells with the joint-learning-based SSE (♯5), which yields an average improvement of 2.1+ BLEU points than the NMT baseline (♯2). It suggests that learning the segmentation model along with the translation process jointly is effective, even without any supervisions of segmentation. The segmentation errors caused by our segmentation model do not present heavy hurt for translation performance.

Compared row 4 with row 5, the joint-learning-based SSE is more effective than simply arranging all the possible n-grams. More interestingly, we find that if we only use the segment-level cells in joint-learning-based SSE (♯8), the translation performance will decrease dramatically. The reason could be that the segment-level cells are sketchy representations of the source sentence, while the more concrete representations are contained in word-level cells.

Consider n-gram-based SSE, when $n = 1$, using the combination of word-level cells with segment-level cells (♯3) outperforms approximate 1.0 BLEU point than using segment-level cells alone (♯6). But it is worth noting that using independent segment-level cells of $n = 1$ (♯6) obtains an almost identical performance compared to the baseline (♯2). A possible explanation is that the encoding of segment with length 1 is different from conventional word-level cells. That is, the segment is represented by the subtraction of adjacent states and the resulting can put more emphasis on the meaning of the independent word. However, the explicit meaning of word is ambiguous in standard encoding procedures as the bidirectional RNN gives a global meaning in each position.

To our surprise, the superiority of combination disappears when $n = 4$ ($\natural 4$ vs. $\natural 7$). It seems that the word-level cells do not play their part and are drowned when mixed with bigger segments. It indicates that using context-sensitive local encoding is comparable to global encoding based on RNN. This finding is consistent with the result in [5]. [5] use a convolution neural network with position embedding, which can be seen as a case of context-sensitive local encoding.

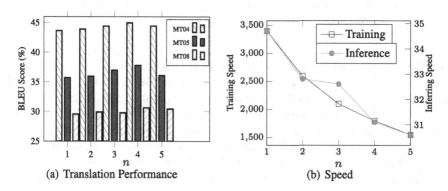

(a) Translation Performance   (b) Speed

Fig. 3. Translation performance and Speed (words/second) with different n using n-gram-based SSE in Chinese→English task. Average length of source sentence in test set for inference is 24 words.

5.2 Impact of n

Figure 3(a) shows the BLEU scores on all the test sets along with different settings of n. We only compare the segment-level cells in order to eliminate the effect of the standard word-level cells. It is obvious that BLEU is improved as n increases. The best result is achieved when $n = 4$, with sharp decreases as n grows bigger. The largest gap is 1.49 BLEU points in all test tests averagely ($n = 4$ vs. $n = 1$). It is an evidence that the translation model can generate better translation results by observing more source contexts. It is also consistent with our intuition that learning and memorizing more source segments are important in the translation process. However, when n is too large ($n > 4$ in our experiments), BLEU starts to drop sharply meaning that current model is saturated and can not benefit from more cells. We also plot the system speed as a function of n (Fig. 3(b)). As expected, a choice of larger n slows down the system. Together with the BLEU results in Fig. 3(a), it suggests that choosing n around 3 is optimal for trade-off of BLEU improvement and speed decrease.

5.3 Samples of Learned Segments

Table 2 presents three samples of the learned source segments by our joint-learning-based SSE in Chinese→English task. An interesting finding is that our

Table 2. Samples of learned source segments in Chinese→English translation task. A red rectangle denotes a segment.

#	Source Sentences			
1	zhōngguó	pàituán fùměi cǎigòu		èrshíduōyì měiyuán gāokējì shèbèi
2	míngnián	rìběn jīngjì zēngzhǎng sùdù		kěnéng fàng huǎn
3	wèishēngbù : quánguó cānyǐnyè		jiāng zhúbù shíxíng	wèishēng jiāndū gōngshì zhìdù

model appears capable of capturing the positions of subject, verb and temporal adverbial. For example, *"zhōngguó"* ("China" in English), *"rìběn jīngjì zēngzhǎng sùdù"* ("the growth rate of Japanese economic" in English) is the subjects of the 1st and 2nd samples respectively; *"cǎigòu"* ("purchase" in English), *"shíxíng"* ("implement" in English) is the verbs of the 1st and 3rd samples respectively; *"míngnián"* ("next year" in English) is the temporal adverbial of the 2nd sample. Note that the segmentation model is trained without any supervisions of segmented sentences, and learned absolutely from the translation procedures.

6   Related Work

The essence of our model is to improve the original encoder network. Apart from those syntax-enhancement methods as introduced in Sect. 1 [4,7,13,16], multi-task learning and deeper network can also enhance the representational power of the encoder.

In multi-task learning framework, the encoder network is shared in different tasks, which can benefit from joint objective function. The translation model can be trained with the source syntax parsing task in [11], or with the source reorder task in [26]. Both these methods require other external resources, such as human-annotated treebanks or source-side monolingual data, whereas our model only needs bitext data for translation model.

Deeper encoder network models have also been successfully employed in NMT. [27] introduce the fast-forward connections method to train a deep Long Short-Term Memory (LSTM) network (18 LSTM layers) as the encoder. Also, [5] propose a deep convolutional encoder with source position embedding. These models have a high cost for training and inference, whereas our model is light and easy to be implemented. It is worth noting that [5] also can be seen as a case of modeling segment in some sense due to the local filters in convolution neural network. However, our model is still based on recurrent neural network.

Another related work is [20], which apply a pre-prepared phrase table to label the source phrases and can directly generate a target phrase at one step. By contrast, our approach learns all segments without any supervision.

7   Conclusion

In this paper, we propose two simple yet effective methods to explicitly model the source segments in the encoder of attention-based NMT. In the first method, we directly encode off-the-shelf n-grams of the source sentence into source memory. In the second method, we jointly learn a segmentation model with translation model in the end-to-end manner. Both of the methods require no external resources (e.g. segmented sentences). Experimental results on the word-based Chinese-to-English translation task show that our method outperforms the baseline significantly. It is observed that using larger linguistic unit helps and gives further improvements on top of the word-based NMT system. In addition, we give an evidence that context-sensitive local encoding is comparable to global encoding based on recurrent neural network. Also, we present that the automatically learned segmentation model is sensitive to some key constituents of a sentence (e.g. subject, verb, temporal adverbial) in some cases.

References

1. Bahdanau, D., Cho, K., Bengio, Y.: Neural machine translation by jointly learning to align and translate. In: Proceedings of the 3rd International Conference on Learning Representations (2015)
2. Cho, K., van Merrienboer, B., Gulcehre, C., Bahdanau, D., Bougares, F., Schwenk, H., Bengio, Y.: Learning phrase representations using RNN encoder-decoder for statistical machine translation. In: Proceedings of the 2014 Conference on Empirical Methods in Natural Language Processing (EMNLP), pp. 1724–1734. Association for Computational Linguistics (2014)
3. Cross, J., Huang, L.: Span-based constituency parsing with a structure-label system and provably optimal dynamic oracles. In: Proceedings of the 2016 Conference on Empirical Methods in Natural Language Processing, pp. 1–11. Association for Computational Linguistics (2016)
4. Eriguchi, A., Hashimoto, K., Tsuruoka, Y.: Tree-to-sequence attentional neural machine translation. In: Proceedings of the 54th Annual Meeting of the Association for Computational Linguistics (Volume 1: Long Papers), pp. 823–833. Association for Computational Linguistics (2016)
5. Gehring, J., Auli, M., Grangier, D., Dauphin, Y.N.: A convolutional encoder model for neural machine translation. arXiv preprint arXiv:1611.02344 (2016)
6. Graves, A., Wayne, G., Danihelka, I.: Neural turing machines. arXiv preprint arXiv:1410.5401 (2014)
7. Hashimoto, K., Tsuruoka, Y.: Neural machine translation with source-side latent graph parsing. arXiv preprint arXiv:1702.02265 (2017)
8. Kim, Y., Denton, C., Hoang, L., Rush, A.M.: Structured attention networks. arXiv preprint arXiv:1702.00887 (2017)
9. Koehn, P., Och, F.J., Marcu, D.: Statistical phrase-based translation. In: Proceedings of the 2003 Conference of the North American Chapter of the Association for Computational Linguistics on Human Language Technology-Volume 1, pp. 48–54. Association for Computational Linguistics (2003)
10. Krizhevsky, A., Sutskever, I., Hinton, G.E.: ImageNet classification with deep convolutional neural networks. In: Advances in Neural Information Processing Systems, pp. 1097–1105 (2012)

11. Luong, M.T., Le, Q.V., Sutskever, I., Vinyals, O., Kaiser, L.: Multi-task sequence to sequence learning. arXiv preprint arXiv:1511.06114 (2015)
12. Luong, T., Pham, H., Manning, D.C.: Effective approaches to attention-based neural machine translation. In: Proceedings of the 2015 Conference on Empirical Methods in Natural Language Processing, pp. 1412–1421. Association for Computational Linguistics (2015)
13. Nadejde, M., Reddy, S., Sennrich, R., Dwojak, T., Junczys-Dowmunt, M., Koehn, P., Birch, A.: Syntax-aware neural machine translation using CCG. arXiv preprint arXiv:1702.01147 (2017)
14. Peng, F., Feng, F., McCallum, A.: Chinese segmentation and new word detection using conditional random fields. In: COLING 2004: Proceedings of the 20th International Conference on Computational Linguistics (2004)
15. Schuster, M., Paliwal, K.K.: Bidirectional recurrent neural networks. IEEE Trans. Sig. Process. **45**(11), 2673–2681 (1997)
16. Sennrich, R., Haddow, B.: Linguistic input features improve neural machine translation. In: Proceedings of the First Conference on Machine Translation: Volume 1, Research Papers, pp. 83–91. Association for Computational Linguistics (2016)
17. Sennrich, R., Haddow, B., Birch, A.: Edinburgh neural machine translation systems for WMT 16. In: Proceedings of the First Conference on Machine Translation: Volume 2, Shared Task Papers, pp. 371–376. Association for Computational Linguistics (2016)
18. Sukhbaatar, S., Weston, J., Fergus, R., et al.: End-to-end memory networks. In: Advances in Neural Information Processing Systems, pp. 2440–2448 (2015)
19. Sutskever, I., Vinyals, O., Le, Q.V.: Sequence to sequence learning with neural networks. In: Advances in Neural Information Processing Systems, pp. 3104–3112 (2014)
20. Tang, Y., Meng, F., Lu, Z., Li, H., Yu, P.L.: Neural machine translation with external phrase memory. arXiv preprint arXiv:1606.01792 (2016)
21. Wang, W., Chang, B.: Graph-based dependency parsing with bidirectional LSTM. In: Proceedings of ACL, vol 1, pp. 2306–2315 (2016)
22. Wu, Y., Schuster, M., Chen, Z., Le, Q.V., Norouzi, M., Macherey, W., Krikun, M., Cao, Y., Gao, Q., Macherey, K., et al.: Google's neural machine translation system: Bridging the gap between human and machine translation. arXiv preprint arXiv:1609.08144 (2016)
23. Xiao, T., Zhu, J., Zhang, H., Li, Q.: NiuTrans: an open source toolkit for phrase-based and syntax-based machine translation. In: Proceedings of the ACL 2012 System Demonstrations, pp. 19–24. Association for Computational Linguistics (2012)
24. Xu, K., Ba, J., Kiros, R., Cho, K., Courville, A.C., Salakhutdinov, R., Zemel, R.S., Bengio, Y.: Show, attend and tell: neural image caption generation with visual attention. In: ICML, vol. 14, pp. 77–81 (2015)
25. Zeiler, M.D.: Adadelta: an adaptive learning rate method. arXiv preprint arXiv:1212.5701 (2012)
26. Zhang, J., Zong, C.: Exploiting source-side monolingual data in neural machine translation. In: Proceedings of EMNLP (2016)
27. Zhou, J., Cao, Y., Wang, X., Li, P., Xu, W.: Deep recurrent models with fast-forward connections for neural machine translation. Trans. Assoc. Comput. Linguist. **4**, 371–383 (2016)

Youdao's Winning Solution to the NLPCC-2018 Task 2 Challenge: A Neural Machine Translation Approach to Chinese Grammatical Error Correction

Kai Fu[✉], Jin Huang, and Yitao Duan

NetEase Youdao Information Technology (Beijing) Co., LTD, Beijing, China
{fukai,huangjin,duan}@rd.netease.com

Abstract. The NLPCC 2018 Chinese Grammatical Error Correction (CGEC) shared task seeks the best solution to detecting and correcting grammatical errors in Chinese essays written by non-native Chinese speakers. This paper describes Youdao NLP team's approach to this challenge, which won the 1st place in the contest. Overall, we cast the problem as a machine translation task. We use a staged approach and design specific modules targeting at particular errors, including spelling, grammatical, etc. The task uses M^2 Scorer [5] to evaluate every system's performance, and our final solution achieves the highest recall and $F_{0.5}$.

Keywords: Grammatical error correction · Machine translation

1 Introduction

Chinese is the most spoken language in the world. With the growing trend in economic globalization, more and more non-native Chinese speakers are learning Chinese. However, Chinese is also one of the most ancient and complex languages in the world. It is very different from other languages in both spelling and syntactic structure. For example, unlike English or other western languages, there is no different forms of plurality and verb tenses in Chinese. Also, reiterative locution is much more common in Chinese than it is in e.g., English. Because of these differences, it is very common for non-native Chinese speakers to make grammatical errors when using Chinese. Effective Chinese Grammatical Error Correction (CGEC) systems can provide instant feedback to the learners and are of great value during the learning process.

However, there are much fewer studies on Chinese grammatical error correction compared with the study of English grammatical error correction. Relevant resources are also scarce. The NLPCC 2018 CGEC shared task provides researchers with both platforms and data to investigate the problem more thoroughly. The goal is to detect and correct grammatical errors present in Chinese essays written by non-native speakers of Mandarin Chinese. Performance is evaluated by computing the overlap between a system's output sequence and the gold standard.

© Springer Nature Switzerland AG 2018
M. Zhang et al. (Eds.): NLPCC 2018, LNAI 11108, pp. 341–350, 2018.
https://doi.org/10.1007/978-3-319-99495-6_29

Youdao's NLP team has been actively studying language learning technologies as part of the Company's greater endeavour to advance online education with AI. Through careful analysis of the problem, we tackle it using a three-stage approach: first we remove the so called "surface errors" (e.g., spelling errors, to be elaborated later) from the input. We then cast the grammatical error correction problem as a machine translation task and apply a sequence-to-sequence model. We build several models for the second stage using different configurations. Finally, those models are combined to produce the final output. With careful tuning, our system achieves the highest recall and $F_{0.5}$, ranking first in the task.

This paper describes our solution. It is organized as follows. Section 2 describes the task, as well as the corresponding data format. Section 3 describes the related research work on grammatical error correction. Section 4 illustrates how our whole system works. Section 5 presents evaluation results. We summarize in Sect. 6.

2   Chinese Grammatical Error Correction

Although Chinese Grammatical Error Diagnosis (CGED) task has been held for a few years, this is the first time correction is introduced into the challenge. The CGEC task aims at detecting and *correcting* grammatical errors in Chinese essays written by non-native Chinese speakers. The task provides annotated training data and unlabeled test data. Each participant is required to submit the corrected text on the test data.

The training data consists of sentences written by Chinese learners and corrected sentences revised by native Chinese speakers. It should be noted that there may be $0 \sim N$ corrected results for the sentences. Specifically, the distribution of original sentences and corrected sentences in the training data is shown in Table 1, and typical examples of the data are shown in Table 2.

Table 1. Overview of the training data.

Corrected Sentence	Sentences Number
0	123,500
1	299,789
2	170,261
3+	123,691
Total	717,241

The task uses M^2 Scorer [5] to evaluate every system's performance. It evaluates correction system at the phrase level in terms of correct edits, gold edits, and use these edits to calculate $F_{0.5}$ for each participant.

Table 2. Samples from training data.

Original Sentence	Corrected Sentences
我从去年3月开始学汉语	
请把我修改一下!	请帮我修改一下
他们是离婚了，所以不一起住	他们今年离婚了，所以不一起住
	他们已经离婚了，所以不住在一起。
	他们离婚了，所以不一起住

3   Related Work

Grammar Error Correction (GEC) task has been attracting more and more attention since the CoNLL 2013–2014 Shared task. Most earlier GEC systems build specific classifiers for different errors and combine these classifiers to form a hybrid system [11]. Later, some researchers begin to treat GEC as a translation problem and propose solutions based on Statistical Machine Translation (SMT) models [2]. Some achieve fairly good results with improved SMT [3]. Recently, with the development of deep learning, Neural Machine Translation (NMT) has emerged as a new paradigm in machine translation, outperforming SMT systems with great margin in terms of translation quality. Yuan and Briscoe [16] apply NMT to the GEC task. Specifically, they use a classical translation model framework: a bidirectional RNN encoder and an RNN decoder with attention. To address the issue of out of vocabulary (OOV) words, Ji [8] presents a GEC model based on hybrid NMT, combining both word and character level information. Chollampatt et al. [4] proposes using convolution neural network to better capture the local context via attention.

Until this year, studies on Chinese grammatical error problem have been focused on diagnosis, spearheaded by Chinese Grammatical Error Diagnosis shared task. Both Zheng [17] and Xie [15] treat CGED as a sequence labeling problem. Their solutions combine the traditional method of conditional random fields (CRF) and long short term memory (LSTM) network.

4   Methodology

In this paper, we regard the CGEC task as a translation problem. Specifically, we aim at letting the neural network learn the corresponding relation between wrong and corrected sentences, and translate the wrong sentence into the correct one. However, unlike in conventional machine translation task, the source sentences in GEC contain numerous types of errors. This is the nature of the GEC problem (otherwise there is no need to perform corrections). As a result, the apparent patterns in the GEC parallel corpus are far more sparse and difficult to learn. On the other hand, grammar is the higher level of abstraction of a language and there are only a few grammatical mistakes language learners tend to make. The traditional Chinese Grammatical Error Diagnosis (CGED) task deals with only four types of grammatical errors: redundant words (R), missing words (M), bad word selection (S) and disordered words (W) [17]. Therefore once the surface errors (e.g., spelling errors) are removed, it becomes relatively easier for the model to learn to identify them. We thus use a three-stage approach: a pre-processing stage aimed to remove most of the surface errors (e.g., spelling and punctuation errors), transformation stage that identifies and corrects grammatical errors and ensemble stage that combines the above two stages to generate the final output. Separating the stages allows us to use different modules targeting at their specific goals and tuned individually. This results in better overall performance.

4.1 Data Preparation

During this task, in addition to the training data NLPCC provides, we make use of two public datasets:

Language Model. Language model is commonly used in the field of grammar correction since it's able to measure the probability of a word sequence. Specifically, a grammatically correct sentence will get a higher probability in language model while a grammatically incorrect or uncommon word sequence will reduce the probability of the sentence. We use a language model as an assistant model to provide features to score the results. The model we use is a character-based 5-gram Chinese language model trained on 25 million Chinese sentences crawled from the Internet.

Similar Character Set. Since Chinese is logographic, the causes of spelling errors are quite different from languages that are alphabetical such as English. For example, Chinese characters with similar shapes or pronunciations are often confused, even for native speakers. Also, since Chinese words are typically shorter (2 to 4 characters), the usual dictionary and edit-distance based spell correction method does not perform well. To this end, we design a specific algorithm for Chinese spell correction. Specifically, we obtain Similar Shape and Similar Pronunciation Chinese character set (generally referred to as the Similar Character Set (SCS)) from the SIGHAN 2013 CSC Datasets [9,14]. The following are some sample entries in the data:

 – Similar Shape: 可， 何呵坷奇河柯苛阿倚寄崎荷椅畸啊婀蚵犄琦轲
 – Similar Pronunciation: 隔， 革格咯骼膈葛鬲蛤

We use SCS to generate candidate spell corrections and the language model to pick the most probable one.

NLPCC Data Processing. Training a machine translation model requires parallel corpus in the form of a collection of (srcSent, tgtSent) pairs where srcSent is the source sentence and tgtSent the target sentence. The NLPCC 2018 CGEC shared task provides training corpus where each sentence is accompanied with 0 or more corrected sentence(s). The original data contains about 0.71 million sentences. We process the data and generate 1.22 million (srcSent, tgtSent) pairs where `srcSent` is the sentence probably containing grammatical mistakes and `tgtSent` the corrected result. If there is no error in `srcSent`, `tgtSent` remains the same as `srcSent`. If there are multiple corrections for an incorrect sentence, multiple pairs are generated. Next, we use the character based 5-gram language model to filter out sentence pairs where the score of `srcSent` is significantly lower than that of `tgtSent`. After the data cleaning step, the data size is reduced to 0.76 million.

4.2   Spelling Error Correction

The main component in the preprocessing stage that removes most of the surface errors is a spelling correction model. For this we use a simple 5-gram language model. The probability of a character sequence W of length n is given by:

$$P(w_1, w_2, ..., w_n) = p(w_1)p(w_2|w_1) \cdots p(w_n|w_1, w_2, ..., w_{n-1}) \tag{1}$$

The perplexity of the sequence is defined as the geometric average of the inverse of the probability of the sequences:

$$PP(W) = P(w_1, w_2, ..., w_n)^{-\frac{1}{n}} \tag{2}$$

We will use $PP(W)$ as the language model score. Higher $PP(W)$ indicates less likely sentence.

To perform spelling error correction, we first divide the sentence x into characters. For each character c in x, we generate candidate substitution character set S_c using SCS. We then try to replace c in the sentence by every $c' \in S_c$. Among the sentences (including the original one) with the lowest perplexity will be selected.

4.3   Grammatical Error Correction Model

After removing the spelling errors, we treat the grammatical error correction task as a translation problem and use a Neural Machine Translation (NMT) model to correct the errors, i.e., "translating" an incorrect sentence into grammatically correct one. Recently, neural networks have shown superior performance in various NLP tasks and they have done especially well in sequence modeling such as machine translation. Generally, most neural translation models are based on the encoder-decoder paradigm, in which the encoder (a neural network) encodes the input sequence $(x_1, x_2, ..., x_n)$ into a sequence of hidden states $(h_1, h_2, ..., h_n)$ and the decoder (also a neural network) generates the output sequence $(y_1, y_2, ...y_m)$ based on hidden state. An obvious advantage of this framework is that it does not need to explicitly extract linguistic features.

There are several variants of NMT models. They can be based on Recurrent Neural Network (RNN) such as Long Short Term Memory (LSTM) or Gated Recurrent Unit (GRU) [1,10], or Convolutional Neural Network (CNN) [6,7]. The recent Transformer model is a new encoder-decoder framework based on the self-attention mechanism, proposed by Google in 2017 [13]. Transformer has shown excellent capability and achieved state of the art performance in machine translation. Thus we adopt it for our task. However, our framework is general and once a new, more advanced MT model emerges, we can easily upgrade the system with the new model.

Specifically, when the transformer reads in a sequence, it encodes it by several self-attentional iterations. Decoding is done in a similar manner. The Attention mechanism (scaled dot-product Attention) is defined as:

$$Attention(Q, K, V) = softmax(\frac{QK^T}{\sqrt{d}})V \qquad (3)$$

where Q represents the query matrix packed from individual queries, K the keys used for processing the query, and V the values to be retrieved. The transformer adopt multiple attention heads:

$$MultiHead(Q, K, V) = [head_1, ..., head_h]W^O \qquad (4)$$

where $head_i = Attention(QW_i^Q, KW_i^K, VW_i^V)$. Besides, a residual connection and hierarchical normalization are added for each attentional layer.

We used the open source tensorflow-based implementation, tensor2tensor [1], to train the transformer model. The hidden size parameter is set to 800. All the other parameters are in the default configuration.

4.4   Models Ensemble and Reranking

NMT models can be configured in different ways to suit different situations. They can be character-based or word-based. To handle rare and out-of-vocabulary words, sub-words can also be used [12]. The general understanding in the machine translation community is that sub-word models perform the best. In the case of CGEC, however, we have to deal with various errors and each may be handled using different tools or configurations. For example, spelling and character level syntax errors in Chinese are not handled well by (sub-)word level models which do a good job at correcting word level grammatical errors. Therefore we take a hybrid approach and build several models using different configurations. We then use a reranking mechanism to select among the model results the best one for each error.

We build the following 5 models, which all take spelling error correction as the first step:

M1: Spelling Checker alone
M2: Spelling Checker + Character NMT
M3: Spelling Checker + Character + Sub-word NMTs
M4: Spelling Checker + Sub-word NMT
M5: Spelling Checker + Sub-word + Character NMTs

---

[1] https://github.com/tensorflow/tensor2tensor

M3 and M5 use the same models but in different order. They may produce different results since the input to a model can be altered by the models preceding it.

For an input sentence x, each of the five models above will output a corrected result. The reranking is simply scoring them using the 5-gram language model. The pipeline of this process is shown in Fig. 1.

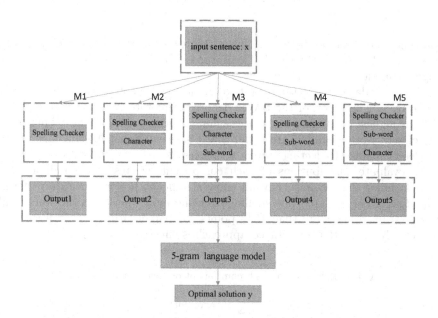

Fig. 1. Ensemble models.

5   Experiment Results

Among the 0.76 million sentence pairs generated according to the method described in Sect. 4.1, we take 3,000 pairs as the valid data set. The remaining are used for training. The validation set is mainly for parameter tuning and model selection.

The NLPCC 2018 CGEC shared task allows 3 submissions. Table 3 shows the performance of our models on the evaluation dataset for the 3 submissions. The differences among the models are mainly due to different selection strategies during the ensemble stage. Specifically, in all the cases, we use the 5-gram language model to score the outputs from the five individual models. S1 selects the result with the lowest perplexity. S2 behaves like S1 when the difference between the two lowest perplexity results is greater than a certain threshold (we set the threshold at a small value), otherwise it chooses the sentence with the *second* lowest perplexity. S3 behaves like S2 except that it makes a random selection

Table 3. Results on evaluation dataset.

Method	Annotator0			Annotator1			Annotator0,1		
	Precision	Recall	$F_{0.5}$	Precision	Recall	$F_{0.5}$	Precision	Recall	$F_{0.5}$
S1	34.17	**17.94**	**28.94**	34.30	**17.79**	**28.93**	35.24	**18.64**	**29.91**
S2	**34.18**	17.78	28.86	**34.40**	17.68	**28.93**	**35.34**	18.52	**29.91**
S3	34.16	17.73	28.82	34.33	17.60	28.85	35.28	18.45	29.83

between the outputs with the two lowest scores when the perplexity difference is less than the certain threshold. The purpose of this perturbation is to test the language model's selection capability.

Gold Standard is annotated by two annotators, denoted Annotator0 and 1 respectively. The union of the two is denoted Annotator0,1. S1 performs best, with the highest recalls and $F_{0.5}$ scores against all three annotations. S2 performs very closely. Both S1 and S2 are better than S3, showing that the language model is indeed capable of selecting correct sentences.

To evaluate contributions of each component, we test them individually. Table 4 shows the results. They all show significant performance drop if run individually. For example, the spelling checking model performs the worst and its $F_{0.5}$ score drops more than 15 % points compared with our overall system. This clearly shows that our staged approach is effective.

Table 4. Results of each component on evaluation dataset.

Method	Annotator0			Annotator1			Annotator0,1		
	Precision	Recall	$F_{0.5}$	Precision	Recall	$F_{0.5}$	Precision	Recall	$F_{0.5}$
Best model	34.17	**17.94**	**28.94**	34.30	**17.79**	**28.93**	35.24	**18.64**	**29.91**
Sub-word level NMT	30.26	10.82	22.26	30.85	10.89	22.57	31.66	11.40	23.36
Char level NMT	32.08	11.00	23.19	32.43	10.96	23.31	33.31	11.52	24.16
Spelling checker	**39.11**	4.17	14.61	**39.21**	4.11	14.49	**39.36**	4.24	14.83

6   Conclusion

This paper describes our solution to the NLPCC 2018 shared task 2. Ours is a staged approach. We first use a spelling error correction model to remove the spelling mistake. This reduces perturbation to later models and allows them to perform better. We then cast the problem into a translation task and use neural machine translation models to correct the grammatical errors. Experiments demonstrate that each stage plays a significant role. Our solution achieves the highest $F_{0.5}$ score and recall rates in all the three annotation files.

There is still plenty of room for improvement and future investigation. Due to the time constraint, we only used a simple 5-gram model for correcting spelling errors. A more sophisticated model such as neural network would perform better.

There are also techniques that we would like to try to improve the effectiveness of the 2nd stage (e.g., data augmentation). Finally, grammatical error correction is only a small initial step into advancing language learning through AI. The current solutions do not handle semantic issues well. This certainly is a challenging research direction that has great potential to change many aspects of language learning. Our goal is to build comprehensive products that could make learning more effective.

References

1. Bahdanau, D., Cho, K., Bengio, Y.: Neural machine translation by jointly learning to align and translate. arXiv preprint arXiv:1409.0473 (2014)
2. Brockett, C., Dolan, W.B., Gamon, M.: Correcting ESL errors using phrasal SMT techniques. In: Proceedings of the 21st International Conference on Computational Linguistics and the 44th annual meeting of the Association for Computational Linguistics, pp. 249–256. Association for Computational Linguistics (2006)
3. Chollampatt, S., Ng, H.T.: Connecting the dots: Towards human-level grammatical error correction. In: Proceedings of the 12th Workshop on Innovative Use of NLP for Building Educational Applications, pp. 327–333 (2017)
4. Chollampatt, S., Ng, H.T.: A multilayer convolutional encoder-decoder neural network for grammatical error correction. arXiv preprint arXiv:1801.08831 (2018)
5. Dahlmeier, D., Ng, H.T.: Better evaluation for grammatical error correction. In: Proceedings of the 2012 Conference of the North American Chapter of the Association for Computational Linguistics: Human Language Technologies, NAACL HLT 2012, pp. 568–572. Association for Computational Linguistics, Stroudsburg (2012). http://dl.acm.org/citation.cfm?id=2382029.2382118
6. Gehring, J., Auli, M., Grangier, D., Dauphin, Y.N.: A convolutional encoder model for neural machine translation. arXiv preprint arXiv:1611.02344 (2016)
7. Gehring, J., Auli, M., Grangier, D., Yarats, D., Dauphin, Y.N.: Convolutional sequence to sequence learning. arXiv preprint arXiv:1705.03122 (2017)
8. Ji, J., Wang, Q., Toutanova, K., Gong, Y., Truong, S., Gao, J.: A nested attention neural hybrid model for grammatical error correction. arXiv preprint arXiv:1707.02026 (2017)
9. Liu, C.L., Lai, M.H., Tien, K.W., Chuang, Y.H., Wu, S.H., Lee, C.Y.: Visually and phonologically similar characters in incorrect chinese words: analyses, identification, and applications. ACM Trans. Asian Lang. Inf. Process. (TALIP) 10(2), 10 (2011)
10. Luong, M.T., Pham, H., Manning, C.D.: Effective approaches to attention-based neural machine translation. arXiv preprint arXiv:1508.04025 (2015)
11. Rozovskaya, A., Chang, K.W., Sammons, M., Roth, D., Habash, N.: The Illinois-Columbia system in the CoNLL-2014 shared task. In: Proceedings of the Eighteenth Conference on Computational Natural Language Learning: Shared Task, pp. 34–42 (2014)
12. Sennrich, R., Haddow, B., Birch, A.: Neural machine translation of rare words with subword units. arXiv preprint arXiv:1508.07909 (2015)
13. Vaswani, A., Shazeer, N., Parmar, N., Uszkoreit, J., Jones, L., Gomez, A.N., Kaiser, L., Polosukhin, I.: Attention is all you need. In: Advances in Neural Information Processing Systems, pp. 6000–6010 (2017)

14. Wu, S.H., Liu, C.L., Lee, L.H.: Chinese spelling check evaluation at SIGHAN bake-off 2013. In: Proceedings of the Seventh SIGHAN Workshop on Chinese Language Processing, pp. 35–42 (2013)
15. Xie, P., et al.: Alibaba at IJCNLP-2017 task 1: Embedding grammatical features into LSTMS for Chinese grammatical error diagnosis task. In: Proceedings of the IJCNLP 2017, Shared Tasks, pp. 41–46 (2017)
16. Yuan, Z., Briscoe, T.: Grammatical error correction using neural machine translation. In: Proceedings of the 2016 Conference of the North American Chapter of the Association for Computational Linguistics: Human Language Technologies, pp. 380–386 (2016)
17. Zheng, B., Che, W., Guo, J., Liu, T.: Chinese grammatical error diagnosis with long short-term memory networks. In: Proceedings of the 3rd Workshop on Natural Language Processing Techniques for Educational Applications (NLPTEA2016), pp. 49–56 (2016)

NLP Applications

Target Extraction via Feature-Enriched Neural Networks Model

Dehong Ma, Sujian Li, and Houfeng Wang[(✉)]

MOE Key Lab of Computational Linguistics, Peking University,
Beijing 100871, China
{madehong,lisujian,wanghf}@pku.edu.cn

Abstract. Target extraction is an important task in target-based sentiment analysis, which aims at identifying the boundary of target in given
text. Previous works mainly utilize conditional random field (CRF) with
a lot of handcraft features to recognize the target. However, it is hard
to manually extract effective features to boost the performance of CRF-
based methods. In this paper, we employ gated recurrent units (GRU)
with label inference, to find valid label path for word sequence. At the
same time, we find that character-level features play important roles
in target extraction, and represent each word by concatenating word
embedding and character-level representations which are learned via
character-level GRU. Further, we capture boundary features of each word
from its context words by convolution neural networks to assist the identification of the target boundary, since the boundary of a target is highly
related to its context words. Experiments on two datasets show that our
model outperforms CRF-based approaches and demonstrate the effectiveness of features learned from character-level and context words.

1 Introduction

Target extraction is a fundamental work in the task of target-based sentiment
analysis, which tries to find all targets (e.g. entity, product...) in open corpus
like tweets, product comments, etc. For example, in the sentence *"I vote to send
Dwyane Wade to the NBA All-Star Game."*, the destination of target extracting is to identify all targets: *person: Dwyane Wade* and *organization: NBA*.
The popular approach is regarding target extraction task as a sequence labeling
problem. The goal of sequence labeling is to assign a label for each element in
the sequence, and we can use Hidden Markov Model (HMM), Max Entropy and
Conditional Random Field (CRF) to tackle sequence labeling task. Generally in
target extraction task, the label set is composed of the three symbols {*B, I, O*},
which stand for the target beginning, the target inside and non-target respectively. In the above example, the labels of the words *Dwyane Wade* and *NBA*
are *"B-PERSON, I-PERSON"* and *"B-ORGANIZATION"* respectively while
all the other words are labeled *O*.

Although CRF-based approaches [19] could achieve good results on target
extraction, they suffer from automatically extracting effective features for boosting system performance. Recently, neural network methods have exhibited their

© Springer Nature Switzerland AG 2018
M. Zhang et al. (Eds.): NLPCC 2018, LNAI 11108, pp. 353–364, 2018.
https://doi.org/10.1007/978-3-319-99495-6_30

ability of feature extraction. [26] study the effect of word embeddings and automatic feature combinations on the task by extending a CRF baseline using neural networks. [23] use recursive neural networks (RNN) to extract features, feed features to CRF and get good performance on target extraction. However, they just use neural networks as a feature extractor and do not make full use of neural networks' ability on sequence labeling. In this paper, we prefer to explore the potentials of neural networks in sequence labeling for the task of target extraction. To make use of neural networks, we take gated recurrent unit (GRU) [3] networks rather than CRF as decoder because GRU is good at modeling long distance dependency which is good for sequence labeling. As we know, there are dependencies between target labels. For example, label I will never follow label O in a sequence of labels. To avoid these illegal transitions between labels, we adopt a transition matrix [5] which measures the probability of jumping from label i to label j to ensure valid paths of labels and discourage all other paths.

In target extraction, we find that character-level features are a key factor for deciding the labels of a sequence. In the example above, the initial characters of the targets are uppercase. In addition, many words have different variants, but with a similar meaning. In such cases, characters can be used to strengthen the word representation. Further, out-of-vocabulary words are hard to be tackled because they have the same representation without distinction. But character-level representation of word could address this problem in some degree. Thus, to incorporate character-level features into our model, we use character-level GRU on word character sequence to obtain character-level representation for each word.

Although GRU can learn long distance dependency of words, the context of a word also plays an important role in target extraction. In the example above, 2-word contexts of *Dwyane* are *'to send'* and *'Wade to'*. From the left context and right context, the label of *Dwyane* tends to be B. From the context of *Wade*, we can infer that the label of *Wade* should be I. To learn context features of word, we employ convolution neural networks (CNN) on the word context.

We evaluate our model on two open-domain datasets [19], and experimental results show that our method achieves the state-of-the-art performance and validate the effectiveness of character-level features and context features.

2   Model

In this section, we first display the details of our model. After that, we introduce the details of our model. The overall architecture of our model is shown in Fig. 1.

Our model consists of three parts which are character-level layer, word-level layer and label inference layer. The character-level layer mainly learns the character-level representation for each word to assist word level layer and address the OOV words problem via stack bidirectional gated recurrent unit networks (SBi-GRU). The word-level layer has two parts which the first part also utilize SBi-GRU to learn the long distance dependencies between words and the second part learn the local feature for each word in its context. The last part finds a valid path via modeling labels dependencies with transition score matrix.

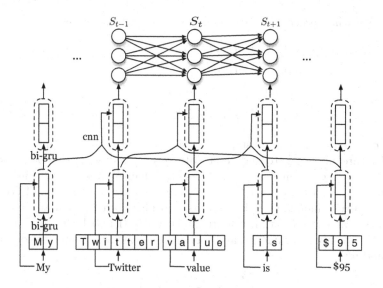

Fig. 1. The overall architecture of our model.

2.1 Embeddings

The first step of neural networks is to transform words and characters into distributed representation, which is also called embeddings [1,16–18].

In our model, word embeddings and characters embedding are used. Formally, we have a word dictionary W of size $|W|$ and a character dictionary C of size $|C|$. W and C are extracted from training data. Word and character will be transformed into corresponding real-value vector if they are in the dictionary. Otherwise, they will be assigned a unique real-value vector. We suppose that a sentence consists of n words and each word i is composed of m characters. The word embeddings of all words are $[W_1, W_2, ..., W_n]$ and the character embeddings of word i is $[C_i^1, C_i^2, ..., C_i^m]$, where $W_i \in R^{|W|*d_w}$ and $C_i^j \in R^{|C|*d_c}$. d_w and d_c are word embedding size and character embedding size. Word embeddings and character embeddings can be learned during training process or pre-trained from large corpus by language model.

2.2 Character-Level Layer

Character-level features are important to target extracting and have positive impact on out-of-vocabulary word problem. To incorporate character-level features, we adopt stack bidirectional gated recurrent units networks (SBi-GRU). The Gated Recurrent Unit (GRU) networks is an extension of recurrent neural networks (RNN) because RNN suffers from the gradient vanishing and exploration when processing long sequence, and GRU is defined by:

$$z_t = \sigma(W_z \cdot x_t + U_z \cdot h_{t-1}) \tag{1}$$

$$r_t = \sigma(W_r \cdot x_t + U_r \cdot h_{t-1}) \tag{2}$$

$$\hat{h}_t = tanh(W_h \cdot x_t + U_h \cdot (r_t \odot h_t)) \tag{3}$$

$$h_t = (1 - z_t) \odot h_{t-1} + z_t \odot \hat{h}_t \tag{4}$$

where x_t is the input at time step t, h_{t-1} is the hidden state at time step $t-1$, z_t and r_t are update gate and reset gate respectively, σ is sigmoid function, W and U are weight metrics, and $\odot$ denotes element-wise multiplication. For simplification, we use $h_t = GRU(x_t, h_{t-1})$ to denotes the definition of GRU.

For sequence modeling, it is useful to consider both forward and backward information at the same time. Bidirectional GRU (Bi-GRU) is good at learning both direction information, and forward and backward information can be computed by:

$$h_t^f = GRU(x_t, h_{t-1}^f); \tag{5}$$

$$h_t^b = GRU(x_t, h_{t-1}^b). \tag{6}$$

The bidirectional hidden state at time t is the concatenation of forward and backward hidden state, which is defined as:

$$\overline{h}_t = h_t^f \oplus h_t^b. \tag{7}$$

where $\oplus$ is a operation that concatenates two tensors alone the last dimension. For simplification, we use $\overline{h}_t = BiGRU(x_t, \overline{h}_{t-1})$ to stand for bidirectional GRU.

Stack bidirectional GRU (SBi-GRU) can learn high level abstract features for sequence. Therefore, we utilize SBi-GRU to learn effective features for each word. Assuming a stack bidirectional GRU has N layer using the same layer function, and the hidden states $\overline{h}^{(n)}$ are iteratively computed from $n = 1$ to $n = N$ by:

$$\overline{h}_t^{(n)} = BiGRU(\overline{h}_t^{(n-1)}, \overline{h}_{t-1}^{(n)}). \tag{8}$$

where $\overline{h}_t^{(x)}$ is the hidden state of layer x at time t, and $\overline{h}^{(0)}$ is the input sequences.

With the character sequence $[C_i^1, C_i^2, ..., C_i^m]$ of word i as inputs, we can obtain the stack bidirectional hidden states $[\overline{h}_1^{(N)}, \overline{h}_2^{(N)}, ..., \overline{h}_m^{(N)}]$ as the hidden states of character sequence. We can get the final character-level features c_i^r for word i by applying max-pooling or mean-pooling operation on its hidden states.

2.3   Word-Level Layer

In word-level layer, there two parts which serve to learn representation for each word. The first part is a deep architecture consisting of a SBi-GRU networks which are able to build up progressively higher level representations of sequence data. The input sequences are the concatenation of word embeddings $[W_1, W_2, ..., W_n]$ and character-level word representations $[c_1^r, c_2^r, ..., c_n^r]$. The hidden state of word i from SBi-GRU is $\overline{h}_t^{(n)}$ according to Eq. 8.

In target extracting, it is beneficial to capture local features around word. We employ the convolution neural networks (CNN) to learn local features on context window $[W_{t-\lfloor \frac{k}{2} \rfloor}, ..., W_t, ..., W_{t+\lfloor \frac{k}{2} \rfloor}]$ for word t, and k is the context window size, which is defined as:

$$c_i = f(W_f \cdot X + b_f). \tag{9}$$

where f is non-linear function, $W_f \in R^{uv}$ is filter used to produce new features, X is the concatenation of context window of word t, and b_f is bias.

The *filter* will produce a feature map $c = [c_1, c_2, ..., c_{k-u+1}]$. We then apply the max-pooling operation on c and get the local features $\hat{c} = max(c)$. The concatenation of the outputs of stack bidirectional GRU and convolution neural networks is word t's features $f(t) = \overline{h}_t^{(n)} \oplus \hat{c}$. The word features are fed into linear transformation layer with activation:

$$f_t = \sigma(W_l \cdot f(t) + b_l). \tag{10}$$

where σ is activation function, W_l and b_l is weight matrix and bias respectively. After that, we use non-linear layer to project f_t into the label space by:

$$y_t = \sigma(W_p \cdot f_t + b_p). \tag{11}$$

where W_p and b_p are weight matrix and bias respectively, and σ is non-linear function.

2.4   Label Inference Layer

In sequence labeling task, there are dependencies between labels, but word level loss discards this kind of label dependency information. To model the dependencies between labels of sequence, [5] introduce a transition score $A_{i,j}$ to measure the probability jumping from label i to label j in a successive words. For an input sequence $x = [x_i, x_2, ..., x_n]$ and its label sequence $y = [y_1, y_2, ..., y_n]$, a sentence level score is the sum of transition score and labeling probability, which is computed by:

$$S(x, y, \theta) = \sum_{t=1}^{n} (A_{i-1,i} + y_t^i). \tag{12}$$

where y_t^i is the score of i_{th} label at t_{th} word, and $\theta = [W_z, W_r, W_h, U_z, U_r, U_h, W_f, b_f, W_l, b_l, W_p, b_p, W_s, b_s]$ are the parameters of our model. We normalize this score over all paths Y via softmax by:

$$p(y|x) = \frac{exp(S(x, y, \theta))}{\sum_{\hat{y} \in Y} exp(S(x, \hat{y}, \theta))}. \tag{13}$$

The word embeddings, character-embeddings and θ will be optimized during training processing, and the training loss is computed by:

$$\text{loss}(x, y) = -\log(p(y|x)). \tag{14}$$

From the formulation above, it is evident that we encourage our model to produce a valid sequence of output labels. While decoding, we predict the output sequence that obtains the maximum score given by:

$$y^* = \arg \max_{\hat{y} \in Y}(S(x, \hat{y}, \theta)). \tag{15}$$

$y*$ can be found by dynamic algorithm like viterbi.

3   Experiments

3.1   Datasets and Evaluate Metric

To verify the effectiveness of our model, we conduct experiments on the data of Mitchell [19][1] which is composed of English and Spanish tweets annotated with entity and sentiment, and we report ten-fold cross-validation results used in [19] and [26].

In order to evaluate the performance of target extracting, we adopt f measure which is defined as:

$$F = 2 * P * R/(P + R);$$

$$P = T/N; R = T/G.$$

where t is the number of correctly predicted targets, n and g are the number of predicted targets and ground truth targets separately.

3.2   Hyperparamters Setting

In our experiments, the character embeddings of two datasets are initialized by Xavier [6]. The word embeddings of English Dataset and Spanish Dataset are from [20][2] and [4][3] respectively. All unknown characters and words, weight matrices and biases are initialized by Xavier. The hidden state size of character-level and word-level stack bidirectional GRU are set to 300 and 600 respectively, and the layer number of character-level and word-level stack bidirectional GRU are all set to 2. We use Adam [11] to optimize all parameters of our model. We also use dropout on word embeddings and character embeddings to avoid overfitting, and the dropout rate is set to 0.5.

---

[1] http://www.m-mitchell.com/code/index.html.
[2] https://nlp.stanford.edu/projects/glove/.
[3] https://spinningbytes.com/resources/embeddings/.

3.3   Model Comparison

To evaluate the effectiveness of our model comprehensively, we list some baselines for comparison. All baselines are introduced as follows.

- **Discrete** is a CRF-based approach, which incorporates many handcraft features including surface features, linguistic features, cluster features and sentiment features [26].
- **Neural** is an extension of *Discrete* with two changes. The discrete features in *Discrete* are replaced by continuous word embeddings, and a hidden neural layer is added between the inputs and outputs [26].
- **Integrated** makes a combination of discrete models and neural models by integrating both types of inputs into a same CRF framework for the reason that both features can complements each other. [26].
- **GRU** is a baseline completely based on neural networks without any handcraft features and CRF components. *GRU* utilizes gated recurrent units networks to model the long distance dependency between words. *GRU* then uses transition matrix to measure the dependencies between labels and obtain the predicted label for each word.
- **Bi-GRU** extends the *GRU* model by model the input sequence with bidirectional gated recurrent networks for both forward information and backward information which play key role in target extracting, and the other components are the same as *GRU*.

Table 1 shows the performances of our model and other baselines. All baselines can be split into two parts. Baselines in first part are based on CRF and the second part baselines take GRU incorporating label inference as decoder.

From the results of CRF-based approaches, we can see that the performance of Neural model is worst in Spanish dataset and Discrete model obtains worst results in English dataset. Neural model outperforms Discrete model about 4.83% on English dataset, while Discrete model improves the performance of Spanish dataset about 1.73% compared with Neural model. This verifies that both discrete features and word embeddings are useful for target extracting.

Table 1. Performances of baselines and our model.

Model/Dataset	English			Spanish		
	P	R	F	P	R	F
Discrete	0.5937	0.3483	0.4384	0.7077	0.4775	0.5700
Neural	0.5364	0.4487	0.4867	0.6559	0.4782	0.5527
Integrated	0.6069	0.5163	0.5567	0.7023	0.6200	0.6576
GRU	0.5649	0.3849	0.4569	0.6157	0.5045	0.5532
Bi-GRU	0.5780	0.4078	0.4772	0.6281	0.5381	0.5794
Our model	0.6245	**0.5185**	0.5658	0.6917	**0.6325**	0.6605
Ensemble	**0.6451**	0.5089	**0.5687**	**0.7201**	0.6189	**0.6654**

The Integrated approach achieves the highest result among CRF-based methods on two datasets. The great improvements of the F-measure demonstrate that it is useful to integrate both discrete and neural features into a framework because both kinds of features can complement each other.

From the performances of approaches in second part, we can observe that GRU performances better than the worst system on two datasets without any handcraft features, which demonstrates that recurrent neural networks with label inference is an alternative approach for target extracting. Bi-GRU outperforms GRU about 2.03% and 2.62% on English and Spanish datasets respectively. Compared with GRU, Bi-GRU incorporates both forward information and backward information, we can infer that bidirectional information plays great important roles in modeling the boundary features in sequence labeling. However, the performances of Bi-GRU are much worse than the Integrated model. This phenomenon implies that only bidirectional information is not enough for target extracting and more useful features should be added to Bi-GRU.

Finally, we can see that our model achieves the state-of-the-art on two datasets, which demonstrates the effectiveness of our model. This validates that character-level features and context features have great influence on target extracting. We also use an ensemble of 6 our models and improve the performance about 0.29% and 0.49% on two datasets respectively. We can also observe that ensemble model can greatly improve the precision but has negative effect on the recall compared with single model.

3.4   Model Variants

In this subsection, we design a series of variants to validate the effectiveness of our model. The first variant is *SBi-GRU* which contains a SBi-GRU and does not contain character-level stack bidirectional GRU and context CNN. The second variant is *SBi-GRU-Context* which incorporates context information of each word by CNN, and the last variant *SBi-GRU+Character* integrates character-level features into SBi-GRU via applying stack bidirectional GRU to character sequence. Table 2 shows the results of our model and its variants.

From the Table 2, we can see that SBi-GRU+Context improves the performances about 0.91% and 0.11% on two datasets compared with SBi-GRU model. This verifies that the context information of each word promotes the performance

Table 2. Performance of the variants of our model.

Model/Dataset	English			Spanish		
	P	R	F	P	R	F
SBi-GRU	0.5728	0.4152	0.4806	0.6458	0.5328	0.5837
SBi-GRU+Context	0.5682	0.4294	0.4897	0.6393	0.5391	0.5848
SBi-GRU+Character	0.5843	**0.5313**	0.5561	0.6773	0.6324	0.6538
Our model	**0.6245**	0.5185	**0.5658**	**0.6917**	**0.6325**	**0.6605**

of target extracting indeed because the boundary of target is highly related to surrounding words. We can also see that context information have positive effect on improving recall and is little harmful to the precision. But higher recall means that system can cover more existing target, good system generally has higher F-measure and similar precision and recall.

Compared with SBi-GRU and SBi-GRU+Context, SBi-GRU+Character improves both precision and recall and outperforms SBi-GRU about 7.55% and 7.01%, which demonstrates that character-level features are very important to target extracting because character-level features include morphological characteristics and grammatical features. Further, character-level features can address OOV words problems in some degree.

Our model integrates character-level features and context information into SBi-GRU and achieve the best performances. From the results, we can see that the improvements from SBi-GRU to our model are greater the accumulation of the improvements from SBi-GRU to SBi-GRU+Context and from SBi-GRU to SBi-GRU+Character (8.52% > 0.91% + 7.55% and 7.68% > 0.11% + 7.01%), and we can infer that the character-level features and context information can complement each other without negative effects in target extracting.

In a word, context information and character-level features play an important role in target extracting, and we can integrate them into SBi-GRU for better performances.

3.5 Error Cases

In this subsection, we will show some error cases in English dataset predicted by our model to show the shortages of our model. Figure 2 shows the error cases.

(1) sergio aguero greets man city fans at the etihad stadium

(2) sergio aguero greets man city fans at the etihad stadium

(3) check out my personal newspaper on the tweeted times

(4) check out my personal newspaper on the tweeted times

Fig. 2. Error cases. Red, yellow, blue and green denote begin word of Person, inside word of Person, begin word of Organization, and inside word of Organization respectively. (1) and (3) are the correct labeling sentence. (2) and (4) are labeled by our model. (Color figure online)

There are main four kinds of errors caused by our model. The first error case is **sergio aguero** which should be a person name but is recognized as organization by our model. From the first error case, we can see that our model has ability to correctly recognizes the boundary of target but does not match the true target type. The reason may be that context information is not enough and we need take the long distance information into account.

The second error case caused by our model is **man city** which should be a organization target while is ignored by our model. In fact, *man* is the abbreviation of *manchester*, and *man* is not very common to be a target word. However, *city* is often regarded as inside word of target. This error case implies that our model is not good at finding new target via obvious clues.

The third error case is **etihad stadium** which should not be a target and is labeled as an organization by our model. *etihad stadium* does not appear in train data and is an existing place, but it should not be regarded as a target. Although our model does not correctly identify this non-target, it shows our model have potential in find new words.

The final error case is **the tweeted times**, our model only recognizes *tweeted*, *times* and misses *the*. This target contains a very common used function word *the* which almost does not be regarded as a part of target. Our model may learn the information about *the* and gives the wrong label. To avoid this kind of error, we need to let our model to associate with the word collection and word context.

From four error cases above, we can observe that our model can achieve comparable results but still has some problems illustrated above. The main reason may be that our model is lack of modeling higher level features like long distance features, word collection, etc., which are key factors for target extracting.

4   Related Work

Target extracting is a fundamental task for target-based sentiment analysis [7,8]. Early works often used unsupervised approaches which rely on predefined rules and handcraft features. For example, [21] introduced an unsupervised information extraction system which mines reviews in order to build a model of important product features, their evaluation by reviewers and their relative quality across products. [14] proposed a unsupervised approach which consists of two forms of recommendations based on semantic similarity and aspect associations respectively for aspect extraction. [25] developed a model to extract aspect term via unsupervised learning of distributed representations of words and dependency paths.

As supervised learning methods, [24] solved the target extraction by introducing a phrase dependency parsing which segments an inputs sentence into phrases and links segments with directed arcs. [15] developed a centering theory to extracting explicit and implicit opinion target from new comments. [22] proposed a double propagation method which propagates information between opinion words and targets to extract target and opinion word. [13] developed a partially-supervised word alignment model to mine opinion relation and used graph-based algorithm to estimate the confidence of candidate.

Recently, Target extracting is often regarded as sequence labeling task. The sequence labeling approaches mainly focus on Hidden Markov Model (HMM) and CRF-based framework. For example, [10] proposed a lexicalized HMM-based framework to extract specific product related entities expressing reviewers' opinion. [12] proposed a CRF-based framework which employs features to extract opinion and object features for review sentence. [9] modeled opinion target extraction as information extraction task and address it by conditional random fields. [19] extracted name entities and their sentiment jointly by CRF.

With the development of neural networks (NN) approaches, NN methods also achieve good performances on target extracting. [26] developed a model which integrates discrete features and word embeddings into CRF to jointly extract target and their opinions. [23] built a joint model which integrates recursive neural networks and conditional random fields into a unified framework to explicit aspect and opinion term extraction. Although above NN approaches achieve good performance, they do not take the advantage of recurrent neural networks on sequence labeling [2]. Therefore, we use stack bidirectional GRU with label inference which integrates character-level features and context features to tackle target extracting, and experimental results on two open-domain datasets validate the effectiveness of our model.

5   Conclusion

In this paper, we propose to use stack bidirectional GRU (SBi-GRU) with label inference to address target extracting for target-based sentiment analysis. The first step of our model is to use SBi-GRU to model each word's character-level features which have great influence on target extracting. Then, SBi-GRU is also used to learn long distance feature for each word on the concatenation of character-level features and word embeddings, and convolution neural networks are adopted to capture local features around word. Finally, local features with outputs from sentence-level SBi-GRU are used to infer the target. Experiments on two datasets show the effectiveness of our model and verify the effectiveness of character-level and local features. Error cases imply the shortages of our model, in future work, we will explore how to learn global features for extracting target.

Acknowledgments. We would like to thank the anonymous reviewers for their insightful suggestions. Our work is supported by National Natural Science Foundation of China (No. 61370117). The corresponding author of this paper is Houfeng Wang.

References

1. Bengio, Y., Ducharme, R., Vincent, P., Jauvin, C.: A neural probabilistic language model. JMLR **3**, 1137–1155 (2003)
2. Chen, X., Qiu, X., Zhu, C., Liu, P., Huang, X.: Long short-term memory neural networks for Chinese word segmentation. In: EMNLP, pp. 1197–1206 (2015)
3. Cho, K., et al.: Learning phrase representations using RNN encoder-decoder for statistical machine translation. arXiv preprint arXiv:1406.1078 (2014)
4. Cieliebak, M., Deriu, J., Egger, D., Uzdilli, F.: A twitter corpus and benchmark resources for German sentiment analysis. In: SocialNLP, p. 45 (2017)
5. Collobert, R., Weston, J., Bottou, L., Karlen, M., Kavukcuoglu, K., Kuksa, P.: Natural language processing (almost) from scratch. JMLR **12**, 2493–2537 (2011)

6. Glorot, X., Bengio, Y.: Understanding the difficulty of training deep feedforward neural networks. In: AISTATS, pp. 249–256 (2010)

7. Hu, M., Liu, B.: Mining and summarizing customer reviews. In: SIGKDD, pp. 168–177. ACM (2004)

8. Hu, M., Liu, B.: Mining opinion features in customer reviews. In: AAAI, vol. 4, pp. 755–760 (2004)

9. Jakob, N., Gurevych, I.: Extracting opinion targets in a single-and cross-domain setting with conditional random fields. In: EMNLP, pp. 1035–1045 (2010)

10. Jin, W., Ho, H.H., Srihari, R.K.: A novel lexicalized hmm-based learning framework for web opinion mining. In: ICML, pp. 465–472 (2009)

11. Kingma, D., Ba, J.: Adam: a method for stochastic optimization. arXiv preprint arXiv:1412.6980 (2014)

12. Li, F., et al.: Structure-aware review mining and summarization. In: ACL, pp. 653–661 (2010)

13. Liu, K., Xu, H.L., Liu, Y., Zhao, J.: Opinion target extraction using partially-supervised word alignment model. In: IJCAI, vol. 13, pp. 2134–2140 (2013)

14. Liu, Q., Liu, B., Zhang, Y., Kim, D.S., Gao, Z.: Improving opinion aspect extraction using semantic similarity and aspect associations. In: AAAI, pp. 2986–2992 (2016)

15. Ma, T., Wan, X.: Opinion target extraction in Chinese news comments. In: Coling, pp. 782–790 (2010)

16. Mikolov, T., Chen, K., Corrado, G., Dean, J.: Efficient estimation of word representations in vector space. arXiv preprint arXiv:1301.3781 (2013)

17. Mikolov, T., Karafiát, M., Burget, L., Cernockỳ, J., Khudanpur, S.: Recurrent neural network based language model. In: Interspeech, vol. 2, p. 3 (2010)

18. Mikolov, T., Sutskever, I., Chen, K., Corrado, G.S., Dean, J.: Distributed representations of words and phrases and their compositionality. In: NIPS, pp. 3111–3119. Curran Associates, Inc. (2013)

19. Mitchell, M., Aguilar, J., Wilson, T., Van Durme, B.: Open domain targeted sentiment. In: ENMLP, pp. 1643–1654 (2013)

20. Pennington, J., Socher, R., Manning, C.: GloVe: global vectors for word representation. In: EMNLP, pp. 1532–1543 (2014)

21. Popescu, A.M., Etzioni, O.: Extracting product features and opinions from reviews. In: Kao, A., Poteet, S.R. (eds.) Natural Language Processing and Text Mining, pp. 9–28. Springer, London (2007). https://doi.org/10.1007/978-1-84628-754-1_2

22. Qiu, G., Liu, B., Bu, J., Chen, C.: Opinion word expansion and target extraction through double propagation. Comput. Linguist. $37(1)$, 9–27 (2011)

23. Wang, W., Pan, S.J., Dahlmeier, D., Xiao, X.: Recursive neural conditional random fields for aspect-based sentiment analysis. arXiv preprint arXiv:1603.06679 (2016)

24. Wu, Y., Zhang, Q., Huang, X., Wu, L.: Phrase dependency parsing for opinion mining. In: EMNLP, pp. 1533–1541 (2009)

25. Yin, Y., Wei, F., Dong, L., Xu, K., Zhang, M., Zhou, M.: Unsupervised word and dependency path embeddings for aspect term extraction. arXiv preprint arXiv:1605.07843 (2016)

26. Zhang, M., Zhang, Y., Vo, D.T.: Neural networks for open domain targeted sentiment. In: EMNLP, pp. 612–621 (2015)

A Novel Attention Based CNN Model for Emotion Intensity Prediction

Hongliang Xie, Shi Feng[(✉)], Daling Wang, and Yifei Zhang

School of Computer Science and Engineering, Northeastern University, Shenyang, China
x123872842@163.com,
{fengshi,wangdaling,zhangyifei}@cse.neu.edu.cn

Abstract. Recently, classifying sentiment polarities or emotion categories of social media text has drawn extensive attentions from both academic and industrial communities. However, limited efforts have been paid for emotion intensity prediction problem. In this paper, we propose a novel attention mechanism for CNN model that associates attention based weights for every convolution window. Furthermore, a new activation function is incorporated into the full-connected layer, which can alleviate the small gradient problem in function's saturated region. Experiment results on benchmark dataset show that our proposed model outperforms several strong baselines and achieves comparable performance with the state-of-the-art models. Unlike the reported models that used different neural network architectures for different emotion categories, our proposed model utilizes a unified architecture for intensity prediction.

Keywords: Emotion intensity prediction · CNN · Attention mechanism

1 Introduction

The social media platforms such as Facebook, Twitter and Instagram have aggregated huge amount of personal feelings and altitudes. In recent years, classifying sentiment polarities (Positive/Negative) or emotion categories (Anger/Fear/Joy/Sadness) of the user generated content has drawn extensive attentions from both academic and industrial communities. However, in text we not only convey the emotion category of our feeling, but also the intensity of that emotion. Therefore, the sentences with the same emotion category may have quite different intensities, as shown in the following example tweets.

Tweet A: *Just got back from seeing @GaryDelaney in Burslem. AMAZING!! Face still hurts from laughing so much #hilarious.*
Tweet B: *What a great training course, lots of photos, fun and laughter. Photo's will be up soon #Boostercourse #fun.*

The above two examples are selected from WASSA-2017 emotion intensity detection dataset, and both of them are associated with emotion label 'Joy'. We can easily observe that Tweet A express Joy much more intensively than Tweet B. In the original dataset, the gold intensity label of Tweet A is 0.980 while Tweet B's gold score is 0.740. The traditional emotion category classification methods could not distinguish the

© Springer Nature Switzerland AG 2018
M. Zhang et al. (Eds.): NLPCC 2018, LNAI 11108, pp. 365–377, 2018.
https://doi.org/10.1007/978-3-319-99495-6_31

intensity difference of these tweets. In this paper, we regard emotion intensity predicting as a regression problem. The goal of the algorithm is to predict real-valued score ranging from 0 to 1, which refers to the degree or amount of this emotion when given the input sentence and the emotion label of this sentence expressed. A score of 1 means that this sentence expresses the highest intensity emotion, and the score of 0 means that this sentence expresses the lowest intensity emotion.

Automatically determining the intensity of emotion felt by speaker has potential applications in commerce, public health, intelligence gathering, and social welfare [1]. The previous studies have achieved promising results for classifying the social media text according to their embedded emotions. However, limited efforts have been paid for tweet emotion intensity prediction problem, which has brought in some brand new challenges. Firstly, emotion intensity prediction is a finer-grained problem than emotion category classification and this new task needs to capture the nuances of different emotional words and provide more effective feature representation for tweets. Secondly, the length limitation of tweets results in sparseness problem in the feature space and sets up obstacles for extracting effective features. Thirdly, Twitter has an extremely large user base which leads to rich textual content, including nonstandard language, creatively spelled words (e.g. happee), and hash-tagged words (e.g. #luvumon) [1]. These informal writing styles of tweets increase the difficulties for understanding the semantics in short text.

To tackle these challenges, in this paper, we propose a novel attention mechanism for Convolutional Neural Network with a revised activation function that are suitable for the tweet emotion intensity prediction problem. Deep learning method has already shown some promising results on this topic. For example, Goel et al. leveraged the ensemble CNN and RNN models to achieve the best performance in WASSA-2017 shared task on emotion intensity [2]. However, they had to train different models for different emotion categories and the basic models they used treated all words equally in the modeling of sentences. Different from the existing methods, our proposed CNN model associates attention based weights for every convolution window. Furthermore, a new activation function in full-connected layer is proposed which can alleviate the small gradient problem in saturated region. In summary, our key contributions are as follows.

- We propose a novel attention mechanism for CNN which makes our model pay more attention to the words contributing to emotion intensity prediction.
- We introduce a novel activation function in full-connected layer for regression problem whose result is real-valued and ranges from 0 to 1.
- We conduct experiment on benchmark dataset. Experiment results show that our proposed model outperforms several strong baselines and achieves comparable performance with the state-of-the-art non-ensemble models. Unlike that reported models used different neural network architectures for different emotion categories, our proposed model utilizes a unified architecture for intensity prediction.

2   Related Work

Modeling emotion intensity has attracted more and more attentions from researchers. The WASSA-2017 EmoInt is a shared task to predict emotion intensity value of the given English tweet with emotion category label, which was held in conjunction with EMNLP-2017. And SemEval-2018 also organizes a shared task to predict emotion intensity value whose dataset has English, Arabic, and Spanish tweets. Continuously holding these shared tasks show that the academic community is gradually aware of the importance of this problem.

Methods for solving emotion intensity prediction can be divided into two categories. One is traditional feature-based methods which rely on a set of features selected from preprocessing steps. And they usually need another external resources. Mohammad et al. created a lot of features includes Word N-grams, Character N-grams, Word Embedding and many Affection Lexicons features [1]. Then they used a L2-regularized L2-loss SVM regression model to predict the emotion intensity. Köper et al. used affective norms and automatically extended resources to build features, and then utilized random forest classifier to predict the emotion intensity [3]. Duppada et al. leveraged the word embedding and emoji embedding [4] to build features, and then used many regression algorithms to predict the emotion intensity [5]. The shortage of those methods are that they rely on manually created affection lexicons and preprocessing steps.

Another category of methods are based on the deep learning technology. Goel et al. used the feed-forward neural network, multitask deep neural network and sequence modeling using CNNs and LSTMS to predict the emotion intensity of tweets [2]. And they also employed the ensemble results of those models to get the state-of-art result in the WASSA-2017 EmoInt shared tasks. John el al. fed affect clues, sentiment polarity and word embedding into the deep neural network to predict the emotion intensity [6]. Lakomkin et al. used the character-level and word-level recurrent neural network models to predict the emotion intensity, and showed that the effectiveness of using the character-level models to model the noisy and short texts [7]. The shortage of these methods are that they treat all word equally in the modeling of sentence. But when we judge the emotional intensity of sentences in reality, different words have quite different effects. Our experiments demonstrate that we can use attention mechanism to associate different words with different weights in CNN framework, and the proposed attention mechanism can indeed help improve the performances of models in predicting emotion intensity.

3   Proposed Method

Given a sentence s and its emotion label, the emotion intensity prediction is a task to get a real value ranges from 0 to 1 which is in proportion to the emotion intensity. The schematic overview of our attention based convolutional neural network is shown in Fig. 1. Given a sentence s, we use the pre-trained word embedding matrix transform it into a matrix $\mathbf{X}$. Secondly, we use a LSTM layer encodes this sentence into a vector representation $\mathbf{v}_s$ from the matrix representation $\mathbf{X}$. Then we use $\mathbf{v}_s$ and $\mathbf{X}$ to compute attention score $\mathbf{v}_a$ by using general method [8]. Each element in the $\mathbf{v}_a$ represents the attention score of each word in s. Next we can scale our word embedding in each convolutional window by the corresponding score. But to make the expectation of convolutional output unchanged, we need to apply the softmax function to the attention scores in that convolutional window to transform it into a probability distributions and multiply the score with the convolutional window size to get the genuine attention score in that convolutional window. Due to the same word in different convolutional windows having different attention scores, it will scale with different factors. So we need transform the **X** into the genuine convolutional input **Z**. After getting the genuine input, we use a CNN layer extracts the features of the sentences which denotes as $\mathbf{v_f}$. Finally, we use a fully-connected layer and a revised activation function to get the predicted emotion intensity.

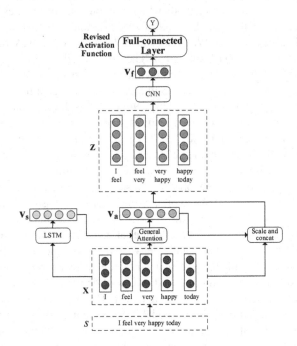

Fig. 1. Schematic overview of attention based Convolutional Neural Network

3.1   Input Representation

Given a sentence $s = (\omega_1, \omega_2, \ldots, \omega_n)$. We transform every word ω_i into a real-valued vector through by looking up work embedding matrix to provide lexical-semantic features. The shape of the word embedding matrix $\mathbf{W_v}$ is $|V| \times d_w$, where $|V|$ is the vocabulary size and d_w is the embedding size. So we map every word ω_i into a row vector $\mathbf{W_v^i} \mathbb{R}^{1 \times d_w}$.

We can get the word embedding matrix from the pretrained word vectors. A common tool for training word embedding is Word2Vec [9, 10]. This tool leverages a lot of unsupervised domain corpus as input, and gets the trained word embedding matrix for every word in the dataset. And the trained word vectors can capture a large number of precise syntactic and semantic word relationships.

Alternatively, we can use the random initialized word embedding matrix. We can fine-tune this word embedding matrix when we train CNN model. Kim demonstrates that random initialized word embeddings can also get a good results [11].

There is a big difference between the vocabulary of WASSA-2017 EmoInt training set and test set. The vocabulary size information of training set and test set is in the Table 1. If we use the random initialized word embedding matrix, many words that only appears in test set cannot be trained in the training phase, which has a bad effect on our experimental results. Therefore, we use pretrained word vectors to bridge the vocabulary gap and alleviate difficulties introduced by the informal writing styles of tweets, such as the creatively spelled words, emojis and so on. We use the public available embeddings [12] which were trained on 400 million tweets for ACL WNUT 2015 shared task.

Table 1. The vocabulary size about training set, test set and the overlap vocabulary size

Emotion	Training set	Test set	Overlap in Training & Test set
Anger	3345	3127	1129
Fear	4253	3664	1445
Joy	3263	3057	1110
Sadness	3605	3102	1189

3.2 Attention Mechanism

We propose a novel attention mechanism for CNN to make our model treat different words with different weights. The LSTM based sentence representation and the word vector is utilized to compute the weight on this word.

Sentence Representation

We use a LSTM layer [13] to encode our input matrix X to a vector v_s which is further used to compute the attention weight of each word in the sentence. The formulas are as follow:

$$i_t = \text{sigmoid}\left(W_i h_{t-1} + U_i X_t + b_i\right) \tag{1}$$

$$f_t = \text{sigmoid}\left(W_f h_{t-1} + U_f X_t + b_f\right) \tag{2}$$

$$o_t = \text{sigmoid}\left(W_o h_{t-1} + U_o X_t + b_o\right) \tag{3}$$

$$\widetilde{c_t} = \tanh\left(W_c h_{t-1} + U_c X_t + b_c\right) \tag{4}$$

$$c_t = f_t * c_{t-1} + i_t * \widetilde{c_t} \tag{5}$$

$$h_t = o_t * \tanh\left(c_t\right) \tag{6}$$

the above formula, $\mathbf{X}_t$ is the t-th word vector in the input matrix $\mathbf{X}$. The weights $\mathbf{W}_i, \mathbf{U}_i, \mathbf{W}_f, \mathbf{U}_f, \mathbf{W}_o, \mathbf{U}_o, \mathbf{W}_c, \mathbf{U}_c$ and bias $\mathbf{b}_i, \mathbf{b}_f, \mathbf{b}_o, \mathbf{b}_c$ are the parameters of LSTM layer. And $\mathbf{c}_t, \mathbf{h}_t$ are the value of cell state and hidden state at timestep t. We use the last hidden state $\mathbf{h}_n$ as the vector representation of that sentence which is also denoted as $\mathbf{v}_s$.

Attention in Convolutional Window

After get the vector representation of sentence, we need to use the sentence vector $\mathbf{v}_s$ and every word vector compute the word attention score. We use the general method [8] to compute the attention weight. In other words this process can be expressed as follow:

$$\mathbf{v}_a = \mathbf{X}\mathbf{W}\mathbf{v}_s \tag{7}$$

In the Eq. 7, we get a vector $\mathbf{v}_a \mathbb{R}^n$. Each element in the vector represents the attention score of each word in the sentence respecting with the sentence vector $\mathbf{v}_s$.

As we all know, convolutional neural networks share weight parameters through the convolution kernel. Due to this fact, we cannot apply attention weight to the convolutional weights directly. And the convolution operation is equal to the sum of each element in the tensor produced by the element-wise product of the input matrix and the convolutional weights. So we can apply the attention weights to the input matrix.

In our model, we use attention mechanism to each convolutional windows. In other words, we multiply every word vector in the convolutional windows by the corresponding attention weight. To make the expectation of the convolution result unchanged, we need to transform the weights in the vector slice corresponding to this convolutional windows to a probability distribution and then we multiply the transformed probability distribution by the convolutional window size to get the genuine attention weight vector. At last, we can multiply the word vector by the genuine attention weight to get the genuine convolution layer input. This process can be expressed as follow:

$$\alpha_i = l * \text{softmax}\left(\mathbf{v}_a[i{:}i+l]\right), i \in \{0, 1, \ldots, n-l\} \tag{8}$$

$$\mathbf{Z}_i = \left[\alpha_{i0} * \mathbf{X}_{(i+0)}, \alpha_{i1} * \mathbf{X}_{(i+1)}, \ldots, \alpha_{i(l-1)} * \mathbf{X}_{(i+l-1)}\right] \tag{9}$$

In the above equations, we transform the input matrix $\mathbf{X}$ to the genuine convolutional layer inputs $\mathbf{Z}$. Each row of the matrix $\mathbf{Z}$ is corresponded to the input of one convolution operation step. Because sentence length is n and the convolutional window size is l. So the number of rows of matrix $\mathbf{Z}$ is $n-l+1$. And $\mathbf{Z}_i$(the i-th row of the matrix $\mathbf{Z}$) is a row vector which dimension is $l \times d_w$. From the above Eq. 8, we can easily know that attention weight of the same word is different when it is in different convolutional operation step. That is the reason why should make the new matrix $\mathbf{Z}$.

CNN uses max pooling to extract the significant feature in that convolutional windows. Inspired by adding attention to word embedding [14], our proposed attention mechanism is able to scale the word embedding in the convolutional window according to their genuine attention score. That will amplify the related word embedding which is more important for predicting emotion intensity and shrink unrelated embedding. So we can extract the significant feature more easily than original CNN architecture.

Convolutional Layer

In the above section, we get the genuine convolutional layer's input matrix **Z**. Then we can use the convolution operation on this matrix. But we need to notice that we have considered the convolutional window size in the process of transforming **X** to the genuine input matrix **Z**. So the convolutional window in this convolutional filter is 1. And the input channel in our model is also 1. Then we get our convolutional filters has the shape $[l \times d_w, 1, 1, k]$. The parameter k is a hyperparameter which controls the number of features extracted for each convolution window.

We use max-pooling over the dimension corresponding to convolutional window after the convolution operation. After the pooling operation, we can get a vector which has k elements for each convolutional window size. Then we add the non-linear operation on that vector. We use the ReLU unit as the activation function [15].

3.3 Fully-Connected Layer and Activation Function

We can get an output vector of k dimensions for each convolutional window size for a sentence. So when we use t different convolutional window sizes to build the final sentence vector representation that is composed by concatenating t output vectors.

$$\mathbf{v} = [\mathbf{v}_{i_1}^T, \mathbf{v}_{i_2}^T, \ldots, \mathbf{v}_{i_t}^T]^T \tag{10}$$

The vector $\mathbf{v}_{i_t}$ is the output of convolution layer which convolutional window size is i_t. Following the convolutional layer, we use a fully-connected layer to transform the sentence vector representation to a scalar value which will be used to get the final predicted value.

One common activation function for output ranging from 0 to 1 is the sigmoid function. Its gradient in linear region is much greater than the gradient in non-linear region. This property is harmful to network learning when the input of activation function is out of its linear region. Because when the input is in the saturated region, the gradient will be very small which can easily result in the increment of parameters underflow when updating parameters and bring in difficulties during model learning. So we need a new activation function which can control its saturated region and alleviate the small gradient problem.

Inspired by the idea of ReLU function, we revised the activation function used for predicting emotion score as follow:

$$o_{max} = \max\left(0, \frac{x}{2 * gap} + 0.5\right) \tag{11}$$

$$o = \min(1, o_{max}) \tag{12}$$

In the revised function, we can use the hyperparameter gap to control the linear region of the activation function. So we can make most of input of the activation function in its linear region. That will give efficient gradient to the former layer which is beneficial to learning parameters in the former layer.

4   Experiments

4.1   Experimental Setup

Dataset and Metrics

We conduct our experiments on the dataset of WASSA-2017 shared task on emotion intensity (EmoInt). The number of instances in this dataset is shown in Table 2. This dataset is composed of four emotion sub-dataset. Each sub-dataset is composed of the samples expressing same emotion.

Table 2. The number of instances in the experimental dataset

Emotion	Training set	Development set	Test set
Anger	857	84	760
Fear	1147	110	995
Joy	823	79	714
Sadness	786	74	673

We evaluate the models using the Spearman Rank Coefficient of the predicted score with the gold score of data and the Pearson Correlation Coefficient of the predicted score with the gold ratings. The correlation scores across all four emotions averaged to get the final coefficient score.

Settings

We use the pretrained word2vec [12] to initialize the word embedding matrix, and keep the word embedding fixed when the CNN model is training. Other weight parameters are initialized using the default Tensorflow initializer. We combine the training set and development set using the 5 fold cross validation on all emotion sub-datasets. And to reduce the influences of the random factor, we run every model 3 times in every cross-validation.

The result of test set is produced by the model that achieved the best result on validation set. But to compare with the result reported by Goel et al. [2], the result of test set in the Sect. 4.3 is produced by training on the dataset which is composed of merging original training dataset and validation dataset with determined number of epochs. Because of this, the result of our proposed model CNN-ATT-RA reported in Table 5 is better than the results reported in Tables 3 and 6.

Table 3. Pearson correlations of emotion intensity predictions with gold score on test dataset

Model	Average	Anger	Fear	Joy	Sadness
CNN	69.39	67.24	71.98	65.57	72.77
LSTM	47.10	41.56	45.75	50.92	50.18
SVM	64.82	63.90	65.20	65.40	64.80
CNN-ATT-RA	**70.01**	**68.06**	**72.62**	**66.32**	**73.05**

The network parameters are learned by minimizing the Mean Absolute Error between the actual and predicted values of emotion intensity. We optimize this loss function by Adam optimization algorithm [16] with the default parameters. And the number of convolutional filters k is set 200 for all model. The convolutional window size are set as 2, 3 and 4. The hidden vector size of LSTM is 256. The hyperparameter *gap* for all emotions are set as 4 except *sadness* which is 10. All hyperparameters are selected by cross-validation.

4.2 Compared with Baselines

To illustrate the performance boost of our proposed attention model and the revised activation function, we compare our model with several strong baseline methods which includes the basic neural network architecture we use in our proposed model.

- **CNN:** Convolutional Neural Network using the pretrained word embeddings.
- **LSTM:** Using LSTM units to encode the sentence.
- **SVM:** Features includes that constructed by a lot of affection lexicon and many other common text features used by emotion classifier.
- **CNN-ATT-RA:** our proposed model.

The experimental results on all sub-dataset are shown in Tables 3 and 4.

Table 4. Spearman coefficient of emotion intensity predictions with gold score on test dataset

Model	Average	Anger	Fear	Joy	Sadness
CNN	68.38	65.61	69.89	65.61	72.40
LSTM	45.63	39.07	43.02	50.47	49.94
CNN-ATT-RA	**69.23**	**66.73**	**70.76**	**66.25**	**73.20**

In the above results, we observe that our proposed model which uses attention weight in every convolutional window and revised activation function can improve both Pearson correlation and Spearman coefficient in all emotion dataset. Specifically, our proposed method outperforms the basic CNN with a relative 0.62% in average Pearson correlation and with a relative 0.85% in average Spearman coefficient. Our proposed method outperforms the basic LSTM with a relative 22.91% in average Pearson correlation and with a relative 23.60% in average Spearman coefficient. Moreover, those results are obtained by the fine-tuned network. And all of results reported above are average of results of 5-fold cross validation and the result of each cross validation are the average of results obtained by repeatedly running models 3 time. That ensure the improvement of results are indeed benefit from the attention mechanism and the revised activation function.

The results obtained by SVM is reported in [2]. And there is no results on the Spearman coefficient in that paper.

We can observe that the results obtained by LSTM is very poor. That may be caused by lacking of training data. Therefore, the basic LSTM model cannot learn particularly effective features that can make our prediction precisely.

4.3   Compared with Results of WASSA-2017 EmoInt

To demonstrate the effective of our models, we also compare our proposed method with the reported models in WASSA-2017 EmoInt shared task.

The Pearson correlation results obtained in test dataset is shown in Table 5. Because those models doesn't report the Spearman coefficient results on their paper and results with Spearman coefficient is largely inline with those obtained using Pearson correlation, we only compare the results on the Pearson correlation.

Table 5. Pearson correlations of emotion intensity predictions with gold scores on test dataset

Approach	Average	Anger	Fear	Joy	Sadness
FFNN	69.58	67.88	72.42	68.26	69.77
Multi-DL	66.20	64.49	67.74	65.37	67.22
CNN + LSTM	**71.79**	70.15	**72.95**	69.14	74.93
Lexicons(BL)	65.00	65.00	66.00	60.00	70.00
Ext.Twitter + BL	70.00	68.00	72.00	66.00	74.00
CNN-LSTM + BL	70.00	69.00	69.00	67.00	**76.00**
Seernet	71.52	67.84	70.52	**72.81**	74.89
CNN-ATT-RA	71.76	**71.64**	72.82	69.24	73.34

In the above results, the model FFNN, Multi-DL and CNN + LSTM are the models used by the best performing system [2]. The model Lexicons, Ext.Twitter + BL and CNN-LSTM + BL are models used by the system ranked second in the shared task [3]. The model Seernet [5] are an ensembles models which ranked third in the shared task.

As the experiment settings reported on [2], we train the model on the dataset which is composed of the merger of training dataset and validation dataset when we use the model to predict the emotional intensity of samples in the test dataset. The number of epochs for each emotion is the average of the epochs which achieved best performance in the validation dataset.

Our proposed model achieves competitive performance with the best single model of the best system (CNN + LSTM) [2]. CNN + LSTM used different architecture for different emotion sub-datasets and the architecture was selected by validation dataset. Different from that, our proposed model use a unified architecture for all emotion dataset. And our proposed model also performs better than the single model (CNN-LSTM + BL, Ext.Twitter + BL) which employed many manually constructed features.

Our proposed model achieves better performance than the ensemble model Seernet [5] which ranks third in the shared task. As we can observe in the results, our proposed model achieves relatively poor results in *joy* than the result achieved by Seernet. That may be caused by predicting emotion intensity of *joy* is harder than the other emotions and the number of training data is not enough. So the ensemble methods can get better performance. The fact that all single model achieves relative poor result than the ensemble model Seernet in *joy* sub-dataset can support our view.

4.4   Ablation Experiments

To better quantify the contribution of the attention mechanism and the revised activation function, we conduct an ablation experiments in our model as follows.

- **CNN-ATT:** Convolutional neural network with our attention mechanism.
- **CNN-RA:** Convolutional neural network with our revised activation function.

The results of the ablation experiments are shown in Tables 6 and 7.

Table 6. Pearson correlation of ablation experiment on test dataset

Approach	Average	Anger	Fear	Joy	Sadness
CNN	69.39	67.24	71.98	65.57	72.77
CNN-ATT	69.69	**68.57**	72.15	65.82	72.23
CNN-RA	69.24	66.96	71.84	65.49	72.67
CNN-ATT-RA	**70.01**	68.06	**72.62**	**66.32**	**73.05**

Table 7. Spearman coefficient of ablation experiment on test dataset

Approach	Average	Anger	Fear	Joy	Sadness
CNN	68.38	65.61	69.89	65.61	72.40
CNN-ATT	68.66	**66.75**	69.95	65.94	72.01
CNN-RA	68.36	65.68	69.80	65.48	72.51
CNN-ATT-RA	**69.23**	66.73	**70.96**	**66.25**	**73.20**

As the results show, our proposed attention mechanism can achieve better performances in most of emotion dataset. And if the revised activation function is used alone, it may cause a slight decrease in performance. But if we combine the revised activation function with the attention mechanism, the revised activation function can improve the performance of the attention mechanism.

The reason why our revised activation function can improve the performance of our proposed attention mechanism is that we use a LSTM layer to encode the sentence which is used to compute the attention score. The neural network is distressed by the gradient vanish. If the input of sigmoid unit is out of its linear region, their gradients are much smaller than gradients at the linear region. Using our revised activation function in the output unit, we can control the range of linear region. So we can propagate bigger gradient to the LSTM unit, which can alleviate the gradient vanish problem in the network. This mechanism makes the LSTM unit to encode the sentence more precisely and brings in more accurate attention scores.

5   Conclusion

In this paper, we propose an attention mechanism which can improve the results for emotion intensity prediction problem. We also present a revised activation function used in the output unit which can make our attention mechanism work better through controlling linear region of activation function. The experiment results obtained by our proposed model are better than several baselines and the ensemble system which rank third in the shared task. And our propose model can achieve competitive performance with the best performance single model in the task. Unlike the single model that selected different neural architectures for different emotion sub-datasets, our proposed model use a unified architecture for all emotion sub-datasets.

Acknowledgements. The work was supported by the National Key R&D Program of China under grant 2018YFB1004700, and National Natural Science Foundation of China (61772122, 61402091).

References

1. Mohammad, S.M., Bravo-Marquez, F.: WASSA-2017 shared task on emotion intensity. arXiv preprint arXiv:1708.03700 (2017)
2. Goel, P., Kulshreshtha, D.: Prayas at EmoInt 2017: an ensemble of deep neural architectures for emotion intensity prediction in tweets. In: Proceedings of the 8th Workshop on Computational Approaches to Subjectivity, Sentiment and Social Media Analysis, pp. 58–65 (2017)
3. Köper, M., Kim, E.: IMS at EmoInt-2017: emotion intensity prediction with affective norms, automatically extended resources and deep learning. In: Proceedings of the 8th Workshop on Computational Approaches to Subjectivity, Sentiment and Social Media Analysis. pp. 50–57 (2017)
4. Eisner, B., Rocktäschel, T.: emoji2vec: Learning emoji representations from their description. arXiv preprint arXiv:1609.08359 (2016)
5. Duppada, V., Hiray, S.: Seernet at EmoInt-2017: tweet emotion intensity estimator. arXiv preprint arXiv:1708.06185 (2017)
6. John, V., Vechtomova, O.: UWat-Emote at EmoInt-2017: emotion intensity detection using affect clues, sentiment polarity and word embeddings. In: Proceedings of the 8th Workshop on Computational Approaches to Subjectivity, Sentiment and Social Media Analysis, pp. 249–254 (2017)
7. Lakomkin, E., Bothe, C.: GradAscent at EmoInt-2017: character- and word-level recurrent neural network models for tweet emotion intensity detection. arXiv preprint arXiv: 1803.11509 (2018)
8. Luong, M.T., Pham, H.: Effective approaches to attention-based neural machine translation. arXiv preprint arXiv:1508.04025 (2015)
9. Mikolov, T., Chen, K.: Efficient estimation of word representations in vector space. arXiv preprint arXiv:1301.3781 (2013)
10. Le, Q., Mikolov, T.: Distributed representations of sentences and documents. In: International Conference on Machine Learning, pp. 1188–1196 (2014)
11. Kim, Y.: Convolutional neural networks for sentence classification. arXiv preprint arXiv: 1408.5882 (2014)
12. Godin, F., Vandersmissen, B.: Multimedia Lab @ ACL W-NUT NER shared task: named entity recognition for Twitter microposts using distributed word representations. In: Proceedings of the Workshop on Noisy User-generated Text, pp. 146–153 (2015)
13. Hochreiter, S., Schmidhuber, J.: Long short-term memory. In: Neural Computation, pp. 1735–1780 (1997)

14. Wang, L., Cao, Z.: Relation classification via multi-level attention CNNs. In: Proceedings of the 54th Annual Meeting of the Association for Computational Linguistics (2016)
15. Nair, V., Hinton, G.E.: Rectified linear units improve restricted boltzmann machines. In Proceedings of the 27th International Conference on Machine Learning (2010)
16. Kingma, D.P., Ba, J.: Adam: a method for stochastic optimization. arXiv preprint arXiv: 1412.6980 (2014)

Recurrent Neural CRF for Aspect Term Extraction with Dependency Transmission

Lindong Guo[1], Shengyi Jiang[1,2(✉)], Wenjing Du[1], and Suifu Gan[1]

[1] School of Information Science and Technology,
Guangdong University of Foreign Studies, Guangzhou, China
1216920263@qq.com, 515056384@qq.com, 657742829@qq.com,
jiangshengyi@163.com
[2] Engineering Research Center for Cyberspace Content Security of Guangdong
Province, Guangzhou, China

Abstract. This paper presents a novel neural architecture for aspect term extraction in fine-grained sentiment computing area. In addition to amalgamating sequential features (character embedding, word embedding and POS tagging information), we train an end-to-end Recurrent Neural Networks (RNNs) with meticulously designed dependency transmission between recurrent units, thereby making it possible to learn structural syntactic phenomena. The experimental results show that incorporating these shallow semantic features improves aspect term extraction performance compared to a system that uses no linguistic information, demonstrating the utility of morphological information and syntactic structures for capturing the affinity between aspect words and their contexts.

Keywords: Aspect term extraction · Dependency transmission
Recurrent neural networks · CRF

1 Introduction

Aspect term extraction [1, 2], also generally named opinion target extraction [3, 4] in some literature, is aimed to identify objects commented in subjective texts. For instance, in a product review of a phone "an average phone with a great screen, but poor battery life", the review is targeted at the phone's "screen" and "battery life", which are aspect terms expected to be extracted. Aspect term extraction is an important prerequisite for fine-grained sentiment analysis. However, sentiment analysis has been based on sentence or paragraph level for years [5, 6], where much accurate information and different opinions towards distinct targets could be missed. To overcome such limitation, aspect-based sentiment analysis has become the focus of a growing number of research.

There are two types of aspects defined in aspect-based sentiment analysis task: explicit aspects and implicit aspects [7]. Explicit aspects are words which appear explicitly in opinioned sentences. In the above example, the opinion targets, "screen" and "battery life" explicitly mentioned in the text are the explicit aspects. On the contrary, implicit aspects are the targets that are not explicitly mentioned in the text but

© Springer Nature Switzerland AG 2018
M. Zhang et al. (Eds.): NLPCC 2018, LNAI 11108, pp. 378–390, 2018.
https://doi.org/10.1007/978-3-319-99495-6_32

can be inferred from the context or opinion words. We can see from the following sentence "My phone is shine but expensive" the appearance and price of the phone are implicit aspects which can be deduced from the opinion words "shine" (corresponds to the appearance of the phone) and "expensive" (corresponds to the price of the phone).

In this paper, we focus on the explicit aspect term extraction task. We propose BiLSTM-DT (Bidirectional LSTM with Dependency Transmission), a novel neural network architecture combining the ability of learning long-term sequential dependencies of LSTM [8] and the guidance of syntactical structural priors provided by dependency transmission. The network takes as input a variable length of embeddings, which consist of character-level embedding, word-level embedding and POS embedding. Specifically, the character-level embedding is the concatenation of the two final state outputs of a bidirectional Recurrent Neural Networks running over a character stream. The word-level embedding transforms word tokens to word vectors given a pre-trained or randomly initialized word embedding lookup table. And the POS embedding is like the word-level embedding, but it is always initialized randomly at the beginning of the training phrase and it provides beneficial lexicological information which is lacking in word-level embedding. The three types of embeddings are then fed into a bidirectional LSTM network with well-designed dependency transmission between recurrent units. To ensure that the network learns label dependencies, a CRF layer is added as the final output to restrict the decoding process. Experimental results on publicly available datasets show that our proposed model achieves new state-of-the-art performance.

2 Related Work

Various approaches have been proposed to tackle aspect term extraction. These approaches can be roughly classified into unsupervised and supervised ones. For unsupervised approaches, most of them are based on statistics or linguistic rules. With the assumption that aspects of products are mostly nouns or noun phrases, Hu and Liu [7] firstly proposed an approach where explicit aspects, with high frequency in corpus, were extracted by association mining and implicit aspects were also detected by minimum distance from opinion words. Though easy to implement, the approach probably obtains low precision because it is vulnerable to frequent noises. For instance, daily expressions, usually with high frequency, could be mistakenly recognized as explicit aspects. Later, point-wise mutual information (PMI) was introduced by Popescu and Etzioni [9]. The precision was improved by computing PMI value between candidate aspect and some entity-related meronymy discriminators. However, the algorithm needs to collect product category expressions in advance since the category indicators are used to compute PMI scores, and at the same time, the computing of PMI score needs the help of search engine, which will be a time-consuming process. Scaffidi et al. [10] proposed an approach based on language model under the assumption that product aspects are more frequently mentioned in product reviews than in general English texts. However, this approach based on statistical model resulted in favorable performance for frequent aspects but instability for infrequent aspects. Different from the above frequency-based methods, syntactic relations between aspects and opinion words for aspect term extraction are exploited in follow-up studies [11–13]. Given a small number of seed opinion words, Qiu et al. [12] proposed an algorithm called Double Propagation (DP) that iteratively expands opinion words and extracts aspects simultaneously with pre-defined syntactic rules from dependency parse trees. However, the rules are often targeted at a specific domain, and often encounter problems such as matching order conflict.

On the other hand, aspect term extraction can be regarded as a sequence labeling problem and many supervised methods such as HMMs-based [14] and CRFs-based [15–17] approaches are developed to solve this problem. Jakob et al. [15] conducted experiments on four different domains (movies, web-services, cars and cameras), in which the CRF model was trained with features defined as token, POS tag, short dependency path, word distance and opinion sentence. Conditional Random Fields based methods, have no problems as the above rules based method, but they also rely on a large number of manual process of feature engineering. These features are decisive for the extraction performance.

Recent studies have found that deep neural networks have the ability to automatically learn feature representations. Therefore, the deep learning based aspect term extraction has become an important research direction in this field. Liu et al. [18] proposed to employ different types of Recurrent Neural Networks to extract aspects and showed that fine-tuning RNNs outperform feature-rich CRF models without any task-specific manual features. However, the proposed method simply employed RNNs in conjunction with word embeddings, and hence many linguistic constraints cannot be learned to deliver beneficial information. Wang et al. [19] proposed an approach combining Dependency-Tree Recursive Neural Networks with CRF for aspect terms and opinion words co-extraction, in which syntactic dependencies and semantic robustness are considered. Our method is inspired by this one, however, we differ in the way we incorporate syntactic relations into neural networks. Instead of simply taking recursive neural networks, we adopt Recurrent Neural Networks with carefully-designed dependency transmission, which are able to take into account the syntactic structural priors while simultaneously reserve the naturally sequential context. Yin et al. [20] used RNNs to learn distributed representation of dependency paths and then fed the learned dependency path embeddings as one of features into CRF model to extract aspect terms. Essentially, we view this method as a CRF-based model since the training of embeddings and the extraction model are segregated, and the modeling capacity of neural networks is not made use of in the supervised phrase.

3   Method

3.1   Overview

In this paper, we investigate the problem of aspect term extraction in opinion mining, as a sequence labeling task. Our model consists of three components: an embedding component including character-level embeddings, word embeddings, and POS tagging embeddings, which capture lexicological and morphological features; a bidirectional LSTM layer with dependency transmission capturing contextual and syntactic correlations among words; a CRF layer leveraging label information to make valid predictions. The main architecture of the full model is shown in Fig. 1.

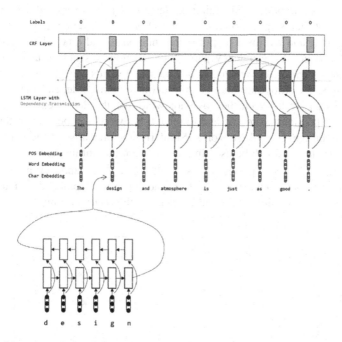

Fig. 1. Architecture of the full model. The left and right output of the character-level bidirectional LSTM are concatenated together to form char embedding. Then char embedding, word embedding, and POS embedding are fed into a sentence-level bidirectional LSTM with dependency transmission. The CRF layer is employed on top of the LSTM to predict BIO labels.

3.2 Embedding Layer

The embedding layer consists of word embedding, character embedding, and POS tag embedding. Word embedding reflecting resemblance between words has been shown effective in a wide range of NLP tasks. Therefore we use it as the basic input of our model. Noting that words themselves of different languages contain rich morphological (e.g. English) or hieroglyphic (e.g. Chinese) information, we can further exploit a neural model to encode words from their own characters in order to make full use of these character-level knowledge. We use Bi-LSTM here since we are tackling English and Recurrent Neural Networks are capable of capturing position-dependent features. We also take advantage of the POS tagging information which guarantees providing strong indicative knowledge for targeted words. Finally, each word in the sentence is associated with a word embedding, a final state output of the forward pass of the character Bi-LSTM, a final state output of the backward pass of the character Bi-LSTM and a POS tagging embedding. These features will be concatenated as a vector and fed to the next layer.

3.3   RNNs Incorporating Dependency Transmission

Given a review sentence $s = \langle w_1, w_2, \ldots, w_T \rangle$ consisting of T words, each represented as an n-dimensional word embedding x_t learned by unsupervised neural nets, a recurrent unit, at time step t, receives the current word embedding x_t and the previous hidden state h_{t-1}, and returns a output representation h_t and a new hidden state.

We first produce a dependency parse for each review sentence using an off-the-shelf dependency parser. In the dependency parse, all words except one are assumed to have a syntactic governor and there is a pre-defined genre of dependency relationship represented as an arc between the governor word and the dependent word. The arcs begin from the previous word and end to the current word will be used in the forward pass of the Recurrent Neural Networks. Similarly, the arcs arise at the rear word and drop at the current word will be used in the backward pass of the network, as depicted in Fig. 2. Each arc is represented as a vector $\mathbf{r} \in \mathbb{R}^d$ and an affine function $(f(\mathbf{r}) = W_r \mathbf{r} + \mathbf{b}_r)$ is introduced to transform the dependency embedding $\mathbf{r}$ to a vector $\mathbf{d}_r$ with the same dimension as the hidden state of the recurrent unit.

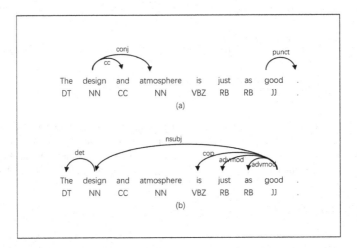

Fig. 2. An illustration of the dependency transmission. (a) represents the dependencies used in the forward pass, and (b) represents the dependencies used in the backward pass.

For each time step t, the input hidden state is now computed from the direct previous hidden state, the vector $\mathbf{d}_r$, and the previous output vector which has a dependency relation with the current token. As an illustration, consider the relation "conj" between words "design" and "atmosphere" in Fig. 2(a), we first calculate the hidden representation of the dependency relation *conj* as follows:

$$\mathbf{d}_{conj} = f\left(W_r \mathbf{r}_{conj} + \mathbf{b}_r\right)$$

where W_r and $\mathbf{b}_r$ denote the weight matrix and the bias vector, respectively, and f is a non-linear activation function. In our experiments, we choose the hyperbolic tangent

$\tanh(\cdot)$ to be the activation function. After this, the new hidden state is now become the summation of the dependently connected output vector o_r, the hidden representation vector $\mathbf{d}_{conj}$ and the directly previous hidden state vector $\mathbf{c}_{t-1}$, i.e.:

$$\mathbf{c}_{t-1} \leftarrow \mathbf{c}_{t-1} + o_r + \mathbf{d}_{conj}$$

The updates for LSTM units now become:

$$i_t = \sigma\left(W_{xi}x_t + W_{hi}h_{t-1} + W_{ci}(c_{t-1} + o_r + d) + b_i\right)$$

$$f_t = \sigma\left(W_{xf}x_t + W_{hf}h_{t-1} + W_{cf}(c_{t-1} + o_r + d) + b_f\right)$$

$$c_t = f_t \cdot c_{t-1} + i_t \cdot \tanh(W_{xt}x_t + W_{hc}h_{t-1} + b_c)$$

$$o_t = \sigma\left(W_{xo}x_t + W_{ho}h_{t-1} + W_{co}(c_t + o_r + d) + b_o\right)$$

$$h_t = o_t \cdot \tanh(c_t)$$

The motivation for this dependency transmission is that the syntactic priors offer beneficial clues about the high-level abstract concepts of the sentence that may help the extraction of the aspect words.

3.4   CRF Layer and Objective Function

The CRF layer is superior to conventional cross entropy loss and has been proved effective in modeling sequence labeling decisions with strong dependencies. In aspect term extraction scenario, for example, it is impossible that a label I is directly followed by a label O. Accordingly, we feed the outputs of the previous bidirectional LSTM layer into a CRF layer as the unary potentials.

Given an input sentence $\mathbf{x} = \langle x_1, x_2, \ldots, x_T \rangle$ and a label sequence $\mathbf{y} = \langle y_1, y_2, \ldots, y_T \rangle$, the score of the label predictions for $\mathbf{x}$ is calculated as:

$$\text{score}(\mathbf{x}, \mathbf{y}) = \sum_{t=0}^{T} T_{y_t, y_{t+1}} + \sum_{t=1}^{T} H_{t, y_t}$$

where T is the transition matrix denoting the probabilities of one tag transiting to another, and H is the matrix stacked by the Bi-LSTM outputs.

The probability for the label sequence $\mathbf{y}$ given $\mathbf{x}$ is then computed from $\text{score}(\mathbf{x}, \mathbf{y})$, using a softmax transformation:

$$p(\mathbf{y}|\mathbf{x}) = \frac{e^{s(\mathbf{x}, \mathbf{y})}}{\sum_{\hat{\mathbf{y}} \in \mathbf{Y}_\mathbf{x}} e^{s(\mathbf{x}, \hat{\mathbf{y}})}}$$

where $\mathbf{Y}_\mathbf{x}$ is the set containing all conceivable assignments of sequence labels for $\mathbf{x}$.

The network parameters are chose to minimize the negative log-likelihood of the gold tag sequence for an input **x**:

$$L(\theta) = -\sum_{\mathbf{x},\mathbf{y}} \log(p(\mathbf{y}|\mathbf{x}))$$

While inferencing, we find the best label sequence $\mathbf{y}^*$ that gives a maximum probability, which can be calculated efficiently by Viterbi algorithm.

$$\mathbf{y}^* = argmax_{\mathbf{y}' \in \mathbf{Y}_x} p(\mathbf{y}'|\mathbf{x})$$

4 Experiments

4.1 Datasets and Evaluation

The experiments are conducted on two publicly available datasets provided by SemEval-2014 Task 4: Aspect Based Sentiment Analysis. Table 1 presents some basic corpus statistics and feature statistics we used in the experiments.

Table 1. Corpus statistics: words are case-insensitive and an off-the-shelf NLP tool was used to tokenize the review sentence and generate the POS and dependencies.

	Laptop		Restaurant	
	Training set	Test set	Training set	Test set
# Sentences	3045	800	3041	800
# Aspect terms	2358	654	2786	818
# Words	4217	1913	4558	2251
# Chars	85	87	80	82
# POS	44	44	44	43
# Dependency	33	33	33	33

Evaluation. We choose the same evaluation metrics suggested in ABSA task.

$$F_1 = \frac{2TP}{2TP + FP + FN}$$

True positives (TP) are defined as the set of aspect terms in gold standard for which there exists a predicted aspect term that matches exactly. In our experiments, t-test is used to evaluate the statistical significance between two models and the corresponding p-value is reported in the table.

4.2 Experimental Settings and Compared Models

Pre-trained Word Embeddings. We used two domain-specific corpus of Amazon Product Data and Yelp Open Dataset for word embedding pre-training. Due to the similarity consideration, we chose the Electronics category provided by the Amazon corpus, which consists of 7,824,482 user reviews. The Yelp Open Dataset consists of 4,736,898 user reviews of various restaurants. All the unlabeled review texts were tokenized by the MBSP system. We trained the word embeddings of 300-dimensions using CBOW architecture with negative sampling.

Experimental Settings. For the two labeled review datasets, we perform tokenization, part-of-speech tagging, and dependency parsing all by Stanford CoreNLP [21]. We build the char vocabulary and word vocabulary from training set and embedding raw corpus by removing low frequency words. This resulted in a vocabulary of approximate 20 K/13 K words for Laptop/Restaurant dataset. In addition, we replace normal number strings, ordinal number, and time expression with \$NUM\$, \$ORD\$, and \$TIM\$, respectively. In test phrase, when encounter an unknown word, we replace it with \$UNK\$. For char level, we just ignore that character. All sentences will be padded to the maximum length with \$PAD\$.

The dimension of the word embedding, character embedding, POS tagging embedding and dependency embedding is 300, 100, 100, and 100, respectively. The size of the hidden state of the character Bi-LSTM is set to 100, while the sentence-level Bi-LSTM is set to 300. We adopt the Adam optimizer default parameters (lr: 0.001, beta1: 0.9, beta2: 0.999) and batch size 20. All hyper-parameters are chosen via cross validation. To further eliminate the influence of random error, we train 10 models with the same hyper-parameters and an average score is calculated on the test set, instead of only 5 times as [22].

Baseline and Comparable Models. To evaluate the effectiveness of our method with dependency transmission, we conduct comparison experiments with the following state-of-the-art models:

ISH_RD_Belarus: The top system for Laptop domain in SemEval 2014 Challenge Task 4. A linear-chain CRF model with a variety of hand-engineered feature sets, including token, part-of-speech, named entity, semantic category, semantic orientation, frequency of token occurrence, opinion target, noun phrase, semantic label and SAO features. The model was trained on a blend of both two domain training sets and used to predict all test sets with the same settings.

DLIREC: The top system for Restaurant domain in SemEval 2014 Challenge Task 4. The system also used a CRF-based model with rich handcrafted features. In addition to general features commonly used in NER systems, voluminous extrinsic resources are exploited to generate word cluster and name list as features in the system.

RNCRF + F (Wang et al. [19]): A Recursive Neural Network with CRF as the output layer. The results reported here are produced by the best setting incorporating hand-crafted features such as name list and sentiment lexicon.

WDEmb_CRF (W + L + D + B Yin et al. [20]): A CRF-based model with embedding features as input, in which the word embedding, the linear context embedding, and the dependency context embedding are trained unsupervisedly.

MIN (Li et al. [22]): A LSTM-based model with memory interactions. The full model is trained with multi-task learning setting.

Giannakopoulos et al. [23]: A regular bidirectional LSTM based model with CRF layer as final output. Additionally, the author conducted experiments on automatically labelled datasets.

4.3   Results and Analysis

In Table 2 we present the extraction performances evaluated by F1 score, compared with those of previous state-of-the-art models. We see that our model significantly outperforms the best systems in SemEval 2014 challenge, by 5.67% and 1.96% absolute gains on Laptop and Restaurant domains respectively, suggesting that deep neural network is capable of memorizing pivotal patterns for aspect term extraction while the latter systems rely on extensive hand-crafted feature engineering and template rules. With consideration of POS tagging information and dependency transmission, our model have surmounted the results of previous published works on each dataset. It clearly demonstrates the effectiveness of leveraging linguistic knowledge and the carefully designed dependency transmissions between recurrent units.

Table 2. Experimental results.

Models	D1	D2
IHS_RD (CRF-based, top system)	74.55	79.62
DLIREC (CRF-based, top system)	73.78	84.01
RNCRF + F(Wang et al., EMNLP, 2016)	78.42	84.93
W + L + D + B (Yin et al., IJCAI, 2016)	75.16	84.97
MIN (Li et al., EMNLP, 2017)	77.58	–
Giannakopoulos et al., EMNLP, 2017	77.96	84.12
Ours (full model)	**80.22**	85.96
Ours (-character embedding)	79.54	**85.97**
Ours (-POS tagging)	79.58	85.38
Ours (-dependency transmission)	79.79	85.41
Ours (only word embedding)	79.77	85.10

Ablation Experiments. To further provide insight into the contribution of the constituent parts of the overall performance, we carry out ablation experiments. Table 2 presents the ablation results in terms of F1 performance. Without character features, the extraction performance declines on Laptop domain but is roughly the same on Restaurant domain. This is because there are more OOVs in the Laptop domain, which proves that character-level embedding helps deal with unknown words. We find that using POS tagging information helps boost the performance since aspect terms usually appear as nominal words or phrases. More importantly, we observed that dependency transmission is significantly contributive to increasing the performance, indicating that it is useful for capturing skip information.

Different Word Embeddings. Besides ablation experiments, we also carry experiments to observe the performance with different pre-trained word embeddings. The experimental results are reported in Table 3. Our domain-specific pre-trained word embedding yields the best performance on all datasets. The result indicates that the pertinence of the corpus is probably more important than its size when it is used to train the word embedding, since the size of Google News corpus is much larger than the Yelp Dataset or the Amazon Review.

Table 3. Experimental results with different pre-trained word embeddings

	D1	D2
word2vec(GoogleNews, 300d)	75.43	79.98
fastText(Wikipedia, 300d)	73.94	81.43
GloVe(Wikipedia 2014 + Gigaword 5, 300d)	71.51	81.73
Ours(Yelp Dataset/Amazon Review, 300d)	**80.22**	**85.96**

Error Analysis. Table 4 presents some examples which are not handled well by our model. Sentence (a) and sentence (b) both introduce external aspects which do not belong to its own domain. The reviewer of sentence (a) expresses his/her opinion by using the "movie" as a metaphor, while sentence (b) refers to "air flow" with strong aspect indicator "good". These linguistic phenomena are not unusual in review text and bear a non-negligible responsibility for interfering the performance of the extraction system. Applying metaphor recognition and introducing domain-specific knowledge may alleviate the interference. We left the verification of this conjecture to future research. Sentence (c) and sentence (d) are examples in which aspect words appear as verbs. It is relatively uncommon since aspect words are usually nouns or noun phrases. Other errors caused by conditions such as human error or unknown words are not discussed here since this can be solved through qualitative and quantitative improvements.

5   Conclusion

In this paper, we have presented a novel architecture leveraging both the sequential message and structural priors for aspect words extraction in opinioned reviews. In addition to fundamental features such as character-level morphology and word-level POS tagging, which have been used extensively to improve performance, we investigate the probability of incorporating structural information into Recurrent Neural Networks. To achieve this, we equip the recurrent unit with the ability to receive information from the dependently connected word, and this ensures that the Recurrent Neural Networks are able to learn features sequentially and structurally. As expected, the comparison of other state-of-the-art models with the experimental results show that

Table 4. Error analysis. For each example, the first line represents the words, and the second and the third denote the gold labels and the predicted labels, respectively.

(a)

it	's	good	but	,	like	the	movie	is	never	as	the	book	,	this	is	same	analogy	.
O	O	O	O	O	O	O	O	O	O	O	O	O	O	O	O	O	O	O
O	O	O	O	O	O	O	B	O	O	O	O	O	O	O	O	O	O	O

(b)

be	sure	to	have	good	air	flow	where	ever	you	put	it	.
O	O	O	O	O	O	O	O	O	O	O	O	O
O	O	O	O	O	B	I	O	O	O	O	O	O

(c)

it	's	just	what	we	were	looking	for	and	it	works	great	.
O	O	O	O	O	O	O	O	O	O	B	O	O
O	O	O	O	O	O	O	O	O	O	O	O	O

(d)

starts	up	in	a	hurry	and	everything	is	ready	to	go	.
B	I	O	O	O	O	O	O	O	O	O	O
O	O	O	O	O	O	O	O	O	O	O	O

the proposed model exhibits a more favorable performance and confirm the effectiveness of the dependency transmission between relational words.

In addition, we have performed error analysis on a few noteworthy sentences, which may provide future directions to implement a more accurate extraction system. One of them is the metaphor recognition in reviews since the increasing number of such sentences used to express analogical feelings about the product, in which the tenor should be recognized as the aspect instead of the vehicle. Furthermore, though the proposed architecture is presented in the context of handling aspect words extraction, it will be a considerable future direction to employ this model to other NLP applications such as fine-grained sentiment classification and stance detection.

Acknowledgments. The research is supported by the National Natural Science Foundation of China (No. 61572145) and the Major Projects of Guangdong Education Department for Foundation Research and Applied Research (No. 2017KZDXM031). We would like to acknowledge the anonymous reviewers for their helpful comments and suggestions.

References

1. Mukherjee, A., Liu, B.: Aspect extraction through semi-supervised modeling. In: Proceedings of the 50th Annual Meeting of the Association for Computational Linguistics (Volume 1: Long Papers), vol. 1, pp. 339–348 (2012)
2. Pavlopoulos, J., Androutsopoulos, I.: Aspect term extraction for sentiment analysis: new datasets, new evaluation measures and an improved unsupervised method. In: Proceedings of the 5th Workshop on Language Analysis for Social Media (LASM), pp. 44–52 (2014)
3. Liu, K., Xu, L., Zhao, J.: Opinion target extraction using word-based translation model. In: Proceedings of the 2012 Joint Conference on Empirical Methods in Natural Language Processing and Computational Natural Language Learning. Association for Computational Linguistics, pp. 1346–1356 (2012)
4. Liu, K., Xu, L., Liu, Y., et al.: Opinion target extraction using partially-supervised word alignment model. In: Proceedings of the Twenty-Third International Joint Conference on Artificial Intelligence, pp. 2134–2140. AAAI Press (2013)
5. Socher, R., Perelygin, A., Wu, J., et al.: Recursive deep models for semantic compositionality over a sentiment treebank. In: Proceedings of the 2013 Conference on Empirical Methods in Natural Language Processing, pp. 1631–1642 (2013)
6. Lin, Y., Lei, H., Wu, J., et al.: An empirical study on sentiment classification of Chinese review using word embedding. In: Proceedings of the 29th Pacific Asia Conference on Language, Information and Computation: Posters, pp. 258–266 (2015)
7. Hu, M., Liu, B.: Mining and summarizing customer reviews. In: Proceedings of the Tenth ACM SIGKDD International Conference on Knowledge Discovery and Data Mining, pp. 168–177. ACM (2004)
8. Hochreiter, S., Schmidhuber, J.: Long short-term memory. Neural Comput. **9**(8), 1735–1780 (1997)
9. Popescu, A.M., Etzioni, O.: Extracting product features and opinions from reviews. In: Kao, A., Poteet, S.R. (eds.) Natural Language Processing and Text Mining, pp. 9–28. Springer, London (2007). https://doi.org/10.1007/978-1-84628-754-1_2

10. Scaffidi, C., Bierhoff, K., Chang, E., Felker, M., Ng, H., Jin, C.: Red Opal: Product-feature scoring from reviews. In: Proceedings of the 8th ACM Conference on Electronic Commerce, pp. 182–191. ACM (2007)

11. Zhuang, L., Jing, F., Zhu, X.Y.: Movie review mining and summarization. In: Proceedings of the 15th ACM International Conference on Information and Knowledge Management, pp. 43–50. ACM (2006)

12. Qiu, G., Liu, B., Bu, J., et al.: Opinion word expansion and target extraction through double propagation. Comput. Linguist. **37**(1), 9–27 (2011)

13. Teng-Jiao, J., Chang-Xuan, W., De-Xi, L.: Extracting target-opinion pairs based on semantic analysis. Chin. J. Comput. **40**(3), 617–633 (2017)

14. Jin, W., Ho, H.H.: A novel lexicalized HMM-based learning framework for web opinion mining. In: Proceedings of the 26th Annual International Conference on Machine Learning, pp. 465–472. ACM (2009)

15. Jakob, N., Gurevych, I.: Extracting opinion targets in a single-and cross-domain setting with conditional random fields. In: Proceedings of the 2010 Conference on Empirical Methods in Natural Language Processing. Association for Computational Linguistics, pp. 1035–1045 (2010)

16. Li, F., Han, C., Huang, M., et al.: Structure-aware review mining and summarization. In: International Conference on Computational Linguistics. Association for Computational Linguistics, pp. 653–661 (2010)

17. Yang, B., Cardie, C.: Joint inference for fine-grained opinion extraction. In: Proceedings of the 51st Annual Meeting of the Association for Computational Linguistics (Volume 1: Long Papers), vol. 1, pp. 1640–1649 (2013)

18. Liu, P., Joty, S., Meng, H.: Fine-grained opinion mining with recurrent neural networks and word embeddings. In: Conference on Empirical Methods in Natural Language Processing, pp. 1433–1443 (2015)

19. Wang, W., Pan, S.J., Dahlmeier, D., et al.: Recursive neural conditional random fields for aspect-based sentiment analysis. In: Proceedings of the 2016 Conference on Empirical Methods in Natural Language Processing, pp. 616–626 (2016)

20. Yin, Y., Wei, F., Dong, L., et al.: Unsupervised word and dependency path embeddings for aspect term extraction. In: International Joint Conference on Artificial Intelligence, pp. 2979–2985. AAAI Press (2016)

21. Manning, C., Surdeanu, M., Bauer, J., et al.: The Stanford CoreNLP natural language processing toolkit. In: Proceedings of 52nd Annual Meeting of the Association for Computational Linguistics: System Demonstrations, pp. 55–60 (2014)

22. Li, X., Lam, W.: Deep multi-task learning for aspect term extraction with memory interaction. In: Proceedings of the 2017 Conference on Empirical Methods in Natural Language Processing, pp. 2886–2892 (2017)

23. Giannakopoulos, A., Musat, C., Hossmann, A., et al.: Unsupervised aspect term extraction with B-LSTM & CRF using automatically labelled datasets. In: Proceedings of the 8th Workshop on Computational Approaches to Subjectivity, Sentiment and Social Media Analysis, pp. 180–188 (2017)

Dependency Parsing and Attention Network for Aspect-Level Sentiment Classification

Zhifan Ouyang and Jindian Su[✉]

School of Computer Science and Engineering,
South China University of Technology, Guangzhou, China
csouzhi@mail.scut.edu.cn, sujd@scut.edu.cn

Abstract. Aspect-level sentiment classification aims to determine the sentiment polarity of the sentence towards the aspect. The key element of this task is to characterize the relationship between the aspect and the contexts. Some recent attention-based neural network methods regard the aspect as the attention calculation goal, so they can learn the association between aspect and contexts directly. However, the above attention model simply uses the word embedding to represent the aspect, it fails to make a further improvement on the performance of aspect sentiment classification. To solve this problem, this paper proposes a dependency subtree attention network (DSAN) model. The DSAN model firstly extracts the dependency subtree that contains the descriptive information of the aspect based on the dependency tree of the sentence, and then utilizes a bidirectional GRU network to generate an accurate aspect representation, and uses the dot-product attention function for the dependency subtree aspect representation, which finally yields the appropriate attention weights. The experimental results on SemEval 2014 Datasets demonstrate the effectiveness of the DSAN model.

Keywords: Aspect-level sentiment classification · Attention network
Dependency tree

1 Introduction

Aspect-level sentiment classification is a fine-grained task in the field of sentiment analysis [13], which aims to identify the sentiment expressions for aspects in their contexts. This task can provide more detailed and in-depth sentiment analysis results, which has been getting much attention recently. For example, given a sentence *"Air has higher resolution, but the fonts are small."*, the polarity is positive for the aspect *"resolution"*, and negative for the aspect *"fonts"*.

Because the sentiment polarity of an aspect needs to consider both the aspect and the contexts, the key point is how to characterize the relationship between the aspect and the contexts [19]. Dependency parsing plays a very important role in the aspect-level sentiment classification task. In some previous work, the

© Springer Nature Switzerland AG 2018
M. Zhang et al. (Eds.): NLPCC 2018, LNAI 11108, pp. 391–403, 2018.
https://doi.org/10.1007/978-3-319-99495-6_33

dependency tree is used to extract aspect-related features to build sentiment classifiers in traditional machine learning methods [7], or to establish aspect-specific recursive structure used for the input in Recursive Neural Network methods [4,11].

Since attention mechanism can help to enforce a model to learn the task-related parts of a sentence [1], some works exploit this advantage and achieve superior performance for aspect-level sentiment classification, i.e. [2,3,15,17]. The attention-based models regard the aspect as the attention calculation goal, which enable the model to learn the association between aspect and its contexts directly. Usually, different aspects in the same sentence might have different attention weights. Despite of the advantages of attention mechanism, previous models simply use the word embedding of the aspect to represent the aspect and calculate the corresponding attention weights, which as a result might lose a lot of aspect information and the aspect representations are not accurate enough, so that the attention model fails to learn the appropriate attention weights for each aspect.

The generation of an accurate aspect representation for each aspect becomes an important factor to make further improvements on the performance of the attention models for aspect-level sentiment classification. Similar to modeling contextual information, utilizing a Recurrent Neural Network (RNN) to model the aspect and generate aspect representations is a very worthwhile try, which is the same as to Ma et al. [9]. However, it is insufficient to simply model the aspect in the form of a noun or a noun phrase.

From the perspective of sentiment expression, we found that when people express their sentiment about some specific target, they tend to use adjectives or adverbs to describe the target and express their inner feelings. These modifiers are closely related with the target, and can form the accurate descriptions of the aspect in the sentence. By using dependency parsing, we can obviously see that these modifiers are generally subject to the aspect. Therefore, we can extract a dependency subtree of the aspect based on the dependency tree of the sentence. For example, we can extract two dependency subtrees from the sentence *"Air has higher resolution, but the fonts are small."*, as shown in Fig. 1, each of which is

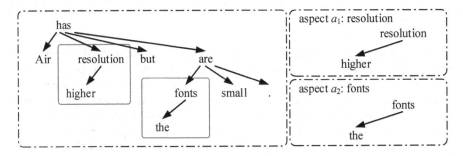

Fig. 1. Dependency parsing tree for sentence: *"Air has higher resolution, but the fonts are small."*, and the dependency subtree of aspect a_1 *"resolution"* and aspect a_2 *"fonts"*.

also a sub-sentence of the sentence and includes some context information about the aspect. So, we can try to model the aspect sub-sentence by RNN networks, and use it to denote the aspect instead of a noun or a noun phrase.

Motivated by the above intuition, we propose a dependency subtree attention network (DSAN) model, which is based on dependency parsing and attention mechanism. DSAN utilizes gated recurrent unit (GRU) to separately modelling the aspect sub-sentence and the contexts, and uses the attention mechanism to generate aspect-related sentiment features based on aspect sub-sentence. We have evaluated our model on Laptop and Restaurant datasets from SemEval 2014 [13]. Experimental results show that our DSAN model achieves the comparable state-of-the-art performance for aspect-level sentiment classification.

2 Model

In this section, we describe the dependency subtree attention network (DSAN) model for aspect-level sentiment classification. The architecture of DSAN model is shown in Fig. 2.

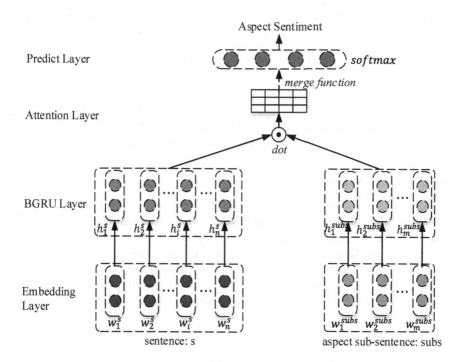

Fig. 2. The architecture of DSAN model

2.1 Embedding Layer

Let $L \in \mathbb{R}^{d_w \times |V|}$ be a word embedding look-up table which is usually trained on an external large corpus [10,12], d_w be the dimensions of word vectors and $|V|$ be the size of the vocabulary. Given a sentence $s = \{w_1^s, w_2^s, ..., w_i^s, ..., w_n^s\}$ and the dependency subtree of an aspect (sub-sentence) $subs = \{w_1^{subs}, w_2^{subs}, ..., w_i^{subs}, ..., w_m^{subs}\}$, the embedding layer project each word into a low dimensional, continuous and real-valued vector, denoted as $X^s = [x_1^s, x_2^s, ..., x_i^s, ..., x_n^s]$ where $x_i^s \in \mathbb{R}^{d_w}$ represents w_i^s, and $X^{subs} = [x_1^{subs}, x_2^{subs}, ..., x_i^{subs}, ..., x_n^{subs}]$ where $x_i^{subs} \in \mathbb{R}^{d_w}$ represents w_i^{subs}.

2.2 BGRU Layer

RNN has already demonstrated its superior performance on variable-length sequences modeling. It can capture long-term dependency information of words and is popularly used in the area of sentiment analysis [18]. In this paper, we use bidirectional GRU (BGRU) to separately model the contexts and the aspect. Denote the hidden states of the forward GRU at time step i as $\overrightarrow{h_i} = \overrightarrow{GRU}(x_i, \overrightarrow{h_{i-1}})$ and the backward GRU as $\overleftarrow{h_i} = \overleftarrow{GRU}(x_i, \overleftarrow{h_{i-1}})$, and $h_i = [\overrightarrow{h_i}; \overleftarrow{h_i}]$ as the output of BGRU at time step i. Then, the output of the contexts and the aspect modeled by BGRU can be defined as:

$$H^s = BGRU^s(X^s) \tag{1}$$

$$H^{subs} = BGRU^{subs}(X^{subs}) \tag{2}$$

where the hidden states $H^s = [h_1^s, h_2^s, ..., h_i^s, ..., h_n^s]$, $h_i^s \in \mathbb{R}^{2d}$ is the representations for the contexts, and the hidden states $H^{subs} = [h_1^{subs}, h_2^{subs}, ..., h_i^{subs}, ..., h_m^{subs}]$, $h_i^{subs} \in \mathbb{R}^{2d}$ is the representations for the aspect.

2.3 Position Weight

The position of the aspect helps the model to distinguish different aspects in the same sentence, and the sentiment expression of the aspect is close to the aspect in the contexts. Therefore, we bring the positional information of the aspect into consideration in the form of position weights during attention calculation. We define the distance as the path of the aspect and context word in the dependency tree. If the aspect is a phrase, then the distance will be simply calculated with the last word in the aspect phrase. The calculation formula of the weight for the context word w_i is:

$$l_i = 1 - \frac{dist_{i,a}}{2dist_{max}} \tag{3}$$

where $dist_{i,a}$ denotes the distance of context word w_i and the aspect a, $dist_{max}$ denotes the max distance of context words in the input sentence s. The range of the position weight is limited to $[0.5, 1]$.

Based on Eq. 3, we can get an aspect-customized hidden states of the sentence s, denoted as $E^s = [e_1^s, e_2^s, ..., e_i^s, ..., e_n^s]$, where $e_i^s = l_i \cdot h_i^s \in \mathbb{R}^{2d}$. The position weights can give higher weight to the context word which is close to the aspect. This can help the model to predict the sentiment of different aspects flexibly and prevent the model from being misled by a strong unrelated sentiment expression.

2.4   Attention Layer

The goal of the attention layer is to allocate appropriate attention weights to the words of the sentence according to the aspect. The attention layer generates aspect-related sentiment features. We employ the attention function which is similarly as Vaswani et al. [16], and the attention weights calculation is based on the hidden states H^{subs} of sub-sentence and aspect-customized hidden states E^s,

$$Q = relu(W_1 H^{subs}) \tag{4}$$

$$K = relu(W_2 E^s) \tag{5}$$

$$Score = \frac{K^T Q}{\sqrt{d_k}} \tag{6}$$

where $W_1, W_2 \in \mathbb{R}^{d_k \times 2d}$ are linear transfer parameters, $Score \in \mathbb{R}^{n \times m}$ is an attention score matrix. $Score_{i,j}$ represents the attention score of the word w_i in the sentence s and the word w_j in the sub-sentence $subs$. In practice, we found that adding rectified linear unit (relu) activation function to filter out negative values can yield a more stable performance. In order to merge the attention contributions to each word of the sub-sentence $subs$, we define a column merging function over score matrix $Score$,

$$\alpha = \begin{cases} softmax\left(\sum_{j=1}^{m} Score_j\right), & \text{if } mode = sum; \\ softmax\left(\frac{1}{m}\sum_{j=1}^{m} Score_j\right), & \text{if } mode = mean; \end{cases} \tag{7}$$

where m is the length of the sub-sentence $subs$, α is the final attention weights of context words. In this work, the merging function includes two different types, sum and $mean$.

After getting the attention weights, we can calculate the final aspect-related sentiment representations r as follows,

$$V = H^s + W_3 X^s \tag{8}$$

$$r = V \cdot \alpha \tag{9}$$

where $W_3 \in \mathbb{R}^{2d \times d_w}$ is a linear transfer parameter for the context word embedding sequence X^s. $V \in \mathbb{R}^{2d \times n}$ is the cumulative result of H^s and X^s, which can be viewed as a key-value memory network, whose keys and values are K and V [5] respectively.

2.5 Sentiment Predict Layer

Finally, we concatenate the last hidden states of BGRU of the sentence s to represent the sentence, and use a nonlinear transfer to get the final representations of a sentence after given an aspect,

$$r^s = [\overrightarrow{h_n^s}; \overleftarrow{h_1^s}] \tag{10}$$

$$h^* = relu\,(W_4 r + W_5 r^s + b_4) \tag{11}$$

where $h^* \in \mathbb{R}^{d_r}$, $W_4, W_5 \in \mathbb{R}^{d_r \times 2d}$ and $b_4 \in \mathbb{R}^{d_r}$ are the parameters of nonlinear layer. Then, we feed the representation h^* into a $softmax$ layer to predict the aspect sentiment polarity.

2.6 Model Training

We use the cross entropy as the objective function, and plus an L_2 regularization term to prevent overfitting,

$$J = \sum_{(x,y)\in D} \sum_{c\in C} P_c^g(x,y)logP_c(x,y;\theta) + \lambda\|\,\theta\,\|^2 \tag{12}$$

where C is the sentiment category set, D is the collection of training instance, $P^g(x,y)$ is a one-hot vector that indicates the true sentiment of aspect, $P(x,y;\theta)$ is the predicted sentiment probability of the model, θ is the parameter set, and λ is the regularization weight. We adapt the ADAM method to update parameters [8].

3 Experiment

3.1 Datasets and Settings

We conduct experiments on two datasets from SemEval task 4 [13], reviews of laptop and restaurant domain respectively. Each aspect with reviews is labeled with three sentiment polarities: positive, negative and neutral. Table 1 shows the final statistics of two datasets.

Table 1. Statistics of two datasets.

DataSet	Positive	Neutral	Negative
Laptop-train	987	460	866
Laptop-test	341	169	128
Restaurant-train	2164	633	805
Restaurant-test	728	196	196

We implement the DSAN model with Keras[1] and TensorFlow[2], and use spaCy[3] to parse the structures of sentences. We use the pre-trained GloVe word embeddings [12] for our experiments. In addition, we use the Amazon electronic dataset [6] for laptop domain, and the Yelp Challenge dataset[4] for restaurant domain to train 300-dimension word embeddings with the Skipgram[5] training method. All parameters except word embeddings are initialized with random uniform distribution $U(-0.05, 0.05)$. The hidden size is set to 120 for both two BGRU and 100 for the nonlinear layer. To prevent overfitting, we set regularization weight λ to be 0.012, and the dropout rate of two BGRU to be 0.5. Evaluation metrics are Accuracy and Macro-F1 because the datasets are unbalanced.

3.2   Experimental Results

In order to evaluate our DSAN model, we compare it with the following methods: **TD-LSTM**[14], **MemNet**[15], **RAM**[2], **ATAE-LSTM**[17], **IAN**[9].

Tables 2 and 3 show the experiment results of our model compared with other related models. We denote our models as **DSAN-sum, DSAN-mean**, which means using *sum* and *mean* mode respectively. In Table 2, all models use pre-trained Glove 300-dimension word embeddings, The difference is that the first group whose vocabulary size is 1.9M and the second one is 2.2M. In Table 3, all models use the skipgram word embeddings that are trained on domain-specific corpus.

From Tables 2 and 3, we can conclude:

(1) Our DSAN model achieves the comparable results with RAM method on both Laptop and Restaurant datasets, exceeding other four benchmark methods except the RAM, which is the state-of-the-art method for aspect-level sentiment classification. Compared with the RAM method, the two merge modes of the DSAN model can achieve better performance than the RAM on the Restaurant dataset, and its performance is slightly lower than the RAM on the Laptop dataset. The biggest difference between the RAM model and the DSAN model is that the RAM implements multiply attention mechanisms based on the GRU network, so it can catch information about different important parts of a sentence by attention layers, and combine the result of each attention layer in a non-linear manner. But RAM still simply uses the word embedding to represent the aspect. The DSAN model utilizes the bidirectional GRU to model the sub-sentence of the aspect that containing the descriptive information of aspect, so the aspect can be represented more accurately than using the way of word embeddings. Therefore, even

---

[1] https://keras.io.

[2] https://www.tensorflow.org.

[3] https://spacy.io/.

[4] https://www.yelp.com/dataset.

[5] https://radimrehurek.com/gensim/models/word2vec.html.

Table 2. Results of different methods on Laptop and Restaurant datasets. The results with '*' are retrieved from RAM paper, and the results with '◊' are retrieved from the papers of compared methods. Best results in each group are in bold.

Word embeddings	Model	Laptop		Restaurant	
		Accuracy	Macro-F1	Accuracy	Macro-F1
Glove (1.9M vocabulary size)	TD-LSTM	0.7183*	0.6843*	0.7800*	0.6673*
	MemNet	0.7033*	0.6409*	0.7816*	0.6583*
	RAM	**0.7449***	**0.7135***	0.8023*	0.7080*
	DSAN-sum	0.7273	0.6878	0.8071	0.7238
	DSAN-mean	0.7382	0.7001	**0.8080**	**0.7273**
Glove (2.2M vocabulary size)	ATAE-LSTM	0.6870◊	NA	0.7720◊	NA
	IAN	0.7210◊	NA	0.7860◊	NA
	DSAN-sum	0.7382	0.6961	**0.8009**	0.7109
	DSAN-mean	**0.7461**	**0.7052**	0.7964	**0.7136**

Table 3. Results of different methods on Laptop and Restaurant datasets with skip-gram word embeddings. Best results in each group are in bold.

Word embeddings	Model	Laptop		Restaurant	
		Accuracy	Macro-F1	Accuracy	Macro-F1
Skipgram	TD-LSTM	0.7179	0.6665	0.7848	0.6812
	MemNet	0.7257	0.6765	0.8027	0.7076
	RAM	0.7445	0.7009	0.7884	0.6835
	ATAE-LSTM	0.7257	0.6833	0.7866	0.6802
	IAN	0.7194	0.6664	0.7991	0.7046
	DSAN-sum	**0.7696**	**0.7265**	**0.8143**	**0.7311**
	DSAN-mean	0.7633	0.7136	0.8134	0.7235

if the DSAN model is calculated with single-layer attention, it also can be achieve better performance.

(2) In the second group of Table 2, the accuracies of the ATAE-LSTM, IAN, and DSAN models gradually increase, which demonstrate the importance of aspect information for aspect-level sentiment classification. All three models use single-layer attention mechanism, but they have shown an increasing trend in the use of aspect information. The ATAE-LSTM method simply uses the word embedding of the aspect as the aspect representations. IAN method utilizes LSTM to model the aspect themselves, and the DSAN model utilizes bidirectional GRU to model the aspect and the descriptive information of aspect. The experimental results of the ATAE-LSTM, IAN and

DSAN models show that the increase in the use of aspect information in the attention model helps the model to achieve better results.

(3) Word embeddings trained on domain-specific corpus are very helpful to the final classification results. After using the skipgram word embeddings, the MemNet, ATAE-LSTM and DSAN models all benefit from the domain knowledge, which have a great performance improvement on the Laptop and Restaurant datasets. The TD-LSTM and IAN models also have improvements on the Restaurant dataset to some extent, and performance is closed on the Laptop dataset. In this paper, the reproducible experimental results of the RAM model are not as good as those of the original author Chen et al. [2] on the published GloVe word vector, which may be related to the hyperparameter setting of the neural network. Chen et al. [2] did not mention their hyperparameter settings about the RAM model.

(4) The results of the two merge modes of DSAN model are similar in the three sets of word embeddings. From the calculation of Formula 7, we can see that the difference between the *sum* mode and the *mean* mode is the scale factor m, which is the length of the aspect sub-sentence. Since the *softmax* function computes each element in the attention weights in an exponential form, the local part of the words will be assigned more attentional weights in the sum model. It re-scales this concentration of the attentional weights, and the distribution of weights is relatively uniform in the mean mode. However, the dimensional factors have been scaled when calculating the attention scores of aspect sub-sentence and the sentence, and the length of most of the aspect sub-sentences are less than 5. So the classification of these two methods exhibit similar performance.

3.3 Effects of Dependency Subtree

In this subsection, we design a set of model comparison experiments to analyze the effect of the dependency subtree in the DSAN model. There are three different models in the comparison experiments. The aspect is represented by the word embedding in the first model, denoted as **W-AN**. The second model, denoted as **A-AN**, uses bidirectional GRU to model the aspect, and it doesn't include the descriptive information of the aspect. The third model is DSAN model. In the A-AN and DSAN model, the merge mode is *sum*, so we refer them as **A-AN-sum** and **DSAN-sum** for short.

From Table 4, the performance of W-AN, A-AN-sum, and DSAN-sum on the Laptop and Restaurant datasets increases sequentially, which reflects their increasing utilization degree of the aspect information. More accurate that the aspect representations are catched, more better that the performance of the attention model in the aspect-level sentiment classification task will achieve. The experimental results fully proves the importance of the dependency subtree of the aspect in the DSAN model.

Table 4. Effects of dependency subtree in DSAN model. Best results in each group are in bold.

Word embeddings	Model	Laptop		Restaurant	
		Accuracy	Macro-F1	Accuracy	Macro-F1
Skipgram	W-AN	0.7429	0.7061	0.7955	0.6992
	A-AN-sum	0.7571	0.7116	0.8054	0.7123
	DSAN-sum	**0.7696**	**0.7265**	**0.8143**	**0.7311**

3.4   Effects of Position Weight

As mentioned in Sect. 2.3, when different aspects are mentioned in a common sentence, the position information of the aspect is a very important feature that can help the model to distinguish different aspects from each other in the same sentence and make an improvement on the performance in the model.

Therefore, we introduce the position information of the aspect in the form of position weights, and define the distance of the aspect and the context word as the path length in dependency tree. We verify the effects of the position weights in two different word embeddings. Table 5 shows the experiment results of the effect of the position weights in different word embeddings. The experimental results prove that the position information of aspect can help model to better identify different aspects in the same sentence, thereby the model can achieve better performance. So, designing more effective ways of using position information is a worthwhile future work, such as position embedding.

Table 5. Effects of position weights in two different word embeddings. The postposition *base* means without position weights, and *position* means with position weights in the model. Best results in each group are in bold.

Word embeddings	Model	Laptop		Restaurant	
		Accuracy	Macro-F1	Accuracy	Macro-F1
Glove (2.2M vocabulary size)	DSAN-sum base	0.7288	0.6872	0.7982	0.7087
	DSAN-sum position	**0.7382**	**0.6961**	**0.8009**	**0.7109**
	DSAN-mean base	0.7351	0.6958	**0.7964**	0.7081
	DSAN-mean position	**0.7461**	**0.7052**	**0.7964**	**0.7136**
Skipgram	DSAN-sum base	0.7665	0.7213	0.8107	0.7209
	DSAN-sum position	**0.7696**	**0.7265**	**0.8143**	**0.7311**
	DSAN-mean base	0.7586	**0.7142**	0.8098	**0.7248**
	DSAN-mean position	**0.7633**	0.7136	**0.8134**	0.7235

3.5 Visualize Attention

In this subsection, we pick a review context *"Great food but the service was dreadful!"* from the Restaurant datasets as an example to visualize the attention weights of DSAN model on different aspect in the same sentence. There are two aspects: *"food"* and *"service"*, whose sentiment polarities are positive and negative respectively. We predict them by Skipgram word embeddings with the DSAN-sum and DSAN-mean model. Figures 3 and 4 show the attention weights of the DSAN-sum and DSAN-mean model on these two different aspects in the same sentence.

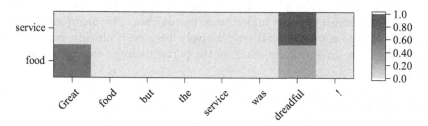

Fig. 3. Attention weights of DSAN-sum model on *"food"* and *"service"*.

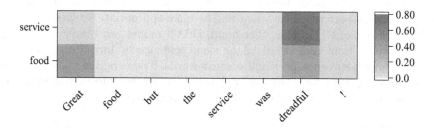

Fig. 4. Attention weights of DSAN-mean model on *"food"* and *"service"*.

From Figs. 3 and 4, we can clearly see the difference between the attention weights of two aspects. Both DSAN-sum and DSAN-mean show a high degree of attention to aspect-related sentiment expression, and assign most of the attentional weights, which helps to correctly predict the sentiment of these two aspects.

In Fig. 3, the DSAN-sum model assigns most of the attention weights to one or two words. When the DSAN-sum model predicts the aspect *"service"*, the word *"dreadful"* gets more than 90% of the attention weights, which helps the sentiment polarity of the aspect *"service"* to be correctly predicted to be negative. When the DSAN-sum model predicts the aspect *"food"*, the words *"Great"* and *"dreadful"* also get higher weights, but the word *"Great"* has much greater attention weights than the word *"dreadful"*, and the sentiment polarity of the aspect *"food"* is correctly predicted to be positive.

In Fig. 4, the DSAN-mean model assigns the attention weights more uniformly than DSAN-sum relatively. The DSAN-mean model also assigns most of the attention weights to the word *"dreadful"* for the aspect *"service"*, but it assigns a little weights to the word *"Great"*. When the DSAN-mean model predicts the aspect *"food"*, the word *"Great"* gets only a little more attention weights than the word *"dreadful"*. Although the DSAN-mean model correctly predicts the aspect *"food"*, the probability of the positive category is only slightly higher than the probability of the negative category.

From the visualization of the attention weights of the DSAN-sum and DSAN-mean model, the performance of the two models is the same as we analyzed in Sect. 3.2. The *sum* merge mode of the attention scores makes the model focus on a local part which gets the major attention weights. The *mean* merge mode makes the attention weights uniform relatively, because it already exists a scale factor $\sqrt{d_k}$ in the attention function, so the performance of two merge mode is similar.

4   Conclusions

In this paper, we propose a dependency subtree attention network (DSAN) model to determine the sentiment polarity of aspect in a sentence. On one hand, DSAN model can extract the dependency subtree that contains the descriptive information of the aspect based on the dependency tree of the sentence, which can offer more accurate aspect representations and assign e appropriate attention weights to the context words. On the other hand, DSAN model can better distinguish multiple aspect from each other in the same sentence by introducing the syntactic distance between aspect and context words. The experimental results on Laptop and Restaurant datasets show that the descriptive information of aspect can be more helpful to correctly predict aspect sentiment.

Acknowledgments. This paper is supported by the Applied Scientific and Technological Special Project of Department of Science and Technology of Guangdong Province (20168010124010); Natural Science Foundation of Guangdong Province (2015A030310318); Medical Scientific Research Foundation of Guangdong Province (A2015065).

References

1. Bahdanau, D., Cho, K., Bengio, Y.: Neural machine translation by jointly learning to align and translate. In: International Conference on Learning Representations 2015 (2015)
2. Chen, P., Sun, Z., Bing, L., Yang, W.: Recurrent attention network on memory for aspect sentiment analysis. In: Proceedings of the 2017 Conference on Empirical Methods in Natural Language Processing, pp. 463–472 (2017)
3. Cheng, J., Zhao, S., Zhang, J., King, I., Zhang, X., Wang, H.: Aspect-level sentiment classification with heat (hierarchical attention) network. In: Proceedings of the 26th ACM International Conference on Information and Knowledge Management, pp. 97–106 (2017)

4. Dong, L., Wei, F., Tan, C., Tang, D., Zhou, M., Xu, K.: Adaptive recursive neural network for target-dependent twitter sentiment classification. In: Proceedings of the 52nd Annual Meeting of the Association for Computational Linguistics (Volume 2: Short Papers), vol. 2, pp. 49–54 (2014)
5. Gehring, J., Auli, M., Grangier, D., Yarats, D., Dauphin, Y.N.: Convolutional sequence to sequence learning. In: International Conference on Machine Learning 2017, pp. 1243–1252 (2017)
6. He, R., Mcauley, J.: Ups and downs: modeling the visual evolution of fashion trends with one-class collaborative filtering. In: International Conference on World Wide Web, pp. 507–517 (2016)
7. Jiang, L., Yu, M., Zhou, M., Liu, X., Zhao, T.: Target-dependent twitter sentiment classification. In: Proceedings of the 49th Annual Meeting of the Association for Computational Linguistics: Human Language Technologies, pp. 151–160 (2011)
8. Kingma, D.P., Ba, J.L.: Adam: a method for stochastic optimization. In: International Conference on Learning Representations 2015 (2015)
9. Ma, D., Li, S., Zhang, X., Wang, H., Ma, D., Li, S., Zhang, X., Wang, H.: Interactive attention networks for aspect-level sentiment classification. In: Proceedings of the 26th International Joint Conference on Artificial Intelligence, pp. 4068–4074 (2017)
10. Mikolov, T., Chen, K., Corrado, G.S., Dean, J.: Efficient estimation of word representations in vector space. arXiv preprint arXiv:1301.3781 (2013)
11. Nguyen, T.H., Shirai, K.: PhraseRNN: phrase recursive neural network for aspect-based sentiment analysis. In: Proceedings of the 2015 Conference on Empirical Methods in Natural Language Processing, pp. 2509–2514 (2015)
12. Pennington, J., Socher, R., Manning, C.D.: Glove: global vectors for word representation. In: Proceedings of the 2014 Conference on Empirical Methods in Natural Language Processing, pp. 1532–1543 (2014)
13. Pontiki, M., Galanis, D., Pavlopoulos, J., Papageorgiou, H., Androutsopoulos, I., Manandhar, S.: Semeval-2014 task 4: aspect based sentiment analysis. In: Proceedings of the 8th International Workshop on Semantic Evaluation (SemEval 2014), pp. 27–35 (2014)
14. Tang, D., Qin, B., Feng, X., Liu, T.: Effective lstms for target-dependent sentiment classification. In: International Conference on Computational Linguistics, pp. 3298–3307 (2016)
15. Tang, D., Qin, B., Liu, T.: Aspect level sentiment classification with deep memory network. In: Proceedings of the 2016 Conference on Empirical Methods in Natural Language Processing, pp. 214–224 (2016)
16. Vaswani, A., Shazeer, N., Parmar, N., Jones, L., Uszkoreit, J., Gomez, A.N., Kaiser, L.: Attention is all you need. In: Neural Information Processing Systems 2017, pp. 6000–6010 (2017)
17. Wang, Y., Huang, M., Zhu, X., Zhao, L.: Attention-based LSTM for aspect-level sentiment classification. In: Proceedings of the 2016 Conference on Empirical Methods in Natural Language Processing, pp. 606–615 (2016)
18. Yin, Y., Song, Y., Zhang, M.: Document-level multi-aspect sentiment classification as machine comprehension. In: Proceedings of the 2017 Conference on Empirical Methods in Natural Language Processing, pp. 2044–2054 (2017)
19. Zhang, M., Zhang, Y., Vo, D.T.: Gated neural networks for targeted sentiment analysis. In: AAAI 2016 Proceedings of the Thirtieth AAAI Conference on Artificial Intelligence, pp. 3087–3093 (2016)

Abstractive Summarization Improved by WordNet-Based Extractive Sentences

Niantao Xie[1], Sujian Li[1(✉)], Huiling Ren[2], and Qibin Zhai[3]

[1] MOE Key Laboratory of Computational Linguistics,
Peking University, Beijing, China
{xieniantao,lisujian}@pku.edu.cn
[2] Institute of Medical Information, Chinese Academy of Medical Sciences,
Beijing, China
ren.huiling@imicams.ac.cn
[3] MOE Information Security Lab, School of Software and Microelectronics,
Peking University, Beijing, China
qibinzhai@ss.pku.edu.cn

Abstract. Recently, the *seq2seq* abstractive summarization models have achieved good results on the CNN/Daily Mail dataset. Still, how to improve abstractive methods with extractive methods is a good research direction, since extractive methods have their potentials of exploiting various efficient features for extracting important sentences in one text. In this paper, in order to improve the semantic relevance of abstractive summaries, we adopt the *WordNet based sentence ranking algorithm* to extract the sentences which are most semantically to one text. Then, we design a *dual attentional seq2seq framework* to generate summaries with consideration of the extracted information. At the same time, we combine *pointer-generator* and *coverage* mechanisms to solve the problems of out-of-vocabulary (OOV) words and duplicate words which exist in the abstractive models. Experiments on the CNN/Daily Mail dataset show that our models achieve competitive performance with the state-of-the-art ROUGE scores. Human evaluations also show that the summaries generated by our models have high semantic relevance to the original text.

Keywords: Abstractive summarization · Seq2seq model
Dual attention · Extractive summarization · WordNet

1 Introduction

For automatic summarization, there are two main methods: extractive and abstractive. Extractive methods use certain scoring rules or ranking methods to select a certain number of important sentences from the source texts. For example, [2] proposed to make use of Convolutional Neural Networks (CNN) to represent queries and sentences, as well as adopted a greedy algorithm combined with *pair-wise ranking algorithm* for extraction. Based on Recurrent Neural Networks (RNN), [10] constructed a sequence classifier and obtained the

© Springer Nature Switzerland AG 2018
M. Zhang et al. (Eds.): NLPCC 2018, LNAI 11108, pp. 404–415, 2018.
https://doi.org/10.1007/978-3-319-99495-6_34

highest extractive scores on the CNN/Daily Mail corpus set. At the same time, The abstractive summarization models attempt to simulate the process of how human beings write summaries and need to analyze, paraphrase, and reorganize the source texts. It is known that there exist two main problems called OOV words and duplicate words by means of abstraction. [16] proposed an improved pointer mechanism named *pointer-generator* to solve the OOV words as well as came up with a variant of coverage vector called *coverage* to deal with the duplicate words. [12] created the *diverse cell* structures to handle duplicate words problem based on query-based summarization. For the first time, a *reinforcement learning* method based neural network model was raised and obtained the state-of-the-art scores on the CNN/Daily Mail corpus [14].

Both extractive and abstractive methods have their merits. In this paper, we employ the combination of extractive and abstractive methods at the sentence level. In the extractive process, we find that there are some ambiguous words in the source texts. The different meanings of each word can be acquired through the synonym dictionary called WordNet. First *WordNet based Lesk algorithm* is utilized to analyze the word semantics. Then we apply the *modified sentence ranking algorithm* to extract a specified number of sentences according to the sentence syntactic information. During the abstractive part based on *seq2seq model*, we add a new encoder which is derived from the extractive sentences and put the *dual attention* mechanism for decoding operations. As far as we know, it is the first time that joint training of sentence-level extractive and abstractive models has been conducted. Additionally, we combine the *pointer-generator* and *coverage* mechanisms to handle the OOV words and duplicate words.

Our contributions in this paper are mainly summarized as follows:

- Considering the semantics of words and sentences, we improve the *sentence ranking algorithm* based on the *WordNet-based simplified lesk algorithm* to obtain important sentences from the source texts.
- We construct two parallel encoders from the extracted sentences and source texts separately, and make use of *seq2seq dual attentional model* for joint training.
- We adopt the *pointer-generator* and *coverage* mechanisms to deal with OOV words and duplicate words problems. Our results are competitive compared with the state-of-the-art scores.

2   Our Method

Our method is based on the *seq2seq attentional model*, which is implemented with reference to [11] and the attention distribution α_t is calculated as in [1]. Here, we show the architecture of our model which is composed of eight parts as in Fig. 1. We construct two encoders (②④) based on the source texts and extracted sentences, as well as take advantage of a *dual attentional* decoder (①③⑤⑥) to generate summaries. Finally, we combine the *pointer-generator* (⑦) and *coverage* mechanisms (⑧) to manage OOV and duplicate words problems.

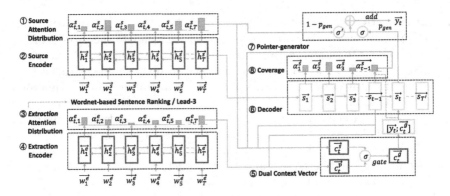

Fig. 1. A *dual attentional encoders-decoder model* with *pointer-generator* network.

2.1 Seq2seq Dual Attentional Model

Encoders-Decoder Model. Referring to [1], we use two single-layer bidirectional Long Short-Term Memory (BiLSTM) encoders including source and extractive encoders, and a single-layer unidirectional LSTM (UniLSTM) decoder in our model, as shown in Fig. 1. For encoding time i, the source texts and the extracted information respectively input the word embeddings $\boldsymbol{w}_i^s$ and $\boldsymbol{w}_i^e$ into two encoders. Meanwhile, the corresponding hidden layer states $\overleftrightarrow{\boldsymbol{h}}_i^s$ and $\overleftrightarrow{\boldsymbol{h}}_i^e$ are generated. At decoding step t, the decoder will receive the word embedding from the step $t-1$, which is obtained according to the previous word in the reference summary during training, or provided by the decoder itself when testing. Next we acquire the state $\boldsymbol{s}_t$ and produce the vocabulary distribution $P(\boldsymbol{y}_t)$.

Here, we are supposed to calculate $\overleftrightarrow{\boldsymbol{h}}_i^s$ by the following formulas:

$$\overrightarrow{\boldsymbol{h}}_i^s = LSTM(\boldsymbol{w}_i^s, \overrightarrow{\boldsymbol{h}}_{i-1}^s) \tag{1}$$

$$\overleftarrow{\boldsymbol{h}}_i^s = LSTM(\boldsymbol{w}_i^s, \overleftarrow{\boldsymbol{h}}_{i+1}^s) \tag{2}$$

$$\overleftrightarrow{\boldsymbol{h}}_i^s = [\overrightarrow{\boldsymbol{h}}_i^s; \overleftarrow{\boldsymbol{h}}_i^s] \tag{3}$$

Also, $\overleftrightarrow{\boldsymbol{h}}_i^e$ could be obtained as follows:

$$\overrightarrow{\boldsymbol{h}}_i^e = LSTM(\boldsymbol{w}_i^e, \overrightarrow{\boldsymbol{h}}_{i-1}^e) \tag{4}$$

$$\overleftarrow{\boldsymbol{h}}_i^e = LSTM(\boldsymbol{w}_i^e, \overleftarrow{\boldsymbol{h}}_{i+1}^e) \tag{5}$$

$$\overleftrightarrow{\boldsymbol{h}}_i^e = [\overrightarrow{\boldsymbol{h}}_i^e; \overleftarrow{\boldsymbol{h}}_i^e] \tag{6}$$

Dual Attention Mechanism. At the tth step, we need not only the previous hidden state $\boldsymbol{s}_{t-1}$, but also the context vector $\boldsymbol{c}_{t-1}^s$, $\boldsymbol{c}_{t-1}^e$, $\boldsymbol{c}_t^s$, $\boldsymbol{c}_t^e$ obtained by the corresponding attention distribution [1] to gain state $\boldsymbol{s}_t$ and vocabulary distribution $P(\boldsymbol{y}_t)$.

Firstly, for source encoder, we calculate the context vector c_t^s in the following way ($\mathbf{V}^s$, $\mathbf{W}_1^s$, $\mathbf{W}_2^s$, $\boldsymbol{b}^s$ are learnable parameters):

$$e_{i,\,t}^s = \mathbf{V}^{sT} \cdot tanh(\mathbf{W}_1^s \cdot \boldsymbol{s}_t + \mathbf{W}_2^s \cdot \overleftrightarrow{\boldsymbol{h}}_i^s + \boldsymbol{b}^s) \tag{7}$$

$$\alpha_{i,\,t}^s = \frac{e_{i,\,t}^s}{\sum_{j=1}^{n_s} e_{j,\,t}^s} \tag{8}$$

$$c_t^s = \sum_{i=1}^{n_s} \alpha_{i,\,t}^s \cdot \overleftrightarrow{\boldsymbol{h}}_t^s \tag{9}$$

Secondly, for extractive encoder, we utilize the identical method to compute the context vector c_t^e ($\mathbf{V}^e$, $\mathbf{W}_1^e$, $\mathbf{W}_2^e$, $\boldsymbol{b}^e$ are learnable parameters):

$$e_{i,\,t}^e = \mathbf{V}^{eT} \cdot tanh(\mathbf{W}_1^e \cdot \boldsymbol{s}_t + \mathbf{W}_2^e \cdot \overleftrightarrow{\boldsymbol{h}}_i^e + \boldsymbol{b}^e) \tag{10}$$

$$\alpha_{i,\,t}^e = \frac{e_{i,\,t}^e}{\sum_{j=1}^{n_e} e_{j,\,t}^e} \tag{11}$$

$$c_t^e = \sum_{i=1}^{n_e} \alpha_{i,\,t}^e \cdot \overleftrightarrow{\boldsymbol{h}}_t^e \tag{12}$$

Thirdly, we get the gated context vector c_t^g by calculating the weighted sum of context vectors c_t^s and c_t^e, where the weight is the *gate network* obtained by the concatenation of c_t^s and c_t^e via *multi-layer perceptron (MLP)*. Details are shown as below (σ is Sigmoid function, $\mathbf{W}^g$, $\boldsymbol{b}^g$ are learnable parameters):

$$\boldsymbol{g}_t = \sigma(\mathbf{W}^g \cdot [c_t^s;\ c_t^e] + \boldsymbol{b}^g) \tag{13}$$

$$c_t^g = \boldsymbol{g}_t \cdot c_t^s + (1 - \boldsymbol{g}_t) \cdot c_t^e \tag{14}$$

In the same way, we can obtain the hidden state $\boldsymbol{s}_t$ and predicte the probability distribution $P(\boldsymbol{y}_t)$ at time t ($\mathbf{W}_1^{in}$, $\mathbf{W}_2^{in}$, $\boldsymbol{b}_{in}$, $\mathbf{W}_1^{out}$, $\mathbf{W}_2^{out}$, $\boldsymbol{b}_{out}$ are learnable parameters).

$$\boldsymbol{s}_t = LSTM(\boldsymbol{s}_{t-1},\ \mathbf{W}_1^{in} \cdot \boldsymbol{y}_{t-1} + \mathbf{W}_2^{in} \cdot c_{t-1}^g + \boldsymbol{b}_{in}) \tag{15}$$

$$P(\boldsymbol{y}_t | \boldsymbol{y}_{<t}, \boldsymbol{x}) = softmax(\mathbf{W}_1^{out} \cdot \boldsymbol{s}_t + \mathbf{W}_2^{out} \cdot c_t^g + \boldsymbol{b}_{out}) \tag{16}$$

2.2   WordNet-Based Sentence Ranking Algorithm

To extract the important sentences, we adopt a *WordNet-based sentence ranking algorithm*. WordNet[1] is a lexical database for the English language, which groups English words into sets of synonyms called synsets and provides short definitions and usage examples. [13] used the *simplified lesk approach* based on WordNet

---

[1] http://www.nltk.org/howto/wordnet.html.

to extract abstracts. We refer to its algorithm and set up our *sentence ranking algorithm* so as to construct the extractive encoder.

For sentence $x = (x_1, x_2, ..., x_n)$, after filtering out the stop words and unambiguous tokens through WordNet, we obtain a reserved subsequence $x' = (x_{i_1}, x_{i_2}, ..., x_{i_m})$. Since some words contain too many different senses which may result in too much calculation, we set a window size n_{win} (default value is 5) and sort x' in descending order according to the number of senses of words, as well as keep the first n_{sav} ($n_{sav} = min(m, n_{win})$) words left to get $x'' = (x_{s_1}, x_{s_2}, ..., x_{s_{n_{sav}}})$. Next, we count the common number of senses of each word as word weight. Finally, we get the sum weights of each sentence and acquire an average sentence weight.

Taking a sentence $x'' = (x_1, x_2, x_3)$ for instance, we make an assumption that x_1 has two senses m_a and m_b, x_2 has two senses m_c and m_d, while x_3 has two senses m_e, m_f. Currently considering x_1 as the keyword, we measure the number of common words between a pair of sentences, which describe the word senses of x_1 and another word.

Table 1 shows all possible matches of the senses of x_1, x_2, x_3. For the two senses of x_1, we can separately obtain the sum of co-occurrence word pairs for each meaning. For m_a, we obtain $count_{m_a} = count_{ac} + count_{ad} + count_{ae} + count_{af}$, for m_b, we gain $count_{m_b} = count_{bc} + count_{bd} + count_{be} + count_{bf}$. The significance corresponding to the higher score $count_{x_1}$ ($count_{m_a}$ or $count_{m_b}$) is assigned to the the keyword x_1.

Table 1. The number of common words between a pair of sentences.

Pair of sentences	Common words in sense description
m_a and m_c	$count_{ac}$
m_a and m_d	$count_{ad}$
m_b and m_c	$count_{bc}$
m_b and m_d	$count_{bd}$
m_a and m_e	$count_{ae}$
m_a and m_f	$count_{af}$
m_b and m_e	$count_{be}$
m_b and m_f	$count_{bf}$

In this way, we're capable of acquiring the average weight of sentence x.

$$weight_{avg} = \frac{1}{n_{sav}} \sum_{i=1}^{n_{sav}} count_{x_i} \tag{17}$$

Let's assume that document $D = (x_1, x_2, ..., x_N)$, which contains a total of N sentences. We sort them in descending order according to the average weights of sentences, and then extract the top n_{top} sentences (default value is 3).

2.3   Pointer-Generator and Coverage Mechanisms

Pointer-Generator Network. *Pointer-generator* is an effective method to solve the problem of OOV words and its structure has been expanded in Fig. 1. We borrow the method improved by [16]. p_{gen} is defined as a switch to decide to generate a word from the vocabulary or copy a word from the source encoder attention distribution. We maintain an extended vocabulary including the vocabulary and all words in the source texts. For the decoding step t and decoder input x_t, we define p_{gen} as:

$$p_{gen} = \sigma(\mathbf{W}_1^p \cdot c_t^s + \mathbf{W}_2^p \cdot s_t + \mathbf{W}_3^p \cdot x_t + b^p) \tag{18}$$

$$P_{vocab} = P(y_t | y_{<t}, x) \tag{19}$$

$$P(w_t) = p_{gen} P_{vocab}(w_t) + (1 - p_{gen}) \sum_{i:w_i = w_t} \alpha_{i,t}^s \tag{20}$$

where w_t is the value of x_t, and $\mathbf{W}_1^p$, $\mathbf{W}_2^p$, $\mathbf{W}_3^p$, b^p are learnable parameters.

Coverage Mechanism. Duplicate words are a critical problem in the *seq2seq model*, and even more serious when generating long texts like multi-sentence texts. [16] made some minor modifications to the *coverage model* [18] which is also displayed in Fig. 1.

First, we calculate the sum of attention distributions from previous decoder steps $(1, 2, 3, ..., t-1)$ to get a coverage vector cov_t:

$$cov_t^s = \sum_{t'=0}^{t-1} \alpha_{t'}^s \tag{21}$$

Then, we make use of coverage vector cov_t to update the attention distribution:

$$e_{i,t}^s = \mathbf{V}^{s^T} \cdot tanh(\mathbf{W}_1^s \cdot s_t + \mathbf{W}_2^s \cdot \overleftrightarrow{h}_i^s + \mathbf{W}_3^s \cdot cov_{i,t}^s + b^s) \tag{22}$$

Finally, we define the coverage loss function $covloss_t$ for the sake of penalizing the duplicate words appearing at decoding time t, and renew the total loss:

$$covloss_t = \sum_i min(\alpha_{i,t}^s, cov_{i,t}^s) \tag{23}$$

$$loss_t = -log(P(w_t^*)) + \lambda\, covloss_t . \tag{24}$$

where w_t^* is the target word at t^{th} step, $-log(P(w_t^*))$ is the primary loss for timestep t during training, hyperparameter λ (default value is 1.0) is the weight for $covloss_t$, $\mathbf{W}_1^s$, $\mathbf{W}_2^s$, $\mathbf{W}_3^s$, b^s are learnable parameters.

3   Experiments

3.1   Dataset

CNN/Daily Mail dataset[2] is widely used in the public automatic summarization evaluation, which contains online news articles (781 tokens on average) paired with multi-sentence summaries (56 tokens on average). [16] provided the data processing script, and we take advantage of it to obtain the non-anonymized version of the data including 287,226 training pairs, 13,368 validation pairs and 11,490 test pairs, though [10,11] used the anonymized version. During training steps, we find that 114 of 287,226 articles are empty, so we utilize the remaining 287,112 pairs for training. Then, we perform the splitting preprocessing for the data pairs with the help of Stanford CoreNLP toolkit[3], and convert them into binary files, as well as get the vocab file for the convenience of reading data.

3.2   Implementation

Model Parameters Configuration. The corresponding parameters of controlled experimental models are described as follows. For all models, we have set the word embeddings and RNN hidden states to be 128-dimensional and 256-dimensional respectively for source encoders, extractive encoders and decoders. Contrary to [11], we learn the word embeddings from scratch during training, because our training dataset is large enough. We apply the optimization technique Adagrad with learning rate 0.15 and an initial accumulator value of 0.1, as well as employ the gradient clipping with a maximum gradient norm of 2.

For the one-encoder models, we set up the vocabulary size to be 50k for source encoder and target decoder simultaneously. We try to adjust the vocabulary size to be 150k, then discover that when the model is trained to converge, the time cost is doubled but the test dataset scores have slightly dropped. In our analysis, the models' parameters have increased excessively when the vocabulary enlarges, leading to overfitting during the training process. Meanwhile, for the models with two encoders, we adjust the vocabulary size to be 40k.

Each pair of the dataset consists of an article and a multi-sentence summary. We truncate the article to 400 tokens and limit the summary to 100 tokens for both training and testing time. During decoding mode, we generate at least 35 words with *beam search algorithm*. Data truncation operations not only reduce memory consumption, speed up training and testing, but also improve the experimental results. The reason is that the vital information of news texts is mainly concentrated in the first half part.

We train on a single GeForce GTX 1080 GPU with a memory of 8114 MiB, and the batch size is set to be 16, as well as the beam size is 4 for *beam search* in decoding mode. For the *seq2seq dual attentional models* without *pointer-generator*, we trained them for about two days. Models with *pointer-generator*

---

[2] https://cs.nyu.edu/~kcho/DMQA/.
[3] https://stanfordnlp.github.io/CoreNLP/.

expedite the training, the time cost is reduced to about one day. When we add *coverage*, the coverage loss weight λ is set to 1.0, and the model needs about one hour for training.

Controlled Experiments. In order to figure out how each part of our models contributes to the test results, based on the released codes[4] of Tensorflow, we have implemented all the models and done a series of experiments.

The baseline model is a general *seq2seq attentional model*, the encoder consists of a biLSTM and the decoder is made up of an uniLSTM. The second baseline model is our *encoders-decoder dual attention model*, which contains two biLSTM encoders and one uniLSTM decoder. This model combines the extractive and generative methods to perform joint training effectively through a *dual attention mechanism*.

For the above two basic models, in order to explain how the OOV and duplicate words are treated, we lead into the *pointer-generator* and *coverage* mechanism step by step. For the second baseline, the two tricks are only related to the source encoder, because we think that the source encoder already covers all the tokens in the extractive encoder. For the extractive encoder, we adopt two methods for extraction. One is the *leading three (lead-3)* sentences technique, which is simple but indeed a strong baseline. The other is the *Modified sentence ranking algorithm* based on WordNet that we explain in details in Sect. 3. It considers semantic relations in words and sentences from source texts.

3.3   Results

ROUGE [7] is a set of metrics with a software package used for evaluating automatic summarization and machine translation results. It counts the number of overlapping basic units including n-grams, longest common subsequences (LCS). We use pyrouge[5], a python wrapper to gain ROUGE-1, ROUGE-2 and ROUGE-L scores and list the $\mathbf{F}_1$ scores in Table 2.

We carry out the experiments based on original dataset, i.e., non-anonymized version of data. For the top three models in table 2, their ROUGE scores are slightly higher than those executed by [16], except for the ROUGE-L score of *Seq2seq + Attn + PGN*, which is 0.09 points lower than the former result. For the fourth model, we did not reproduce the results of [16], ROUGE-1, ROUGE-2, and ROUGE-L decreased by an average of 0.41 points.

For the four models in the middle, we apply the *dual attention* mechanism to integrate extraction with abstraction for joint training and decoding. These model variants own a single *PGN* or *PGN* together with *Cov*, achieve better results than the corresponding vulgaris attentional models simultaneously. We conclude that the extractive encoders play a role, among which we obtained higher ROUGE-1 and ROUGE-2 scores based on the *Lead-3 + Dual-attn +*

---

[4]   https://github.com/tensorflow/models/tree/master/research/textsum.
[5]   https://pypi.org/project/pyrouge/0.1.3/.

Table 2. ROUGE F_1 scores on CNN/Daily Mail non-anonymized testing dataset for all the controlled experiment models mentioned above. According to the official ROUGE usage description, all our ROUGE scores have a 95% confidence interval of at most ±0.25. *PGN, Cov, ML, RL* are abbreviations for *pointer-generator, coverage, mixed-objective learning* and *reinforcement learning*. Models with subscript $_a$ were trained and tested on the anonymized CNN/Daily Mail dataset, as well as with * are the state-of-the-art extractive and abstractive summarization models on the anonymized dataset by now.

Models	ROUGE F_1 scores		
	1	2	L
Seq2seq + Attn	31.50	11.95	28.85
Seq2seq + Attn (150k)	30.67	11.32	28.11
Seq2seq + Attn + PGN	36.58	15.76	33.33
Seq2seq + Attn + PGN + Cov	**39.16**	**16.98**	**35.81**
Lead-3 + Dual-attn + PGN	37.26	16.12	33.87
WordNet + Dual-attn + PGN	36.91	15.97	33.58
Lead-3 + Dual-attn + PGN + Cov	**39.41**	**17.30**	35.92
WordNet + Dual-attn + PGN + Cov	39.32	17.15	**36.02**
Lead-3 ([16])	40.34	17.70	36.57
Lead-3 ([10])$_a$	39.20	15.70	35.50
SummaRuNNer ([10])$_a^*$	**39.60**	**16.20**	**35.30**
RL + Intra-attn ([14])$_a^*$	**41.16**	15.75	**39.08**
ML + RL + Intra-attn ([14])$_a$	39.87	**15.82**	36.90

PGN + Cov model, and achieve a better ROUGE-L score on *WordNet + Dual-attn + PGN + Cov* model.

Let's take a look at the five models at the bottom, two of which give the state-of-the-art scores for the extractive and generative methods our scores are already comparable to them. It is worthy to mention that based on the *dual attention*, our models related to both *Lead-3* and *WordNet* with *PGN* and *Cov* have exceeded the previous best ROUGE-2 scores. When in fact, previous *SummaRuNNer, RL* related models are based on anonymized dataset, these differences may cause some deviations in the comparison of experimental results.

We give some generated summaries of different models for one selected test article. From Fig. 2, we can see that the red words represent key information about *who, what, where* and *when*. We can match the corresponding keywords in the remaining seven summaries to find out whether they cover all the significant points, and check if they are expressed in a concise and coherent way. It can be discovered from Fig. 2 that most of the models have lost several vital points, and the model *Lead-3 + Dual-attn + PGN* has undergone fairly serious repetition. Our model *WordNet + Dual-attn + PGN + Cov* holds the main key information as well as has better readability and semantic correctness reliably.

Reference: six people , including one civilian , are killed when a car bomb explodes near a police station . six others are killed when their armored vehicle is attacked on a highway in northern sinai . ansar beit al-maqdis , an isis affiliate , claims responsibility .
Seq2seq + Attn: an attack on a police station , a guard is killed in a car bomb . the attack comes hours after another operation that the group claims responsibility . the attacks come as the military announced a departure of several senior military positions .
Seq2seq + Attn + PGN: six people were killed when a car bomb exploded near the police station in al-arish , capital of north sinai . a first lieutenant , a sergeant and four conscripts were killed when their armored vehicle was attacked on the highway from al-arish to sheikh zuweid in northern sinai . ansar beit al-maqdis has claimed many attacks against the army and police in sinai .
Seq2seq + Attn + PGN + Cov: six people were killed when a car bomb exploded near the police station in al-arish , capital of north sinai . ansar beit al-maqdis , a sergeant and four conscripts were killed when their armored vehicle was attacked on the highway from al-arish to sheikh zuweid in northern sinai . a third attack sunday on a checkpoint in rafah left three security personnel injured , after unknown assailants opened fire .
WordNet + Dual-attn + PGN: six people killed when a car bomb exploded near the police station in al-arish , capital of north sinai . a first lieutenant , a sergeant and four conscripts were killed when their armored vehicle was attacked on the highway . a third attack sunday on a checkpoint in rafah left three security personnel injured .
Lead-3 + Dual-attn + PGN: six people , including one civilian , were killed when a car bomb exploded near the police station in al-arish , capital of north sinai . six people , including one civilian , were killed when a car bomb exploded near the police station in al-arish , capital of north sinai , authorities say .
WordNet + Dual-attn + PGN + Cov: six people , including one civilian , were killed when a car bomb exploded near the police station . a first lieutenant , a sergeant and four conscripts were killed when their armored vehicle was attacked on the highway . ansar beit al-maqdis has claimed many attacks against the army and police in sinai .
Lead-3 + Dual-attn + PGN + Cov: six people , including one civilian , were killed when a car bomb exploded near the police station . ansar beit al-maqdis , an isis affiliate , claimed responsibility for the attack . ansar beit al-maqdis has claimed many attacks against the army and police .

Fig. 2. Summaries for all the models of one test article example.

4   Related Work

Up to now, automatic summarization with extractive and abstractive methods are under fervent research. On the one hand, the extractive techniques extract the topic-related keywords and significant sentences from the source texts to constitute summaries. [3] proposed a *seq2seq model* with a hierarchical encoder and attentional decoder to solve extractive summarization tasks at the word and sentence levels. Currently [10] put forward *SummaRuNNer*, a RNN based sequence model for extractive summarization and it achieves the previous state-of-the-art performance. On the other hand, abstractive methods establish an intrinsic semantic representation and use natural language generation techniques to produce summaries which are closer to what human beings express. [1] applied the combination of *seq2seq model* and *attention* mechanism to machine translation tasks for the first time. [15] exploited *seq2seq model* to sentence compression to lay the groundwork for subsequent summarization with different granularities. [8] used *encoder-decoder with attention* method to generate news headlines. [20] added a *selective gate network* to the basic model in order to control which part of the information flowed from encoder to decoder. [17] raised a model based on graph and *attention* mechanism to strengthen the positioning of vital information of source texts.

So as to solve rare and unseen words, [5,6] proposed the *COPYNET model* and *pointing* mechanism, [19] created read-again and copy mechanisms. [11] made a combination of the basic model with *large vocabulary trick (LVT)*, *feature-rich encoder, pointer-generator*, and *hierarchical attention*. In addition to *pointer-generator*, other tricks of this paper also contributed to the experiment results. [16] presented an updated version of *pointer-generator* which proved to be better. As for duplicate words, for sake of solving problems of over or missing translation, [18] came up with a *coverage* mechanism to avail oneself of historical information for attention calculation, while [16] provided a progressive version. [12] introduced a series of *diverse cell* structures to solve the duplicate words.

So far, few papers have considered about the structural or sementic issues at the language level in the field of summarization. [4] presented a novel unsupervised method that made use of a pruned dependency tree to acquire the sentence compression. Based on a Chinese short text summary dataset (LCSTS) and the *attentional seq2seq model*, [9] proposed to enhance the semantic relevance by calculating the cos similarities of summaries and source texts.

5   Conclusion

In our paper, we construct a *dual attentional seq2seq model* comprising source and extractive encoders to generate summaries. In addition, we put forward the *modified sentence ranking algorithm* to extract a specific number of high weighted sentences, for the purpose of strengthening the semantic representation of the extractive encoder. Furthermore, we introduce the *pointer-generator* and *coverage* mechanisms in our models so as to solve the problems of OOV and duplicate words. In the non-anonymized CNN/Daily Mail dataset, our results are close to the state-of-the-art ROUGE F_1 scores. Moreover, we get the highest abstractive ROUGE-2 F_1 scores, as well as obtain such summaries that have better readability and higher semantic accuracies. In our future work, we plan to unify the *reinforcement learning* method with our abstractive models.

Acknowledgments. We thank the anonymous reviewers for their insightful comments on this paper. This work was partially supported by National Natural Science Foundation of China (61572049 and 61333018). The correspondence author is Sujian Li.

References

1. Bahdanau, D., Cho, K., Bengio, Y.: Neural machine translation by jointly learning to align and translate. arXiv preprint arXiv:1409.0473 (2014)
2. Cao, Z., Li, W., Li, S., Wei, F., Li, Y.: Attsum: Joint learning of focusing and summarization with neural attention. arXiv preprint arXiv:1604.00125 (2016)
3. Cheng, J., Lapata, M.: Neural summarization by extracting sentences and words. arXiv preprint arXiv:1603.07252 (2016)
4. Filippova, K., Strube, M.: Dependency tree based sentence compression. In: Proceedings of the Fifth International Natural Language Generation Conference, pp. 25–32. Association for Computational Linguistics (2008)
5. Gu, J., Lu, Z., Li, H., Li, V.O.: Incorporating copying mechanism in sequence-to-sequence learning. arXiv preprint arXiv:1603.06393 (2016)
6. Gulcehre, C., Ahn, S., Nallapati, R., Zhou, B., Bengio, Y.: Pointing the unknown words. arXiv preprint arXiv:1603.08148 (2016)
7. Lin, C.Y.: Rouge: A package for automatic evaluation of summaries. In: Text Summarization Branches Out (2004)
8. Lopyrev, K.: Generating news headlines with recurrent neural networks. arXiv preprint arXiv:1512.01712 (2015)
9. Ma, S., Sun, X., Xu, J., Wang, H., Li, W., Su, Q.: Improving semantic relevance for sequence-to-sequence learning of chinese social media text summarization. arXiv preprint arXiv:1706.02459 (2017)
10. Nallapati, R., Zhai, F., Zhou, B.: Summarunner: A recurrent neural network based sequence model for extractive summarization of documents. In: AAAI, pp. 3075–3081 (2017)
11. Nallapati, R., Zhou, B., Gulcehre, C., Xiang, B., et al.: Abstractive text summarization using sequence-to-sequence rnns and beyond. arXiv preprint arXiv:1602.06023 (2016)
12. Nema, P., Khapra, M., Laha, A., Ravindran, B.: Diversity driven attention model for query-based abstractive summarization. arXiv preprint arXiv:1704.08300 (2017)
13. Pal, A.R., Saha, D.: An approach to automatic text summarization using wordnet. In: Advance Computing Conference (IACC), 2014 IEEE International, pp. 1169–1173. IEEE (2014)
14. Paulus, R., Xiong, C., Socher, R.: A deep reinforced model for abstractive summarization. arXiv preprint arXiv:1705.04304 (2017)
15. Rush, A.M., Chopra, S., Weston, J.: A neural attention model for abstractive sentence summarization. arXiv preprint arXiv:1509.00685 (2015)
16. See, A., Liu, P.J., Manning, C.D.: Get to the point: Summarization with pointer-generator networks. arXiv preprint arXiv:1704.04368 (2017)
17. Tan, J., Wan, X., Xiao, J.: Abstractive document summarization with a graph-based attentional neural model. In: Proceedings of the 55th Annual Meeting of the Association for Computational Linguistics (Volume 1: Long Papers), vol. 1, pp. 1171–1181 (2017)
18. Tu, Z., Lu, Z., Liu, Y., Liu, X., Li, H.: Modeling coverage for neural machine translation. arXiv preprint arXiv:1601.04811 (2016)
19. Zeng, W., Luo, W., Fidler, S., Urtasun, R.: Efficient summarization with read-again and copy mechanism. arXiv preprint arXiv:1611.03382 (2016)
20. Zhou, Q., Yang, N., Wei, F., Zhou, M.: Selective encoding for abstractive sentence summarization. arXiv preprint arXiv:1704.07073 (2017)

Improving Aspect Identification
with Reviews Segmentation

Tianhao Ning[1,2], Zhen Wu[1,2], Xin-Yu Dai[1,2(✉)], Jiajun Huang[1,2],
Shujian Huang[1,2], and Jiajun Chen[1,2]

[1] National Key Laboratory for Novel Software Technology,
Nanjing University, Nanjing 210023, China
{ningth,wuz,huangjj}@nlp.nju.edu.cn,
{daixinyu,yincy,huangsj,chenjj}@nju.edu.cn
[2] Collaborative Innovation Center of Novel Software Technology
and Industrialization, Nanjing 210023, China

Abstract. Aspect identification, a key sub-task in Aspect-Based Sentiment Analysis (ABSA), aims to identify aspect categories from online user reviews. Inspired by the observation that different segments of a review usually express different aspect categories, we propose a reviews-segmentation-based method to improve aspect identification. Specifically, we divide a review into several segments according to the sentence structure, and then automatically transfer aspect labels from the original review to its derived segments. Trained with the new constructed segment-level dataset, a classifier can achieve better performance for aspect identification. Another contribution of this paper is extracting alignment features, which can be leveraged to further improve aspect identification. The experimental results show the effectiveness of our proposed method.

Keywords: Aspect identification · Reviews segmentation
Alignment features

1 Introduction

Sentiment analysis and opinion mining have drawn increasing attention in recent years because of the rapid growth of user-generated reviews on the Internet. For a product, users usually evaluate it from multiple aspects in a review. For example, a review *"Get this computer for portability and fast processing!!!"* of laptop domain contains two aspects, namely *portability* and *cpu operation performance*. So instead of classifying the overall sentiment of a review into binary polarity (positive or negative), a finer-grained task, known as Aspect-Based Sentiment Analysis (ABSA) [16], is proposed to discover more detailed entities, attributes, and emotions of users towards various aspects from reviews. In ABSA, a key sub-task is to identify aspect categories from reviews before sentimental polarity can be predicted towards each aspect.

Supported by the NSFC (No. 61472183, 61672277, 61772261).

© Springer Nature Switzerland AG 2018
M. Zhang et al. (Eds.): NLPCC 2018, LNAI 11108, pp. 416–428, 2018.
https://doi.org/10.1007/978-3-319-99495-6_35

For a specific domain of product or service, the set of aspect categories is usually predefined as E#A, where E is an entity and A is an attribute of E [10] such as *LAPTOP#PRICE*. Users usually express opinions toward multiple aspect categories in a review. Thus aspect identification can be formulated as a multi-label classification problem. Some previous works focus on designing the classification models and feature representations [17–19] and obtain some competitive results.

Different from previous works, we observe that different segments of a review usually express different aspect categories. For example, as shown in Table 1, the review *"Fantastic for the price, but the keys were not illuminated."* can be divided into two segments, namely *"Fantastic for the price."* and *"But the keys were not illuminated."*. These two segments are mutually independent, the former segment expresses aspect category *LAPTOP#PRICE* and the latter describes aspect category *KEYBOARD#DESIGN_FEATURES*. The example shows that each segment and its aspect categories have finer-grained mapping relation than the whole review and overall aspect categories have. Therefore, we claim that classification performance can be improved if we obtain the finer-grained mapping dataset, because we do not need to consider the interference from other segments when dealing with current segment.

To address the issues, we propose a reviews-segmentation-based method to divide a review into multiple segments, and then transfer aspect labels from the original review to corresponding segments automatically. These two steps will help us construct a review-segment-level labeled dataset with finer-grained mapping relation. After reviews segmentation and labels transferring, like solving other classification problems, we train a classifier on the constructed dataset for predicting aspect categories of new reviews. In this paper, we use Long Short-Term Memory (LSTM) [5] as classifier.

In addition, we also observe that in reviews some words have strong indication for aspects. For example, in the review of Table 1, the word *"price"* expresses the aspect category *LAPTOP#PRICE*, the words *"keys"* and *"illuminated"* indicate the aspect category *KEYBOARD#DESIGN_FEATURES*. However, due to the sparseness of the training data, it is hard to learn some sparse words like *"illuminated"* as important features to identify some aspect categories. Therefore, we introduce alignment algorithm in machine translation to extract alignment

Table 1. Examples of aspect categories identification.

Review: Fantastic for the price, but the keys were not illuminated **Aspects:** LAPTOP#PRICE, KEYBOARD#DESIGN_FEATURES
Segment_1: Fantastic for the price **Aspects:** LAPTOP#PRICE **Segment_2:** But the keys were not illuminated **Aspects:** KEYBOARD#DESIGN_FEATURES

features between words and aspect categories, which are used for further improving the aspect identification performance.

The main contributions of our work can be summarized as follows:

1. We improve the performance of aspect identification with reviews segmentation. Especially, we propose an effective method to divide a review into multiple segments and transfer the aspect labels from reviews to the corresponding segments.
2. We introduce the alignment algorithm in machine translation to extract the alignment features to further improve aspect identification.

2   Related Work

The ABSA task was added to the SemEval challenges since 2014 [11]. The subtask aspect identification of ABSA predefines aspect categories for a specific domain, so it can be regarded as a multi-label classification problem. Some early works employ traditional features and classification algorithms for aspect identification. [6] follows the one-vs-all strategy and build a binary Support Vector Machine (SVM) [2] classifier with ngrams and lexicon features for each aspect category. However, if a token implying an aspect, e.g., "expensive", is not taken as a feature, the SVM classifier cannot correctly identify its corresponding category. Therefore, [21] enhances the results from the SVM classifier by using implicit aspect indicators [4]. In addition, Maximum Entropy is also adopted for aspect identification with bag-of-words-like features (e.g. words, lemmas) [14].

Recently, neural network based models are explored to solve this problem. [17] extracts lexicon, syntax and word cluster as features, and trains a binary single layer feedforward network for each aspect category. [18] enhances the system of [17] by adding neural network features learned from a Deep Convolutional Neural Network system [15]. Different from previous works, [13] does not use traditional hand-crafted features, and directly train a convolutional neural network to output probability distributions over all aspect categories.

However, the above works all pay attention to designing hand-crafted features and classification models, but ignore the phenomenon that different segments of a review usually expressed different aspect categories, which motivates our work.

3   Method

3.1   Overview of Our Method

From our observation mentioned in Sect. 1, we have strong motivation to train a segment-level classifier to capture finer-grained mapping relation. To achieve the goal, we propose an effective method to build the corresponding segment-level dataset from an original review-level dataset.

Firstly, we use reviews segmentation method according to punctuations or dependency parsing tree to divide a long review into multiple segments. As Fig. 1

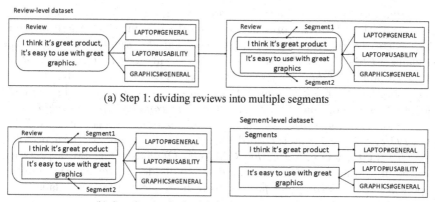

(a) Step 1: dividing reviews into multiple segments

(b) Step 2: transferring labels from reviews to segments

Fig. 1. Reviews segmentation and labels transferring.

shows, in the step 1 the review "*I think it's great product, it's easy to use with great graphics.*" will be divided into two segments, namely "*I think it's great product.*" and "*It's easy to use with great graphics.*".

Secondly, we train an LSTM classifier on the original review dataset and design some conservative rules (refer to Algorithm 1) to transfer the aspect labels from reviews to the corresponding segments. In above example, the label *LAPTOP#GENERAL* will be transferred to the segment "*I think it's great product.*". The labels *LAPTOP#USABILITY* and *GRAPHICS#GENERAL* will be transferred to the other segment "*It's easy to use with great graphics.*". After labels transferring, we will have a segment-level dataset.

Finally, a new LSTM classifier will be trained on the constructed segment-level dataset for predicting aspect labels of new reviews.

3.2 Reviews Segmentation

Reviews Segmentation with Punctuations. In linguistics, a clause is the smallest grammatical unit that can express a complete proposition. Sometimes some sentences themselves are clauses. The simplest reviews segmentation approach is to divide a review into several clauses according to punctuations. However, this approach does not work when there is no punctuation in the sentence. For instance, there is no punctuation in the review "*It's more expensive but well worth it in the long run*", whereas it has two clauses "*It's more expensive*" and "*well worth it in the long run*". These two clauses express different aspects. The similar sentences are quite common in real reviews.

Reviews Segmentation with Dependency Parsing Tree. The above example shows that we cannot divide a review into multiple segments when there is no punctuation in the sentence. Therefore, we need to consider more structural information. Here we present two typical cases in which there are multiple aspects. In the review *"I like the food but the waiter was rude."*, the first clause *"I like the food"* describes the aspect category *FOOD#GENERAL*, and the second one *"the waiter was rude"* expresses the aspect category *SERVICE#QUALITY*. In this example, the two aspects are expressed in two independent clauses. For another review *"Get this computer for portability and fast processing!"*, the word *"portability"* indicates the aspect category *LAPTOP#PORTABILITY*, and the word *"fast processing"* expresses the aspect category *CPU#OPERATION_PERFORMANCE*. Obviously, the two aspects are dispersed in syntactic coordinate structures. To divide this review, the common component *"Get this computer for"* needs to be replicated for each clause.

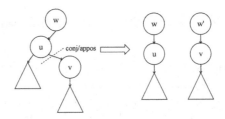

Fig. 2. Sentence segmentation with dependency parsing tree.

Fortunately, dependency parsing can address the above issues. Figure 2 shows how a sentence is divided into multiple relatively complete clauses with dependency parsing tree. We break dependency relations denoted by $\langle u, v \rangle$, whose dependency type is *conj* (conjunt) or *appos* (appositional modifier). More specifically, we denote all ancestors of u and their other descendants as w, except the subtree rooted by u. Let w' be the clone of w, then we append v and its descendants to w'. One exception is those sentences with compound predicates, like *"subject verb1 object1 and verb2 object2"*. In this case, *verb1* is u and *verb2* is v, but *subject* is not an ancestor of *verb1* and needs to be appended to *verb2*. In our work, this sentence is divided into *"subject verb1 object1"* and *"subject verb2 object2"*.

3.3   Labels Transferring from Reviews to Segments

After reviews segmentation, a complex review is divided into multiple segments. The next step is to transfer the aspect labels from original review to the corresponding segments. One way for labels transferring is to train a classifier (LSTM) on the original review-level dataset, and then predict the aspect labels of each segment. However, with this method, the quality of the constructed segment-level dataset is not guaranteed and heavily depends on the trained classifier. Therefore, based on the classifier trained on the review-level dataset, we add

Algorithm 1. Labels Transferring Algorithm

Require: A review r, its label bag b, and trained classifier f on review-level labeled dataset

Ensure: A set s of segments, and corresponding label bags B'

1: $s = divided(r)$, dividing r into segments according to the method in Section 3.2
2: **if** $|s| = 1$ **then**
3:     set $B' := \{b\}$
4: **else**
5:     set $B' := [\varnothing].* |s|$ {Initialize the label bags B' whose size is $|s|$.}
6:     **for** l in b **do**
7:         set $flag := False$ {The $flag$ means whether the label l is transferred to segments.}
8:         **for** $i = 1$ to $|s|$ **do**
9:             **if** $f(s_i)_l >= f(r)_l$ **then**
10:                 add l to B'_i {Transfer label l to i-th segment.}
11:                 set $flag := True$
12:             **end if**
13:         **end for**
14:         **if** $flag == False$ **then**
15:             **if** r not in s **then**
16:                 add r to s {If label l is not transferred to segments, make sure that review r is in s.}
17:                 $B'_{|s|+1} := [\varnothing]$ {$|s| + 1$ is the index of r in s}
18:             **end if**
19:             add l to $B'_{|s|+1}$ {Return the label l to the review r.}
20:         **end if**
21:     **end for**
22:     **return** s, B'
23: **end if**

---

some constraints (refer to Algorithm 1) for labels transferring to improve the quality of segment-level dataset.

Algorithm 1 demonstrates the pseudocode of Labels Transferring algorithm. Firstly, we train an LSTM classifier f on review-level labeled dataset. Let $f(x)_y$ be the predicted probability of aspect label y towards the input x. Then, to ensure the reliability of labels transferring, we set some constraints during the transferring. Specifically, for each aspect label l of a review r and each segment s_i divided from r, l would be transferred from r to s_i on condition that $f(s_i)_l >= f(r)_l$. This condition means that the segment s_i has stronger indication for aspect label l compared with the whole review r. If none of the segments satisfies the condition, we will return the label l to the review r and add the original review r and the label l to the new dataset. We use a classifier and fallback strategy to ensure that the labels are transferred to review segments as accurately as possible.

After labels transferring, we have a segment-level labeled dataset, with which we can train a more powerful classifier for aspect identification.

3.4 Alignment Feature

In reviews, some words or phrases have strong indication for expressed aspect categories. For example, for the review *"Fantastic for the price, but the keys were not illuminated."* in Fig. 3, the word *"illuminated"* is quite significant for the identification of aspect category *KEYBOARD#DESIGN_FEATURES*. However, it is hard to learn the word *"illuminated"* as an effective feature for a classifier due to the sparseness of training data. Therefore, we employ alignment algorithm to extract alignment features between the words in reviews and expressed aspects to improve aspect identification.

Fig. 3. An alignment example between a review and corresponding parallel data.

Firstly, we build the parallel data for extracting alignment features. The construction process is illustrated with an example. For the review in Fig. 3, we can obtain its paired labels (LAPTOP#PRICE, positive) and (KEYBOARD#DESIGN_FEATURES, negative) (The polarity of aspect is provided by original datasets). Then we rewrite the paired labels as *"LAPTOP#PRICE is positive; KEYBOARD#DESIGN_FEATURES is negative;"*. In fact, the original review and the rewritten text are parallel and express the same meaning towards expressed aspects. With the parallel training data, we remove those stop-words and punctuations, and use Giza++ [9] to train IBM model 4 [3] to obtain bidirectional alignment probabilities, which contains the probability from words to aspects and the probability from aspects to words. We add these probabilities as alignment features to improve the performance of aspect identification.

3.5 Sequence Encoder and Aspect Identification

In this work, we adopt Long Short-Term Memory (LSTM) to encode reviews or review segments because of its excellent performance on sequence modeling. For a review or review segment consisting of n words $\{w_1, w_2, \cdots, w_n\}$, each word w_i is mapped to its embedding $\mathbf{w}_i \in \mathbb{R}^d$. LSTM network receives $[\mathbf{w}_1, \mathbf{w}_2, \cdots, \mathbf{w}_n]$ and generates hidden states $[\mathbf{h}_1, \mathbf{h}_2, \cdots, \mathbf{h}_n]$. Then we concatenate the last hidden state $\mathbf{h}_n$ and alignment features in Sect. 3.4 as the final representation $\mathbf{r}$ for aspect identification. We use a linear layer to project representation $\mathbf{r}$ into the target space of C aspect categories. Since aspect identification is a multi-label classification problem, we add no-linear function sigmoid rather than softmax before calculating cross entropy loss:

$$p = \sigma\left(\mathbf{W}_r \mathbf{r} + \mathbf{b}_r\right), \tag{1}$$

where $\mathbf{W}_r$ and $\mathbf{b}_r$ are weight matrix and bias vector respectively, every dimension p_a of p is in $[0, 1]$ and corresponds to the predicted probability of aspect category a. We set a threshold θ and make a prediction that a sample has a aspect label a when p_a exceeds the threshold θ. The loss function for optimization when training is defined as:

$$L = -\frac{1}{M} \sum_{m=1}^{M} \sum_{a=1}^{C} (p_a^g(d_m) \cdot \log(p_a(d_m)) + (1 - p_a^g(d_m)) \cdot \log(1 - p_a(d_m))), \quad (2)$$

where p_a^g is the gold probability of aspect label a with ground truth being 1 and others being 0, M denotes the number of training data, d_m represents the m-th sample of training data. Finally, for a review we merge all the predicted results of its segments as the aspects of the whole review.

4   Experiments

4.1   Setup

- **Dataset:** We evaluate the effectiveness of the proposed method on SemEval-2015 task-12 from two domains (laptop and restaurant)[1]. Statistics of the original datasets are shown in Table 2. In these two datasets, all aspect categories in testing set exist in training set.
- **Preprocessing:** We use NLTK [1] to tokenize reviews and keep a vocabulary of 1500 most frequent words excluding stop-words. We use the dependency parser in Stanford CoreNLP [7] for reviews segmentation. Word embeddings are pretrained with skip-gram model [8] on the Yelp Phoenix Academic Dataset, which includes eighteen million user reviews[2] in restaurant domain.
- **Hyper-parameters selection:** We set word vector size to be 300. The dimensions of hidden states in LSTM are set to be 256. We train all models with AdaDelta [20]. The predicting thresholds θ are obtained via grid search in $[0.1, 0.3]$ with increments of 0.01.
- **Metrics:** We use the precision and recall to compute F1-score as evaluation metrics of the performance of aspect identification.

Table 2. Statistics of the original datasets.

Dataset	Laptop-Train	Laptop-Test	Restaurant-Train	Restaurant-Test
Reviews	1739	761	1315	685
Aspects	81	58	12	12

---

[1] http://alt.qcri.org/semeval2015/task12/.
[2] https://www.yelp.com/dataset/challenge/.

4.2   Validating the Performance of Labels Transferring

It is obvious that the quality of new constructed datasets has a significant effect on the performance of aspect identification. To evaluate the performance of labels transferring, we randomly select 500 samples from segment-level dataset in restaurant domain, and invite three experience-rich annotators to manually annotate the aspect categories of every segment. The results are as the Table 3 shows.

Table 3. Performance of labels transferring from reviews to segments in restaurant domain.

	Precision	Recall	F1-score
Dependency tree	0.9268	0.9172	0.9220
Punctuation	0.9451	0.8498	0.8949

We validate the performance of labels transferring based on two different reviews segmentation methods, namely punctuations and dependency tree. From Table 3, we can observe that even the simple review segmentation method based on punctuations achieves around 90% F1-score, which proves most of aspect labels are correctly transferred to corresponding segments. In addition, we obtain better results of labels transferring with dependency parsing. The precision, recall and F1-score are all above 91%. The results is reasonable because we consider more structural information with dependency parsing.

4.3   The Statistics of Reviews Segmentation

In our observation, different segments of a review usually express different aspect categories, which means we can obtain finer-grained mapping datasets after reviews segmentation and labels transferring. In order to validate the rationality of review segmentation, we count the number of data containing n aspects before and after reviews segmentation respectively. As Fig. 4(a) shows, after reviews segmentation and labels transferring, on both two domains, we have more samples with only one aspect, and fewer samples with two or more aspects. More expected results can be achieved after reviews segmentation with dependency parsing, as shown in Fig. 4(b). Compared to segmentation with punctuations, segmentation with dependency parsing achieves more segments with only one aspect.

Overall, Table 3 and Fig. 4 validate our observation that finer-grained correspondence between aspects and review segments exists. Therefore, we can build high-quality segment-level datasets with less interference from other segments after reviews segmentation and labels transferring.

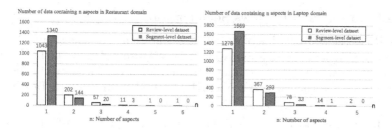

(a) Reviews segmentation with punctuations.

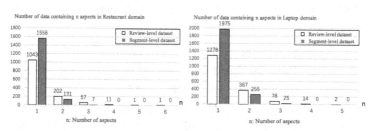

(b) Reviews segmentation with dependency parsing.

Fig. 4. Number of data containing n aspects on two review datasets of SemEval-2015 before and after reviews segmentation respectively.

4.4 Baselines

We compare our method with several baseline methods for aspect identification:

- **TJUdeM:** TJUdeM [21] combines a SVM classifier with implicit aspect indicators. The SVM classifier uses words as features to determine the aspect categories. Additionally, they identify the implicit aspect indicators manually by setting a set of indicators for several aspects.
- **Sentiue:** Sentiue [14] uses a separate Maximum Entropy classifier with bag-of-words-like features (e.g. words, lemmas) for each entity and each attribute. Subsequently, heuristics are applied to the output of the classifiers to determine which aspect categories will be assigned to each sentence.
- **NLANGP:** NLANGP [17] is the winning system of SemEval-2015 task-12 and achieved the best performance in two domains. They train a sigmoid feedforward network as a classifier respectively for each aspect category. They use features containing bag-of-word, n-grams, parsing, and word embeddings learnt from Amazon and Yelp data [12].
- **LSTM:** We train an LSTM classifier on original datasets as one of our baselines.

4.5   Effectiveness of Reviews Segmentation

We use RS as the abbreviation of reviews segmentation. The experimental results
are shown in Table 4. We can observe that in restaurant domain LSTM without
any feature engineering outperforms the traditional classification model TJU-
deM, Sentinue and the winning model NLANGP of SemEval-2015 Task 12.

Table 4. Effectiveness of reviews segmentation for aspect identification on two
datasets. On the basis of LSTM, RS_Punc represents reviews segmentation with punctuations, and RS_Tree denotes reviews segmentation with dependency parsing. "**"
means that LSTM+RS_Punc and LSTM+RS_Tree are significantly better than LSTM
with 99% t-test.

Models	Restaurant			Laptop		
	Precision	Recall	F1-score	Precision	Recall	F1-score
TJUdeM	0.4782	0.5806	0.5245	0.4489	0.4821	0.4649
Sentiue	0.6330	0.4720	0.5410	0.5770	0.4410	0.5000
NLANGP	0.6386	0.6155	0.6268	0.6425	0.4209	0.5086
LSTM	0.6919	0.5809	0.6315	0.6009	0.4241	0.4972
LSTM+RS_Punc	0.6895	0.6219	0.6540**	0.6135	0.4420	0.5138**
LSTM+RS_Tree	0.6895	0.6361	**0.6617****	0.5899	0.4673	**0.5215****

In addition, compared to the baseline methods, LSTM+RS_Punc and
LSTM+RS_Tree achieve significant improvements on the two datasets. Especially, LSTM+RS_Tree improves the performance over the LSTM by 3.02% on
Restaurant dataset and 2.43% on Laptop dataset in F1-score. The results show
that reviews segmentation is effective for aspect identification.

Compared with LSTM+RS_Punc, the model LSTM+RS_Tree achieves better
performance on two datasets. The comparison shows that reviews segmentation
with dependency parsing is more reasonable because more structural information
is considered.

4.6   Effectiveness of Alignment Features

To validate the effectiveness of alignment features, we also report the
results of our model incorporating alignment features into representation
of review segments. As shown in Table 5, the LSTM+RS_Punc+align and
LSTM+RS_Tree+align are our models containing alignment features. When we
add alignment features to representation, more promising results are achieved in
both two domains. The results show that alignment features can strengthen the
connection between some key words and aspect categories. For example, in the
review segment "*excellent speed for processing data.*", the alignment probability
from "*speed*" to *LAPTOP#PERFORMANCE* is 0.9533. With the probability,
it is quite possible that the review segment is assigned with the aspect category
LAPTOP#PERFORMANCE, which can help improve classifier's performance.

Table 5. Effectiveness of alignment features for aspect identification on two datasets. The "+align" represents the model using alignment features. "*" means that LSTM+RS_Punc+align and LSTM+RS_Tree+align are significantly better than other no alignment features methods with 95% t-test.

Models	Restaurant			Laptop		
	Precision	Recall	F1-score	Precision	Recall	F1-score
LSTM+RS_Punc	0.6895	0.6219	0.6540	0.6135	0.4420	0.5138
LSTM+RS_Punc+align	0.7076	0.6245	**0.6635***	0.6667	0.4346	**0.5262***
LSTM+RS_Tree	0.6895	0.6361	0.6617	0.5899	0.4673	0.5215
LSTM+RS_Tree+align	0.7074	0.6335	**0.6685***	0.6376	0.4620	**0.5358***

5 Conclusion

In this work, we propose a reviews-segmentation-based method to improve aspect identification. Specifically, we firstly divide a review into multiple segments, then propose an algorithm to transfer the aspects from the original reviews to the corresponding segments. With the segment-level dataset, we can train a more powerful classifier for aspect identification. For better identification, we also introduce the alignment algorithm in machine translation to extract alignment probabilities. With our proposed method and novel alignment features, promising results are achieved on two benchmark datasets.

References

1. Bird, S.: NLTK: the natural language toolkit. In: Proceedings of the COLING/ACL on Interactive Presentation Sessions, pp. 69–72. Association for Computational Linguistics (2006)
2. Boser, B.E., Guyon, I.M., Vapnik, V.N.: A training algorithm for optimal margin classifiers. In: Proceedings of the 5th Annual Workshop on Computational Learning Theory, pp. 144–152. ACM (1992)
3. Brown, P.F., Pietra, V.J.D., Pietra, S.A.D., Mercer, R.L.: The mathematics of statistical machine translation: parameter estimation. Comput. Linguist. **19**(2), 263–311 (1993)
4. Cruz, I., Gelbukh, A.F., Sidorov, G.: Implicit aspect indicator extraction for aspect based opinion mining. Int. J. Comput. Linguist. Appl. **5**(2), 135–152 (2014)
5. Hochreiter, S., Schmidhuber, J.: Long short-term memory. Neural Comput. **9**(8), 1735–1780 (1997)
6. Kiritchenko, S., Zhu, X., Cherry, C., Mohammad, S.: NRC-Canada-2014: detecting aspects and sentiment in customer reviews. In: Proceedings of the 8th International Workshop on Semantic Evaluation (SemEval 2014), pp. 437–442 (2014)
7. Manning, C.D., Surdeanu, M., Bauer, J., Finkel, J.R., Bethard, S., McClosky, D.: The stanford CoreNLP natural language processing toolkit. In: ACL (System Demonstrations), pp. 55–60 (2014)
8. Mikolov, T., Yih, W.t., Zweig, G.: Linguistic regularities in continuous space word representations. In: HLT-NAACL, pp. 746–751 (2013)

9. Och, F.J., Ney, H.: Giza++: training of statistical translation models (2000)

10. Pontiki, M., Galanis, D., Papageorgiou, H., Manandhar, S., Androutsopoulos, I.: SemEval-2015 task 12: aspect based sentiment analysis. In: Proceedings of the 9th International Workshop on Semantic Evaluation (SemEval 2015), pp. 486–495. Association for Computational Linguistics, Denver (2015)

11. Pontiki, M., Galanis, D., Pavlopoulos, J., Papageorgiou, H., Androutsopoulos, I., Manandhar, S.: SemEval-2014 task 4: aspect based sentiment analysis. In: Proceedings of the 8th International Workshop on Semantic Evaluation (SemEval 2014), pp. 27–35. Association for Computational Linguistics (2014)

12. Qiu, G., Liu, B., Bu, J., Chen, C.: Opinion word expansion and target extraction through double propagation. Comput. Linguist. **37**(1), 9–27 (2011)

13. Ruder, S., Ghaffari, P., Breslin, J.G.: INSIGHT-1 at SemEval-2016 task 5: deep learning for multilingual aspect-based sentiment analysis. arXiv preprint arXiv:1609.02748 (2016)

14. Saias, J.: Sentiue: target and aspect based sentiment analysis in SemEval-2015 task 12. In: Proceedings of the 9th International Workshop on Semantic Evaluation (SemEval 2015), pp. 767–771. Association for Computational Linguistics, Denver, June 2015

15. Severyn, A., Moschitti, A.: UNITN: training deep convolutional neural network for twitter sentiment classification. In: Proceedings of the 9th International Workshop on Semantic Evaluation (SemEval 2015), pp. 464–469 (2015)

16. Thet, T.T., Na, J.C., Khoo, C.S.: Aspect-based sentiment analysis of movie reviews on discussion boards. J. Inf. Sci. **36**(6), 823–848 (2010)

17. Toh, Z., Su, J.: NLANGP: supervised machine learning system for aspect category classification and opinion target extraction. In: Proceedings of the 9th International Workshop on Semantic Evaluation (SemEval 2015), pp. 496–501 (2015)

18. Toh, Z., Su, J.: NLANGP at SemEval-2016 task 5: improving aspect based sentiment analysis using neural network features. In: Proceedings of SemEval, pp. 282–288 (2016)

19. Xenos, D., Theodorakakos, P., Pavlopoulos, J., Malakasiotis, P., Androutsopoulos, I.: AUEB-ABSA at SemEval-2016 task 5: ensembles of classifiers and embeddings for aspect based sentiment analysis. In: International Workshop on Semantic Evaluation, pp. 312–317 (2016)

20. Zeiler, M.D.: ADADELTA: an adaptive learning rate method. arXiv preprint arXiv:1212.5701 (2012)

21. Zhang, Z., Nie, J.Y., Wang, H.: TJUdeM: a combination classifier for aspect category detection and sentiment polarity classification. In: SemEval-2015, p. 772 (2015)

Cross-Lingual Emotion Classification
with Auxiliary and Attention Neural Networks

Lu Zhang, Liangqing Wu, Shoushan Li$^{(\boxtimes)}$, Zhongqing Wang,
and Guodong Zhou

Natural Language Processing Lab, School of Computer Science and Technology,
Soochow University, Suzhou, China
{lzhang0107, lqwu}@stu.suda.edu.cn,
{lishoushan, wangzq, gdzhou}@suda.edu.cn

Abstract. In the literature, various supervised learning approaches have been adopted to address the task of emotion classification. However, the performance of these approaches greatly suffers when the size of the labeled data is limited. In this paper, we tackle this challenge from a cross-lingual sensoria where the labeled data in a resource-rich language (i.e., English in this study) is employed to improve the emotion classification performance in a resource-poor language (i.e., Chinese in this study). Specifically, we first use machine translation services to eliminate the language gap between Chinese and English data and then propose a joint learning framework to leverage both Chinese and English data, which develops auxiliary representations from several auxiliary emotion classification tasks. Furthermore, in our joint learning approach, we introduce an attention mechanism to capture informative words. Empirical studies demonstrate the effectiveness of the proposed approach to emotion classification.

Keywords: Sentiment analysis · Emotion classification · Attention mechanism

1 Introduction

Emotion classification aims to determine the involving emotion within a piece of text. With the tremendous growth of social media, such as Twitter and Facebook, emotion classification has drawn more and more attention. In the last decade, emotion classification has been proved to be invaluable in many applications, such as stock markets [1], online chat [2] and news classification [3].

Conventional approaches to emotion classification mainly conceptualize the task as a supervised learning problem where sufficient labeled data is essential for training the model. However, in most scenarios, the annotated corpus for emotion classification is scarce, and to obtain such labeled data is extremely costly and time-consuming. Some previous studies tackle this challenge by applying semi-supervised technique to make use of unlabeled data. For instance, Liu et al. [3] propose a co-training algorithm to improve the performance of emotion classification by leveraging the information in the unlabeled data. Li et al. [4] propose a two-view label propagation approach to emotion

© Springer Nature Switzerland AG 2018
M. Zhang et al. (Eds.): NLPCC 2018, LNAI 11108, pp. 429–441, 2018.
https://doi.org/10.1007/978-3-319-99495-6_36

E1:
Original: 今天大甩卖！我们去逛街吧~
Translation: *There's a big sale on today! Let's go shopping.*
E2:
Original: 最近总是七上八下。
Translation: *Recently, I'm always in an unsettled state of mind.*

Fig. 1. Some examples in Chinese emotion corpus with their English translations

classification by exploiting two views, namely source text and response text in a label propagation algorithm (Fig. 1).

Instead of semi-supervised learning, we focus on addressing this issue from a cross-lingual view. On one hand, Chinese emotion corpus is limited but many English emotion corpora are freely available. On the other, the emotion involved in a given text may not be learned in an exact manner with the representation in Chinese. However, if we translate it into English, it becomes easier to determine the emotion. For instance, in **E1**, due to the lack of Chinese emotion corpus, "大甩卖" may not exist in the training set so that it cannot be correctly classified. But if we translate this word into English, i.e., "*big sale*", then we can leverage English emotion corpus to make up for this. Similarly, in **E2**, "七上八下" is a Chinese idiom, which is difficult for machine to understand. However, if we translate it into English, i.e., "*an unsettled state of mind*", the emotion expressed by this phrase can be understood more easily.

In this study, we propose a joint learning framework, namely, Aux-LSTM-Attention, which learns simultaneously from the labeled data from both resource-poor and resource-rich languages. First, machine translation services are used to translate Chinese emotion corpus into English corpus and also translate English emotion corpus into Chinese corpus. Then, we view the emotion classification task with original Chinese emotion corpus as a main task and the emotion classification tasks with additional corpora as auxiliary tasks. To perform joint learning, we share neural network layers from the auxiliary tasks into the main task. Consequently, the main task learns the emotion classification by using the knowledge from both the main and auxiliary tasks through the layer sharing. Furthermore, we utilize an attention mechanism [5, 6] to aggregate the representation of informative words into a vector for emotion prediction. Empirical studies demonstrate that the proposed joint learning approach significantly outperforms several baseline approaches to emotion classification.

The remainder of this paper is organized as followed. Section 2 gives a brief overview of related work. Section 3 proposes our joint learning framework on emotion classification with both resource-poor and resource-rich corpora. Section 4 evaluates the proposed approach before presenting the concluding remarks in Sect. 5.

2 Related Work

2.1 Cross-Lingual Sentiment Classification

Sentiment analysis is the field of analyzing people's opinions, sentiments, attitudes and emotions from the text they have published [7]. In previous studies, conventional approaches to sentiment analysis mainly focus on sufficient labeled data [8]. However, in most scenarios, there is insufficient labeled data and to manually label reliable corpus is not a trivial task. Cross-lingual addresses this issue in sentiment classification from a cross-language view.

Over the last decades, there has been a proliferation of work exploring various aspects of cross-lingual sentiment classification. Mihalcea et al. [9] generate resources for subjectivity annotations for a new language, by leveraging resources and tools available for English. Wan [10] uses machine translation services to eliminate the language gap between Chinese corpus and English corpus. Chinese features and English features are considered to be two independent views of the classification problem and a co-training algorithm is employed to make use of unlabeled Chinese data. Balamurali et al. [11] use WordNet synset identifiers as features of a supervised classifier. They leverage the linked WordNets of two languages to bridge the language gap. Prettenhofer and Stein [12] introduce the structural correspondence learning algorithm to learn a map between the source language and the target language. More recently, Zhou et al. [13] propose a bilingual document representation learning method for cross-lingual sentiment classification which directly learns the vector representation for documents in different languages.

Unlike all above studies, this work focuses on cross-lingual emotion classification. Compared to cross-lingual sentiment classification, cross-lingual emotion classification is more challenging due to the fact that the sentiment categories in two languages are the same while the emotion taxonomies in two languages might be different.

2.2 Emotion Classification

Our work is also related to emotion classification. Tokuhisa et al. [14] propose a data-oriented method for inferring the emotion of an utterance sentence in a dialog system. Bhowmick et al. [15] present a method for classifying news sentences into multiple emotion categories using multi-label KNN classification technique. Xu et al. [16] propose a coarse-to-fine analysis strategy for emotion classification which takes similarities to sentences in training set as well as adjacent sentences in the context into consideration. Yang et al. [17] introduce an Emotion-aware LDA model to build a domain-specific lexicon for predefined emotions. Felbo et al. [18] show how millions of readily available emoji occurrences on Twitter can be used to pertrain models to learn a richer emotional representation than traditionally obtained through distant supervision.

Unlike all above studies, our work is the first attempt to apply cross-lingual in emotion classification.

3   Our Approach

3.1   Machine Translation

In this section, we propose our joint learning approach to perform joint learning with both resource-poor and resource-rich data. In this study, we assume that Chinese is the resource-poor language and it has only a few labeled samples. English is the resource-rich language and it has many more labeled samples. In order to overcome the language gap, we translate one language into the other language with a machine translation tool. Specifically, we adopt *Baidu Translate*[1] for both English-to-Chinese translation and Chinese-to-English translation. Figure 2 shows the general framework of the machine translation in the training phase. We translate the labeled Chinese emotion corpus into English to set up a translated English view and translate English emotion corpus into Chinese to establish a translated Chinese view.

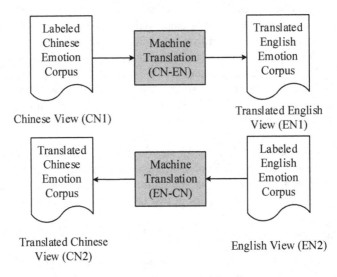

Fig. 2. Framework of the machine translation in the training phase

3.2   The Main Emotion Classification Task

Figure 3 illustrates the overall architecture of our Aux-LSTM-Attention approach which contains a main task and three auxiliary tasks. Specially, we consider the emotion classification task with original Chinese emotion corpus as the main task and the emotion classification tasks with other corpora as auxiliary tasks. The main idea of the proposed approach is to employ some auxiliary representations learned from the auxiliary tasks to assist the performance of the main task. Note that not all words contribute equally to representing the meaning of a post. Hence, instead of simply

---

[1] http://fanyi.baidu.com/translate.

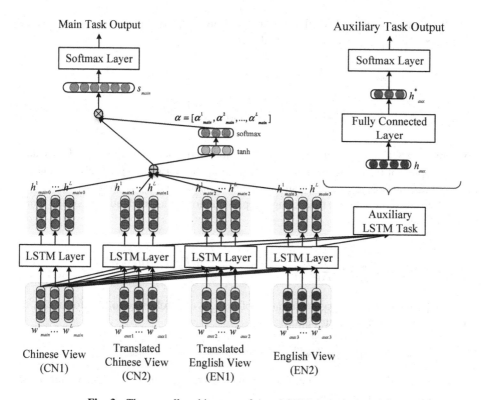

Fig. 3. The overall architecture of Aux-LSTM-Attention model

concatenating the representations from the main encoder layer and auxiliary sharing layers, we introduce an attention mechanism to produce an attention weight vector α and a weighted hidden representation s. To obtain the auxiliary representations, we adopt standard LSTM [19] layers as sharing layers between the networks from the main and auxiliary tasks.

Formally, the middle representation of the main task is generated from both the main encoder layer and auxiliary sharing layers, i.e.

$$h_{main0} = \text{LSTM}_{main}(T_{main}) \tag{1}$$

$$h_{main1} = \text{LSTM}_{aux1}(T_{main}) \tag{2}$$

$$h_{main2} = \text{LSTM}_{aux2}(T_{main}) \tag{3}$$

$$h_{main3} = \text{LSTM}_{aux3}(T_{main}) \tag{4}$$

where $T_{main} = \{w_{main}^1 \cdots w_{main}^L\}$ represents the input sequence from the Chinese emotion corpus (CN1). h_{main0} means the representation for the classification model via the main encoder layer. While h_{main1}, h_{main2} and h_{main3} mean the representations for the classification model via the auxiliary sharing layers.

With these main and auxiliary representations, we compute an attention weight vector $\alpha = [\alpha^1_{main}, \alpha^2_{main}, \ldots, \alpha^L_{main}]$ as follows:

$$m^i_{main} = \tanh(W_m \cdot [h^i_{main0} \oplus h^i_{main1} \oplus h^i_{main2} \oplus h^i_{main3}] + b_m) \tag{5}$$

$$\alpha^i_{main} = softmax(m^i_{main}) = \frac{\exp(m^i_{main})}{\sum_{t=1}^{L} m^t_{main}} \tag{6}$$

where $1 \leq i \leq L$ and L is the length of the input sequence. h^i_{main0}, h^i_{main1}, h^i_{main2} and h^i_{main3} represent the i-th word in h_{main0}, h_{main1}, h_{main2} and h_{main3} respectively. $\oplus$ denotes the concatenate operator. W_m is an intermediate matrix and b_m is an offset value.

Then, we compute the final sample representation as a weighted sum of the word annotations:

$$s_{main} = \sum_{i=1}^{L} \alpha^i_{main} \cdot [h^i_{main0} \oplus h^i_{main1} \oplus h^i_{main2} \oplus h^i_{main3}] \tag{7}$$

To perform emotion classification, a softmax layer is followed to transform s_{main} to conditional probability distribution:

$$p(y_{main}|T_{main}) = softmax(W_{main} \cdot s_{main} + b_{main}) \tag{8}$$

where $p(y_{main}|T_{main})$ is the output of the main task, W_{main} is the weight vector to be learned and b_{main} is the bias term.

3.3   The Auxiliary Emotion Classification Task

The representations of three auxiliary tasks are generated from corresponding auxiliary sharing layers respectively, i.e.,

$$h_{aux1} = LSTM_{aux1}(T_{aux1}) \tag{9}$$

$$h_{aux2} = LSTM_{aux2}(T_{aux2}) \tag{10}$$

$$h_{aux3} = LSTM_{aux3}(T_{aux3}) \tag{11}$$

where $T_{aux1} = \{w^1_{aux1} \ldots w^L_{aux1}\}$, $T_{aux2} = \{w^1_{aux2} \ldots w^L_{aux2}\}$, $T_{aux3} = \{w^1_{aux3} \ldots w^L_{aux3}\}$ mean the input sequences from English-to-Chinese (CN2), Chinese-to-English (EN1) and English (EN2) emotion corpora respectively. h_{aux1}, h_{aux2} and h_{aux3} are the outputs from three auxiliary sharing layers respectively.

Then, a fully-connected layer followed by a dropout layer is leveraged to gain a feature vector for classification, i.e.,

$$h^*_{aux1} = dense(h_{aux1}) \cdot D(p^*_{aux1}) \tag{12}$$

$$h^*_{aux2} = dense(h_{aux2}) \cdot D(p^*_{aux2}) \tag{13}$$

$$h^*_{aux3} = dense(h_{aux3}) \cdot D(p^*_{aux3}) \tag{14}$$

where $dense(\cdot)$ denotes the output of the fully-connected layer. D defines the dropout operation and p^* is the dropout probability.

Once obtaining the representations of these auxiliary tasks, we feed them into a softmax layer respectively to perform emotion classification:

$$p(y_{aux1}|T_{aux1}) = \text{softmax}(W_{aux1} \cdot h^*_{aux1} + b_{aux1}) \tag{15}$$

$$p(y_{aux2}|T_{aux2}) = \text{softmax}(W_{aux2} \cdot h^*_{aux2} + b_{aux2}) \tag{16}$$

$$p(y_{aux3}|T_{aux3}) = \text{softmax}(W_{aux3} \cdot h^*_{aux3} + b_{aux3}) \tag{17}$$

where $p(y_{aux1}|T_{aux1})$, $p(y_{aux2}|T_{aux2})$ and $p(y_{aux3}|T_{aux3})$ are the outputs of the auxiliary tasks respectively. W_{aux1}, b_{aux1}, W_{aux2}, b_{aux2}, W_{aux3} and b_{aux3} are the parameters f or softmax layers.

3.4 Joint Learning

The model can be trained in an end-to-end manner where the objective loss function is a linear combination of the main task and auxiliary tasks:

$$
\begin{aligned}
J(\theta) = &- \lambda_1 \cdot \sum_{i=1}^{N} \sum_{j=1}^{C} y^j_{main} \cdot \log p(y^j_{main}|T^i_{main}) - \lambda_2 \cdot \sum_{i=1}^{N} \sum_{j=1}^{C} y^j_{aux1} \cdot \log p(y^j_{aux1}|T^i_{aux1}) \\
&- \lambda_3 \cdot \sum_{i=1}^{N} \sum_{j=1}^{C} y^j_{aux2} \cdot \log p(y^j_{aux2}|T^i_{aux2}) - \lambda_4 \cdot \sum_{i=1}^{N} \sum_{j=1}^{C} y^j_{aux3} \cdot \log p(y^j_{aux3}|T^i_{aux3}) \\
&+ \frac{l}{2} \|\theta\|^2_2
\end{aligned}
\tag{18}
$$

where y^j_{main}, y^j_{aux1}, y^j_{aux2} and y^j_{aux3} are the ground-truth labels from the main task and auxiliary tasks. N is the total quantity of training samples. C is the category number. l is a L_2 regularization to bias parameters and θ denotes all parameters. λ_1, λ_2, λ_3 and λ_4 are the weight parameters to balance the importance of losses between the main task and auxiliary tasks and $\lambda_1 + \lambda_2 + \lambda_3 + \lambda_4 = 1$. We take Adadelta [20] as the optimizing algorithm with a learning rate of 1.0. All the matrix and vector parameters are initialized with a uniform distribution in $\left[-\sqrt{6/(r+c)}, \sqrt{6/(r+c)}\right]$, where r and c are the rows and columns of the matrices.

4 Experiment

In this section, we systematically evaluate the performance of our approach to emotion classification.

4.1   Experimental Settings

- **Data Settings:** In order to assess the performance of the proposed approach, we use the Chinese emotion corpus constructed by Yao et al. [21]. This corpus consists of 14,000 instances, of which 7,407 instances express emotions. Seven basic emotions are defined as candidate categories, namely *anger, happiness, sadness, fear, like, surprise* and *disgust*. In addition, we use the dataset of SemEval 2018 Task1 as the English emotion corpus. It contains a lot of tweets and corresponding emotion categories, i.e., *anger, joy, sadness* and *fear*. Table 1 illustrates the distribution of these two datasets. As to enlarge the corpora mentioned above, we can translate one language into the other language. The Chinese data is much imbalanced and we extract a balanced dataset for each emotion category in Chinese. Due to the fact that the number of instances in *fear* category is too small, we decide to set the number of instances in *surprise* category as the basis to avoid contingency. We use 80% of instances as training data and the remaining 20% as test data. Furthermore, we set aside 10% of the training data as development data to fine tune the parameters in learning algorithm.

Table 1. Emotion categories and distribution on two corpora

Emotion	#Sentences in Chinese Corpus	#Sentences in English Corpus
anger	669	1901
happiness	1460	1816
sadness	1173	1733
fear	148	2452
like	2203	–
surprise	362	–
disgust	1392	–

- **Word Segmentation and Representations:** FudanNLP[2] is employed to segment each Chinese post into words and we learn distributed representation of each word with word2vec[3] (The skip-gram model is used) on each dataset. The vector dimension is set to be 100 and the window size is set to be 5.
- **Hyper-parameters:** The hyper-parameters in our approach are tuned according to the performance in the development data. The size of units in LSTM layer is 128 and all models are trained by mini-batch of 32 instances. λ_1 is set to be 0.5, λ_2, λ_3 and λ_4 are the same as each other.
- **Evaluation Metric:** We use *Macro-F1 (F)* and *Accuracy* to measure the divergences between predicted labels and ground-truth labels. Besides, *t*-test is used to determine whether the performance difference is statistically significant.

---

[2] https://github.com/FudanNLP/fnlp/.

[3] https://github.com/dav/word2vec/.

4.2 Experimental Results

In this section, we report the experimental results of our joint learning approach to emotion classification. For thorough comparison, we provide selected baseline approaches. In addition, we also implement some state-of-the-art approaches in sentiment classification to emotion classification.

- **LSTM (CN1):** This method applies the standard LSTM model using only the Chinese emotion corpus for emotion classification.
- **CNN-Tensor (CN1) [22]:** This is a state-of-the-art approach to sentiment classification, which appeals to tensor algebra and uses low-rank n-gram tensors to directly exploit interactions between words already at the convolution stage. It applies only the Chinese emotion corpus for emotion classification.
- **Attention-LSTM (CN1) [23]:** This is a state-of-the-art approach to aspect-level sentiment classification, which leverages the attention mechanism to concentrate on different parts of a sentence. Note that we ignore aspect embedding and use sentence representations from LSTM to yield an attention weight vector directly. It applies only the Chinese emotion corpus for emotion classification.
- **LSTM (CN1 + EN1):** This method combines the results of LSTM (CN1) and LSTM (EN1) by averaging the probabilities. It applies both Chinese and Chinese-to-English emotion corpora for emotion classification. This is an ensemble approach by Wan [24] which is proposed to deal with cross-lingual sentiment classification. Since the categories in the Chinese corpus and the English corpus are different, this approach could not be directly applied to combine all corpora (i.e., CN1 + CN2 + EN1 + EN2).
- **LSTM (CN1 + CN2):** This method simply merges Chinese and English-to-Chinese emotion samples in corresponding categories and applies the standard LSTM model for emotion classification.
- **Aux-LSTM (CN1 + CN2):** It applies the Aux-LSTM model with both Chinese and English-to-Chinese emotion corpora for emotion classification. It simply concatenates the representations from the main and auxiliary task.
- **Aux-LSTM (CN1 + EN1):** It applies the Aux-LSTM model with both Chinese and Chinese-to-English emotion corpora for emotion classification. It simply concatenates the representations from the main and auxiliary task.
- **Aux-LSTM (CN1 + EN2):** It applies the Aux-LSTM model with both Chinese and English emotion corpora for emotion classification. It simply concatenates the representations from the main and auxiliary task.
- **Aux-LSTM (CN1 + CN2 + EN1 + EN2):** It applies the Aux-LSTM model with all Chinese, English-to-Chinese, Chinese-to-English and English emotion corpora for emotion classification. It simply concatenates the representations from the main and auxiliary tasks.
- **Aux-LSTM-Attention (CN1 + CN2 + EN1 + EN2):** It applies the Aux-LSTM model with attention on all Chinese, English-to-Chinese, Chinese-to-English and English emotion corpora for emotion classification.

Table 2 shows the results of different approaches to Chinese emotion classification. From the table, we can see that all Aux-LSTM models consistently outperform the

Table 2. Performance comparison of different approaches to emotion classification

	Macro-F1	Accuracy
LSTM (CN1)	0.39087	0.36255
CNN-Tensor (CN1)	0.41468	0.40125
Attention-LSTM (CN1)	0.44048	0.41573
LSTM (CN1 + EN1)	0.40278	0.37669
LSTM (CN1 + CN2)	0.42063	0.40619
Aux-LSTM (CN1 + CN2)	0.45238	0.43472
Aux-LSTM (CN1 + EN1)	0.45833	0.44773
Aux-LSTM (CN1 + EN2)	0.45040	0.43806
Aux-LSTM (CN1 + CN2 + EN1 + EN2)	0.46429	0.43795
Aux-LSTM-Attention (CN1 + CN2 + EN1 + EN2)	**0.49802**	**0.49355**

baseline approaches whichever corpus is employed, which verifies the effectiveness of the proposed Aux-LSTM model. These results encourage to incorporate other-language labeled data to improve the performance of emotion classification. However, with the increase in number of additional corpora, it could not bring about remarkable results any more. Hence, instead of simply concatenating the representations from the main encoder layer and auxiliary sharing layers, we introduce an attention mechanism to provide insight into which words contribute to the emotion classification decision. Among all these approaches, our Aux-LSTM-Attention model performs best, which suggests sharing additional corpora and utilizing attention mechanism to capture the informative words. Significance test shows that the improvement of Aux-LSTM-Attention model over the other approaches is significant ($p\text{-value} < 0.05$).

To better understand why our joint learning approach is so effective, we calculate the standard precision (P), recall (R) and F-score (F) in each category. Table 3 demonstrates these specific results. For clarity, we only report the results of LSTM (CN1) and Aux-LSTM-Attention (CN1 + CN2 + EN1 + EN2). From Table 3, we can see that our joint learning approach is obviously superior to LSTM (CN1) in almost every category, especially in the *like* category and *surprise* category. The performance of our approach

Table 3. Comparative results with standard precision (P), recall (R) and F-score (F) in each category

	LSTM (CN1)			Aux-LSTM-Attention (CN1 + CN2 + EN1 + EN2)		
	P	R	F	P	R	F
anger	0.551	0.681	0.609	0.610	0.694	0.649
happiness	0.393	0. 611	0.478	0.603	0.528	0.563
sadness	0.337	0. 486	0.398	0.494	0.542	0.517
fear	0.409	0. 500	0.450	0.419	0.431	0.425
like	0.346	0. 125	0.184	0.524	0.458	0.489
surprise	0.233	0. 139	0.174	0.408	0.556	0.471
disgust	0.333	0. 194	0.246	0.444	0.278	0.342

is consistent in each category while LSTM (CN1) fluctuates widely. It indicates that our approach is more effective for predicting emotion in Chinese emotion text.

4.3 Case Study

In Fig. 4, we list two examples from the test set which have not been correctly inferred by LSTM (CN1) model due to the limitation of Chinese emotion corpus. In **E3**, "骗子 *(liar)*" expresses a strong sentiment signal to the emotion *anger* but it does not exist in the training set so that this sample could not be correctly classified by LSTM (CN1). However, when we translate it into English, i.e., *liar* and it can be found in the additional English emotion corpus, our Aux-LSTM model can work well. In **E4**, "炫耀 *(show off)*" is not contained in the training set either. But if the corresponding translation *"show off"* can be found in the additional English emotion corpus, then we can leverage our Aux-LSTM model to predict the correct emotion *(happiness)* easily.

E3: Ground-truth label: *anger*	
Original: *刚刚收到短信…这绝对是骗子!*	
Translation: *Just received a text message…This is definitely a liar!*	
LSTM (CN1)	Aux-LSTM (CN1+CN2+EN1+EN2)
✕ *(surprise)*	√ *(anger)*

E4: Ground-truth label: *happiness*	
Original: *来炫耀! 入手了表哥千里迢迢捎来的…哪怕惧怕昆虫的人也该去入一本。*	
Translation: *To **show off**! Started the cousin's journey…even those who are afraid of insects should go to a book.*	
LSTM (CN1)	Aux-LSTM (CN1+CN2+EN1+EN2)
✕ *(like)*	√ *(happiness)*

Fig. 4. Examples of emotion classification

4.4 Visualization of Attention

Figure 5 shows the attention visualization for a post in the test set. The color depth indicates the importance degree of corresponding word - the darker the shade, the more important the word. Obviously, the attention mechanism obtains the important elements which carry strong sentiment signals from the whole post dynamically, such as *"terrible"* and *"exaggerated"*.

E5: Ground-truth label: *fear*   Predict label: *fear*

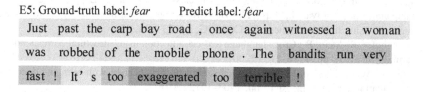

Just past the carp bay road , once again witnessed a woman was robbed of the mobile phone . The bandits run very fast ! It's too exaggerated too terrible !

Fig. 5. Attention visualization

5 Conclusion

In this paper, we address the corpus scarce challenge in emotion classification from a cross-lingual view and propose a joint learning framework, namely Aux-LSTM-Attention, to perform emotion classification when both resource-poor and resource-rich corpora exist. Specially, we employ sharing layers to develop auxiliary representations for the main task. Furthermore, an attention mechanism is utilized to capture the informative words. Empirical studies show that our joint learning approach successfully improves the performance of emotion classification by using the labeled data from a different language. Moreover, empirical studies demonstrate that our approach outperforms several strong baseline approaches to emotion classification.

In our future work, we would like to explore tackling the corpus scarcity challenge by using the labeled data from multiple languages. Furthermore, we will attempt to apply our approach to other natural language processing tasks in which the annotated corpus is limited.

Acknowledgements. This research work has been partially supported by two NSFC grants, No. 61331011 and No. 61672366.

References

1. Bollen, J., Mao, H., Zeng, X.: Twitter mood predicts the stock market. J. Comput. Sci. **2**(1), 1–8 (2011)
2. Galik, M., Rank, S.: Modeling emotional trajectories of individuals in an online chat. In: MATES, pp. 96–105 (2012)
3. Liu, H., Li, S., Zhou, G., Huang, C., Li, P.: Joint modeling of news reader's and comment writer's emotions. In: ACL, pp. 511–515 (2013)
4. Li, S., Xu, J., Zhang, D., Zhou, G.: Two-view label propagation to semi-supervised reader emotion classification. In: COLING, pp. 2647–2655 (2016)
5. Bahdanau, D., Cho, K., Bengio, Y.: Neural machine translation by jointly learning to align and translate. CoRR, abs/1409.0473, 2014
6. Yang, Z., Yang, D., Dyer, C., He, X., Smola, A., Hovy, E.: Hierarchical attention networks for document classification. In: NAACL-HLT, pp. 1480–1489 (2016)
7. Liu, B.: Sentiment analysis and opinion mining. Synthesis Lectures on Human Language Technologies. Morgan & Claypool Publishers. 1–167 (2012)
8. Pang, B., Lee, L., Vaithyanathan, S.: Thumbs up? sentiment classification using machine learning techniques. In: EMNLP, pp. 79–86 (2002)
9. Mihalcea, R., Banea, C., Wiebe, J.: Learning multilingual subjective language via cross-lingual projections. In: ACL, pp. 976–983 (2007)
10. Wan, X.: Co-training for cross-lingual sentiment classification. In: ACL, pp. 235–243 (2009)
11. Balamurali, A., Aditya, J., Pushpak, B.: Cross-lingual sentiment analysis for indian languages using linked wordnets. In: COLING, pp. 73–82 (2012)
12. Prettenhofer, P., Stein, B.: Cross-language text classification using structural correspondence learning. In: ACL, pp. 1118–1127 (2007)
13. Zhou, X., Wan, X., Xiao, J.: Cross-lingual sentiment classification with bilingual document representation learning. In: ACL, pp. 1403–1412 (2016)

14. Tokuhisa, R., Inui, K., Matsumoto, Y.: Emotion classification using massive examples extracted from the web. In: COLING, pp. 881–888 (2008)
15. Bhowmick, P., Basu, A., Mitra, P., Prasad, A.: Multi-label text classification approach for sentence level news emotion analysis. In: PReMI, pp. 261–266 (2009)
16. Xu, J., Xu, R., Lu, Q.: Coarse-to-fine sentence-level emotion classification based on the intra-sentence features and sentential context. In: CIKM, pp. 2455–2458 (2012)
17. Yang, M., Peng, B., Chen, Z., Zhu, D., Chow, K.: A topic model for building fine-grained domain-specific emotion lexicon. In: ACL, pp. 421–426 (2014)
18. Felbo, B., Mislove, A., SØgaard, A., Rahwan, I., Lehmann, S.: Using millions of emoji occurrences to learn any-domian representations for detecting sentiment, emotion and sarcasm. In: EMNLP, pp. 1615–1625 (2017)
19. Graves, A.: Generating sequences with recurrent neural networks. CoRR, abs/1308.0850, 2013
20. Zeiler, M.: ADADELTA: an adaptive learning rate method. CoRR, abs/1212.5701 (2012)
21. Yao, Y., Wang, S., Xu, R., Liu, B., Gui, L., Lu, Q., Wang, X.: The construction of an emotion annotated corpus on microblog. J. Chin. Inf. Process. **28**(5), 83–91 (2014)
22. Lei, T., Barzilay, R., Jaakkola, T.: Modeling CNNs for text: non-linear, non-consecutive convolutions. In: EMNLP, pp. 1565–1575 (2015)
23. Wang, Y., Huang, M., Zhao, L., Zhu, X.: Attention-based LSTM for aspect-level sentiment classification. In: EMNLP, pp. 606–615 (2016)
24. Wan, X.: Using bilingual knowledge and ensemble techniques for unsupervised Chinese sentiment analysis. In: EMNLP, pp. 553–561 (2008)

Are Ratings Always Reliable?
Discover Users' True Feelings
with Textual Reviews

Bin Hao, Min Zhang$^{(\boxtimes)}$, Yunzhi Tan, Yiqun Liu, and Shaoping Ma

Department of Computer Science and Technology,
Beijing National Research Center for Information Science and Technology,
Tsinghua University, Beijing 100084, China
{haob15,tyz13}@mails.tsinghua.edu.cn, {z-m,yiqunliu,msp}@tsinghua.edu.cn

Abstract. In e-commerce systems, users' ratings play an important role in many scenarios such as reputation and trust mechanisms and recommender systems. A general assumption in these techniques is that users' ratings represent their true feelings. Although it has long been adopted in previous work, this assumption is not necessarily true.

In this paper, we first present an in-depth study of the inconsistency between users' ratings and their reviews. Then we propose an approach to mine users' "true ratings" which better represent their real feelings, from textual reviews based on Gated Recurrent Unit (GRU) and hierarchical attention techniques. One major contribution is that we are about the first, to the best of our knowledge, to investigate this new problem of discovering users' true ratings, and to provide direct solutions to revise ratings that are insincere and inconsistent.

Comparative experiments on a real e-commerce dataset have been conducted, which show that the "true ratings" learned by the proposed model is significantly better than the original ones in terms of consistency with the reviews in three sets of crowdsourcing-based evaluations. Furthermore, leveraging different state-of-art recommendation approaches based on the learned "true ratings", more effective results have been achieved at all times in rating prediction task.

Keywords: Rating revision · Review to score
Deep learning for recommendation

1 Introduction

In recommendation systems, users' ratings are widely used as a basis to learn user's preferences and make recommendations in most of the classical algorithms, such as [2]. In recent years, simultaneously exploiting ratings and reviews for recommendation attracts more and more attention for its ability to mitigate

This work is supported by Natural Science Foundation of China (Grant No. 61672311, 61532011).

© Springer Nature Switzerland AG 2018
M. Zhang et al. (Eds.): NLPCC 2018, LNAI 11108, pp. 442–453, 2018.
https://doi.org/10.1007/978-3-319-99495-6_37

data sparsity and build more accurate models [3,4]. In these models, however, ratings still play important roles in recommendation task or even been taken as the targets of learning.

For these ratings-based techniques, the fundamental assumption is that ratings are valid and reliable, which honestly indicate overall feelings of users towards items. While unfortunately, this assumption is not always true [1,5]. Apart from spam users which have been widely studied previously [6,7], a considerable number of ordinary users also give inconsistent ratings with the corresponding reviews in opinion expression. There have been some observations on this phenomenon [8] but little work has been done to overcome it, which is the major topic of this paper. Examples of ratings and their corresponding reviews given by real ordinary users, who have consumed the products or services, from a large-scale e-commerce website are shown in Table 1.

Table 1. Examples from real e-commerce dataset, where ratings fail to represent users' true feelings

No.	Ratings	Reviews
1	5.0	There was a hole on the socks. The user service wasn't very friendly
2	5.0	It was too large, especially the sleeves. The styles was Okay
3	5.0	Feel a bit hard, maybe because I just start to use
4	1.0	The clothes are of good quality, and look beautiful
5	3.0	Very beautiful! I like it. Praise!

In this paper, we first investigate the problem where ratings fail to represent users' opinion expressed in the reviews. Generally, the user's review shows more information on his experience with the item after he has consumed them, which takes more reliable information than a single rating in many cases. Therefore, a deep-learning based approach is then proposed for mining users' true feelings from textual reviews. Furthermore, we design three experiments for evaluation. For recommendation scenario, performances of the learned "true ratings" in rating prediction task, which has been examined via different classical recommendation algorithms. Compared with original ratings, significant improvements are achieved in all the experiments using the learned ratings.

The main contributions of this work are as follows: (1) To the best of our knowledge, it is about the first work on detecting biased user ratings and building models to directly mine users' "true ratings". (2) We show how deep-learning-based approach help in this new problem, representing an inherent connection between textual information and score ratings. Furthermore, various state-of-art recommendation algorithms achieved significant performance improvements by using the learned ratings. (3) The "true ratings" learned by our models can be applied as groundwork in many scenarios where user ratings are adopted.

2   Related Work

Although ratings in e-commerce systems are widely used in many scenarios as important feedback, they are not always reliable. Some sellers do encourage buyers to provide positive feedback and avoid negative feedback to show that consumers are satisfied [8]. The problem of spam users, wherein users promote or degrade targeted items intentionally through fraudulent ratings and reviews, is one of the reasons [7]. There has been a lot of work focusing on detecting spam users [9–11].

Moreover, a growing body of work in recent years has paid attention to simultaneously exploiting user ratings and reviews in order to improve the performance of recommendations [3,4,12]. In such algorithms, information from reviews is introduced to better model user preferences and item features. However, the assumption that the users' ratings are reliable still serves as a basic assumption and ratings are directly used in these approaches. Hence it is essentially different from the basic problem of this work, in which the ratings are supposed to be not necessarily reliable and are not encouraged to be used directly.

Differing markedly from previous work, our work focuses on the reliability of original ratings when they are used to represent the users' "true feelings" towards the items. This problem is crucial but has not been well studied. In this work, we propose a method to mine users' true feelings from their reviews in which they describe in detail their experiences of products and services after consuming them. To the best of our knowledge, this is about the first direct solutions that are proposed for this problem. We also show that properly revised ratings also help the previous recommendation algorithms which make use of rating information.

In general, reviews are written in a free text format with natural language. To mine users' "true ratings" accurately, we need to understand semantic and structural information in reviews, to which deep learning techniques are shown to be helpful. Recurrent neural networks (RNNs) are able to process arbitrary sequences of inputs and form some kind of short-term memory using their internal memories, which make them applicable to tasks such as speech recognition or text processing [13]. Theoretically, RNNs can efficiently represent more complex patterns. The Gated Recurrent Unit network (GRU) [14] is a special kind of RNNs and capable of learning long-term dependencies. Attention mechanism has already been shown effective in many areas, such as machine translation [15] and sentiment analysis [16]. [17] propose a hierarchical attention network for document classification. The main difference between us is that they calculate word attention to find the critical words in the sentence and we calculate it to find the critical words in the whole review.

3   Are Ratings Always Reliable?

As described above, ratings may sometimes fail to represent users' true feelings towards items (i.e., products, services and so on). To investigate this problem with a real-world dataset, we collected a real dataset of clothing and accessories category from a popular e-commerce website. The dataset is a collection of feedback, where each piece of feedback consists of 4 factors: anonymized "UserID" and "ItemID", a numerical "rating" (from 1 to 5 stars) and a corresponding textual "review". In summary, there are 324,925 pieces of feedbacks provided by 284,848 users towards 27,370 items in the dataset (i.e. on average 11.8 feedback per item).

To measure users' true feelings towards items, we randomly selected 15,424 reviews (approximately 5% of the whole dataset) for crowdsourcing-based labeling. For each textual review, we hide its original rating and annotators were randomly selected to manually label five-level ratings according to the given corresponding reviews. Labeling quality was monitored in real time and unreliable annotators were interrupted in the task immediately by the crowdsourcing platform. Finally, 144 annotators provided valid labels and for each review, 3 labels were received. We also publish this dataset for the convenience of related researches.[1]

Some statistics of the labels are shown in Table 2. The average label variance is 0.2237, which indicates a good labeling consistency. We took the arithmetic average of the 3 labels as the review's labeled rating r^{lab}, which is used to represent user's true feelings towards an item in this work. It shows that the mean of all labeled ratings is 4.15, which is much lower than that of original ratings.

Table 2. Statistics of original and labeled ratings

Original mean	Labeled mean	Averaged labeled variance
4.46	4.15	0.2237

The distributions of the labeled and original ratings are shown in Fig. 1(a). 57.44% of original ratings are 5 stars, 35.78% are 4 stars, and less than 7% are 3 stars or less. If we treat 5 and 4 stars as positive feedback, more than 93% of feedback is positive. This percentage is similar to that is discovered in [8]. However, for labeled ratings, 40.06% of ratings are 5 stars and 41.52% are 4 stars, which shows about 12% less on positive ratings. Hence more diverse distribution is observed compared with the original one.

Another observation is, generally speaking, the differences between labeled ratings and original ones are mostly 1 or 2 levels. For example, the original rating of the review "The price of this hat is too high!" was 5; however, it's average label is 3.67. The original rating for the review "This looks good, and I like it very much." was 4, but was re-labeled as 5. Detail information of rating differences distribution is shown in Fig. 1(b). Here the rating difference is defined as the labeled ratings minus the corresponding original ratings. There are 33.85%

---

[1] It can be downloaded at https://pan.baidu.com/s/1O9r1S5ojGnrraivWwqT42w.

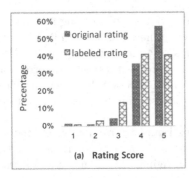 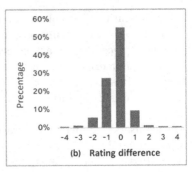

Fig. 1. Labeled ratings statistics. (a) the distribution on rating scores, (b) differences between ratings (labeled minus original). (Round labeled ratings to the nearest integer)

and 10.51% of reviews for which the labeled ratings are lower and higher than the original ratings, respectively, when we round the average labeled ratings to the nearest integer. If we compare the average labeled rating directly with the original one without round operation, then the proportions of the ratings by crowdsourcing labeling are respectively 46.62% lower than and 11.43% higher than the corresponding original ratings, which in total approaches to sixty percent of the data.

Table 3. Number of labeled ratings (columns, rounded to the nearest integer) v.s. that of original ratings (rows)

Ratings	1	2	3	4	5	Total
1	58	74	62	30	4	228
2	8	49	71	33	7	168
3	7	54	235	303	54	653
4	26	196	1206	3100	986	5514
5	28	125	650	2926	5132	8861
Total	127	498	2224	6392	6183	

Moreover, we analyze the distribution of labeled ratings relative to original ones, as shown in Table 3. What's interesting, from the table, some ratings express completely opposite opinions from what is described in the reviews. For example, the original rating of the review "The quality is really bad! I will never buy it again!" was 5; however, it was labeled as 1 by all the three annotators on account of the strong dissatisfaction it expressed. The original rating for the review "Good commodity. Its fabrics, colors and other aspects are also relatively satisfactory." was 1, which was re-labeled as 5 to reflect the user's satisfaction with the item. Similar examples can be found in ratings which are revised by 3 levels (e.g. 1 to 4, 5 to 2, etc.). It does happen in real scenarios when users make misunderstanding on the meaning of the rating stars.

4   Model for Mining Users' True Feelings from Reviews

To understand users' true feelings on the items, we propose to analyze the information expressed in textual reviews, from which we learn revised scores as the users' "true ratings". In general, reviews are written in a free text format. To mine users' "true ratings", we need to analyze the semantic and the structural information expressed in them, which is the strong point of deep learning techniques. As a result, we propose a **Hierarchical Total Attention (HTA)** model based on deep learning techniques to mine users' "true ratings" from their reviews.

4.1   Formalizations

We treat the review as a document d containing n sentences $\{S^1, S^2, \cdots, S^n\}$. The length of the k-th sentence S^k is l_k. The embeddings of the words in sentence S^k are $\{w_1^k, w_2^k, \cdots, w_{l_k}^k\}$.

4.2   Overview of HTA

The goal of our model is to predict a rating given its corresponding review. We treat each review as a document and use a hierarchical structure to capture the relation between sentences in one review and between words in the review. And we utilize the attention mechanism to automatically assign weights to each word and sentence. The structure of the HTA model is shown in Fig. 2. First we use word attention mechanism to get the vector representation of each sentence as $\{s_1, s_2, \cdots, s_n\}$. Then we use sentence attention mechanism to get the vector representation of the review as d. After this, we use a fully connected layer to get the prediction value of the review r.

4.3   From Word to Sentence Vector

The embeddings of the words in each sentence are inputted and processed by bi-directional Gated Recurrent Unit(GRU)[14]. The k-th word embedding of the i-th sentence w_k^i is encoded as h_k^i. Then we use the attention mechanism to calculate the attention value of **the total words** in the review as follows:

$$score(h_i^j) = v_w^{\mathrm{T}} tanh(W_w h_i^j + b_w) \tag{1}$$

$$a_i^j = \frac{exp(score(h_i^j))}{\sum_{j=1}^{n} \sum_{k=1}^{l_j} exp(score(h_k^j))} \tag{2}$$

$$s_i = \sum_{j=1}^{l_i} a_i^j h_i^j \tag{3}$$

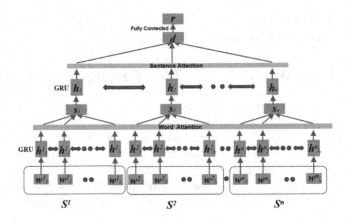

Fig. 2. The neural network architecture of HTA model.

First we calculate each word's importance of the review in Eq. 1, where score(h_i^j) is a score function which scores the importance of each word in the review, v_w is a word level context vector and v_w^{T} denotes its transpose, W_w is the weight matrix, b_w is the bias. The next step we use softmax function to calculate the attention weight of each word a_i^j in Eq. 2, which is the main difference from [17]. Then we aggregate all the word encoded vector h_i^j of a sentence to get its vector representation in Eq. 3.

4.4   From Sentence to Review Vector

We input each sentence vector to bi-directional Gated Recurrent Unit(GRU) encoder. The i-th sentence vector of review s_i is encoded as h_i. Then we use the attention mechanism similar to the word attention to select critical sentences to form the document representation. The document vector representation is formed via:

$$d = \sum_{i=1}^{n} a_i h_i \tag{4}$$

where a_i is the attention weight of sentence's encoded GRU vector h_i, which can be calculated similar to the word attention.

4.5   Regression and Learning

As d is extracted from words and sentences from the review, it can be treated as the feature vector of the review. We use a fully connected layer and a non-linear transformation (Relu) to get the final rating r of the review:

$$r = Relu(W_d d + b_d) \tag{5}$$

In this model, all of the parameters are learnt by minimizing the sum of squared errors between ratings labeled manually and ratings mined from reviews, which is shown as follows.

$$L(R) = \sum_{r^{lab} \in R} (r - r^{lab})^2 + \lambda_1 \|W_d\|^2 + \lambda_2 \|b_d\|^2 \qquad (6)$$

where R denotes the training set of the labeled rating dataset, $\|\cdot\|$ denotes the l_2-norm. And the component $\lambda_1 \|W_d\|^2 + \lambda_2 \|b_d\|^2$ is used for regularization to avoid over-fitting. We also use a *dropout* technique to avoid over-fitting.

5   Experiment and Discussion

5.1   Dataset and Experimental Settings

As described in Sect. 3, we collected a dataset D of clothing and accessories from a popular e-commerce website. Then 15,424 feedbacks are randomly selected L from D which annotators from crowdsourcing platform manually labeled their ratings according to the feelings users expressed in the textual reviews.

We treated the labeled ratings from the reviews as the representation of users' true feelings towards items, i.e., the "true ratings", and designed HTA model to learn the connection between the textual reviews and the "true ratings". The labeled dataset is randomly divided the labeled dataset L into a training set R, a validation set V, and a testing set T. Specifically, 70% of the labeled dataset was used for training, 10% for validation and the remaining 20% for testing. We use the word embeddings from Google trained from Word2Vector model, each word is presented as a 200 dimension vector. We use five models as our baselines: Linear Regression (LR), SVM with linear kernel (LinearSVR), Multi-layer Perceptron with one hidden layer of 100 nodes (MLP), SVM with RBF kernel (SVR), HAN model depicted in [17]. For the top four models, we use the average vector of all words in a review as its feature and for the HAN model, we use the same setting as our model.

To evaluate the performance of our models, we adopted the commonly used metrics root mean squared error (RMSE) and mean absolute error (MAE), which are defined as below:

$$RMSE = \sqrt{\frac{1}{n} \sum_{j=1}^{n} (f_j - y_j)^2}, MAE = \frac{1}{n} \sum_{j=1}^{n} |f_j - y_j| \qquad (7)$$

where f_j and y_j denote the prediction value and the true value, respectively.

5.2   Effectiveness of "True Ratings"

Evaluation of Rating Score Revision. This evaluation task is to measure the effectiveness of direct rating score revision by estimating the proper rating given a piece of related review. Crowdsourcing labeled "true ratings" are taken as the ground truth. Performances on RMSE on the testing set T by HTA model are given in Table 4. Here, the "Ori" denotes the original ratings and the "HTA" denote the ratings learned by HTA model, etc.

Table 4. Performance of HTA Model

Metric	Ori	SVR	LR	LinearSVR	MLP	HAN	HTA
RMSE	0.8163	0.6650	0.5936	0.5879	0.5711	0.4836	**0.4803**
MAE	0.5393	0.5078	0.4567	0.4323	0.4291	0.3497	**0.3431**

Evaluation with Pairwise Preference. To make a quantitative evaluation, we conducted experiments of pairwise preference to examine the consistency of the learned ratings with the feelings expressed in the reviews by crowdsourcing-based labeling. In detail, we randomly select 1,000 textual reviews. For easily distinguishing, we only selected the reviews of which the ratings learned by the HTA model are at least 1 level different from their original ratings. We then rounded the rating to the nearest integer to make the two types of ratings have the same appearances, hence no bias was introduced to annotators. Finally, for each review, 5 randomly selected annotators were asked to discern which rating was more consistent with the feeling expressed by the review. If an annotator found that the learned rating was more consistent, we added an "R2S" tag to the review; otherwise, an "Ori" tag was added. Then we calculated the percentages of reviews with different numbers of "R2S" labels, and the results are shown in Table 5. We can see that almost half of the scores from our model get 5 "R2S" and 86.9% scores of our model better than the original ones.

Table 5. The percentages of reviews with different numbers of "R2S" labels

	# "R2S" = 5	# "R2S" >= 3	# "R2S" < 3	# "R2S" = 0
HTA	**48.8%**	**86.9%**	13.1%	3.2%

Evaluation with Preferences on Recommend-Ability. From the above two sets of experiments, it is shown that given the users' review information, ratings from the proposed model work better than the original ones in terms of theirs consistencies to the users' feelings that are expressed in reviews. But there might be some other concerns: it is possible that a user does want to rate the item with the exact score he gives, while his comments are incomplete in expressing his opinion to the item. Here we call it review bias.

We design the third experiments that measure the quality of the revised ratings by crowdsourcing platform and try to reduce review bias. For this time the URLs of the items in the e-commerce system are shown to the crowdsourcing annotators directly. They are asked to click the link and browse the full information of the item, including the item descriptions and the corresponding complete reviews and ratings. Then the annotators are asked to label the recommend-ability, i.e. "whether the item is worthy of being recommended" in 5 levels, say "must not be recommended", "not a good candidate", "just so-so", "good recommendation candidate" and "strongly recommended". Each item is evaluated by three annotators. Data is removed if the labels are not reliable, for example, the annotation procedure is too short to give a careful evaluation.

Finally, 978 items are annotated, and in total 2,459 (rating, review) pairs of these items are found in the previous experimental dataset. Taking the average score of the crowdsourcing labels on items recommend-ability as the ground truth, we evaluate the gap between the average user rating and the recommend-ability of an item with RMSE. Three types of ratings are measured: the original ratings (noted as "Original"), HTA model revised ratings ("R2S-HTA"), and the manually revised ratings according to the reviews by crowdsourcing experiments.

The Results are shown in Table 6. It verifies revised ratings achieved by the proposed models are significantly better on indicating whether the item is worthy of being recommended. Encouragingly, the performances of the HTA model is similar to that of the manually revised scores by the crowds.

Table 6. The gap between items recommend-ability and the ratings

	Original	R2S-HTA	Manual
RMSE	1.5294	1.2317**	1.2110**

$^{**}p < 0.01$ compared with the Original one.

5.3 Effectiveness for Rating Prediction

Rating prediction is an important research topic in the field of personalized recommendation. The aim of rating prediction is to estimate the rating score a user, say u, will rate for any item i. The rating prediction task differs greatly from the previous rating score revision task because the textual reviews are not available and historical information is taken into consideration here.

We evaluate the effectiveness of ratings learned by our model in rating prediction task with two of the state-of-art rating prediction methods, i.e., MF and BMF, which have been popularly used in recommendation tasks in recent years.

The matrix factorization (MF) model [2]: this model maps user preference distributions q_u and item feature distributions p_i to a joint latent factor space of dimensionality f. It estimates the rating a user u will give to any item i by using $\hat{r}_{u,i} = q_u^T p_i$.

The biased matrix factorization (BMF) model [18]: differing from the MF model, the BMF model explains the full rating value not only by the interaction of users and items but also by the biases associated with either users or items. It predicts the rating of item i by user u by using $\hat{r}_{u,i} = \mu + b_u + b_i + q_u^T p_i$, where u, b_u, b_i denote the global average, user bias and item bias, respectively.

The MF and BMF models exploit the past ratings for training, and there are two kinds of ratings: original ratings ("Ori"), ratings learned by the HTA model ("HTA"). Whether using the MF or BMF models, we keep all settings the same except for the sources of ratings. Specifically, we use ratings associated with the user-item pairs in the whole dataset D except the test set in the labeled dataset T for training and try to predict the ratings of the user-item pairs in dataset T.

The crowdsourcing labeled ratings of dataset T are treated as the ground truth. For fair comparison, the information of dataset T is never used in any training procedure.

Table 7 is shown that relative to different rating prediction methods, we are able to achieve better performance by adopting the ratings learned from textual reviews. The result verifies our idea that ratings learned from reviews are more consistent with users' true feelings; hence we can model users' preferences more accurately based on them compare with the original ratings.

Table 7. Performance of rating prediction

Rating	MF		BMF	
	RMSE	MAE	RMSE	MAE
Ori	0.7470	0.5329	0.7339	0.5560
SVR	0.7185	0.5701	0.7242	0.5813
LR	0.6377	0.4959	0.6498	0.5202
LinearSVR	0.6622	0.5023	0.6078	0.4837
MLP	0.6198	0.4745	0.6304	0.5045
HAN	0.6021	0.4611	0.6036	0.4792
HTA	**0.5691**	**0.4326**	**0.5807**	**0.4582**

6   Conclusions and Futurework

In this work, we investigate the basic problem where users' ratings fail to represent their true feelings. We conduct extensive empirical analysis on real-world dataset and propose a deep-learning based approach to recover users' "true ratings" from textual reviews. The ratings learned by our models are able to provide a more reliable basis for the rating-based tasks. Experimental results on "true rating" revision task show that the learned ratings are more consistent with the reviews compared with the original ratings, and the performance of rating prediction task has also been improved by adopting the learned ratings.

The main contributions of this work are: (1) To the best of our knowledge, this is about the first work to detect the unreliability of user ratings and build models so as to recover users' "true ratings"; (2) We show the power of deep learning approach to learn users' "true ratings" from their reviews, which reveals the inherent connection between textual information and score ratings; (3) Our learned "true ratings" also help significantly on rating prediction task. This work can be taken as the foundation in varied scenarios.

References

1. Jøsang, A., Ismail, R., Boyd, C.: A survey of trust and reputation systems for online service provision. Decis. Support Syst. **43**(2), 618–644 (2007)
2. Koren, Y., Bell, R., Volinsky, C.: Matrix factorization techniques for recommender systems. Computer **8**, 30–37 (2009)
3. Almahairi, A., Kastner, K., Cho, K., Courville, A.: Learning distributed representations from reviews for collaborative filtering. In: Recsys, pp. 147–154 (2015)
4. Bao, Y., Fang, H., Zhang, J.: Topicmf: Simultaneously exploiting ratings and reviews for recommendation. In: 28th AAAI Conference, pp. 2–8 (2014)
5. Zhang, Y., Zhang, H., Zhang, M., Liu, Y., Ma, S.: Do users rate or review? Boost phrase-level sentiment labeling with review-level sentiment classification. In: SIGIR, pp. 1027–1030 (2014)
6. Gunes, I., Kaleli, C., Bilge, A., Polat, H.: Shilling attacks against recommender systems: a comprehensive survey. Computer **42**(4), 767–799 (2014)
7. Zhang, Y., Tan, Y., Zhang, M., Liu, Y., Ma, S.: Catch the black sheep: unified framework for shilling attack detection based on fraudulent action propagation. In: IJCAI, pp. 2408–2414 (2015)
8. Resnick, P., Zeckhauser, R.: Trust among strangers in internet transactions: empirical analysis of eBay's reputation system. the economics of the internet and e-commerce. Adv. Appl. Microeconomics **11**, 127–157 (2002)
9. Bhaumik, R., Mobasher, B., Burke, R.D.: A clustering approach to unsupervised attack detection in collaborative recommender systems. In: ICDM, pp. 181–187 (2011)
10. Burke, R., Mobasher, B., Williams, C., Bhaumik, R.: Classification features for attack detection in collaborative recommender systems. In: KDD, pp. 542–547 (2006)
11. Hurley, N., Cheng, Z., Zhang, M.: Statistical attack detection. In: Recsys, pp. 149–156 (2009)
12. McAuley, J., Leskovec, J.: Hidden factors and hidden topics: understanding rating dimensions with review text. In: 7th ACM Conference on Recommender Systems, pp. 165–172. ACM, Hong Kong (2013)
13. Goodfellow, I., Bengio, Y., Courville, A.: Deep Learning. MIT press, Cambridge (2016)
14. Chung, J., Gulcehre, C., Cho, K.H., Bengio, Y.: Empirical evaluation of gated recurrent neural networks on sequence modeling. arXiv preprint arXiv:1412.3555 (2014)
15. Bahdanau, D., Cho, K., Bengio, Y.: Neural machine translation by jointly learning to align and translate. arXiv preprint arXiv:1409.0473 (2014)
16. Chen, H., Sun, M., Tu, C., Lin, Y., Liu, Z.: Neural sentiment classification with user and product attention. In: EMNLP, pp. 1650–1659 (2016)
17. Yang, Z., Yang, D., Dyer, C., He, X., Smola, A., Hovy, E.: Hierarchical attention networks for document classification. In: Proceedings of the 2016 Conference of the North American Chapter of the Association for Computational Linguistics: Human Language Technologies, pp. 1480–1489 (2016)
18. Su, X., Khoshgoftaar, T.M.: A survey of collaborative filtering techniques. In: Advances in Artificial Intelligence, vol. 4 (2009)

The Sogou Spoken Language Understanding System for the NLPCC 2018 Evaluation

Neng Gong, Tongtong Shen, Tianshu Wang, Diandian Qi, Meng Li, Jia Wang, and Chi-Ho Li[✉]

Sogou Inc., Beijing, China
{gongneng,shentongtong,wangtianshu,qidiandian,
lizhihao}@sogou-inc.com, lm168260@antfin.com, cdwangjia5@jd.com

Abstract. This report analyzes the problem of spoken language understanding, how the problem is simplified in the NLPCC shared task, and the properties of the official datasets. It also describes the system we developed for the shared task and provides experimental analysis that explains how promising results could be achieved by careful usage of standard machine learning and natural language processing techniques and external resources.

1 Introduction

This paper describes the architecture and the technical details for our system used for the NLPCC 2018 shared task on spoken language understanding, as well as provides experimental analysis for the choices behind our system architecture.

The structure of the paper is as follows. Section 2 examines the very problem of spoken language understanding and also the datasets used in the shared task. Two important insights found in the survey would become the guiding principles in building our SLU system, which is elaborated in Sect. 3. In Sect. 4 experimental results are provided to show the effectiveness of the various techniques used in our system. Further discussions are provided in the Sects. 5 and 6.

2 Problem Definition and Data Analysis

Spoken language understanding (SLU) comprises two tasks, intent identification and slot filling. That is, given the current query along with the previous queries in the same session, an SLU system predicts the intent of the current query and also all slots (entities or labels) associated with the predicted intent. The significance of SLU lies in that each type of intent corresponds to a particular service API and the slots correspond to the parameters required by this API. SLU helps the dialog system to decide how to satisfy the user's need by calling the right service with the right information.

The authors Meng Li and Jia Wang left the company after the shared task evaluation.

© Springer Nature Switzerland AG 2018
M. Zhang et al. (Eds.): NLPCC 2018, LNAI 11108, pp. 454–463, 2018.
https://doi.org/10.1007/978-3-319-99495-6_38

In real use cases of dialog systems, the difficulty of SLU depends on many factors:

1. Complexity of the intent categorization. Taking the query pattern "我要去X" ("I want to go to X") as example. In some simple design this kind of queries can only be assigned the intent of navigation or map search. But for more complicated design the query pattern may also refer to the need of booking flight ticket or train ticket. The difficulty of disambiguation goes up with the complexity of the categorization.
2. World knowledge. The intent of the last example depends, to certain extent, on whether there is an airport or train station at the place X.
3. User's situation. The intent of the last example also depends, to certain extent, on whether it is too far to drive to X from the user's current location.

In the NLPCC shared task, the last two factors are removed as it is almost infeasible to provide comprehensive world knowledge and user's situation information in a 'closed' dataset. The correct interpretation of a query depends solely on linguistic factors, i.e. the query itself and its preceding queries in the same session. Moreover, the intent categorization is confined to a system of 11 types of intent (which can be further grouped into 4 domains: MUSIC, NAVIGATION, PHONE_CALL, and OTHERS) and 15 types of slots, thereby enabling the shared task with a relatively small training dataset.

As in real use cases of dialog systems, the queries in the shared task can be roughly divided into two kinds, viz. queries with intent-indicating salient phrases and queries without. By intent-indicating salient phrase (IISP) it is meant a phrase in the query that shows the intent of the query. E.g. the phrase "我要去" in the query "我要去上海" and the phrases "打电话" in the query "打电话给李小明" are IISPs.

Our examination on the shared task training dataset shows two important insights:

A. For a query with IISP, the intent can be determined by the query itself, without considering context (the preceding queries). That is a simplification of the use case of many dialog system products, where the existence of IISP simply limits the range of possible intents, as shown by the example on last page, "我要去X", of which the correct interpretation depends not only on the IISP but also on context, world knowledge, etc. Such simplification is due to the low complexity of the intent categorization used in the shared task.
B. For a query without IISP, or 'entity-only' queries, such as "五道口", "李小明", the intent can be determined solely by the domain labels of the preceding queries in the same session. That is also a simplification of the real use case of many dialog system product, where, for example, whether the mention of a person name should be interpreted as a phone call request partly depends on whether the name exists in the user's phonebook. As all such factual knowledge cannot be provided in the shared task datasets, the organizing committee released a handful of 'data annotation principles' to interpret entity-only queries.

These two insights imply an SLU architecture for the shared task: divide queries into those with IISPs and those without; the former ones are handled by a 'context-independent' SLU model (i.e. it does not consider the preceding queries in the same session at all), while the latter ones are handled by some context-dependent rules. This SLU architecture is explained in details in the next section.

3   System Details

Figure 1 shows the framework of our SLU system, which consists of the context-dependent rules for entity-only queries and the context-independent model for queries with IISPs. The entire system feeds the query to the rules first. If the rule-based component returns null result, that means the query is judged to contain IISPs and the model-based component will continue to process it. Otherwise, it means the query is regarded as entity-only and the result of the rules is returned as the final output.

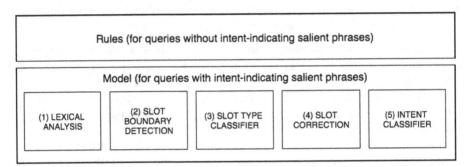

Fig. 1. The Sogou SLU system architecture for the shared task.

3.1   Rule-Based SLU

Our SLU rules apply to only six types of single entities, viz. song titles, singer names, person names, location names, phone numbers, and short commands. Song titles and singer names are defined by the lists provided by the organization committee, and phone numbers are defined by the official annotation principles. There is no official guideline on person names and location names. We used a person name list of around 9 million entries and a location name list of around 70 million entries provided by Sogou Input Method Editor (搜狗输入法) as the definitions of person names and location names respectively.[1] As to short commands like "取消" ("cancel"), we observed the training dataset and collected a dozen of short commands whose correct interpretations depend, in accordance with the annotation principles, on the domain labels of preceding queries.

We imposed altogether 12 rules based on the organizing committee's data annotation principles and our inspection on the training dataset. Each rule is of the form "if the query q is listed in a particular lexicon L, and the preceding queries and their predicted domain labels satisfy certain conditions, then q is assigned a certain intent label and, with the exception of short commands, the entire q is regarded as a slot of the type corresponding to L." The rules are arranged in sequential order in accordance with their priorities. If a query does not fire any rule then it will be processed by the subsequent

---

[1] If there were no such name lists, the rules could still rely on some lexical analyzer to identify person names and location names, since these two kinds of names are represented by special part-of-speech labels in most lexical analyzers. Section 4.1 will compare the value of the name lists with that of using a lexical analyzer.

model-based SLU. If a query fires a rule then the whole process ends with the result returned by the rule.

3.2   Model-Based SLU Pipeline

The SLU model comprises five components that work in a sequential manner, illustrated in Fig. 2.

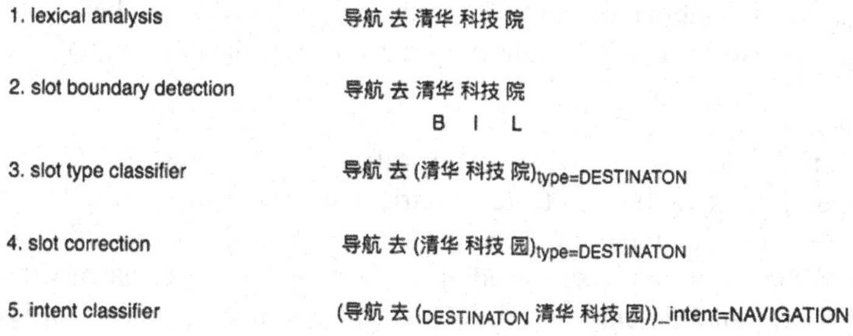

1. lexical analysis	导航 去 清华 科技 院
2. slot boundary detection	导航 去 清华 科技 院 　　　　B　I　L
3. slot type classifier	导航 去 (清华 科技 院)$_{type=DESTINATON}$
4. slot correction	导航 去 (清华 科技 园)$_{type=DESTINATON}$
5. intent classifier	(导航 去 $_{(DESTINATON}$ 清华 科技 园))_intent=NAVIGATION

Fig. 2.   An example illustrating the mechanism of the model-based SLU.

The first step is lexical analysis, i.e. word segmentation and part-of-speech (POS) tagging. The words and POS labels are used as features in the subsequent models. For the shared task we used HanLP [1] as our Chinese lexical analyzer.

Slot Boundary Detection

The second step is slot boundary detection, i.e. to find out the start and end positions of each slot in the query. This task is considered as sequence labeling using the BILOU scheme[2]. We used both character-based and word-based sequence labeling. The character-based version is a Conditional Random Field (CRF) model with window size 7, using lexical features (the characters themselves) and dictionary features, whereas the word-based version is CRF model with window size 5, using lexical features, POS features, and dictionary features. By dictionary features we mean features of the form "whether the current character/word is the prefix/infix/suffix of some entry in a dictionary of certain entity type". The dictionaries used are the same as those in the rule-based SLU. Each CRF model return n outputs[3], and all these $2n$ outputs are passed to the next step. The rationale behind the use of character-based sequence labeling in addition to the word-based version is to reduce the risk due to word segmentation errors. In Sect. 4 we will see the value of this 'combination' strategy in sequence labeling over the standard practice of doing sequence labeling on word tokens only.

---

[2] B stands for beginning position, I for inside, L for last, O for outside, and U for unit, i.e. both as beginning and last position.

[3] $n=3$ in our usage.

Slot Type Classification

The third step is slot type classification, i.e. to identify the type of slot found by the preceding step. Here we used logistic regression with L2-regularizer as the classifier, and the predicted slot, its context characters/words, and the POS labels of its context words (for the word-based input only) as features. As there are several hypotheses from slot boundary detection, where each hypothesis may contain a different number of predicted slots, the slot type classifier calculates the average score of the slots in each hypothesis[4] and chooses the hypothesis with the highest average score as output. This simple strategy of combination is based on the assumption that the best slot boundary detection hypothesis leads to the highest scored slot type classifier output.

Slot Correction

The fourth step is slot correction, i.e. to identify the 'spelling errors' of a slot due to incorrect speech recognition result, etc. A retrieval based method is used. As there is a dictionary for each slot type (c.f. Sect. 3.1), if a slot s with predicted type T is not registered in the dictionary corresponding to T, then s will be matched against the dictionary entries and the entries with lowest edit distance with s are retrieved. This process is carried out twice, one representing s and the entries as Chinese character strings and another representing s and the entries as pinyin strings. The best match is selected by a re-ranker from these two sets of retrieval results.

Intent Classifier

The last step is the intent classifier. It is based on gradient boosting, using the well-known XGBoost tool [2] with its default settings. The features used include the word tokens, query length, and the predicted slots in the preceding steps. Note that the query length feature is proposed by the observation that in the training set queries with more than 20 characters are mainly of the intent OTHERS. That is, the query length feature is a simple hack to distinguish OTHERS from other kinds of intent.

Training Sample Mining

The official training dataset contains about 21,352 samples only and it is tempting to add more samples from other sources. Although we do have two millions labeled queries for Sogou voice interface products, the categorization behind these queries are very different from that used in the shared task, and therefore these assets cannot be directly applied.

Instead we used a cautious method for introducing more samples. For intent identification, the trained classifier is applied to the training dataset itself and the error cases are believed to be under-represented by the training set. For each under-represented query q, it is then matched against our own query set using similarity metrics like ngram overlap ratio, edit distance, etc. Very harsh thresholds on the metrics are used. Those queries that satisfy the thresholds are taken as new samples and they are labeled with the same intent as q.

---

[4] That is, if a hypothesis contains N slots, where the i-th slot is assigned a score s_i by the slot type classifier, then the score of the hypothesis is $\frac{1}{n} \sum_i s_i$.

New samples for slot filling are introduced in a similar but a bit more complicated way. Two queries of similar structure may not be similar on their surface forms, as the slot values in the two queries could be very different from each other. Therefore, when two queries q and q' are compared their slot values should first be replaced by the slot types, thereby converting the queries into query patterns[5]. If the similarity between query patterns satisfy a very harsh threshold then q' would be introduced as a new sample, and the slot types in q can be projected to those in q' via the query patterns.

Eventually around 500 samples are chosen as new training instances for intent classifier and around 1000 samples for slot filling.

4   Experiments

In this section we present the results of some experiments showing the effectiveness of the techniques described in the last section. The training and test datasets are the ones released by the organizing committee. The evaluation criterion for intent classifier is F1, which is the same criterion for subtask 2. The evaluation criterion used in all other experiments is precision, the same criterion for subtask 4. In the experiments on slot filling only, the precision is slightly modified so as not to take the intent label into account.

4.1   Rule-Based SLU

Among the 5350 queries in the test set, our context-dependent SLU rules are fired by 1265 of them; that is, the rules are responsible for about 23.6% of the queries. The precision of the rules for both intent identification and slot filling is 96.60%, and the F1 for intent identification only is 96.56%. Moreover, if the identification of person names and location names in our rules are not based on our own name lists but on the lexical analyzer HANLP, the F1 of intent identification drops to 93.03%, and the precision for the entire SLU drops to 92.89%. That is, our name lists reduce nearly half of the errors.

There are two kinds of errors made by the rules. The first kind is due to the incorrect domain labels predicted by SLU model for the preceding queries. The second kind is more difficult to analyze. On the one hand these errors may be explained as the noises in the dictionaries of person names and location names, yet on the other hand it may also be said that the errors are not of the rules but of the test set itself. For example, the query "楼兰" is labeled in the test set as a destination/location name while our system judges it to be a person name. It seems to be rather arbitrary to say that "楼兰" is less likely to be a person name than a location name.

---

[5] For example, the query in the official training set "我想去北海公园转转" and the query in extra resources "我想去蓝色港湾转一转" are not very similar to each other on their surface forms. Yet because the correct slots "北海公园" and "蓝色港湾" are already labeled in both sets, we could convert the queries into patterns "我想去<slot>转转" and "我想去<slot>转一转". Similarity can be measured on such query patterns.

4.2 Slot Filling Models

Table 1 shows the experiment results on slot filling (combining the steps of boundary detection, type classification, and correction) of the queries that cannot be handled by rule-based SLU. In the baseline setting, the boundary detector is based only on word-based sequence labeling and it does not use any dictionary features; no extra training samples are used either. The other three settings test the values of three techniques, viz. the combination strategy in boundary detection, the dictionary features in boundary detection, and more training samples. The results show that both the combination strategy and the dictionary features are very effective techniques.

Table 1. Experiments on slot filling.

Setting	Precision (before slot correction)	Precision (after slot correction)
1: Baseline	90.58%	91.70%
2: (1) + combination strategy	91.31%	92.43%
3: (2) + dictionary features	92.39%	93.51%
4: (3) + more samples	92.73%	93.85%

Besides, our slot correction technique consistently makes an absolute improvement of about 1.1%. The errors of slot correction are mainly due to the noises in the person and location name lists.

4.3 Intent Classifier

Table 2 shows the experiment results on intent identification of the queries that cannot be handled by rule-based SLU. The baseline setting is about classifier using word token features only. The other three setting test the values of three techniques, viz. adding the slot types predicted by the slot filling models as extra features, adding query length as extra feature, and adding more training samples. As shown in Table 2, the slot features are the most useful technique, leading to an absolute improvement of nearly 4%. Moreover, only about 500 new training samples (2% of the official training set size) reduced 9% of errors. That shows the importance of data and the effectiveness of our mining techniques.

Table 2. Experiments on intent identification

Setting	F1
1: Baseline	91.36%
2: (1) + slot features	95.26%
3: (2) + query length feature	95.46%
4: (3) + more samples	95.97%

4.4 The Complete SLU System

As to intent identification only, the complete system produces an F1 score of 96.11%. Table 3 shows the experiment results of the complete system on both intent identification and slot filling. The three experiment settings are designed to show the contribution of the external resources (dictionaries and extra training samples).

Table 3. Experiments on the entire SLU system.

Setting	Precision
Rules + Model w/o extra dictionaries and samples	93.42%
Rules + Model with extra dictionaries but not samples	94.24%
Rules + Model with extra dictionaries and samples	94.49%

There are three major types of errors on queries with IISPs. The first type is about very long queries that contain multiple intent, e.g. "唱一首等你下课打电话给六三". This type of errors need a means to split up the query into shorter ones, and/or a multi-label classifier to select the prominent intent. The second type of errors are due to both speech recognition errors and data sparsity, such as the query "放手下课别走", for which the correct form should be "放首下课别走". Although we proposed a mechanism to mine similar queries from our own query log, the speech recognition errors in our log are different from those in the official training set. For this example, there are very few samples for the pattern "放手<歌曲>", and our speech recognition system seldom makes this particular mistake. It is not always successful to mine rare query patterns with certain speech recognition error.

Last but not least, the third type of errors are not of the SLU system but of the labeling consistency of the datasets. For example, it is difficult to explain why "打电话不干" is labeled as PHONE_CALL whereas "打电话差不多" is not. Similarly, it is difficult to explain, for the query "打 电 话 给 我 听 王 蓉", why the slot of CONTACT_NAME should be "我听王蓉" instead of "王蓉".

5 Related Works

There are many reviews on SLU as such [3] or as a component in a complete dialogue system [4]. The various techniques can be divided into two camps, viz. the end-to-end approaches and the pipelined approaches. An end-to-end approach completes both intent identification and slot filling in one process, using techniques like [5–7]. In contrast, a pipelined approach tackles the two tasks one by one, and there are a few choices of the pipeline architecture and various techniques in implementing the components. We decided not to use any of the advanced deep learning techniques because it is found in our pilot experiments that the training data size is too small to produce satisfying results from any deep learning technique. Similar to [8], we found that traditional methods like CRF give the optimal results.

Our slot correction method follows a general spelling correction framework using the noisy channel model [9]. That is, given a misspelled string s, the task is to seek s'

which maximizes the translation model of $P(s|s')$ and the language model of $P(s')$. In our system, the translation model is the edit distances between s and s' on both Chinese character and Pinyin character levels, while the language model is determined by entity lexicons.

6   Summary and Discussions

In this report, we analyzed the problem of spoken language understanding, how the problem is simplified in the NLPCC shared task, and the properties of the official datasets. The data survey inspired us to adopt a hybrid approach to SLU, viz. to deal with entity-only queries with context-dependent rules whereas to deal with queries with intent-indicating salient phrases with context-independent models. A pipelined framework is used to integrate individual models to produce the final output. Although the techniques we used for the model-based SLU are all very 'old-school' ones, and yet our system achieved very promising results. The remaining problems are essentially about quality of dictionaries, long queries with multiple intent, rare query patterns with speech recognition errors, and data labeling consistency issues.

The datasets in the shared task seem to be extracted from the query log of some in-car voice interface product. The SLU problem in industrial products has three characteristics. First, users' commands/requests are mostly entity-specific. Therefore, entity knowledge (or at least lexical resources) is very important to the performance of an SLU system, and our experimental analysis proves the value of entity knowledge.

Secondly, data annotation is the key. Nowadays most speech assistant or voice interface products do not provide user feedbacks as effective as users' clicks in search engines. So, the only way to judge whether an SLU output is correct solely depends on annotators' judgments. We also showed that a few labeled samples in the official datasets are dubious.

The most important one is that the complexity of the techniques required by SLU is proportional to the complexity of intent categorization. For the categorization of only 11 intents in the shared task, we have shown that (for queries with IISPs) SLU can be reliably done with information about the current query only. For a more complicated categorization which contains, following the example in Sect. 2, navigation intent and flight/train-booking intent, the SLU models must appeal at least to context (previous queries) and perhaps even more factors, and will therefore require more complicated techniques. In general, if there are more than one intents which correspond to similar query patterns then more advanced techniques are required.

Based on these observations, we hope that the next shared task would be a more challenging one by having a more fine-grained intent categorization, and a larger and more accurately labeled dataset.

References

1. HanLP: Han Language Processing. https://github.com/hankcs/HanLP
2. Chen, T., Guestrin, C.: XGBoost: a scalable tree boosting system. arXiv:1603.02754 (2016)

3. Tur, G., De Mori, R.: Spoken Language Understanding: Systems for Extracting Semantic Information from Speec. Wiley, Hoboken (2011)
4. Chen, H., Liu, X., Yin, D., Tang, J.: A survey on dialogue systems: recent advances and new frontiers. arXiv:1711.01731 (2017)
5. Jeong, M., Lee, G.G.: Triangular-chain conditional random fields. IEEE Trans. Audio Speech Lang. Process. **16**(7), 1287–1302 (2008)
6. Xu, P., Sarikaya, R.: Convolutional neural network based triangular CRF for joint intent detection and slot filling. In: ASRU (2013)
7. Zhang, X., Wang, H.: A joint model of intent determination and slot filling for spoken language understanding. In: IJCAI (2016)
8. Vukotic, V., Raymond, C., Gravier, G.:. Is it time to switch to word embedding and recurrent neural networks for spoken language understanding? In: Interspeech (2015)
9. Kernighan, M., Church, K., Gale, W.: A spelling correction program based on a noisy channel model. In: COLING (1990)

Improving Pointer-Generator Network with Keywords Information for Chinese Abstractive Summarization

Xiaoping Jiang, Po Hu[(✉)], Liwei Hou, and Xia Wang

School of Computer Science, Central China Normal University,
Wuhan 430079, China
{jiangxp,houliwei}@mails.ccnu.edu.cn,
phu@mail.ccnu.edu.cn, wangx_cathy@163.com

Abstract. Recently sequence-to-sequence (Seq2Seq) model and its variants are widely used in multiple summarization tasks e.g., sentence compression, headline generation, single document summarization, and have achieved significant performance. However, most of the existing models for abstractive summarization suffer from some undesirable shortcomings such as generating inaccurate contents or insufficient summary. To alleviate the problem, we propose a novel approach to improve the summary's informativeness by explicitly incorporating topical keywords information from the original document into a pointer-generator network via a new attention mechanism so that a topic-oriented summary can be generated in a context-aware manner with guidance. Preliminary experimental results on the NLPCC 2018 Chinese document summarization benchmark dataset have demonstrated the effectiveness and superiority of our approach. We have achieved significant performance close to that of the best performing system in all the participating systems.

Keywords: Abstractive summarization · Sequence to sequence model
Pointer-generator network · Topical keywords · Attention mechanism

1 Introduction

Automatic summarization aims to simplify the long text into a concise and fluent version while conveying the most important information. It can be roughly divided into two types: extraction and abstraction. Extractive methods usually extract important sentences from the original document to generate the summary. However, abstractive methods often need to understand the main content of the original document first and then reorganize even generate the new summary content with natural language generation (NLG). Compared with extractive methods, abstractive methods are more difficult but are closer to human summarization manner.

Recently, the development of deep neural network makes abstractive summarization viable among which attention-based sequence-to-sequence (Seq2Seq) models have increasingly became the benchmark model for abstractive summarization task (Hou et al. 2018a). Seq2Seq models have encoder-decoder architecture with recurrent neural

© Springer Nature Switzerland AG 2018
M. Zhang et al. (Eds.): NLPCC 2018, LNAI 11108, pp. 464–474, 2018.
https://doi.org/10.1007/978-3-319-99495-6_39

network (RNN) configuration. The attention mechanism is recently added to generate more focused summary by referencing salient original context when decoding.

However, the existing models usually face two shortcomings: one is the content inaccuracy and repetition mainly caused by out-of-vocabulary (OOV) words (See et al. 2017), and another is that the existing attention mechanism does not consider topic information of the original document explicitly which may lead to insufficient decoding.

In this work, we propose a novel approach to improve the summary's informativeness by explicitly incorporating topical keywords information from the original document into a pointer-generator network via a new attention mechanism so that a topic-oriented summary can be generated in a context-aware manner with guidance. Specifically, we first adopt a pointer-generator network proposed by See et al. (2017) to improve accuracy of the generated content and alleviate the problem of OOV words. Meanwhile, the coverage mechanism is used to solve content duplication problem. Second, we put topical keywords extracted from the original document into the attention mechanism and incorporate it into the pointer-generator network so that decoder will pay more attention to topic information to better guide the generation of informative summary. In general, our contributions are as follows:

- We adopt a pointer-generator network model to alleviate the problems of inaccurate detail description caused by OOV words (Sect. 3.2).
- We encode topical keywords extracted from the original document into the attention mechanism and incorporate it into the pointer-generator network to enhance the generated summary's topic coverage (Sect. 3.3).
- We applied our proposed method to the Chinese document summarization benchmark dataset provided by NLPCC 2018 shared task3 and ranked the third among all the participating systems (Sect. 4).

2 Related Work

Automatic summarization has always been a classic and hot topic in the field of natural language processing (NLP). Significant progress has been made recently from traditional extractive summarization to more abstractive summarization (Yao et al. 2017).

Earlier research in the last decade is dominated by extractive methods. Extractive methods first score each sentence in the original document. Unsupervised sentence scoring approaches mostly rely on frequency, centrality and probabilistic topic models. Sentence classification, sentence regression and sequence labeling are the supervised approaches commonly used to evaluate the importance of sentences. Having predicted sentences importance score, the next step is to select sentences according to their information richness, redundancy and some constraint rules (such as total summary length, etc.). The popular approaches for sentence selection include maximum marginal relevance (MMR), integer linear programming (ILP) and submodular function maximization. Recently, it has been shown effective to use neural network to directly predict the relative importance of a sentence given a set of selected sentences, under the consideration of importance and redundancy simultaneously (Cao et al. 2017; Narayan et al. 2018).

Although extractive methods have the advantage of preserve the original information more complete, especially ensuring the fluency of each sentence, one of the problems is that they suffer from the secondary or redundant information. More importantly, there is often a lack of coherence between adjacent sentences in an extractive summary. With the rapid development of deep learning technology in recent years, abstractive summarization has gradually become the current research focus.

RNN-based encoder-decoder structure is proposed by Bahdanau et al. (2014) and used in machine translation successfully. Subsequently, this structure has also been successfully applied to other fields of NLP, including but not limited to syntactic parsing (Vinyals and Le 2015), text summarization (Rush et al. 2015) and dialogue systems (Serban et al. 2016). Rush et al. (2015) first introduce the encoder-decoder structure and the attention mechanism into summarization task and achieve good results on DUC-2004 and Gigawords datasets. Later Nallapati et al. (2016) extend their work and combine additional features to the model, which get better results than Rush on DUC-2004 and Gigawords datasets. The graph-based attention mechanism is proposed by Tan and Wan (2017) to improve the adaptability of the model to sentence saliency. Paulus et al. (2017) combine supervised learning with reinforcement learning while training, and their work not only keeps the readability but also ensures the flexibility of summary. For latent structure modeling, Li et al. (2017) add historical dependencies on the latent variables of Variational Autoencoder (VAEs) and propose a deep recurrent generative decoder (DRGD) to distill the complex latent structures implied in the target summaries. Nallapati et al. (2016); Gu et al. (2016); Zeng et al. (2016), See et al. (2017) and Paulus et al. (2017) use copy mechanism to solve the problem of OOV words in the decoding phase. Moreover, See et al. (2017) propose a coverage mechanism to alleviate words repetition. In a recent work, Wang et al. (2018) incorporate topic information into the convolutional sequence-to-sequence (ConvS2S) model.

3 Model

3.1 Attention-Based Seq2Seq Model

Attention-based Seq2Seq model is first used for machine translation tasks. It is also used to generate abstractive summary due to the resemblance between abstractive summarization and machine translation. Attention-based Seq2Seq model mainly consists of three parts: encoder, decoder and the attention mechanism connecting them.

In the encoding phase, the word embedding sequences of the original document are fed into a single bidirectional LSTM to get the encoder hidden states sequence $h = \{h_1, h_2, \ldots, h_n\}$. At each decoding time step, a single unidirectional LSTM reads the previous word embedding to obtain the decoder hidden state s_t, which is used for the output prediction of the current time step. The hidden states of encoder and decoder pass through a linear layer and a softmax function to get the attention distribution a^t. The attention distribution corresponds to a probability distribution of each word in the original document, that tells which words are more important in the current prediction process. It is calculated as follows:

$$e_i^t = v^T \tan h(W_h h_i + W_s s_t + b_{attn}) \tag{1}$$

$$a^t = softmax(e^t) \tag{2}$$

Where v^T, W_h, W_s and b_{attn} are learnable parameters. Once a^t has been computed, it is used to produce a weighted sum of the encoder hidden states, which is a dynamic representation of the original document called the context vector h_t^*:

$$h_t^* = \sum_i a_i^t h_i \tag{3}$$

Finally, the decoder hidden state s_t and the context vector h_t^* pass through two linear layer and a softmax function to produce the vocabulary distribution P_{vocab} of the current time step. The concrete formulas are as follows:

$$P_{vocab} = softmax(V'(V[s_t, h_t^*] + b) + b') \tag{4}$$

$$P(w) = P_{vocab}(w) \tag{5}$$

Where V', V, b and b' are learnable parameters. $P(w)$ represents the probability of the current prediction for word w. Loss function of the model uses negative log likelihood:

$$loss = \frac{1}{T} \sum_{t=0}^{T} -log(P(w_t^*)) \tag{6}$$

Where w_t^* is the target word for the current time step, and T is the total length of the target summary.

3.2 Pointer-Generator Network

The pointer-generator network proposed by See et al. (2017) is a hybrid model combining both an attention-based Seq2Seq model and a pointer network. It allows the model to generate new words from a fixed vocabulary or copy words from the original document. Therefore, for each original document, it will add the words in it to the fixed vocabulary and get the extended vocabulary. Furthermore, they also adopt the coverage mechanism to solve repetition problem.

The context vector h_t^*, the decoder hidden state s_t and the decoder input x_t pass through a linear layer and a sigmoid function to produce the generation probability P_{gen}, which indicates the probability to generate a new word from the fixed vocabulary:

$$P_{gen} = \sigma(w_{h^*}^T h_t^* + w_s^T s_t + w_x^T x_t + b_{ptr}) \tag{7}$$

Where w_{h^*}, w_s, w_x and b_{ptr} are learnable parameters. P_{gen} is used to produce a weighted sum of the vocabulary distribution P_{vocab} (referring to formula 4) and the attention distribution a^t (referring to formula 2):

$$P(w) = p_{gen}P_{vocab}(w) + (1 - p_{gen}) \sum_{i:w_i} a_i^t \tag{8}$$

The coverage mechanism is introduced into the pointer-generator network to alleviate repetition problem. It maintains a coverage vector c^t (i.e., the sum of attention distributions), which records the coverage degree of those words which have received from the attention mechanism so far. The coverage vector in turn affects the attention distribution for the current time step. The coverage mechanism also calculates additional coverage loss to punish repeated attention. The whole computation process is implemented as follows:

$$c^t = \sum_{t'=0}^{t-1} a^{t'} \tag{9}$$

$$e_i^t = v^T tanh(W_h h_i + W_s s_t + W_c c_i^t + b_{attn}) \tag{10}$$

$$covloss_t = \sum_i min(a_i^t, c_i^t) \tag{11}$$

$$loss = \frac{1}{T} \sum_{t=0}^{T} \left[-logP(w_t^*) + \lambda \sum_i covloss_t \right] \tag{12}$$

Where W_c is a new learnable parameter and λ represent the weight of the coverage loss.

3.3   Keywords Attention Mechanism

Recent studies show that the traditional attention mechanism only considers the relationship between the current target word and the original document, so it fails to grasp the main gist of the original document and leads to insufficient information (Lin et al. 2018). Imagine that when people write a summary, they usually have a clear understanding of the topic content, and the summary they write is often centered around topic. Therefore, we propose to adopt topical keywords as the guidance information for the original document and encode them into the attention mechanism to generate summary with better topic coverage.

In this work, TextRank (Mihalcea 2004) algorithm is used to extract topical keywords. We extract d topical keywords from the original documents and get their word embeddings $k = \{k_1, k_2, \ldots, k_d\}$. As shown in Fig. 1, we calculate the sum of word embeddings for d keywords and use it as a part of input for the attention distribution.

$$t = \sum_{i=1}^{d} k_i \tag{13}$$

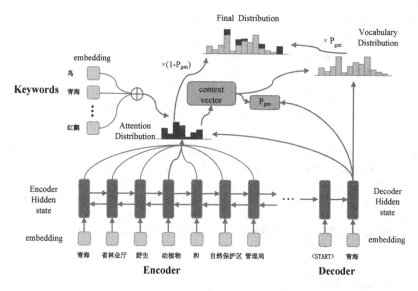

Fig. 1. Pointer-generator model with expanded keywords attention

$$e_i^t = v^T tanh\left(W_h h_i + W_s s_t + W_c c_i^t + W_t t + b_{attn}\right) \qquad (14)$$

Where W_t is a new learnable parameter. We change Eqs. (10) to (14).

4   Experiments

4.1   Dataset

We conduct experiments on the open-available dataset provided by NLPCC 2018 shared task3, which is a Chinese document summarization task. The training set contains 50000 document-summary pairs, the validation set contains 2000 document-summary pairs, and the testing set contains 2000 documents without corresponding golden-standard summaries. We compute the length (i.e., the number of words after segmentation using jieba[1] toolkit) of the original documents and the standard summaries whose statistics are shown in Table 1.

4.2   Evaluation

In this work, we use ROUGE[2] toolkit for evaluation. ROUGE is widely used in the summarization research community for content evaluation, which often calculates the recall rate of N-grams or words between the machine-generated summary and the golden-standard human summary (Chin 2004). Character-based ROUGE-F score is

---

[1] https://pypi.python.org/pypi/jieba/

[2] https://github.com/andersjo/pyrouge/tree/master/tools/ROUGE-1.5.5

Table 1. The statistics of the NLPCC 2018 document summarization benchmark dataset.

Length	Training set		Validation set		Testing set	
	Document	Summary	Document	Summary	Document	Summary
Min	32	6	35	8	7	–
Max	13186	85	8871	44	7596	–
Average	579.4	25.9	579.9	25.8	426.2	–

used as evaluation metric in this work. We conduct evaluation on both the validation set and the testing set due to the lack of golden-standard summaries on the testing set. For the validation set, we calculate and report ROUGE-1, ROUGE-2 and ROUGE-L scores respectively. For the testing set, we present results provided by the NLPCC 2018 official organizer and the results represent the average score of multiple ROUGE metrics including ROUGE-1, ROUGE-2, ROUGE-3, ROUGE-4, ROUGE-L, ROUGE-W-1.2 and ROUGE-SU4.

4.3   Implementation

In our experiments, we first convert all the datasets into plain texts and use jieba toolkit to conduct word segmentation on both news articles and corresponding summaries. Then, we use TextRank4ZH toolkit[3] to extract keywords from each news articles with the number of keywords set to 8. Furthermore, we use a vocabulary of 50 k words for both original documents and target summaries.

Unlike previous work proposed by Hou et al. (2018b), we do not pretrain the word embeddings in advance and all are learned from scratch during training, and the dimension of the word embeddings is set to 128. In our proposed approach, the encoder is a bidirectional LSTM and the decoder is a unidirectional LSTM with both the hidden layer dimension set to 256. And we set the maximum encoding length (i.e., the maximum length of the input sequence) to 1000, and the decoding length (i.e., the output sequence length) is adjusted from 8 to 40. Our model is trained by Adagrad (Kingma et al. 2014) with learning rate of 0.15 and initial accumulator value of 0.1. We implement all our experiments with Tensorflow on an NVIDIA TITAN XP GPU and the batch size is set to 16.

During the training phase, the initial loss value is 10, and then it drops to 7 after the first 300 times of training, and further drops to 3.5 after 15000 times of training. Finally, with the increase of training time, the loss value gradually becomes stable and converges to 3. When the loss value of the model is stable on the training set, the parameters learned from training phase are used for validation. During the testing phase, we use beam search to get the target summary and set beam size to 6.

---

[3] https://github.com/letiantian/TextRank4ZH

4.4   Experimental Results and Analysis

The experimental results are shown in Tables 2 and 3. We choose the pointer-generator network model as the baseline, and we also compare with other representative state-of-the-art extractive and abstractive summarization approaches.

Table 2. Comparison results on the NLPCC 2018 validation dataset using F-measure of ROUGE

Models	R-1	R-2	R-L
TextRank	35.78	23.58	29.60
Pointer-Generator	43.44	29.46	37.66
Pointer-Generator (character)	38.01	24.43	32.32
TextRank + Pointer-Generator	42.79	28.93	37.33
Our Model	**44.60**	**30.46**	**38.83**

Table 3. Examples of the generated summaries set of our approach and the pointer-generator baseline. Here **bold** denotes richer information generated by our model.

Document: 显示图片切尔西官方宣布费利佩离队北京时7月28日，切尔西官方宣布边卫费利佩-路易斯离队 ，巴西人将重返西甲劲旅马德里竞技，马竞官方随后也确认了这一消息。据悉，费利佩的转会费为1500万镑。新浪体育稍后为您带来详细报道
Golden summary: 切尔西官方宣布边卫费利佩 • 路易斯离队 ，巴西人将1500万镑重返 西甲劲旅马德里竞技
Pointer-generator： 显示，切尔西官方宣布边卫费利佩-路易斯离队，巴西人将重返西甲劲旅马德里竞技
Our model: 切尔西官方宣布费利佩离队，巴西人将重返西甲劲旅马德里竞技，**转会费为1500万镑**

Document: 据美国媒体报道，美国总统奥巴马7月7日在白宫椭圆形办公室会见了越南共产党总书记阮富仲。白宫发布"美国—越南联合愿景声明"声明说，两国共同关注南中国海最近提升的紧张局势。声明表示，越南共产党总书记阮富仲对美国进行了历史性的访问，两国于2015 年7月1日制定了这份联合愿景声明 ……
Golden summary: 奥马巴会见越共总书记阮富仲，发布美越联合愿景声明，声称共同关注南海紧张局势
Pointer-generator: 两国再次强调将继续"美越防务关系联合愿景声明"中所体现的防务和安全双边合作。
Our model: 奥巴马会见越南共产党总书记阮富仲，发布越南联合愿景声明，**两国共同关注南中国海最近提升的紧张局势。**

TextRank. This approach adopts the open-source TextRank4zh toolkit to extract the most important sentences with highest informatives scores from the original news article to generate the target summary.

Pointer-Generator. This model has been described in Sect. 3.2.

TextRank + Pointer-Generator. Inspired by the work of Tan et al. (2017), we combine the TextRank approach with the pointer-generator model to generate the summary. First, we obtain an 800 words summary extracted by TextRank (Mihalcea 2004). Then, the summary is used as the input of the pointer generator network to generate the final summary.

Pointer-Generator (character). Basically, there are two typical approaches to pre-process Chinese document: character-based and word-based (Wang et al. 2018). In this work, we adopt the word-based approach as we believe that words are more relevant to latent topic of document than characters. Since the official evaluation metrics are Character-based ROUGE F score, we also evaluate pointer-generator network using character-based approach to obtain a comprehensive comparison.

Our Model. The hybrid method combining keywords attention and pointer-generator network introduced in Sect. 3.3.

According to the results shown in Table 2, we can find that our proposed approach outperforms all other methods on ROUGE-1, ROUGE-2 and ROUGE-L. The key-words attention-based pointer-generator network model exceeds the basic pointer-generator network significantly. In addition, word-based approach achieves higher ROUGE performance than character-based approach, and abstractive methods always achieve higher ROUGE performance than TextRank. Furthermore, the method directly Combining TextRank with pointer-generator does not achieve obviously better results.

Table 3 gives two running cases from which it can be observed that our keywords attention-based pointer-generator network model produces more coherent, diverse, and informative summary than the basic pointer-generator network approach.

The testing results are shown in Table 4. The scores are the average of ROUGE-1, ROUGE-2, ROUGE-3, ROUGE-4, ROUGE-L, ROUGE-SU4 and ROUGE-W1.2 provided by official organizer. Our team ranked the third among all the participating teams, which is 0.11 points slightly below than the first.

Table 4. Official evaluation results for the formal runs of top-10 participating teams

Team	The average score of ROUGE-1, 2, 3, 4, L, SU4, W-1.2
WILWAL	0.29380
Summary++	0.28533
CCNU_NLP (Our Team)	0.28279
Freefolk	0.28149
Kakami	0.27832
Casia-S	0.27395
Felicity_Dream_Team	0.27211
dont lie	0.27078
CQUT_301_1	0.25998
lll_go	0.25611

5   Conclusion

In this work, we propose a novel abstractive summarization approach which incorporates topical keywords information from the original document into a pointer-generator network via a new attention mechanism. The experimental results show that our proposed approach can reduce wording inaccuracy while improve the summary's informativeness. In the future, we will conduct more experiments on other larger-scale datasets like CNN/Daily Mail to verify the effectiveness of our method. Besides, we will also try more advanced keyword extraction algorithm to discover and embed topical keywords information more effectively.

Acknowledgments. This work was supported by the National Natural Science Foundation of China (No. 61402191), the Self-determined Research Funds of CCNU from the Colleges' Basic Research and Operation of MOE (No. CCNU18TS044), and the Thirteen Five-year Research Planning Project of National Language Committee (No. WT135-11).

References

Bahdanau, D., Cho, K., Bengio, Y.: Neural Machine Translation by Jointly Learning to Align and Translate. arXiv preprint arXiv:1409.0473 (2014)

Serban, I.V., Sordoni, A., Bengio, Y., Courville, A.C., Pineau, J.: Building end-to-end dialogue systems using generative hierarchical neural network models. In: AAAI (2016)

Vinyals, O., Le, Q.: A Neural Conversational Model. arXiv preprint arXiv:1506.05869 (2015)

See, A., Liu, P.J., Manning, C.D.: Get to the Point: Summarization with Pointer-Generator Networks. ACL (2017)

Yao, J.-G., Wan, X., Xiao, J.: Recent advances in document summarization. Knowl. Inf. Syst. **53**, 297–336 (2017)

Rush, A.M., Chopra, S., Weston, J.: A Neural Attention Model for Abstractive Sentence Summarization. arXiv preprint arXiv:1509.00685 (2015)

Nallapati, R., Xiang, B., Zhou, B.: Abstractive Text Summarization Using Sequence-to-Sequence RNNs and Beyond. arXiv preprint arXiv:1602.06023 (2016)

Tan, J., Wan, X.: Abstractive Document Summarization with a Graph-Based Attentional Neural Model. ACL (2017)

Paulus, R., Xiong, C., Socher, R.: A Deep Reinforced Model for Abstractive Summarization. arXiv preprint arXiv:1705.04304 (2017)

Li, P., Lam, W., Bing, L., Wang, Z.: Deep recurrent generative decoder for abstractive text summarization. In: Proceedings of the Conference on Empirical Methods in Natural Language Processing. EMNLP (2017)

Gu, J., Lu, Z., Li, H., Li, V.O.K.: Incorporating Copying Mechanism in Sequence-to-Sequence Learning. ACL (2016)

Zeng, W., Luo, W., Fidler, S., Urtasun, R.: Efficient Summarization with Read-Again and Copy Mechanism. arXiv preprint arXiv:1611.03382 (2016)

Wang, L., Yao, J., Tao, Y., Zhong, L., Liu, W., Du, Q.: A reinforced topic-aware convolutional sequence-to-sequence model for abstractive text summarization. In: IJCAI-ECAI (2018)

Hou, L., Hu, P., Bei, C.: Abstractive document summarization via neural model with joint attention. In: Huang, X., Jiang, J., Zhao, D., Feng, Y., Hong, Yu. (eds.) NLPCC 2017. LNCS (NLAI), vol. 10619, pp. 329–338. Springer, Cham (2018a). https://doi.org/10.1007/978-3-319-73618-1_28

Cao, Z., Wei, F., Li, W., Li, S.: Faithful to the Original: Fact Aware Neural Abstractive Summarization arXiv:1711.04434 (2017)

Narayan, S., Cohen, S.B., Lapata, M.: Ranking Sentences for Extractive Summarization with Reinforcement Learning. arXiv preprint arXiv:1802.08636 (2018)

Mihalcea, R.: TextRank: bringing order into texts. In: EMNLP (2004)

Chin, Y.L.: Rouge: A Package for Automatic Evaluation of Summaries. ACL (2004)

Kingma, D.P., Ba, J.: Adam: a method for stochastic optimization. Computer Science (2014)

Tan, J., Wan, X., Xiao, J.: From neural sentence summarization to headline generation: a coarse-to-fine approach. In: IJCAI (2017)

Hou, L.-W., Hu, P., Cao, W.-L.: Chinese abstractive summarization with topical keywords fusion. Acta Automatica Sinica (2018b)

Lin, J., Sun, X., Ma1, S., Su, Q.: Global Encoding for Abstractive Summarization. ACL (2018)

Author Index

Printed in the United States
by Baker & Taylor

Printed in the United States
By Bookmasters